Atmospheric Rivers

F. Martin Ralph • Michael D. Dettinger
Jonathan J. Rutz • Duane E. Waliser
Editors

Atmospheric Rivers

Springer

Editors
F. Martin Ralph
Center for Western Weather and Water Extremes
Scripps Institution of Oceanography
University of California–San Diego
La Jolla, CA, USA

Jonathan J. Rutz
Science and Technology Infusion Division
National Weather Service
Salt Lake City, UT, USA

Michael D. Dettinger
Retired, U.S. Geological Survey
Carson City, NV, USA

Duane E. Waliser
Jet Propulsion Laboratory
California Institute of Technology
Pasadena, CA, USA

ISBN 978-3-030-28905-8 ISBN 978-3-030-28906-5 (eBook)
https://doi.org/10.1007/978-3-030-28906-5

This Springer imprint is published by the registered company Springer Nature Switzerland AG
The registered company address is: Gewerbestrasse 11, 6330 Cham, Switzerland

This book is dedicated to those who came before us, who discovered these rivers in the sky, and to those who will ride them into the future.

Foreword

I agreed to write this foreword because I was motivated by the scope and breadth of the research related to atmospheric rivers (ARs) in this volume. It is my pleasure to recommend this scholarship, which represents the first comprehensive collection of research on the increasingly important phenomenon of ARs. It is both a benchmark for the field now and a springboard for future discoveries.

ARs are "increasingly important" because they are basic to extratropical dynamics of weather and climate and they are increasingly recognized as causes of precipitation totals and extremes in many regions of the world. This volume describes the observations, models, and analyses that are the basis of current understanding of ARs; their global distributions and impacts; the roles of ARs in extratropical meteorology and climatology; forecasting issues and the likely effects of climate change on future ARs; and some nascent applications of AR science.

As a result of research over the past 10 to 15 years, and rapidly advancing and heightened scientific and public awareness, we now know that a significant fraction of the annual precipitation on the western side of continents in the Northern and Southern Hemispheres, as well as extreme precipitation events, occurs in conjunction with landfalling ARs. As a result of those extreme precipitation events, from a practical and operational perspective, it is critical to be able to distinguish the smaller subset of ARs that may be associated with dangerously high-impact precipitation events from the larger and weaker group of ARs that pose no immediate danger to public safety.

For example, based on reliable records dating back to 1921, the all-time wettest water year (1 Oct–30 Sep) on record in northern California occurred during 2016–2017, when the northern Sierra's record rainfall occurred in conjunction with multiple landfalling ARs between December 2016 and April 2017. The record rainfall in February 2017 contributed to concerns about the safety of Oroville Dam on the Feather River. Water-level heights behind the Oroville Dam rapidly increased, damaging the main spillway as excessive water began to overtop it. As the rains continued, the emergency spillway was additionally damaged by erosion. This resulted in heightened concerns that a concrete weir around the dam could fail—which could have caused a devastating 10-meter-high wall of water to surge down into the Feather River and all the way to the Central Valley, potentially flooding communities downstream (see Chap. 7).

To facilitate the identification of this kind of high-impact AR, Ralph et al. (2019) have constructed an AR impact scale based on the magnitude and duration of the integrated water vapor transport (IVT) along ARs that should facilitate the identification of and communications about these ARs. This sort of scale is important because duration and IVT matter. I usually pay attention to ARs when the associated IVT first becomes >250 kg m^{-1} s^{-1}—and I give them my undivided attention when the IVT becomes >1000 kg m^{-1} s^{-1} (see detailed information on this AR scale in Chap. 8).

The last 10 to 15 years of AR research have also heightened the scientific community's understanding of important synoptic-scale and meso-scale aspects of ARs and have resulted in better knowledge of the relationships among ARs, tropical moisture exports (TMEs), and warm conveyor belts (WCBs) (see Chap. 2). This overall increased knowledge about ARs has culminated in the formal approval of an AR definition that appeared in the glossary of the

American Meteorological Society in 2017 (see Chap. 3, Sect. 3.1). This integration of AR science into these other, more traditional aspects of mid-latitude meteorology puts AR science on a firmer footing for future research and is described here by several of the leaders in the fields of AR, TME, and WCB science.

Of course, in the end, given these newly recognized risks of extreme precipitation and hazards that ARs can pose, and the key role in extratropical climate dynamics that they play, forecasting ARs is of growing importance. Currently, we are not able to properly forecast the global and regional distribution of ARs beyond a few days. This limitation is one likely source of the current limitations on predictability of precipitation amounts or types (e.g., snow, sleet, freezing rain, and rain). Verification studies of forecast models have shown that they can better predict the probability of precipitation rather than precipitation amounts or types. The practical implication of this is that precipitation amounts remain hard to forecast, so that forecasters are better able to distinguish between the occurrence of wet and dry days, and are somewhat less able to predict how much precipitation will fall, given the occurrence of a wet day. In an increasing range of settings globally, it is now recognized that forecasts of cold-season precipitation amount and type are limited by uncertainties about the following:

- the strength and location of upstream low- and upper-level jets
- the extent of the coupling between the low-level and upper-level jets
- where the nose of the low-level jet that transports AR moisture poleward will intersect a surface boundary
- the overall structure and configuration of the horizontal and vertical precipitation-producing circulations associated with a progressive upstream upper-level trough

AR representations and forecast ability in modern weather and climate models are described in Chap. 6.

As a result of these (and other) key findings and issues covered within, this book should appeal to a broad spectrum of readers interested in both basic and applied research opportunities, and in undergraduate and graduate education; operationally oriented readers; resource managers; and federal, state, and local emergency management officials as well as technically oriented public officials. Among these readers may be the next generation of AR researchers.

Lance Bosart
Distinguished Professor,
Department of Atmospheric and Earth Sciences,
University at Albany, SUNY

Preface

This book is intended to summarize the state of the science of atmospheric rivers (ARs) and its application to practical decision-making and broader policy topics. It is the first book on the subject and is intended to be a learning resource for professionals, students, and indeed anyone new to the field, as well as a reference source for all.

We first envisioned the book during the heady days of 2013 when the Center for Western Weather and Water Extremes was being planned and established. However, right from the start, we recognized that the effort required would exceed that of any single or couple of authors, and that the book would surely benefit from a broad range of perspectives and knowledge from a variety of leaders of atmospheric-river science from around the world. Consequently, the first step toward this book was to organize workshops addressing various aspects of AR science that we were able to co-opt, in part, for recruitment of, and discussions among, possible contributing authors. This led to the diverse authorship team that ultimately wrote this book, as well as our engagement of an experienced publication and book editing team. Among the strategies agreed to by the contributing authors, one key decision was that the book would focus mostly on results that have already been published and would emphasize figures and references from those formal publications. Where vital, new information has been developed and incorporated. Each chapter was led by a few expert lead authors recruited by the four of us, and those chapter leads recruited contributions from other experts on the chapter topic. Each chapter was reviewed by other specialists who were not part of its authorship team, generally including one highly technical expert and one reviewer intended to represent members of a broader audience. This helped ensure the accuracy of interpretations as well as high standards and accessibility of presentation. We, the editors of the book, reviewed all chapters at various stages of composition and layout.

Given currently high levels of interest in ARs in the scientific community as well as by the public, we hope that the book will be a useful starting place for many readers. Writing a book about a topic that is as new and that is advancing as quickly as AR science is today (in 2018) poses many difficult challenges but, with the help of the large team of expert authors who have contributed, we believe that, with this book, we are providing a firm foundation for future expansion and advances in this important field.

La Jolla, CA, USA — F. Martin Ralph
Carson City, NV, USA — Michael D. Dettinger
Salt Lake City, UT, USA — Jonathan J. Rutz
Pasadena, CA, USA — Duane E. Waliser

Acknowledgements

Co-Editors

- F. Martin Ralph, Editor-in-Chief (Scripps Institution of Oceanography, CW3E)
- Michael D. Dettinger (United States Geological Survey; Retired) *MDD's contributions in chapters 5 and 7 were undertaken as a research hydrologist for the U.S. Geological Survey Water Mission Area.*
- Jonathan J. Rutz (National Oceanic & Atmospheric Administration/National Weather Service)
- Duane E. Waliser (National Aeronautics and Space Administration/Jet Propulsion Laboratory) *DEW's contribution to this study was carried out on behalf of the Jet Propulsion Laboratory, California Institute of Technology, under a contract with the National Aeronautics and Space Administration.*

The co-editors acknowledge the diligent efforts of all the authors and co-authors to the chapters of this book, without which—of course—this book would not exist. Thanks for sticking with us.

Chapter authors/contributing authors: Michael Alexander, Michael L. Anderson, Deniz Bozkurt, Gilbert Compo, Jason M. Cordeira, Dale A. Cox, Francina Dominguez, Bin Guan, Huancui Hu, Jay Jasperse, David A. Lavers, Alexander Gershunov, Irina Gorodetskaya, Bin Guan, Peter Knippertz, Paul J. Neiman, Kelly M. Mahoney, Benjamin J. Moore, William Neff, Florian Pappenberger, Alexandre M. Ramos, David S. Richardson, Lawrence J. Schick, Harald Sodemann, Ryan Spackman, Hans Christian Steen–Larsen, Andreas Stohl, Maria Tsukernik, Raúl Valenzuela, Maximiliano Viale, Andrew J. Wade, Heini Wernli, Allen B. White, Gary A. Wick, Ervin Zsoter

We acknowledge the generous support from the following entities that allowed this book to be conceived of, developed, and produced.

Funders/sponsors: This book would not have been possible without funding from the "Water Operations Technical Support: Research to Investigate Atmospheric Rivers (AR) and the Feasibility of Developing and Using AR Forecast Capabilities to Inform Reservoir Operations Within the USACE" project led by the Center for Western Weather and Water Extremes at UC San Diego's Scripps Institution of Oceanography. This Forecast-Informed Reservoir Operations program is sponsored by and collaboratively executed with researchers from the US Army Engineer Research and Development Center (ERDC) and water management professionals from the US Army Corps of Engineers (USACE).

We also acknowledge the persistence and efforts of the publication team who polished our contributions into this final form.

Publication team: Lauren Muscatine, Managing Editor (UC Davis), Mary Beth Sanders, Designer (Metography), Sheila Chandrasekhar, Developmental Editor (Persuasive Pages)

The co-editors would also like to acknowledge those individuals who provided guidance, critique, and mentorship throughout their careers.

Finally, the co-editors would like to acknowledge the original advance made by Yong Zhu and Reginald E. Newell (1998) who not only identified an intriguing and important phenomena for study, but were insightful enough to also provide a descriptive name that is well posed scientifically as well as attractive and intuitive for the broader community and public.

Contents

List of Figures

List of Tables

Contributors

Michael L. Anderson State of California Department of Water Resources, Sacramento, CA, USA

Deniz Bozkurt Department of Geophysics, Center for Climate and Resilience Research, University of Chile, Santiago, Chile

Gilbert P. Compo Cooperative Institute for Research in Environmental Sciences, University of Colorado–Boulder, Boulder, CO, USA

Physical Sciences Laboratory, National Oceanic and Atmospheric Administration (NOAA), Earth System Research Laboratory, Boulder, CO, USA

Jason M. Cordeira Plymouth State University, Plymouth, NH, USA

Michael D. Dettinger Retired, U.S. Geological Survey, Carson City, NV, USA

Francina Dominguez University of Illinois, Urbana–Champaign, Champaign, IL, USA

Alexander Gershunov Center for Western Weather and Water Extremes, Scripps Institution of Oceanography, University of California–San Diego, La Jolla, CA, USA

Irina V. Gorodetskaya Department of Physics, Centre for Environmental and Marine Studies, University of Aveiro, Aveiro, Portugal

Bin Guan University of California, Los Angeles, Los Angeles, CA, USA

Huancui Hu Pacific Northwest National Lab, Richland, WA, USA

Peter Knippertz Institute of Meteorology and Climate Research, Karlsruhe Institute of Technology, Karlsruhe, Germany

David A. Lavers European Centre for Medium-Range Weather Forecasts (ECMWF), Reading, UK

Kelly M. Mahoney Physical Sciences Laboratory, National Oceanic and Atmospheric Administration (NOAA), Earth System Research Laboratory, Boulder, CO, USA

Benjamin J. Moore Cooperative Institute for Research in Environmental Sciences, University of Colorado–Boulder, Boulder, CO, USA

Physical Sciences Laboratory, National Oceanic and Atmospheric Administration (NOAA), Earth System Research Laboratory, Boulder, CO, USA

William Neff Physical Sciences Laboratory, National Oceanic and Atmospheric Administration (NOAA), Earth System Research Laboratory, Boulder, CO, USA

Cooperative Institute for Research in Environmental Sciences, University of Colorado–Boulder, Boulder, CO, USA

Paul J. Neiman National Oceanic and Atmospheric Administration (NOAA), Earth System Research Laboratory, Boulder, CO, USA

Florian Pappenberger European Centre for Medium-Range Weather Forecasts (ECMWF), Reading, UK

F. Martin Ralph Center for Western Weather and Water Extremes, Scripps Institution of Oceanography, University of California–San Diego, La Jolla, CA, USA

Alexandre M. Ramos Instituto Dom Luiz, Faculdade de Ciências, Universidade de Lisbon, Lisbon, Portugal

David S. Richardson European Centre for Medium-Range Weather Forecasts (ECMWF), Reading, UK

Jonathan J. Rutz Science and Technology Infusion Division, National Weather Service, Salt Lake City, UT, USA

Lawrence J. Schick Retired, U.S. Army Corps of Engineers, Seattle, WA, USA

Harald Sodemann Geophysical Institute, University of Bergen and Bjerknes Centre for Climate Research, Bergen, Norway

Hans Christian Steen-Larsen Geophysical Institute, University of Bergen, Bergen, Norway
Bjerknes Centre for Climate Research, University of Bergen, Bergen, Norway

Andreas Stohl NILU - Norwegian Institute for Air Research, Kjeller, Norway

Maria Tsukernik Brown University, Providence, RI, USA

Raúl Valenzuela Department of Geophysics, Center for Climate and Resilience Research, University of Chile, Santiago, Chile

Maximiliano Viale Argentinean Institute for Snow, Glaciology and Environmental Science, (IANIGLA - CONICET), Mendoza, Argentina

Andrew J. Wade University of Reading, Reading, UK

Duane E. Waliser Jet Propulsion Laboratory, California Institute of Technology, Pasadena, CA, USA

Heini Wernli Institute for Atmospheric and Climate Science, ETH Zürich, Zürich, Switzerland

Allen B. White National Oceanic and Atmospheric Administration (NOAA), Earth System Research Laboratory, Boulder, CO, USA

Gary A. Wick National Oceanic and Atmospheric Administration (NOAA), Earth System Research Laboratory, Boulder, CO, USA

Ervin Zsoter European Centre for Medium-Range Weather Forecasts (ECMWF), Reading, UK

Introduction to Atmospheric Rivers

1

F. Martin Ralph, Michael D. Dettinger, Lawrence J. Schick, and Michael L. Anderson

1.1 A Brief History of AR Science

The term "atmospheric river" (AR) has been evolving from its first mention in the scientific literature to the present day—becoming not only ubiquitous throughout the scientific community, but also garnering enough public attention that a Lake Tahoe casino recently introduced an "atmospheric river" cocktail. After years of differing opinions on the term, the Glossary of Meteorology formed a committee to develop a formal definition, which was vetted broadly with the science community and converged on the definition published in the Glossary in 2018. Ralph et al. (2018) provides the definition and describes the process of developing it. Chapter 2 of this book provides a deep analysis of the relationship between ARs and related phenomena of warm conveyor belts and tropical moisture exports, confusion over which had underlain some of the earlier debate over the atmospheric river concept. Most recently, a scale for AR intensity and an impact-based categorization scheme (Ralph et al. 2019) have been created. This section offers a brief description of the progress of AR science from the 1970s to the present.

- 1970s and 1980s—A number of underlying concepts about frontal systems and airflow through cyclones came into being.
- 1990s—The term "atmospheric river" is coined and the phenomenon is observed in global data sets. ARs over the northeastern Pacific are observed during a field campaign.
- 2000s—AR science is reinvigorated and benefits from Special Sensor Microwave Imager/Sounder (SSM/I) integrated water vapor (IWV) observations. ARs are cataloged, and this data is leveraged to examine AR climatology in the western USA, effects on precipitation events, and hydroclimate. It becomes increasingly clear that ARs produce extreme events over parts of the western USA, and that this represents unique challenges to this region.
- 2010s—AR science branches out and inter-connects with a number of other diverse fields such as hydrology, ecology, and atmospheric aerosols, among others.
- TODAY—This community includes participants from many disciplines: meteorology, climatology, hydrology, civil engineering, but also ecosystems biology, paleoclimatology, and even arctic and polar science. In fact, as 2016's first International Atmospheric Rivers Conference (IARC; Ralph et al. 2017) showed, presentations addressed ARs on six continents and the island of Greenland.

1.1.1 The 1970s

The seminal work by Browning and Pardoe (1973) on the low-level jet (LLJ) does not use the term "atmospheric rivers," but describes many of the physical characteristics that have come to be associated with them. Their schematic cross-section of the LLJ, shown in Fig. 1.1, highlights the upper-level front, warm and moist air ahead of the front, as well as convection and vertical motions associated with frontal dynamics. The LLJ itself, indicated by the "J," is located where recent studies have placed the maxima of water vapor transport associated with ARs. Their study highlights the ver-

F. M. Ralph
Center for Western Weather and Water Extremes, Scripps Institution of Oceanography, University of California–San Diego, La Jolla, CA, USA
e-mail: mralph@ucsd.edu

M. D. Dettinger (✉)
Retired, U.S. Geological Survey, Carson City, NV, USA
e-mail: mdettinger@ucsd.edu

L. J. Schick
Retired, U.S. Army Corps of Engineers, Seattle, WA, USA

M. L. Anderson
State of California Department of Water Resources, Sacramento, CA, USA
e-mail: Michael.L.Anderson@water.ca.gov

F. M. Ralph et al. (eds.), *Atmospheric Rivers*, https://doi.org/10.1007/978-3-030-28906-5_1

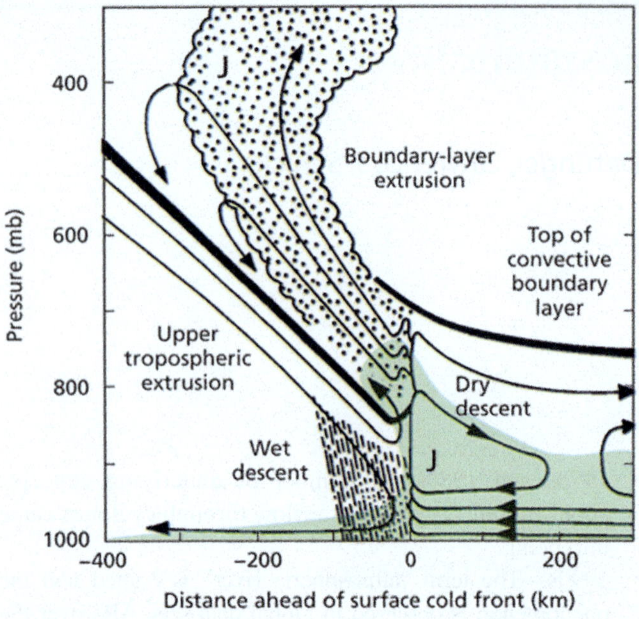

Fig. 1.1 The low-level jet (Browning and Pardoe 1973; image courtesy of Jay Cordeira)

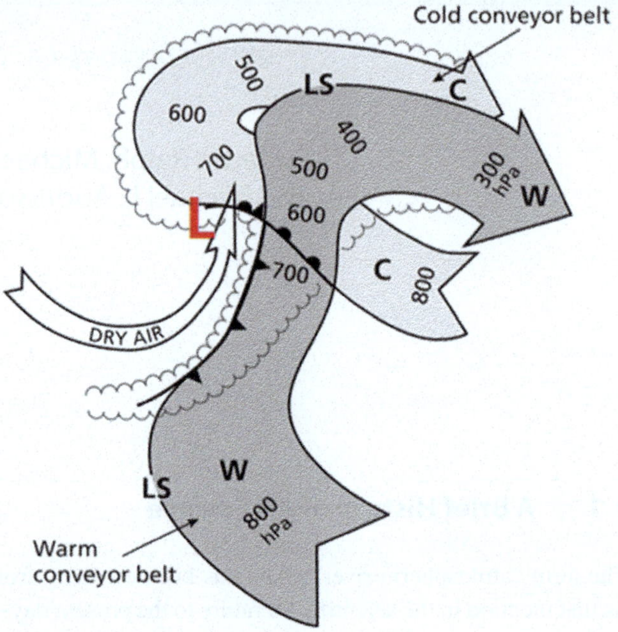

Fig. 1.2 Warm and cold conveyor belt concepts (Carlson 1980; image courtesy of Jay Cordeira)

tical circulations, a major topic of meteorology at that time (and today), but less so the horizontal water vapor transport.

Studies conducted during this time also pointed to the role of orographic precipitation, particularly over the UK (e.g., Browning 1980), which is precipitation enhanced by vertical lifting which increases in surface elevation induce. Note that many of these initial studies on orographic precipitation took place in the UK, where topographic barriers are less dramatic than in many other parts of the globe.

1.1.2 The 1980s

Another key step toward understanding ARs was provided via a schematic produced by Carlson (1980; Fig. 1.2), which highlights airflow through mid-latitude cyclones. It shows both the warm conveyor belt (WCB) and cold conveyor belt (CCB) air streams, and their respective motions relative to the cyclone. Heini Wernli, while doing his dissertation on the WCB (Wernli 1995, 1997; Wernli and Davies 1997), came to define the WCB as, specifically, the region of ascent. The WCB concept also further blended the combination of latent and sensible heat, whereas later work on ARs emphasized the water vapor component (latent heat).

1.1.3 The 1990s

In a now-recognized-as-landmark study for AR science, Newell et al. (1992) described "tropospheric rivers"—narrow plumes of intense, vertically integrated water vapor transport (IVT) observed meandering through the atmosphere. However, it was several more years before Zhu and Newell (1998) coined the term "atmospheric river" while building upon results from Newell et al. (1992). Their study used a global analysis at 2-degree horizontal resolution over a 2-year period, and objectively identified IVT values by grid cell, laying the groundwork for current techniques of AR identification. Furthermore, Zhu and Newell concluded that these ARs were located along the cold fronts associated with mid-latitude cyclones, meaning they were dynamically linked to the extratropical storm track. Their work not only provided scientists with visuals of what this transport looked like on a global scale (Fig. 1.3), but also compiled statistics on the global transport of water vapor. One such enduring statistic is that ARs are responsible for ~90–95% of poleward water vapor flux across mid-latitudes, yet occupy only ~10% of global circumference in doing so, which offers a sense of the narrowness and efficiency of transport that occurs along ARs.

Figure 1.4, from Zhu and Newell (1998), highlights this fact and shows more generally the meridional (i.e., in the north-south direction) water vapor flux associated with ARs as a function of latitude. From these results, it is clear that ARs play a major role in the global water cycle.

The work by Zhu and Newell (1998) encountered some resistance, and a few key questions emerged: How are ARs related to the LLJ, the WCB, and other conceptual models? Why are we calling it a river, given that it meanders and has no fixed position? These concerns were legitimate steps in the progress of AR science: new ideas emerge, are interrogated, and tested—ultimately improving our understanding.

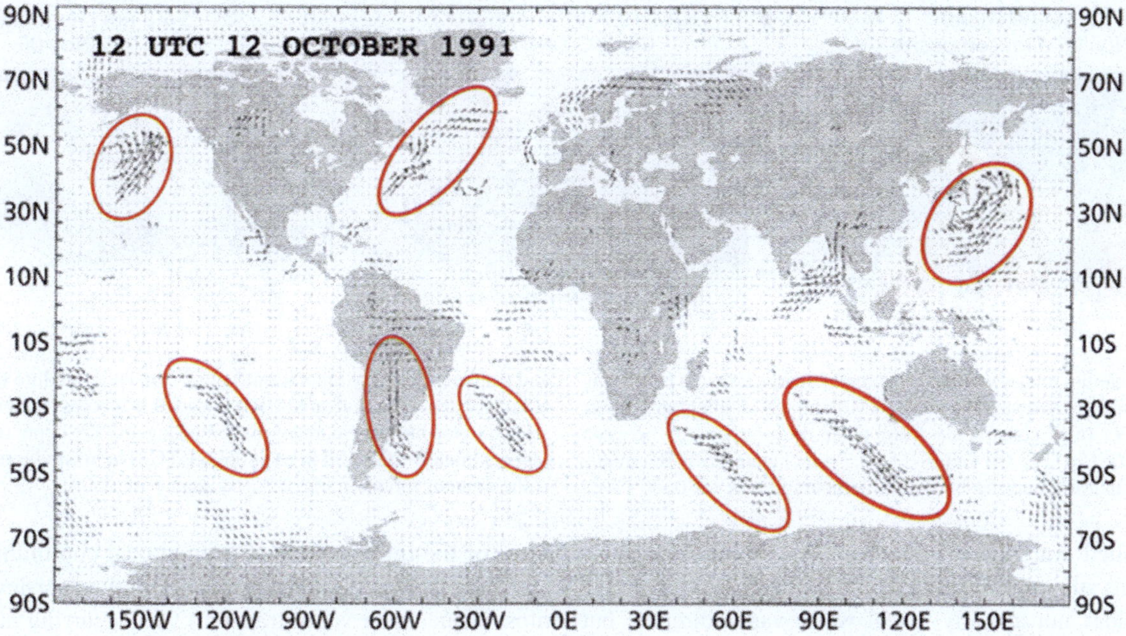

Fig. 1.3 Global analysis of a 2-degree grid (Zhu and Newell 1998). Atmospheric rivers are outlined in red

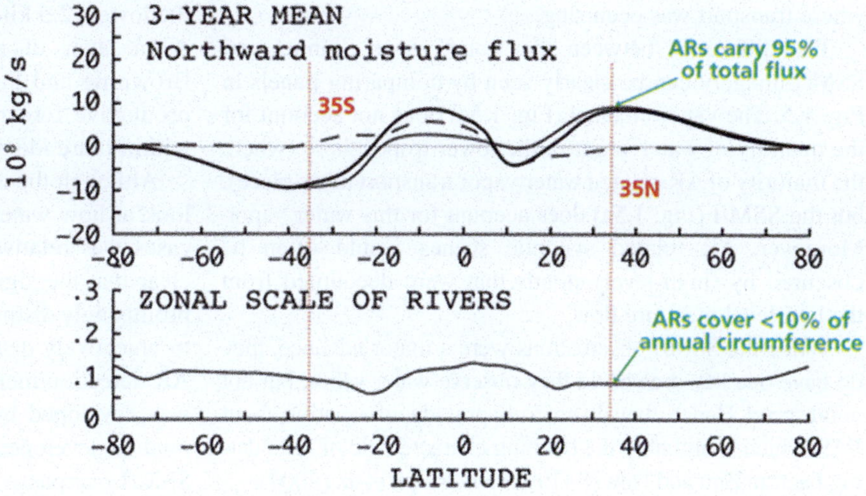

Fig. 1.4 ARs are responsible for 90–95% of the total global meridional water vapor transport at mid-latitudes, yet constitute <10% of the earth's circumference at those latitudes (Zhu and Newell 1998)

1998 can be considered a pivotal year in the emerging science of ARs based on four major developments in that year, or very nearly in that year:

- The work of Zhu and Newell (1998) was published.
- The California Land-Falling Jets (CALJET) field campaign of 1997–1998 was conducted that flew aircraft into and through ARs (described in Ralph et al. 2004).
- The number of polar-orbiting SSM/I satellites had grown large enough to stitch their individual observing swaths together to create a nearly synoptic view of IWV over the oceans globally.
- The National Centers for Atmospheric Prediction and the National Center for Atmospheric Research (NCEP/NCAR) reanalysis was published and became readily available for the first time.

Although these happened in the closing part of the 1990s, the lessons to be learned from them would emerge in the 2000s, and are described below.

1.1.4 The 2000s

The advent of available satellite-based water vapor imagery via SSM/I marked another watershed resource for AR science. The SSM/I enabled weather forecasters and researchers to analyze IWV in near real time. Although IWV presents only the moisture component of IVT (which also depends on the wind profile), meteorologists could now visualize the long and narrow plumes of water vapor associated with the reanalysis-based ARs from Zhu and Newell (1998). Previously, meteorologists often used satellite imagery from

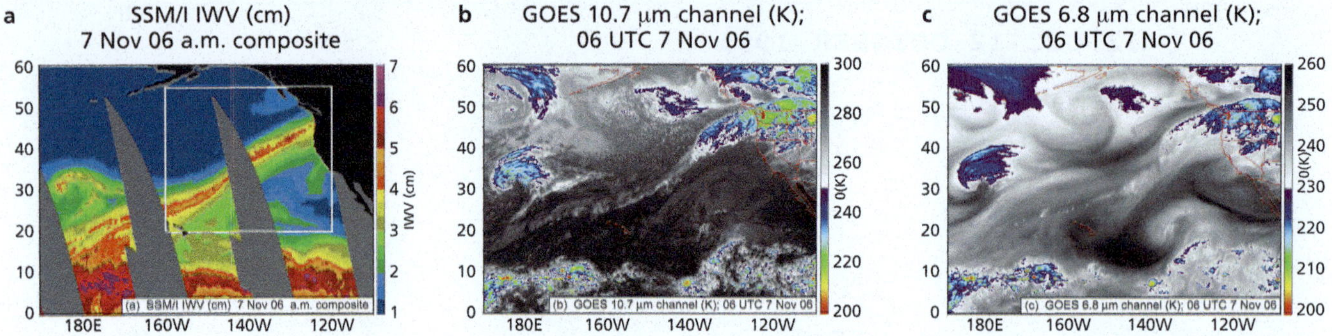

Fig. 1.5 Satellite imagery on the morning of 7 Nov 2006 (**a**) composite Special Sensor Microwave/Imager (SSM/I) image of integrated water vapor (IWV) (cm) constructed from polar orbiting swaths between ~0200 and 0615 UTC; (**b**) GOES-11 10.7 μm channel (i.e., IR) image of surface and/or cloud-top brightness temperature (K) at 0600 UTC; and (**c**) Geostationary Operational Environmental Satellite (GOES)-11 6.7 μm channel (i.e., infrared [IR]) image of brightness temperature (K) related to the moisture content of a broad layer of the upper troposphere (approximately 200–500 hPa) at 0600 UTC. The *white inset box* in (**a**) is the domain shown in Fig. 1.3. (Neiman et al. 2008)

the so-called water vapor channel (Fig. 1.5) to diagnose movements of water vapor, but this was difficult since this channel does not actually reveal water vapor directly, but only the temperature at a level above which 1 mm of water vapor exists. This could reveal circulation patterns, but did not really inform the user about where water vapor existed or where transport was occurring.

The difference between the water vapor channel and SSM/I imagery can be clearly seen by comparing panels in Fig. 1.5. The vapor channel (Fig. 1.5c) does not account for the quantity of water vapor in the lower troposphere (where the majority of AR-related water vapor transport takes place), but the SSM/I (Fig. 1.5a) does account for this water vapor. Moreover, AR related synoptic scenes would often be obscured by (high-level) clouds that were decoupled from the low-level moisture flow.

While the SSM/I observations were a major advance, they do have one key drawback: they observe water vapor, but not wind speed. Hence, they do not truly provide information about IVT, which is arguably the defining characteristic of ARs dating back to Zhu and Newell (1998) and Ralph et al. (2004).

AR science took another major step forward during the CALJET field campaign of 1997–1998 (Ralph et al. 2004). CALJET used one of NOAA's P-3 research aircraft, and the strong El Niño provided not only an abundance of storms, but also an abundance of public interest in the field campaign and the role of ARs in driving major landfalling storms and impacts (Fig. 1.6).

During CALJET, the research team, led by F. Martin Ralph, conducted a number of flights through mid-latitude cyclones and across LLJs, releasing dropsondes to collect data, which enabled the production of cross-sectional analyses through the storms. These analyses revealed that the strongest water vapor transport was indeed taking place near the LLJ, in a filamentary region along and ahead of the cold front, where the team expected it based on satellite imagery. Whereas the strongest winds (and previously, most of the interest from the dynamical community) were associated with the upper-level jet and upper-level front, most of the water vapor transport was occurring in the lowest few kilometers. In fact, Ralph et al. (2005) showed, after compositing observational data from 17 storms, that 75% of the horizontal water vapor flux occurred in about the lowest 2.5 kilometers (kms)—and that they had about a 3-km layer of moist neutrality (Fig. 1.7). No wonder Browning and Pardoe had found that the LLJ was such a productive forcer for orographic precipitation: conditions within it are ideal.

Although the aircraft data in and of itself provided a new look at how water vapor transport occurred, the number of cases was relatively limited. At the same time, it became clear that the signature of ARs in SSM/I satellite data was promisingly distinct. To explore this potential of using SSM/I to objectively detect ARs, Ralph et al. (2004) developed an AR detection method (ARDM) based on IWV. The concept was developed by examining many images of IWV via a method developed by Wick et al. (2013) to combine multiple SSM/I overpasses at different times to show larger areas of IWV. Outside the tropics, it was clear that regions of large IWV sometimes took on a long, narrow structure. This led to a simple set of criteria, described below, that were applied by eye to the entire winter of satellite images. This method was then validated against the in situ aircraft data, confirming that the method worked.

The AR detection criteria applied to SSM/I-observed IWV, developed by Ralph et al. (2004), worked by searching the SSM/I IWV fields for regions of large IWV (>20 mm) that met the following criteria:

- 2000 km long
- <1000 km wide
- >20 mm of IWV, used as a proxy for water vapor transport, along the axis of (or in some cases, throughout) the feature

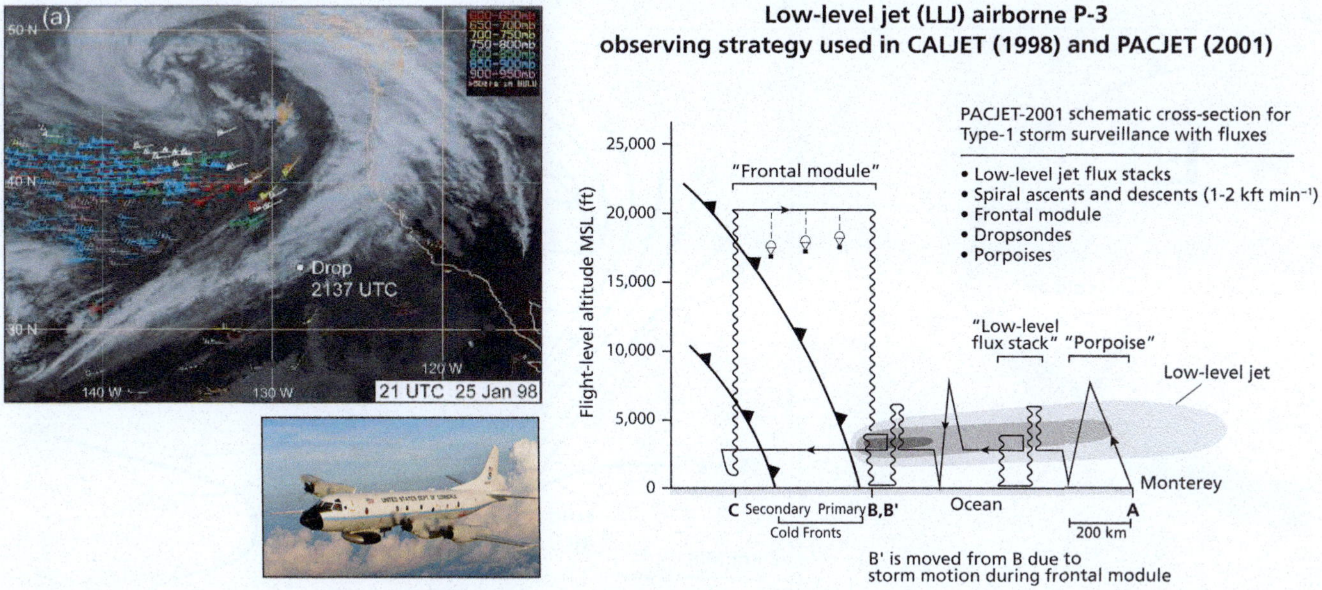

Fig. 1.6 Low-level jet airborne P-3 observing strategy used in CALJET (1998) and the Pacific Land-falling Jets Experiment (PACJET) (2001) (Ralph et al. 2004)

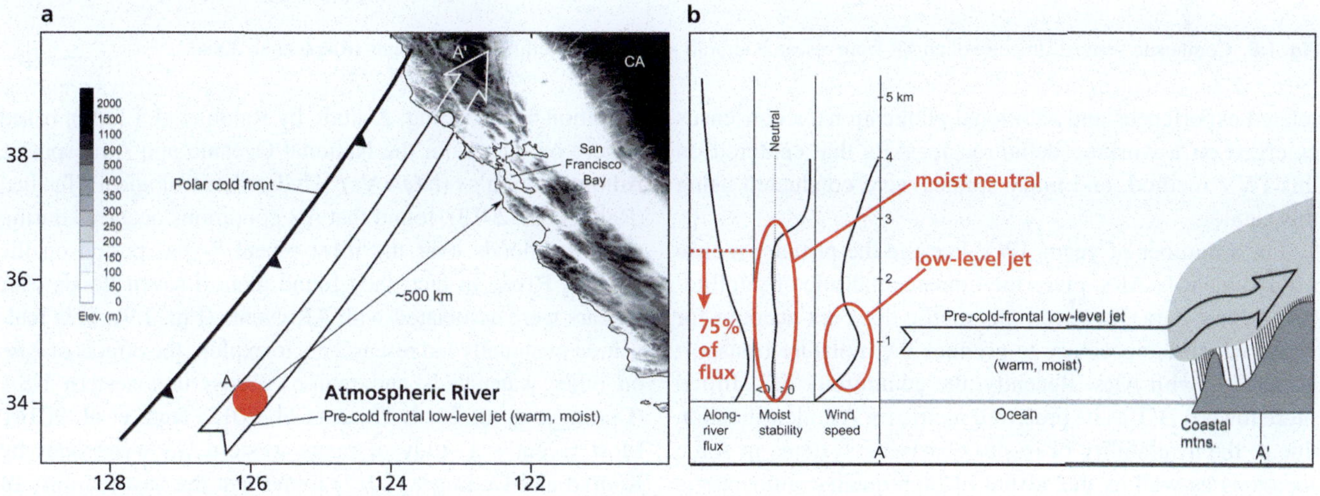

Fig. 1.7 (**a**) Plan view of the mean location of the vertical profile shown in (**b**) and of major weather features. (**b**) Composite vertical structure of an AR from 17 cases observed by research aircraft. (Ralph et al. 2005)

The approach of using SSM/I was combined with the aircraft data to provide the first full picture of AR structure (Ralph et al. 2004; Fig. 1.8).

In a pair of follow-on papers, Ralph et al. (2005) showed how ARs led to heavy coastal precipitation, partly because they represented meteorological conditions that were ideal for creating heavy orographic precipitation upon landfall on the mountainous US West Coast. This was followed by a study that was the first to show how ARs were associated with all flooding events on the Russian River (located just north of the San Francisco Bay Area) during an 8-year period when special ground-based data were available there (Ralph et al. 2006). Adding to the historical context, this finding about the flooding happened when the first-ever multi-year catalog of AR landfalls (based on SSM/I) in northern California was compared with independent streamflow data from the Russian River. In one meeting during a visit to Scripps Institution of Oceanography in La Jolla (Scripps), Marty Ralph and Paul Neiman asked Mike Dettinger for the dates of flood events on the Russian River during those years. An hour later they found a 100% match. Neiman et al. (2008) added further credibility to these criteria by showing that ARs identified using SSM/I IWV are typically consistent with IVT based on using reanalysis. Based on these and

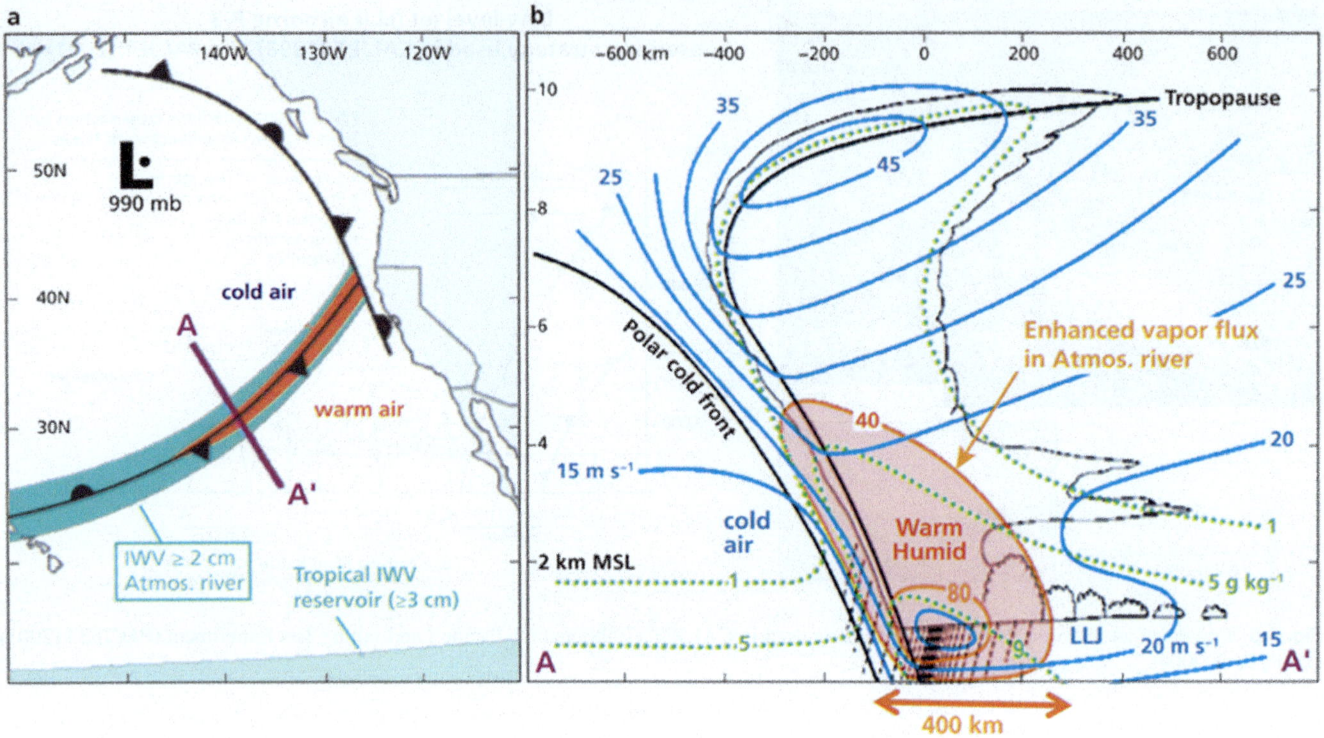

Fig. 1.8 Composite vertical structure of an AR from research aircraft and SSM/I satellite measurements (Ralph et al. 2004)

related experiences and additional publications, a consensus emerged on a working definition for ARs that centered on this IWV method, and many studies were conducted using this catalog.

For a number of years, IWV became the primary metric used to identify ARs, given the fundamental role of satellite-based IWV data. However, this metric does not account for wind, which is necessary to produce the moisture transport associated with ARs. Recently, the community has shifted more toward IVT as its preferred metric for AR identification due to the availability of in situ observing systems in some locations as well as the advent of high-quality atmospheric reanalysis products. Note that while IWV fields often differ from IVT fields (especially in areas of strong wind speeds), IWV is still regarded as a typically useful proxy over the ocean and away from the moist tropics.

One practical result of CALJET, from an operational forecasting perspective, was that accurately forecasting the LLJ—which transports water vapor along and ahead of the cold front associated with the mid-latitude cyclone—was recognized as being the key to accurately forecasting extreme precipitation (Morss and Ralph 2007).

AR Impacts: Precipitation, Flooding, and Water Supply

As the characteristics of ARs and processes associated with them were becoming increasingly well understood, researchers began turning their attention to impacts, particularly precipitation and flooding. A study by Ralph et al. (2006), based largely on data from the National Oceanic and Atmospheric Administration's (NOAA's) Hydrometeorological Testbed (Ralph et al. 2013), found that AR conditions occurred during all seven floods over the most recent 8-year period on the Russian River. In fact, they found that all seven floods over 8 years were associated with AR events (Fig. 1.9). This realization eventually led researchers to explore the effects of ARs on other watersheds and regions across the western USA (Dettinger et al. 2011; Guan et al. 2013; Rutz et al. 2014). Most recently, a study of many western US watersheds by Konrad and Dettinger (2017) found that the vast majority of high flows in watersheds in the region were associated with ARs. The results of Ralph et al. (2006), Neiman et al. (2008), and Dettinger et al. (2011), taken together, showed that ARs, which could generally be identified via SSM/I data, were critical not only to flooding, but also to water supply and hydroclimate across the western USA. Based on a global surface hydrology modeling study, this association seems extensible to many regions across the globe (Paltan et al. 2017).

The AR identification criteria developed by Ralph et al. (2004) were used, with SSM/I data, to develop a multi-year, observations-based catalog of landfalling ARs on the US West Coast (Neiman et al. 2008). Researchers were able to leverage this data set to examine the amount of resulting precipitation that was attributable to ARs, finding it to be substantial across many parts of the western USA (e.g., Dettinger et al. 2011). Late in the 2000s, AR perspectives from

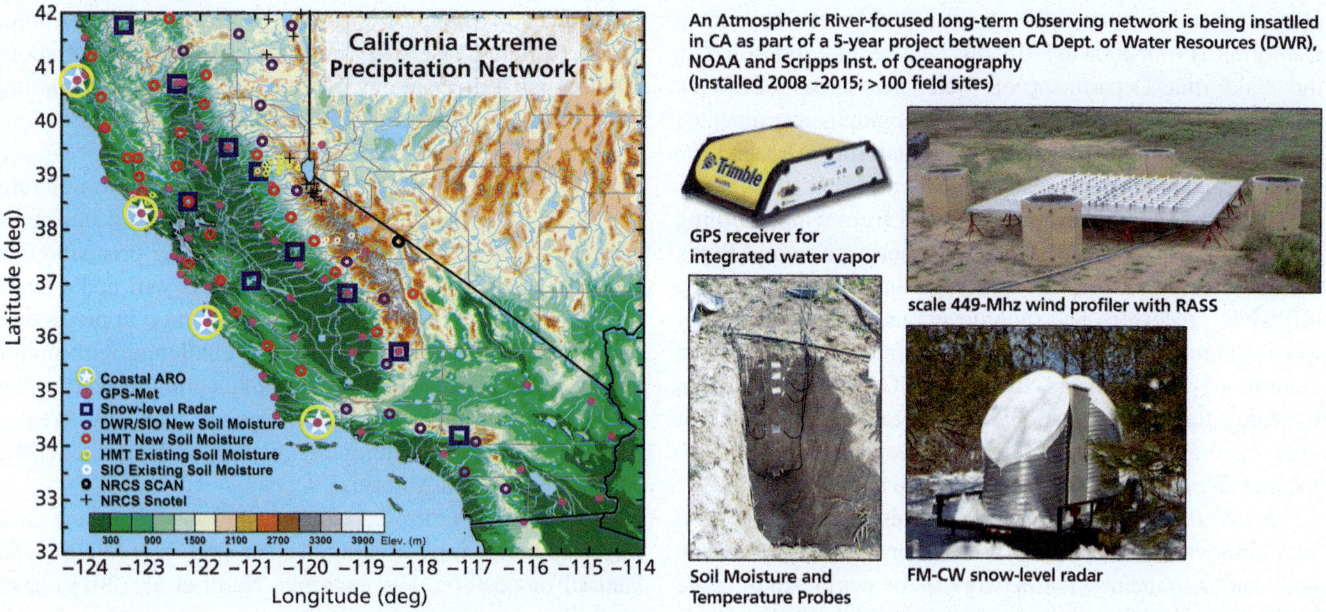

Fig. 1.9 California extreme precipitation network (White et al. 2013; Ralph et al. 2014)

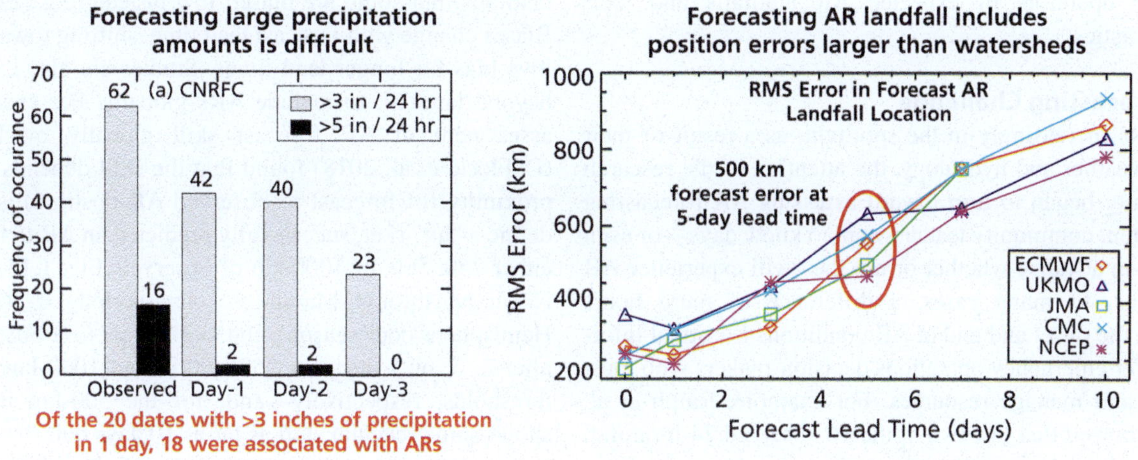

Fig. 1.10 The forecasting challenge (left panel: Ralph et al. 2010; right panel: Wick et al. 2013)

European scientists appeared in a key publication by Stohl et al. (2008), which also found major hydrometeorological impacts of an AR in Norway).

An evaluation of AR-driven, extreme precipitation totals over the western USA relative to primarily tropical storm-driven, extreme precipitation totals over the eastern USA. Ralph and Dettinger (2012) categorization of 72-h events based on accumulated precipitation marked the next key development in AR science history. Their results found that there are approximately two to three events that meet the Rainfall Category (RCAT) 3 or 4 threshold (>400 mm) each year across the USA (Fig. 1.10), which is similar to the number of Enhanced Fujita Scale (EF)–4 and EF–5 tornadoes or Category 4 and 5 hurricanes. In other words, these events are quite rare, nationally. Surprisingly, the occurrence of these extreme events was shown to be as frequent in California as anywhere east of the Rocky Mountains. Since most of these US West Coast events had been shown to be associated with ARs, these results highlighted the need for greater focus on AR science.

1.1.5 2010 and Beyond

The California AR Observation Network

By 2011, based on the results of a number of studies linking ARs to precipitation, flooding, and water supply (e.g., Ralph et al. 2006; Neiman et al. 2008; Dettinger et al. 2011), the state of California began investing in what is now a nationally recognized meso-scale network. It consists of various observational platforms to monitor ARs at over 100 sites throughout California. (From 2008 to 2015, this joint venture sponsored by the California Department of Water Resources (CDWR), NOAA, and Scripps Institution of Oceanography instrumented the region with sensors tailored to monitoring ARs and the extreme precipitation associated with them; Fig. 1.9).

The network includes measurements from wind-profiling radars on the coast (also called AR observatories or AROs), soil moisture, Global Positioning Satellite/Meteorology (GPS/Met) receivers, and snow-level radar, which—via very low-cost innovative design that utilizes frequency-modulated continuous-wave sensors from the NOAA Earth System Research Laboratory in Boulder, Colorado—gives the altitude of snow levels above the reservoirs where they are located. White et al. (2013) describe the network.

New York, Oklahoma, and other states are now starting to adopt innovations that the CDWR first invested in, and which now comprise the best AR network in the world. An ultimate applications-level goal is to improve the state's ability to predict extreme precipitation events; that is, to be able to inform water managers, emergency preparedness personnel, and reservoir operators of expected AR landfalls and their characteristics.

The Forecasting Challenge

With ARs increasingly in the spotlight, as a result of their role in weather and hydrology, the attention of the research community began to turn toward assessing AR forecasting skill. Often, community leaders want to know days—or even weeks—in advance whether or not they will experience AR conditions. In many cases, a difference of mere hours between the onset and end of AR conditions is critical information for emergency operations decision-makers who must prepare and manage resources. For example, Ralph et al. (2010) showed that just two of the 16 observed 24-h rainfall events with precipitation totals of 5 in. or greater were predicted to be severe at that location just 1 day ahead of the storm. Most of these extreme precipitation events were associated with landfalling ARs. This represented quite a challenge for the scientific (and operational) communities involved in predicting ARs and other extreme weather events (Fig. 1.10, left panel).

Wick et al. (2013) used analyses and forecasts from global modeling centers such as the NCEP and the European Centre for Medium-Range Weather Forecasts (ECMWF), among others, to assess how well they predicted[1] AR landfall position. They found that for events with short lead times (1–3 days), root-mean-square (RMS) errors are substantial (200–400 km; Fig. 1.10, right panel). Furthermore, these errors increase with time, generally at a rate of about 100 km per additional day of forecast lead time. Given that ARs are often about 500 km in width, that translates into forecasters with 5 days of lead time having to say, "We see an AR headed toward the West Coast, but we are not sure yet where the greatest impacts will be." From a water-management perspective, this lack of forecasting precision is a key issue. Errors in predicting where ARs will make landfall are similar to the errors from decades ago in predicting hurricanes; *that* community met the challenge with major advances in skill. The AR research community has the opportunity to meet this challenge as well, by greatly improving forecasts along the west coasts of the USA, Europe, and South America.

Very recent studies have continued to explore how well forecasts perform by looking at their skill in predicting AR landfall or position. For example, Nardi et al. (2018) have quantified the bias in frequency of occurrence of predicted AR landfalls from re-forecasting data from nine global models (Figs. 1.11 and 1.12). Most models predict too many AR landfalls, but some are biased low instead (e.g., ECMWF). Biases change with forecast lead time, shifting toward a positive bias for longer lead times. Studies are also expanding beyond landfall, to include ARs globally. For example, an assessment of AR forecast skill globally by ECMWF (DeFlorio et al. 2018) found that the skill depends on what proximity (of forecast to observed AR position) is used to define a "hit" (i.e., successfully predicted an AR to be within either 250, 500, or 1000 km of observed; Fig. 1.13). For the November-through-March time-period (Northern Hemisphere cool season), the skill drops to a 50% hit rate after 2, 5, or 7 days for 250-, 500-, and 1000-km distance thresholds, respectively—and, for the 500-km threshold, shows some residual skill as far as 10 days out.

Jim Hoke, then-Director of the Hydrometeorological Prediction Center,[2] observed that it was the community's work on ARs that gave their office the confidence to start issuing Day 6 and Day 7 forecasts for the West Coast, which eventually expanded to the entire USA. Growing knowledge about ARs thus enabled the WPC to create a valuable new product by extending the lead time out from 5 days to 7 days. Recent work by DeFlorio et al. (2019) has examined the viability of extended-range forecasts of AR activity across the globe, showing that in some regions, seasons, and climate mode modulations, AR activity can be forecast at 3 weeks or more with more skill than using climatology.

[1] In the fields of forecasting and prediction, "skill" is any measure of the accuracy and/or degree of association of prediction to an observation or estimate of the actual value of what is being predicted.

[2] One of the nine national centers of the NOAA-backed National Weather Service; known since 2013 as the Weather Prediction Center (WPC), but still called "Hydro-Met" by long-time weather scientists.

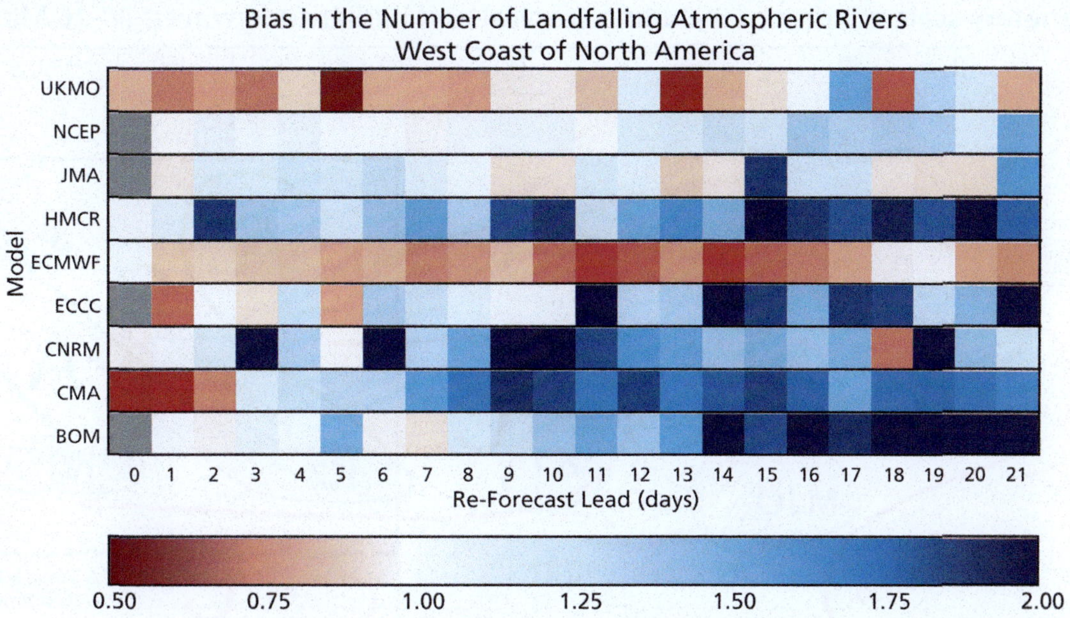

Fig. 1.11 Evaluation of biases in the number of AR landfalls in re-forecasts for the North American West Coast. (Nardi et al. 2018)

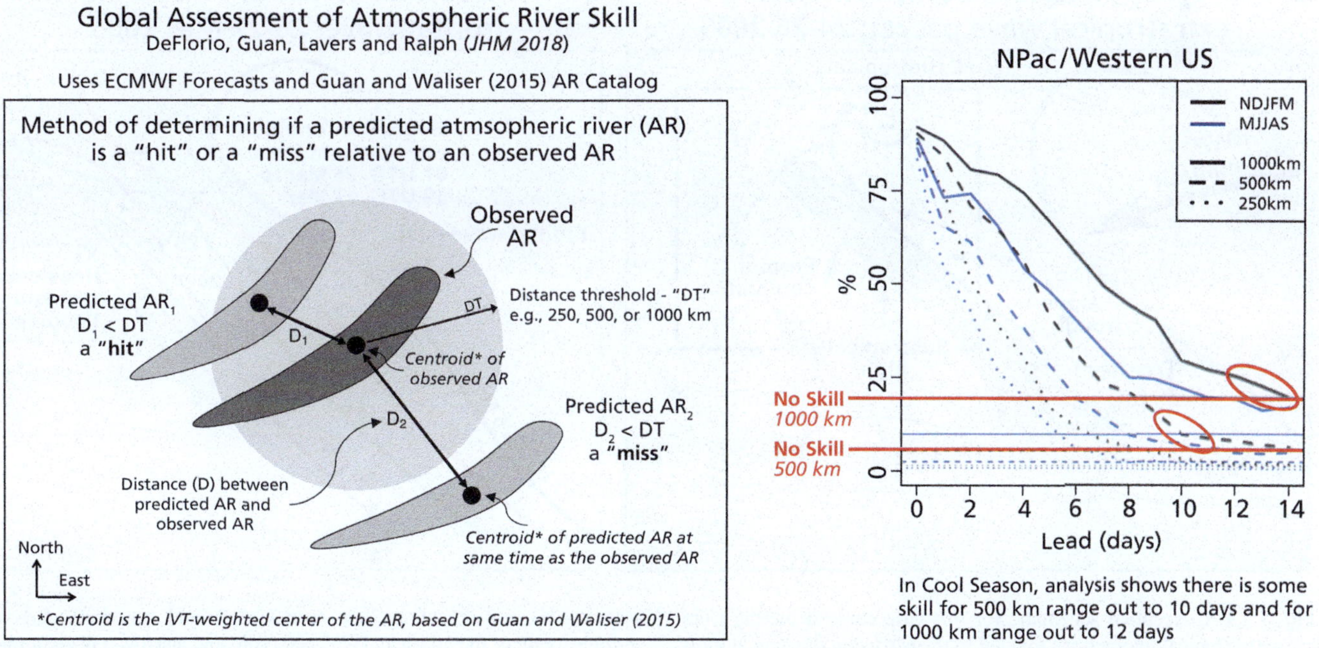

Fig. 1.12 Evaluation of the skill of AR forecasts globally, but shown for the North Pacific/Western USA. (DeFlorio et al. 2018)

AR Duration Found to Help Modulate AR Impacts

Although the maximum intensity of an AR in terms of IVT is a key characteristic, it is possible for a strong AR to last only briefly over an individual site. In contrast, another AR of similar maximum IVT could stall over one location, adding greatly to the storm-total precipitation and impacts. An early paper by Ralph et al. (2011) documents this and the dynamical origins of long-duration AR conditions at a single location. Through observations and diagnostics, this case study showed that a meso-scale frontal wave developed offshore and caused the landfalling AR to stall its progress over a portion of the coast. The AR lasted over 30 h at one site; that location experienced the largest storm-total precipitation in the event. Essentially, this case study showed how the stalling of an AR is a key process that determines the exact location of the most extreme precipitation in a landfalling AR, and that the development of a meso-scale frontal wave can cause this to occur. The study also illustrated the connection

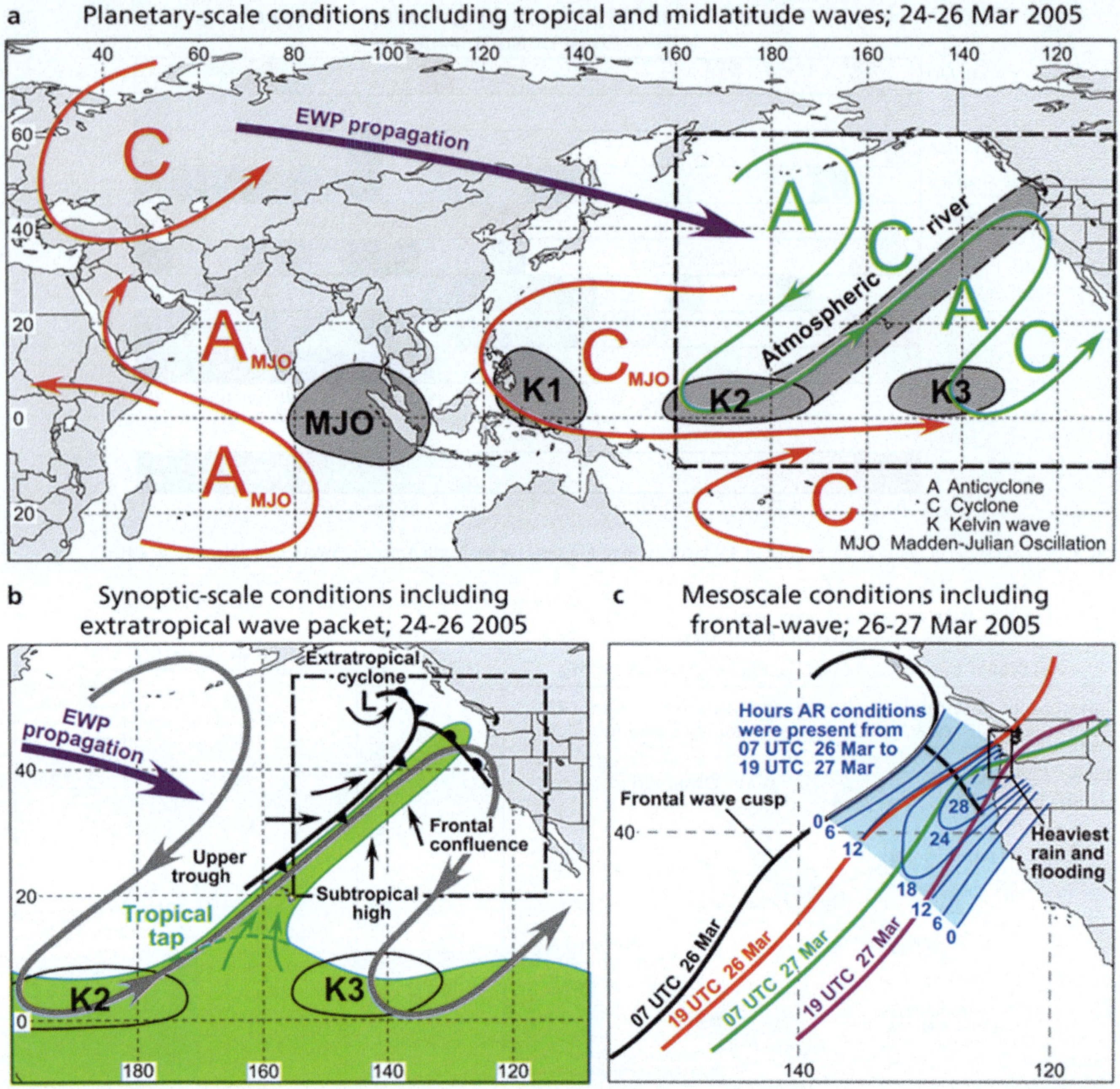

Fig. 1.13 Conceptual schematic for this case study depicting tropical–extratropical interactions that led to the extrusion of tropical moisture into an AR over the eastern Pacific on 24–26 March 2005: (**a**) Large-scale depiction of 150-hPa streamline anomalies (*red*: planetary-scale circulations; *green*: baroclinic wave packet [BWP] tied to the AR). The "A" and "C" labels refer to anticyclonic and cyclonic circulation centers, respectively. The *purple arrows* show the mean direction of BWP energy dispersion. *Gray shading* depicts Cloud Archive User Service (CLAUS) observations of coherent cold cloud tops associated with the MJO, 3 Kelvin waves (K1, K2, and K3), and the AR (enclosed within a *dashed line*). (**b**) Regional-scale depiction of the BWP (*thick gray-shaded arrow*; *purple arrow* shows propagation direction) and associated extratropical cyclone (standard frontal notation). *Green shading* depicts the tropical IWV reservoir and narrow IWV plume associated with the AR, and the *green arrows* depict the tapping of tropical water vapor into the AR. Kelvin waves 2 and 3 are enclosed with *thin, black lines*. The lower-tropospheric flow pattern is shown with *black arrows*. *Dashed inset boxes* in (**a**, **b**) correspond to the domains in the follow-on panels. Panel (**c**) is a frontal isochrone analysis on 26–27 March 2005 that shows a frontal wave propagating across the eastern Pacific and making landfall in northwestern Oregon where heavy rain and flooding occurred. The *blue isopleths* represent the number of hours of AR conditions, based on the isochrone analysis, and an assumption that the AR was 500 km wide. (Ralph et al. 2011)

among the Madden–Julian Oscillation (MJO), a tropical moisture export, a breaking Rossby wave, and the meso-scale frontal wave (Fig. 1.13).

Building on this case study, Ralph et al. (2013) pursued this further by looking at many cases. They used several years of ARO observations in northern California (see Sect. 1.5) to show that the average duration of ARs was 20 h, and that 75% of the variance in storm-total precipitation in the Russian River area could be explained by variance in the storm-total upslope component of water vapor transport (Fig. 1.14). This is partly because orographic forcing is a key mechanism, and ARs are ideally structured to produce orographic precipitation. Although this core connection that exists between the rate of water vapor supplied to the amount of precipitation largely holds true, significant spread remains in the relationship, especially for heavy precipitation events. This spread represents the influence of other physical processes such as dynamically forced vertical motions (frontal), convection, microphysics, and even aerosols. Exploring these factors is an active area of research. Additionally, the study showed that of the 91 cases studied, the top 10% longest-duration ARs averaged 40 h in duration. Most important, even though the duration doubled, the storm total runoff from these cases averaged seven times the average of all cases. This was because not only was the AR of longer duration, but the IWV and winds (and thus IVT) in these ARs were also larger, and they fell on soils that were already saturated by previous storms.

Since these first studies of AR duration, others have explored the topic, including Rutz et al. (2013). Although they used a different method to quantify duration, they found exactly the same result (20 h) as had Ralph et al. (2013). Moore et al. (2012) used 2-day, time-integrated IVT to study the role of a stalled AR in causing extreme flooding in Tennessee. Neiman et al. (2016) combined airborne and ground-based observations to show the role of a meso-scale frontal wave in causing a long-duration AR landfall event.

The duration of an AR is being used as one of the two AR characteristics required to define a scale for AR intensity and impacts AR categories (Ralph et al. 2019). The other is the maximum IVT in the AR. The AR scale ranges from 1 to 5 and can be applied to observations, reanalysis, or forecasts; it is based on the time-series of IVT at any individual geographic location (see Chap. 8).

1.1.6 ARs and Global Climate Change

Looking forward from these discoveries and advances, another question that has been explored increasingly through the 2010s is the future of ARs in a time of global climate change.

As noted earlier, in chronicling Zhu and Newell's (1998) contributions to introducing the concept—and coining the term for—"atmospheric rivers," over 90% of all the poleward water vapor transports in the atmosphere outside of the tropics are concentrated into a relative handful of ARs of intense and narrow bands of vapor transport. Taken together on the average day, they reported, the ARs conducting this overwhelming majority of the atmospheric water cycle out-

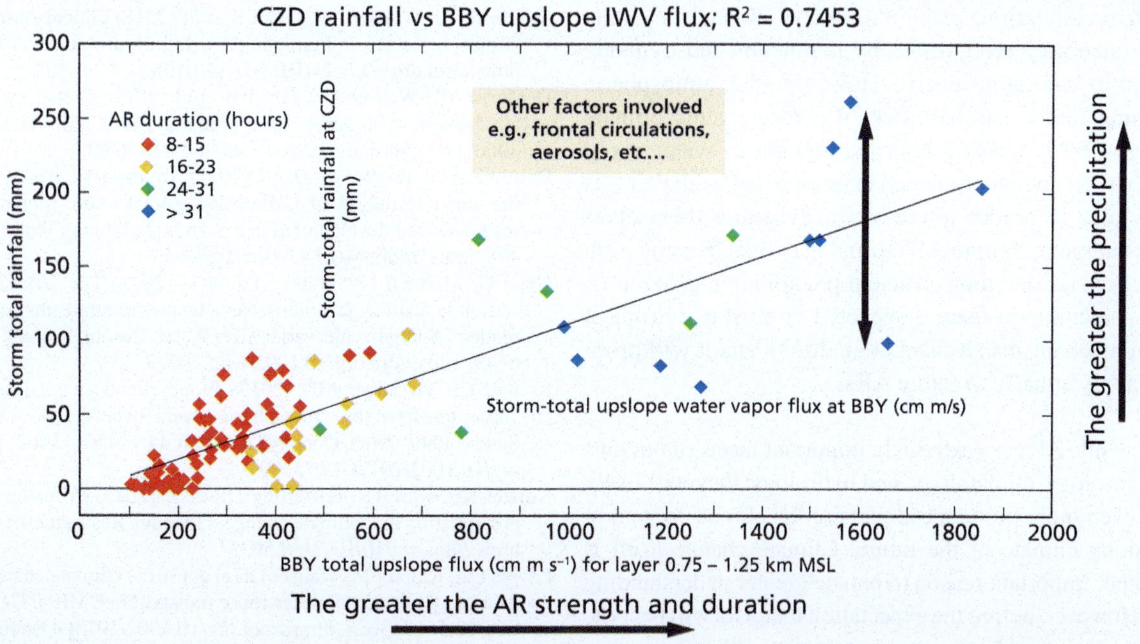

Fig. 1.14 The greater the AR strength and duration, the greater the precipitation (Ralph et al. 2013)

side the tropics spanned only about 5–10% of the Earth's total circumference. The following chapters will discuss at length many important consequences of ARs. But one of the more subtle consequences has been this realization that nearly all of the large-scale transports of water vapor in the (extratropical) atmosphere occur through the mechanism of these narrow ARs. Thus the large-scale transports that make up much of the global water cycle are almost entirely along and within these narrow AR corridors.

Given their importance, it is natural to wonder how ARs will change as the global climate changes in response to increasing greenhouse-gas concentrations in the atmosphere. This is not just an idle possibility. Rather, it is very natural for two reasons that ARs are closely tied to climate change:

1. Under climate change, the atmosphere will warm significantly and, because warmer air can hold and transport more water vapor, the amounts of vapor transport will increase dramatically (e.g., Lavers et al. 2015). Because ARs are the primary mechanism by which water vapor is transported through the atmosphere poleward of the tropics, and because there is no reason to suspect that this central role of ARs in global-scale vapor transports will change, it will be surprising if ARs are not modified as global climate changes. Given the larger quantities of water vapor that will be transported globally, overall amplification of future ARs is the most likely outcome and, indeed, is what current climate projections predict (see Chap. 6).
2. Amplification of future ARs is, of course, a great concern, given their central role in floods, healthy ecosystems, and water resources in many extratropical settings (see Chap. 5). The more vapor that future ARs transport, the more water, precipitation, and—ultimately—floods they may yield once they are uplifted by mountains and dynamic forces in the atmosphere. However, the atmospheric warming that is the hallmark of anthropogenic climate change—by increasing both the amount of water vapor and thermal energy contained in and carried by the air—is likely also to render future storm dynamics themselves more energetic. Future AR storms may then become even more extreme and more efficient precipitation generators. This is a common feature expected of most extratropical storm mechanisms (Kunkel et al. 2013), and it will probably apply equally to future ARs.

Thus, while ARs are extremely important facets of our current meteorology, climatology, and hydrology, they will likely become even more pronounced parts of the climate system in the warming climate of the future. Climate change itself is thus another important reason to pursue greater understanding of ARs. However, before the expectation that ARs will become even more important is accepted, much more about the nature, distributions, and impacts of ARs in today's world need to be understood, including how they are modeled and predicted. Much of the rest of this book offers that foundation.

1.2 Structure of This Book

Examples and details of all these findings of recent AR research comprise the rest of this text. Chapter 2 describes the structure and workings of ARs and their place among other key features of the global water cycle. Chapter 3 details the methods now used to observe, detect, and characterize ARs. Chapter 4 presents the global distribution and characteristics of ARs and Chapter 5 discusses their many global effects, both adverse impacts and vital benefits. Chapter 6 discusses both forecasting of ARs and projections of their futures under climate change. Chapter 7 presents some emerging applications of AR science, and Chapter 8 presents some questions that AR research may address in the coming decade. Ideally, this book will provide a foundation for the AR research and applications to come.

References

Browning KA (1980) Structure, mechanism and prediction of orographically enhanced rain in Britain. In: Hide H, White PW (eds) Orographic effects in planetary flows. WMO GARP Publications Series No. 23, pp 85–114

Browning KA, Pardoe CW (1973) Structure of low level jet streams ahead of mid-latitude cold fronts. Q J Roy Meteorolog Soc 99(422):619–638. https://doi.org/10.1002/qj.49709943304

Carlson TN (1980) Airflow through mid-latitude cyclones and the common cloud patterns. Mon Wea Rev 108:1498–1509. (Figure 9?)

DeFlorio MJ, Waliser DE, Guan B et al (2018) Global Assessment of Atmospheric River Prediction Skill. J Hydrometeor 19:409–426. https://doi.org/10.1175/JHM-D-17-0135.1

DeFlorio MJ, Waliser DE, Guan B et al (2019) Global evaluation of atmospheric river subseasonal prediction skill. Clim Dyn 52:3039–3060. https://doi.org/10.1007/s00382-018-4309-x

Dettinger MD, Ralph FM, Das T (2011) Atmospheric rivers, floods and the water resources of California (special issue: Managing water resources and development in a changing climate). Water 3(2):445–478. https://doi.org/10.3390/w3020445

Guan G, Molotch NP, Waliser DF et al (2013) The 2010/2011 snow season in California's Sierra Nevada: role of atmospheric rivers and modes of large-scale variability. Water Resour Res 49(10):6731–6743. https://doi.org/10.1002/wrcr.20537

Konrad CP, Dettinger MD (2017) Flood runoff in relation to water vapor transport by atmospheric rivers over the Western United States, 1949–2015. Geophys Res Lett 44:11456–11462. https://doi.org/10.1002/2017GL075399

Kunkel KE, Karl TR, Easterling DR et al (2013) Probable maximum precipitation and climate change. Geophys Res Lett 40:1402–1408. https://doi.org/10.1002/grl.5033

Lavers DA, Ralph FM, Waliser DE et al (2015) Climate change intensification of horizontal water vapor transport in CMIP5. Geophys Res Lett 42:5617–5625. https://doi.org/10.1002/2015GL064672

Morss RE, Ralph FM (2007) Use of information by national weather service forecasters and emergency managers during CALJET and PACJET-2001. Wea Forecast 22:539–555. https://doi.org/10.1175/Waf1001.1

Moore BJ, Neiman PJ, Ralph FM et al (2012) Physical processes associated with heavy flooding rainfall in Nashville, Tennessee, and vicinity during 1–2 May 2010: the role of an atmospheric river and mesoscale convective systems. MonWea Rev 140(2):358–378

Nardi KM, Barnes EA, Ralph FM et al (2018) An assessment of numerical weather prediction model re-forecasts of the occurrence, intensity, and location of atmospheric rivers along the west coast of North America. J Hydrometeorol 146:3343–3362

Neiman PJ, Ralph FM, Wick GA et al (2008) Meteorological characteristics and overland precipitation impacts of atmospheric rivers affecting the west coast of North America based on eight years of SSM/I satellite observations. J Hydrometeorol 9(1):22–47. https://doi.org/10.1175/2007JHM855.1

Newell RE, Newell NE, Yong Z et al (1992) Tropospheric rivers? – a pilot study. Geophys Res Lett 19(24):2401–2404. https://doi.org/10.1029/92GL02916

Paltan H, Waliser D, Lim W-H et al (2017) Global floods and water availability driven by atmospheric rivers. Geophys Res Lett44(20):10387–10395. https://doi.org/10.1002/2017GL074882

Ralph FM, Dettinger MD (2012) Historical and national perspectives on extreme west-coast precipitation associated with atmospheric rivers during December 2010. Bull Am Meteorol Soc 93(6):783–790. https://doi.org/10.1175/BAMSOD-11-001881

Ralph FM, Neiman PJ, Wick GA (2004) Satellite and CALJET aircraft observations of atmospheric rivers over the eastern North Pacific Ocean during the winter of 1997/98. Mon Wea Rev 132(7):1721–1745

Ralph FM, Neiman PJ, Rotunno R (2005) Dropsonde observations in low-level jets over the northeastern Pacific Ocean from CALJET-1998 and PACJET-2001: mean vertical-profile and atmospheric-river characteristics. Mon Wea Rev 133:889–910. https://doi.org/10.1175/MRW2896.1

Ralph FM, Neiman PJ, Wick GA et al (2006) Flooding on California's Russian River: role of atmospheric rivers. Geophys Res Lett 33(13):L13801. https://doi.org/10.1029/2006GLO26689

Ralph FM, Neiman PJ, Kiladis GM et al (2010) A multiscale observational case study of a Pacific atmospheric river exhibiting tropical– extratropical connections and a mesoscale frontal wave. Mon Wea Rev 139(4):1169–1189. https://doi.org/10.1175/2010MWR3596.1

Ralph FM, Neiman PJ, Kiladis GN et al (2011) A multiscale observational case study of a Pacific atmospheric river exhibiting tropical–extratropical connections and a mesoscale frontal wave. Mon Wea Rev 139(4):1169–1189. https://doi.org/0.1175/2010MWR3596.1

Ralph FM, Coleman T, Neiman PJ et al (2013) Observed impacts of duration and seasonality of atmospheric-river landfalls on soil moisture and runoff in coastal northern California. J Hydrometeor 14(2):443–459

Ralph FM, Dettinger M, White A et al (2014) A vision for future observations for Western U.S. extreme precipitation and flooding. Special Issue of J. *Contemporary Water Resources Research and Education*, Universities Council for Water Resources, 153:16–32.

Ralph FM, Dettinger M, Lavers D et al (2017) Atmospheric Rivers Emerge as a Global Science and Applications Focus. Bull. Amer. Meteor. Soc., 98, 1969–1973, https://doi.org/10.1175/BAMS-D-16-0262.1

Ralph FM, Dettinger MD, Cairns MM et al (2018) Defining "atmospheric river": how the *Glossary of Meteorology* helped resolve a debate. Bull Amer Meteor Soc 99(4):837–839. https://doi.org/10.1175/BAMS-D-17-0157.1

Ralph FM, Rutz JJ, Cordeira JM et al (2019) A scale to characterize the strength and impacts of atmospheric rivers. Bull Am Met Soc 100(2):269–289

Rutz JJ, Steenburgh WJ, Ralph FM (2014) Climatological characteristics of atmospheric rivers and their inland penetration over the western United States. Mon Weather Rev 42(2):905–921. https://doi.org/10.1175/MWR-D-13-00168.1

Stohl A, Forster C, Sodemann H et al (2008) Remote sources of water vapor forming precipitation on the Norwegian west coast at 60°N—a tale of hurricanes and an atmospheric river. J Geophys Res Atmos 113(D5):1–13. https://doi.org/10.1029/2007JD009006

Wernli JH (1995) Lagrangian perspective of extratropical cyclogenesis. Dissertation No. 11016. Swiss Federal Institute of Technology (ETH), Zürich. 157 pp

Wernli H (1997) A Lagrangian-based analysis of extratropical cyclones. II: a detailed case-study. Q J Roy Meteorol Soc 123(1):1677–1706

Wernli H, Davies HC (1997) A Lagrangian-based analysis of extratropical cyclones. I: the method and some applications. Q J Roy Meteorol Soc 123:467–489

White AB, Anderson AL, Dettinger MD et al (2013) A twenty-first-century California observing network for monitoring extreme weather events. J Atmos Ocean Technol 30(8):1585–1603. https://doi.org/10.1175/JTECH-D-12-00217.1

Wick GA, Neiman PJ, Ralph FM et al (2013) Evaluation of forecasts of the water vapor signature of atmospheric rivers in operational numerical weather prediction models. Weather Forecast 28(6):1337–1352. https://doi.org/10.1175/WAF-D-13-00025.1

Zhu Y, Newell R (1998) A proposed algorithm for moisture fluxes from atmospheric rivers. Mon Wea Rev 126(3):725–735. https://doi.org/10.1175/1520-0493(1998)126<0725:APAFMF>2.0.CO;2

Structure, Process, and Mechanism

2

Harald Sodemann, Heini Wernli, Peter Knippertz, Jason M. Cordeira, Francina Dominguez, Bin Guan, Huancui Hu, F. Martin Ralph, and Andreas Stohl

2.1 Introduction

This chapter describes the 3-dimensional (3-D) structure of ARs and explains the relevant processes and mechanisms involved in their formation, maintenance, and decay. This undertaking is ambitious because there are potential misconceptions about ARs, and, more fundamentally, because achieving a detailed understanding of AR dynamics is complex, given the many processes involved and their interactions. Misconceptions that require clarification include the following:

ARs transport tropical moisture toward the poles

Evaporation occurs in the entrance zone of an AR and precipitation near the exit

ARs replace the earlier concept of warm conveyor belts

In particular, Sects. 2.3 and 2.4 address these points by illustrating the complexity of the atmospheric water cyclone along ARs, and by comparing the AR concept to the alternative flow concepts of warm conveyor belts (WCBs) and tropical moisture exports (TMEs).

Section 2.2 first provides a concise description of ARs, including their definition and horizontal and vertical structure. The definition starts with "an AR is a long, narrow, and transient corridor of strong horizontal water vapor transport that is typically associated with a low-level jet (LLJ) stream ahead of the cold front of an extratropical cyclone …" and includes elements about their geometry (long and narrow shape), their key characteristic (strong horizontal water vapor transport), and their dynamical embedding (LLJ ahead of an extratropical cyclone's cold front).

Section 2.5 illustrates and explains in more detail these linkages to classical concepts of dynamical meteorology.

Overall, this chapter contains fairly technical discussions, but they serve to offer a more complete picture of the 3-D structure and transient nature of ARs; their relationship to cyclones, fronts, WCBs, and TMEs; as well as the complex interplay of AR dynamics with associated moisture transport and patterns of ocean evaporation and precipitation.

H. Sodemann
Geophysical Institute, University of Bergen and Bjerknes Centre for Climate Research, Bergen, Norway
e-mail: harald.sodemann@uib.no

H. Wernli
Institute for Atmospheric and Climate Science, ETH Zürich, Zürich, Switzerland
e-mail: heini.wernli@env.ethz.ch

P. Knippertz (✉)
Institute of Meteorology and Climate Research, Karlsruhe Institute of Technology, Karlsruhe, Germany
e-mail: peter.knippertz@kit.edu

J. M. Cordeira
Plymouth State University, Plymouth, NH, USA
e-mail: j_cordeira@plymouth.edu

F. Dominguez
University of Illinois, Urbana–Champaign, Champaign, IL, USA
e-mail: francina@illinois.edu

B. Guan
University of California, Los Angeles, Los Angeles, CA, USA
e-mail: bin.guan@jpl.nasa.gov

H. Hu
Pacific Northwest National Lab, Richland, WA, USA
e-mail: huancui.hu@pnnl.gov

F. M. Ralph
Center for Western Weather and Water Extremes, Scripps Institution of Oceanography, University of California–San Diego, La Jolla, CA, USA
e-mail: mralph@ucsd.edu

A. Stohl
NILU - Norwegian Institute for Air Research, Kjeller, Norway
e-mail: ast@nilu.no

F. M. Ralph et al. (eds.), *Atmospheric Rivers*, https://doi.org/10.1007/978-3-030-28906-5_2

2.2 Structure of ARs

This section provides a concise description of ARs, including their structure and how some of their key characteristics vary over many samples, as well as brief background on the origins of this information. The term's definition was developed for the *Glossary of Meteorology* in 2017 (Ralph et al. 2018). The definition includes a summary of the vertical and horizontal structure of ARs as informed by 21 aircraft-based measurements of ARs over the eastern North Pacific from several field campaigns (Ralph et al. 2017a). The representativeness of these difficult-to-collect observations—and the typical range of key AR characteristics across many samples—are summarized here based on a comparison of those 21 measurements with examination of several thousand ARs in reanalyses by Guan et al. (2018). (Reanalyses are gridded global meteorological data sets produced by numerical weather prediction centers based on all routinely available observations, to create a composite plan view.)

2.2.1 Definition of the Term "Atmospheric River"

The development of a formal definition was triggered by requests received by the GoM Editor from the meteorological community. These requests reflected a dramatic increase in publications that referred to the term, from 10 papers before 2004, to over 600 between then and 2016 (Ralph et al. 2017b). A process for community input was established. It included formation of a committee comprising four experts: including chairs of three standing American Meteorological Society (AMS) technical committees (mesoscale, hydrology, water resources) and the GoM Editor. The committee created a draft definition, building on a draft created earlier at the International Atmospheric Rivers Conference (IARC) in 2016—the first conference specifically addressing research on ARs. The draft was then discussed at two dedicated Town Hall meetings, one at the 2016 American Geophysical Union (AGU) Fall Meeting, and one at the 2017 AMS Annual Meeting. Each Town Hall included five or six diverse panelists who offered their opinions. Roughly 250 people attended these open forums, many of whom asked questions or offered comments. Because there had been diverse opinions about the appropriateness of the term "atmospheric river"—including how it related to the LLJ studied earlier (e.g., Browning and Pardoe 1973) and about what an "atmospheric river" was—the process of creating the definition was summarized in a brief article in the Bulletin of the American Meteorological Society (Ralph et al. 2018).

The committee settled on the following definition, which was then published in the GoM in May 2017: An atmospheric river is

> a long, narrow and transient corridor of strong horizontal water vapor transport that is typically associated with a LLJ stream ahead of the cold front of an extratropical cyclone. The water vapor in atmospheric rivers is supplied by tropical and/or extratropical moisture sources. Atmospheric rivers frequently lead to heavy precipitation where they are forced upward, e.g., by mountains or by ascent in the warm conveyor belt. Horizontal water vapor transport in the mid-latitudes occurs primarily in atmospheric rivers and is focused in the lower troposphere.

2.2.2 Water Vapor Transport and the Vertical and Horizontal Structure of ARs

Direct Observations of Water Vapor Transport

Between 2005 and 2017, several field campaigns used research or weather reconnaissance aircraft to release dropsondes across ARs over the northeast Pacific during winter (Ralph et al. 2017a). The campaigns were "Ghost Nets" in 2005, "WISPAR" in 2011, "CalWater" in 2014 and 2015, and "AR Recon" in 2016. The National Oceanic and Atmospheric Administration (NOAA) P-3 and G-IV, National Aeronautics and Space Administration (NASA) Global Hawk, and Air Force C-130 aircraft were used. In each case, an aircraft flew at high altitude across an AR (Fig. 2.1), and released dropsondes—14 on average—at about a 100-km horizontal spacing over 1000–1500-km baseline across an AR. Dropsondes measured vertical profiles of water vapor, temperature, wind, and pressure with high vertical resolution as they descended to the ocean surface over 10–20 min. These data allowed calculation of vertically integrated water vapor (IWV) and vertically integrated water vapor transport (IVT). The campaigns were all in January, February, or March and spanned from roughly 20–50°N.

Each dropsonde transect across an AR could be used to calculate the width of the AR, its strength in terms of maximum IVT, and its total IVT (TIVT), which represents the total horizontal flux of water vapor in the AR, integrated horizontally from one edge of the AR to the other, and vertically from the surface to 300 hPa. Because the width and TIVT—which determines where an AR's edges are laterally—depend on the definition of an AR, many cases were analyzed to explore the relative importance of using IWV or IVT as a basis for defining an AR. (IWV was used initially because it was available directly from satellite; it was introduced by Ralph et al. 2004). Ralph et al. (2017a) showed just a few percent difference in width and TIVT in terms of the average of 21 cases. The use of IVT as the basis was also shown as most robust because at lower latitudes IWV sometimes never dropped below the standard

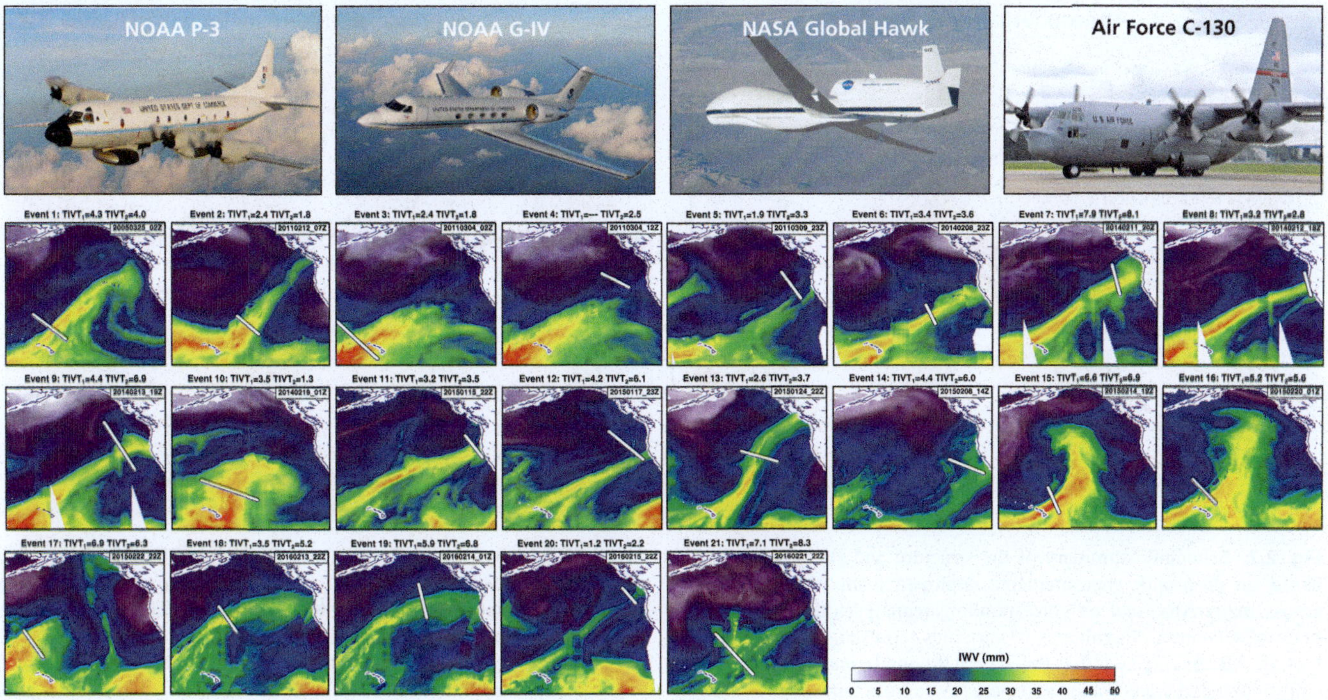

Fig. 2.1 (Top row) Aircraft used in collecting dropsonde data between 2005 and 2016 to develop the observations-based composite of the cross-AR vertical structure (see Fig. 2.2). (Bottom rows) Satellite images of vertically integrated water vapor (IWV) including the AR transect baseline for each of the flights used in the study. The values of TIVT ($\times 108$ kg s^{-1}) are shown as text atop each IWV panel for each case based on both the IWV (*subscript 1*) and IVT (*subscript 2*) AR threshold methods. (Adapted from Ralph et al. 2017a)

2-cm threshold, although IVT did. Also, at higher latitudes, some ARs were well defined by IVT, but, because the air was cooler than at lower latitudes, IWV sometimes remained below the 2-cm threshold. Not only is IVT more robust as a basis, it is a more appropriate parameter than IWV for a phenomenon focused on horizontal water vapor transport. Thus, keeping in mind that shape requirements (e.g., long and narrow) are often used in addition to the IVT threshold, and that researchers have had success using a variable IVT threshold that is scaled to local climatology and varies with season, IVT = 250 kg s^{-1} m^{-1} was recommended as the most useful threshold for defining an AR (as also used in Sect. 2.1).

A key result of the compositing and comparisons was the conclusion that an average AR is roughly 850 ± 250 km wide, and transports $4.7 \pm 2.0 \times 10^8$ kg s^{-1} of water vapor, equivalent to roughly 25 times the Mississippi River's (or 2.5 times the Amazon River's) discharge into the ocean.

Observations of Vertical and Horizontal Structure

The dropsonde transects were then used to create a composite vertical cross-section and combined with reanalysis data to create a composite plan view. These are shown in Fig. 2.2. To create the composite, the center point of each AR was identified as the position of maximum IVT. Each AR was then adjusted in terms of position and orientation to enable spatial compositing.

Figure 2.2a shows that ARs are typically centered over 1000 km from their parent extratropical cyclone (L), and their position is aligned with and overlaps the cold front and trailing stationary front. The AR intersects the warm front, but is depleted through precipitation from ascent in the warm conveyor belt (WCB; e.g., Madonna et al. 2014). The average maximum value of IVT is 700 kg s^{-1} m^{-1}. The vertical structure is shown in Fig. 2.2b and shows its position relative to the jet-front system and tropopause, including an upper-level jet of over 60 m s^{-1} and a LLJ of about 30 m s^{-1}. Although upper-level winds are roughly twice as strong as the LLJ, the air is cold and dry at high altitudes. The water vapor mixing ratio in the LLJ averages over 9 g kg^{-1}, roughly 10 times the value in the upper jet. The result is that water vapor flux is focused in the lower troposphere, with roughly 75% of IVT located below 3 km mean sea level (MSL), and <1% occurring in the upper jet (i.e., above about 400 hPa or 8 km MSL; Ralph et al. 2017a).

The plan-view composite (Fig. 2.2a) includes IWV, which shows an anomaly of large IWV extending from lower (closer to the equator) latitudes into the AR, a feature that is evident in the IWV fields shown in Fig. 2.2 for each AR. Although the composite IVT does not show a clear connection to the lower latitudes, the IWV signature may

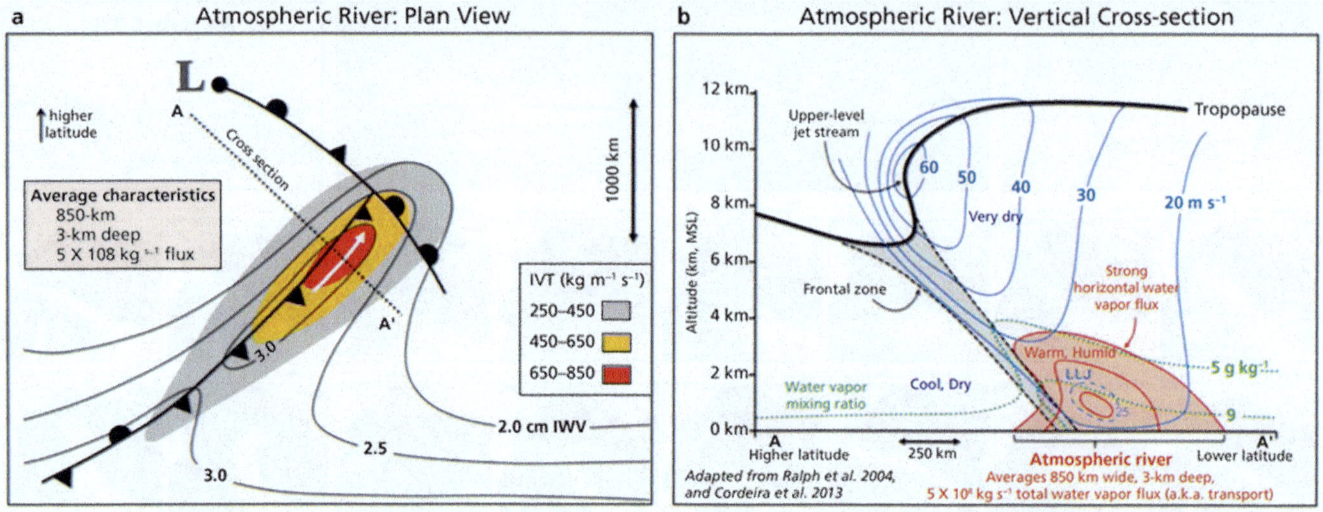

Fig. 2.2 Schematic summary of the structure and strength of an AR based on dropsonde measurements deployed from research aircraft across many ARs, and on corresponding reanalyses that provide the plan-view context. Magnitudes of variables represent an average mid-latitude AR. Average width is based on AR boundaries defined by vertically integrated water vapor transport (IVT; from surface to 300 hPa) lateral boundary threshold of 250 kg m^{-1} s^{-1}. Depth corresponds to the altitude below which 75% of IVT occurs. The total water vapor transport (AKA flux) corresponds to the transport along an AR, bounded laterally by the positions of IVT = 250 kg m^{-1} s^{-1}, and vertically by the surface and 300 hPa. (**a**) Plan view including parent low pressure system, and associated cold, warm, stationary, and warm-occluded surface fronts. IVT is shown by *color fill* (magnitude, kg m^{-1} s^{-1}) and direction in the core (*white arrow*). Vertically integrated water vapor (IWV, cm) is *contoured*. A representative length scale is shown. The position of the cross-section shown in panel (**b**) is denoted by the *dashed line* A–A′. (**b**) Vertical cross-section perspective, including the core of the water vapor transport in the AR (*orange contours* and *color fill*) and the pre-cold-frontal low-level jet (LLJ), in the context of the jet-front system and tropopause. Water vapor mixing ratio (*green dotted lines*, g kg^{-1}) and cross-section-normal isotachs (*blue contours*, m s^{-1}) are shown. (From Ralph et al. (2017a), including elements from Cordeira et al. (2013) and others; Schematic prepared by F. M. Ralph, J. Cordeira, and P. J. Neiman)

represent a link to TMEs (e.g., Knippertz et al. 2013) that are characterized by poleward transport of tropical water vapor into the extratropics. Although many ARs can form without a tropical connection, it is clear that in some cases an AR can entrain significant water vapor from the tropics (e.g., Sodemann and Stohl 2013), and that this entrained water vapor may largely be rained out before reaching mid-latitudes (e.g., Cordeira et al. 2013)

Representativeness of Airborne Observations and Typical Range of Key Characteristics

As much as the 21 aircraft-observed ARs represent a solid basis for calculating the mean and creating a composite, examining many more ARs would be helpful. This would help document the fuller range of AR conditions, as well as the representativeness of the 21 ARs. However, this was impractical in terms of aircraft, so Guan et al. (2018) turned to the European Centre for Medium-Range Weather Forecasts (ECMWF's) European Reanalysis (ERA)-Interim and Modern-Era Retrospective Analysis for Research and Application (MERRA)-2 reanalysis products. This allowed examination of ARs in the northeast Pacific during January–March each year from 1979 to 2016, including the cases documented by aircraft in Ralph et al. (2017a).

Guan et al. (2018) used the AR Detection Method (ARDM) Guan and Waliser (2015) developed, which uses a spatially-variable IVT threshold that represents the 85th to 95th percentile of IVT observed at a given grid cell in the reanalysis product. The threshold is also allowed to vary by season. AR "objects" then must meet shape requirements in terms of length and length-to-width ratio. The ARDM identified almost 6000 ARs during this period and in this large area. Figure 2.3a, b shows the spectrum of AR widths and AR TIVTs for these ARs.

The mean width of all ~6000 ARs in the reanalysis was 847 km. The mean of the reanalysis version of the 21 aircraft cases was 868 km. Both are remarkably close (within 2–5%) to the value of 886 km from the 21 aircraft observations. Similarly, the mean TIVT of all ~6000 ARs in the reanalysis was 4.49 × 10^{8} kg s^{-1}. The mean of the reanalysis version of the 21 aircraft cases was 4.86 × 10^{8} kg s^{-1}. Both are remarkably close (within 5–10%) to the value of 4.71 × 10^{8} kg s^{-1} from the 21 aircraft observations. These similarities are even more remarkable in that they used significantly different methods than in Ralph et al. (2017a, b). This agreement between aircraft observations and reanalyses adds robustness to the mean AR widths and TIVTs presented in Fig. 2.2 for the northeast Pacific in January–March.

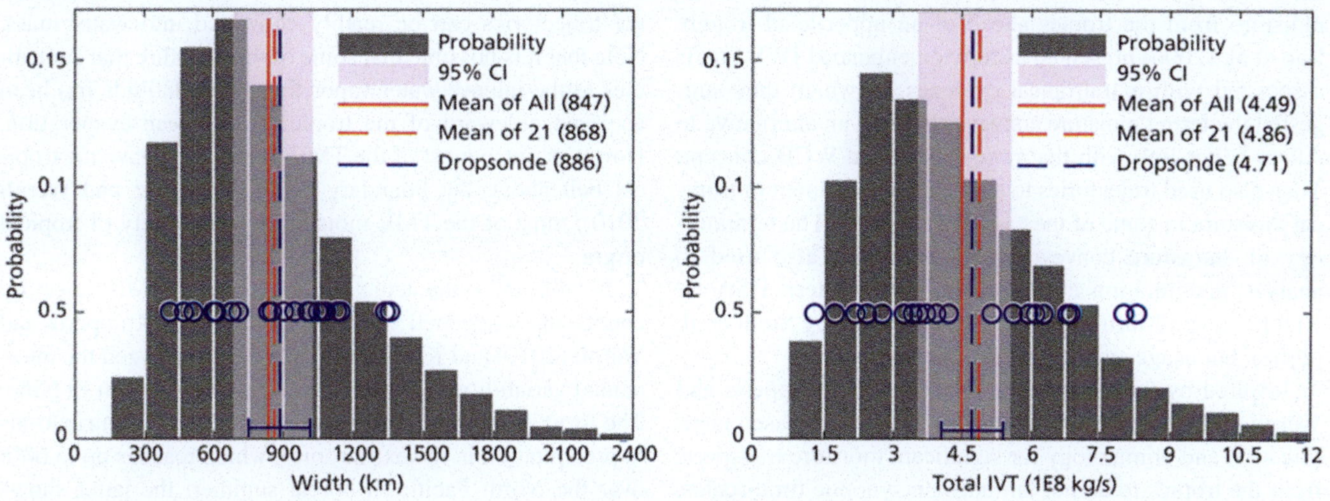

Fig. 2.3 (left) Frequency of occurrence of AR widths based on ERA-Interim and MERRA2 reanalysis data using the Guan and Waliser (2015) ARDT applied to the northeast Pacific basin in January–March each year from 1979–2016. Means are shown of all ~6000 ARs (*solid red line*), the reanalysis representations of the 21 aircraft-observed ARs (*dashed red line*), and the aircraft observations (*dashed blue line*). The 95% confidence range for the mean is shown (*red shading*), as is the distribution of the actual aircraft-measured cases (*blue circles*). (right) Same as (left), but for TIVT (Guan et al. 2018)

2.3 WCBs and TMEs and Their Relationship to ARs

In the meteorological literature, different concepts exist that describe meso- to synoptic-scale filamentary features of enhanced absolute moisture content or moisture transport (sometimes vertically integrated), clouds, and precipitation. In addition to ARs, these include WCBs, moisture conveyor belts, moisture bursts, TMEs, tropical plumes, and tropical intrusions (see, e.g., Knippertz (2007) for an overview of the tropical phenomena). These concepts—usually promoted by different research groups—deviate from each other by emphasizing different meteorological aspects. For some of these features, clear objective identification criteria exist, but others are more loosely defined phenomenologically. Naturally, the boundaries between these concepts are not sharp, leading—at least partly—to coincidences in time and space or temporal succession of features. The emphasis on different meteorological parameters leads to differing geographical, seasonal, and inter-annual variations in the frequency of these features.

This multitude of concepts and the lack of sharp boundaries and systematic comparisons have led to a certain level of confusion in the scientific community about each concept's specific definition, purpose, and usefulness. This section focuses on two concepts that have explicitly been linked to ARs in past literature and for which objective identification methods exist: WCBs and TMEs.

TMEs consist of trajectories characterized by an intense horizontal export of tropical moisture into subtropical and mid-latitudes.

WCBs are trajectories with a strong ascent in the vicinity of extratropical cyclones and their fronts.

Excluded are other concepts that mostly rely on cold clouds identified from infrared satellite imagery, such as tropical plumes; they often occur at upper-tropospheric levels only, and their identification is sensitive to the satellite data set used (Fröhlich et al. 2013). This section:

1. portrays the key characteristics of the three concepts: AR, WCB, and TME—and how they are objectively identified (including a short illustrative case study)
2. quantifies how often the three features partially overlap in space and time (including a comparison of global climatologies for the two solstice seasons and statistical linkages between them)
3. provides a physical interpretation of the findings

All analyses are based on 6-hourly ERA-Interim reanalysis data (Dee et al. 2011) for the time-period 1979–2014 interpolated to a regular 1°-by-1° grid with 60 vertical levels. By systematically clarifying and quantifying the relationship of ARs to TMEs and WCBs for the first time, the key characteristics of ARs can be better understood.

2.3.1 Concepts of TMEs, ARs, and WCBs

The TME concept was developed between 2005 and 2010 on the basis of a number of case studies. Knippertz and Martin (2005) analyzed three unusual boreal winter rainfall events in northern Africa associated with tropical cloud plumes, using trajectories to demonstrate the anticyclonic inflow of moisture from the tropics ahead of an upper-level trough. Bao et al. (2006) presented a study on enhanced IWV bands associated with extratropical cyclones, in which they suggested the term "moisture conveyor belt" as an alternative to AR to better link with the more established WCB concept. They also used trajectories to show the involvement of tropical moisture in some of the cases investigated. The terminology of "moisture conveyor belts" was then also used to analyze precipitation events in the southwestern USA by Knippertz and Martin (2007), and in Norway by Stohl et al. (2008), but usage since then has been sparse.

Inspired by these and some other results, Knippertz and Wernli (2010) developed an objective trajectory-based identification and climatology for significant moisture transports from the tropics to higher latitudes on synoptic time-scales, which they termed TMEs. Knippertz et al. (2013) updated and expanded the climatology to the Southern Hemisphere. Figure 2.4a schematically illustrates a basic TME. One-day forward trajectories are calculated daily, starting at 0000 UTC from every 100 × 100 km × 30-hPa box within the tropical belt (20°S–20°N) and between 1000 and 490 hPa. Out of the ~300,000 trajectories per day, only those that cross 20°N or 20°S are continued for another 6 days. Trajectories that reach a water vapor flux of at least 100 g kg^{-1} m s^{-1} somewhere poleward of 35° latitude are defined as TME trajectories. Every trajectory represents the same atmospheric mass of ~3 × 10^{12} kg, and thus specific humidity interpolated along the trajectories can be readily converted into water mass. Note that it is possible that some of the moisture that contributes to the intense water vapor flux at 35° latitude has been acquired poleward of the tropics from ocean evaporation. However, since most of the TME trajectories leave the tropical belt above the boundary layer (Knippertz and Wernli 2010), most of the TME moisture is most likely of tropical origin.

In addition to the actual climatologies of TME trajectory counts and associated water vapor transport, Knippertz and Wernli (2010) and Knippertz et al. (2013) analyzed the interannual variability of TMEs and its association with El Niño. The former paper also analyzed climatological contributions to precipitation in the extratropics, which reaches up to 60% over the North Pacific in boreal summer; the latter shows that, statistically, 90% of the ARs reaching the west coast of North America during December to May are connected to TMEs. Recently, the TME concept has been used to analyze precipitation in France (Lu et al. 2013) and in North America (e.g., Steinschneider and Lall 2016).

ARs, their history, and structure are explained in detail elsewhere in this book (Chap. 1 and, in particular, Sect. 2.1 earlier in this chapter). Here, a quantitative definition of ARs has been implemented that was agreed upon as "reasonable" at the 2015 AR workshop at the University of California, San Diego's Scripps Institution of Oceanography (UCSD Scripps), and which is broadly consistent with the range of

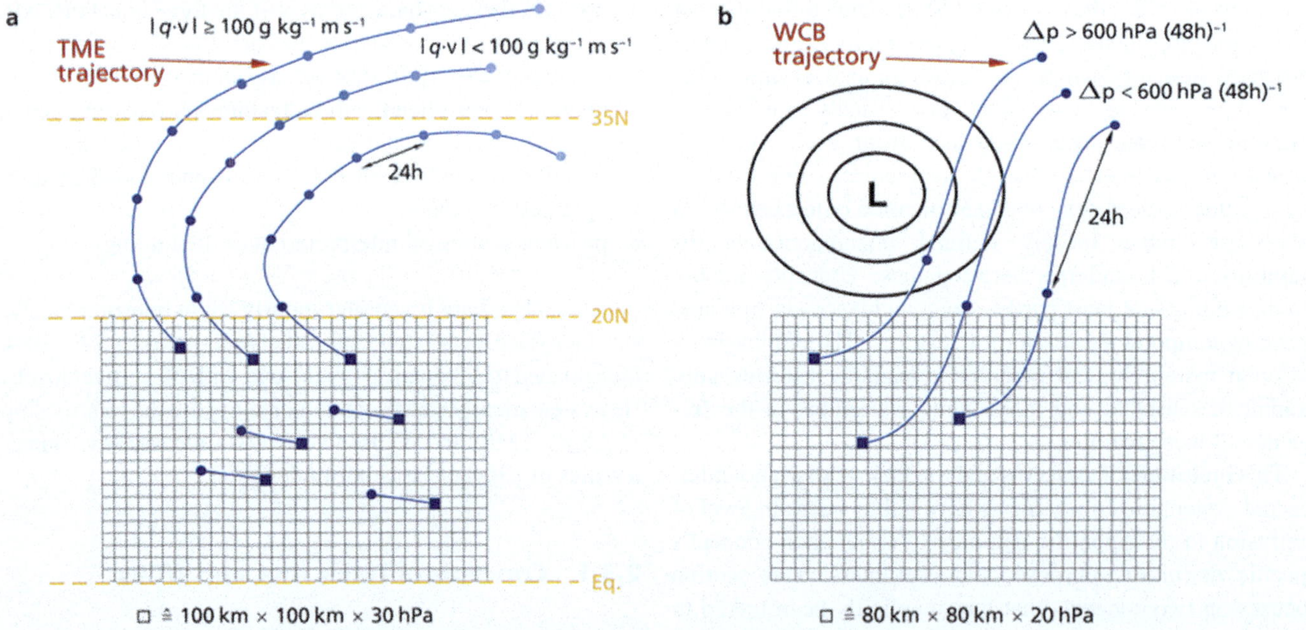

Fig. 2.4 Schematic depiction of the objective identification algorithms for (**a**) TMEs and (**b**) WCBs. *Blue lines* show example trajectories started from pre-defined 3-D atmospheric volumes marked by *blue boxes* with 24-h periods marked by *blue circles*. (**a**) To classify as TME, trajectories need to start equatorward of 20° latitude and reach a water vapor flux of at least 100 g kg^{-1} m s^{-1} somewhere poleward of 35° latitude (example for Northern Hemisphere is shown here). (**b**) WCB trajectories need to cross the objectively identified area of a cyclone and ascend by more than 600 hPa in 48 h. (Panel (**a**) is modified from Fig. 2 in Knippertz and Wernli 2010)

methods used in the literature and with the general GoM description of ARs given in Sect. 2.2.1. It is as follows:

> In contrast to TMEs and WCBs, a Eulerian approach to define ARs is used here: ARs are identified here as 2-dimensional (2-D) objects poleward of 20° latitude with length scales of at least 2000 km. Every grid point in this object fulfills the criteria that (1) IWV is larger than 20 kg m^{-2} (equivalent to 20 mm), and (2) IVT is larger than 250 kg m^{-1} s^{-1}.

According to this definition, ARs are typically, but not always, filamentary synoptic-scale features of both enhanced moisture and strong horizontal moisture transport. Conceptually, ARs differ from TMEs mainly because ARs need not contain moisture of tropical origin, and they focus on vertically coherent structures; whereas, in principle, a TME can be shallow with intense moisture transport only at a particular level.

In contrast to ARs and TMEs, WCBs have a history that can be traced back for longer than a century. Central to WCBs is the idea of coherent air flows in extratropical cyclones. Despite the chaotic nature of atmospheric flows, early synoptic analyses and later satellite imagery revealed the impression that cyclones are characterized by coherent poleward-moving air streams of warm and moist air. Shaw and Lempfert (1906) showed very early illustrations of surface air-parcel trajectories calculated from synoptic weather maps. Decades later, Rossby and Namias pioneered the concept of isentropic analyses and considered the motion of moist and dry air masses on sloping isentropes over the USA (e.g., Rossby et al. 1937). Green et al. (1966) then used moist isentropes to identify zones of saturated ascent along cold fronts. They emphasized that in certain situations subtropical air masses from the Caribbean Sea may rise over a front into the jet stream. Browning and Harrold (1969) in the UK, and later Carlson (1980) in the USA, noted the usefulness of these air streams for understanding the precipitation pattern in extratropical cyclones. Browning (1971) was the first to use the term WCB to denote these moist ascending and precipitation-producing air streams. Since the work by Rossby et al., the ascent of moist air was an essential aspect of what was later termed as WCB, in contrast to ARs and TMEs, which primarily emphasize the horizontal moisture transport. Clearly, the ascent of WCBs leads to condensation, and the formation of clouds and precipitation, and this is why Browning (1990) stated that WCBs are the primary source of precipitation in extratropical cyclones.

With the advent and subsequent availability of reanalysis data, validating the concept of coherently ascending air streams with trajectory calculations became possible. Wernli and Davies (1997) defined WCBs as strongly ascending trajectories in the vicinity of extratropical cyclones and used a criterion of 600-hPa ascent within 2 days (see Fig. 2.4b). They showed that the strong ascent goes along with intense latent heating (typically 20 K over the 2 days) and precipitation, consistent with Browning's earlier definition. Since then, WCBs have been objectively identified by calculating very large numbers of trajectories and selecting those that fulfill the ascent criterion (e.g., to produce WCB climatologies; Stohl 2001; Eckhardt et al. 2004; Madonna et al. 2014). In recent years, WCBs have been shown to be highly relevant for the dynamics of upper-tropospheric Rossby waves and anticyclonic blockings (e.g., Grams et al. 2011; Pfahl et al. 2015), and to be associated with a complex microphysical evolution from warm-phase clouds in the WCB inflow to cirrus clouds in the WCB outflow (Joos and Wernli 2012).

2.3.2 Climatologies

All three features (TMEs, ARs, and WCBs) have 2-D fields of occurrence frequency constructed every 6 h for ERA-Interim climatologies for the December–January–February (DJF) and June–July–August (JJA) seasons. For each feature, a value of 1 is assigned to grid points that belong to this feature, and a value of 0 to all other grid points. In this way, simple time averaging of these binary fields leads to climatological frequency fields. For instance, an averaged value of 0.2 indicates that at this grid point the feature occurs in 20% of all time instants. The calculation of the 6-hourly binary fields is straightforward for the ARs: a value of 1 is simply attributed to all grid points that fulfill the AR criteria set out in the previous section. For TMEs and WCBs, which are trajectory-based features, binary fields are constructed by asking: *for every 6-hour time-step and for every 1°-by-1°grid column, is at least one 5-day TME trajectory or 2-day WCB trajectory present?* For these Lagrangian features, a climatological frequency value of 20% indicates how often at least one trajectory of the considered category is somewhere in the vertical column represented by the grid point—but it does not provide information about how many TME or WCB trajectories are present. Similarly, the AR frequencies just indicate how often the criteria are met, but not how far they are exceeded. Given the criteria outlined above, TMEs and ARs are only defined outside the 20°S, 20°N tropical belt, which can lead to sharp gradients in the frequency diagrams at 20° latitude.

The information provided here complements to some extent the more detailed climatologies presented in Knippertz and Wernli (2010) and Knippertz et al. (2013) for TMEs, and Madonna et al. (2014) for WCBs. For ARs, many more aspects of their climatology and effect on heavy precipitation are discussed in other parts of this book. Climatologies of TMEs, ARs, and WCBs are calculated for the first time here with the same underlying data set and compared directly to one another.

The resulting plots of the frequency of occurrence (hereafter just frequencies) for all three features are shown in Fig. 2.6

for the solstice seasons DJF (left) and JJA (right). In boreal winter, TMEs in the Northern Hemisphere show two clear occurrence maxima: one over the central–eastern North Pacific and one over the western Atlantic (Fig. 2.6a). Values are largest in the subtropics and then stretch across the mid-latitude parts of the ocean basin, finally reaching the adjacent continents: North America and Europe (respectively termed “Pineapple Express” and “Gulf Stream” in Knippertz and Wernli 2010). Mid-latitude maxima reach values of about 30%, although near-20°N frequencies of more than 60% are found. In the Southern Hemisphere, overall values are larger, and two strong maxima are found again over the Pacific and Atlantic, with the latter particularly pronounced just off the coast of Brazil (Fig. 2.6a). These are associated with the South Pacific and South Atlantic Convergence Zones, respectively. Weaker maxima occur east of South Africa near Madagascar and over Western Australia (see also Knippertz et al. 2013). For JJA, patterns in the Southern Hemisphere change rather little but overall values decrease, although activity shifts into the South American continent and over southern Africa (Fig. 2.6b). In the Northern Hemisphere in contrast, dramatic seasonal changes are revealed. The maximum over the Pacific shifts westward to the Asian coast, reaching the largest values of any region and season of more than 70%. Only a small fraction of TME trajectories reach the North American continent during this season. In the Atlantic sector, a sharp narrow band of TMEs connects the Caribbean and the Gulf of Mexico with the US Great Plains, although the “Gulf Stream” maximum is strongly reduced in zonal extent compared to DJF.

The climatology of ARs (Fig. 2.6c, d) shares a number of similarities with the TME occurrence frequencies, despite the overall somewhat smaller frequencies. There are geographical matches for most of the maxima discussed for TMEs above, but AR maxima tend to be more elongated—mostly in the WSW–ENE direction in the Northern Hemisphere and in the WNW–ESE direction in the Southern Hemisphere, respectively—and therefore generally have larger values farther away from the equator, relatively speaking. Given the strong emphasis on moisture and moisture transport, both TMEs and ARs tend to show larger values in the summer hemispheres. (This seasonality aspect will be discussed further in Sect. 2.3.4.) The most striking differences between the two features are: (1) the relatively weak AR signature over the North American Great Plains in summer and over South America in winter (Fig. 2.6d) and (2) AR maxima aligned along 20°N during Northern Hemisphere summer (Fig. 2.6d). The former is most likely a consequence of the length criterion used in the AR definition, which filters out those TMEs where trajectories ascend and moisture condenses and rains out at relatively low latitudes. The latter is related to the warm sea-surface temperatures over the western Atlantic, Indian Ocean, and West Pacific, which allow for enhanced atmospheric moisture contents in combination with relatively stable circulations associated with the Asian and American monsoon systems. This type of AR, which is aligned westward with the trade winds, has not been discussed in the literature.

Another remarkable difference between the AR and the TME climatologies is the large seasonal cycle in areas around South Africa and Australia (cf. Fig. 2.6c, d). This points to a substantial reduction in moisture here from summer to winter, which in the TME climatology may be compensated for by an increase in meridional transport (Fig. 2.6a, b), but this requires further study to be confirmed. Also noteworthy is that ARs as defined here show hardly any penetration into the North American continent from the west in both seasons (e.g., Rutz et al. 2014, 2015), which is probably related to the strong rain-out of moisture by the steep and high topography close to the coast. For Europe, penetration is weak in winter, but reaches far into Russia in summer, which contrasts with TMEs, where the restriction to 5-day trajectories and the weaker summertime flow does not allow an extension that far to the northeast from the main sources.

Finally, the climatology for WCBs is given in the bottom panels of Fig. 2.6. Compared to the other two features, WCBs show a significantly larger extension into higher latitudes. Given the emphasis on vertical motion—and, therefore, storm dynamics—there is a tendency toward greater frequencies in winter, in contrast to the other two features with their emphasis on moisture and largest frequencies in summer.

In Northern Hemisphere winter, there is qualitative agreement in the location of the two maxima of WCBs in the Atlantic and Pacific storm track areas (Fig. 2.6e) with the climatologies of TMEs and ARs (Fig. 2.6a, c). However, in summer, contrasts among the three features are stark, with generally much lower frequencies of WCBs and no Great Plains maximum (Fig. 2.6f). The weak summertime maximum of WCBs over Asia is related to strong orographic ascent caused by monsoonal meridional flow toward the Himalayas. All three features agree, at least, that in JJA the Pacific storm track is more active than its counterpart in the Atlantic. In JJA, the locations of the North Pacific maxima of WCBs, TMEs, and ARs are remarkably similar; which is in line with the large contribution of tropical moisture sources for North Pacific WCBs in summer (see Fig. 3 in Pfahl et al. 2014).

In the Southern Hemisphere, activity maxima in the areas of the South Pacific and South Atlantic Convergence Zones agree well among the three features in terms of geographical position, but not in magnitude. The maximum off the coast of South America shows some agreement with the respective maximum in TME frequency, but, for WCBs, the maximum is extended southeastward in JJA. Contrasts to ARs are, again, large. For the Australian region, WCB activity is concentrated to the southeast and southwest of the continent, similar to ARs but in contrast to TMEs. Differences in WCB activity are small between summer and winter, again in stark contrast to

ARs. The implication here is that baroclinic instability (i.e., extratropical cyclogenesis) remains active at about the same level year-round in the Southern Hemisphere, whereas in the Northern Hemisphere it is most prominent in winter.

2.3.3 Linkages Among the Three Feature Categories

Both the case study (Fig. 2.5) and climatologies (Fig. 2.6) reveal an interesting and rather complex relationship among the three features. Moisture-transport events simultaneously fulfill the criteria of ARs, TMEs, and WCBs; however, more often it appears that the three feature categories emphasize different aspects of the flow and/or different events. Here, this linkage is quantified in detail, again making use of the 6-hourly 2-D fields of AR, TME, and WCB occurrence. For the Northern Hemisphere winter and summer seasons and the region poleward of 20°N, every 6 hours the areas where (1) none of the features is present, (2) where one of them occurs in isolation, and (3) where two or all three of them overlap are determined. The seasonal mean values (in % of the area poleward of 20°N) are then displayed in the form of a Venn diagram in Fig. 2.7.

Focusing first on ARs in winter, the sum of the four values in the red circle in Fig. 2.7a indicates that, on average, 6.71% of the area was covered by an AR. Often, ARs occur in isolation (2.65% of the area, corresponding to about 40% of all ARs). In a similar number of cases (2.51%), ARs co-occur with TMEs but not WCBs. About 23% of all ARs overlap with WCBs, split almost evenly into occasions with (0.79%) and without (0.76%) additional occurrence of TMEs. Similarly, the values in the green and blue circles provide information about TMEs and WCBs, respectively, and their overlap with other features. In agreement with Fig. 2.6, the overall frequency of TMEs (11.7% of the area) is larger than that of ARs, while the frequency of WCBs is smaller (5.85% of area). The most frequent overlap is between ARs and TMEs, which again agrees with the similarity of their climatological distributions (Fig. 2.6a, b). TMEs and WCBs occur more often in isolation than with a mutual overlap. Coincidences of all three features are relatively rare (0.79% of the area), which in relative terms indicate that this situation occurs for about 12% of all ARs, 7% of all TMEs, and 14% of all WCBs, respectively.

In Northern Hemisphere summer (Fig. 2.7b) some values change strongly compared to winter, in agreement with the climatologies shown in Fig. 2.6. ARs become more frequent, now covering 17.05% of the area, TMEs have a similar overall frequency (11.04% of the area), and WCBs become less frequent (only 2.4% of the area). Most of the ARs (61%) occur in isolation, although in summer 50% of all TME grid points coincide with ARs. The overlap between ARs and TMEs is again the most frequent type, even more strongly than in winter. Interestingly, despite the strong seasonality of the AR and WCB frequencies and the geographical distributions of all three features, the frequency of 3-fold coincidence is almost the same in summer as in winter (0.73% and 0.79% of the area, respectively). In summer, this means that only slightly more than 4% of all ARs simultaneously overlap with both TMEs and WCBs, and 32% of all ARs overlap with a TME only. The interpretation is that most ARs do not have a tropical origin (in particular the ones over the North Atlantic, cf. Fig. 2.6b, d) and that only a few of them are characterized by the strong ascent required for WCBs. The first of these results is in line with the findings of Dacre et al. (2015), who composited the moisture budget for 200 intense extratropical cyclones and found that local sources of water vapor are essential for the formation and

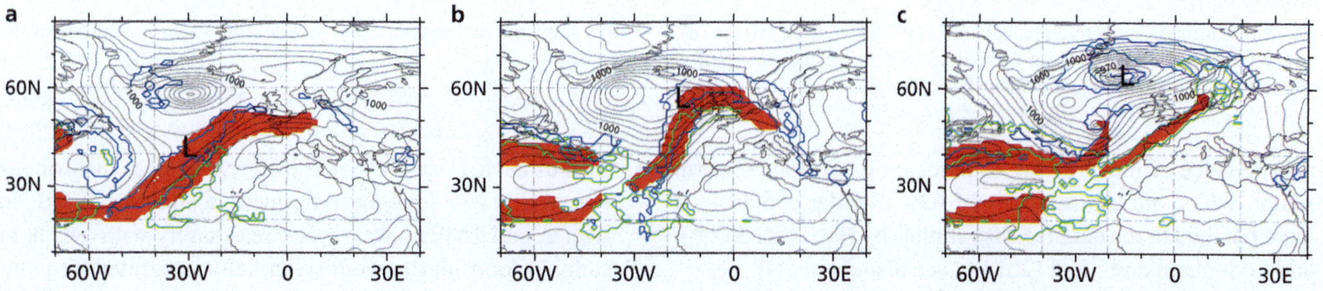

Fig. 2.5 Illustrative case study of a North Atlantic cyclone (position marked with label "L") at (**a**) 0000 UTC 22 Nov, (**b**) 0000 UTC 23 Nov, and (**c**) 0000 UTC 24 Nov 1992. *Black* contours show mean sea-level pressure (in hPa), *red* shading marks the identified ARs, and *green* and *blue* contours are the identified TMEs and WCBs, respectively

[1]It is important to note that this percentage value (and similar values of overlap mentioned below) refers to grid points (and not entire features); i.e., 23% of all AR grid points are at the same time also WCB grid points. Such an overlap value results if, for instance, a full overlap occurred for 23% of all ARs, or if for every AR there was a 23% areal overlap with a WCB. Very likely, there is large variability, with few AR features having an almost perfect overlap with WCBs (see, e.g., Fig. 2.5a) and others with almost no overlap (e.g., Fig. 2.5c).

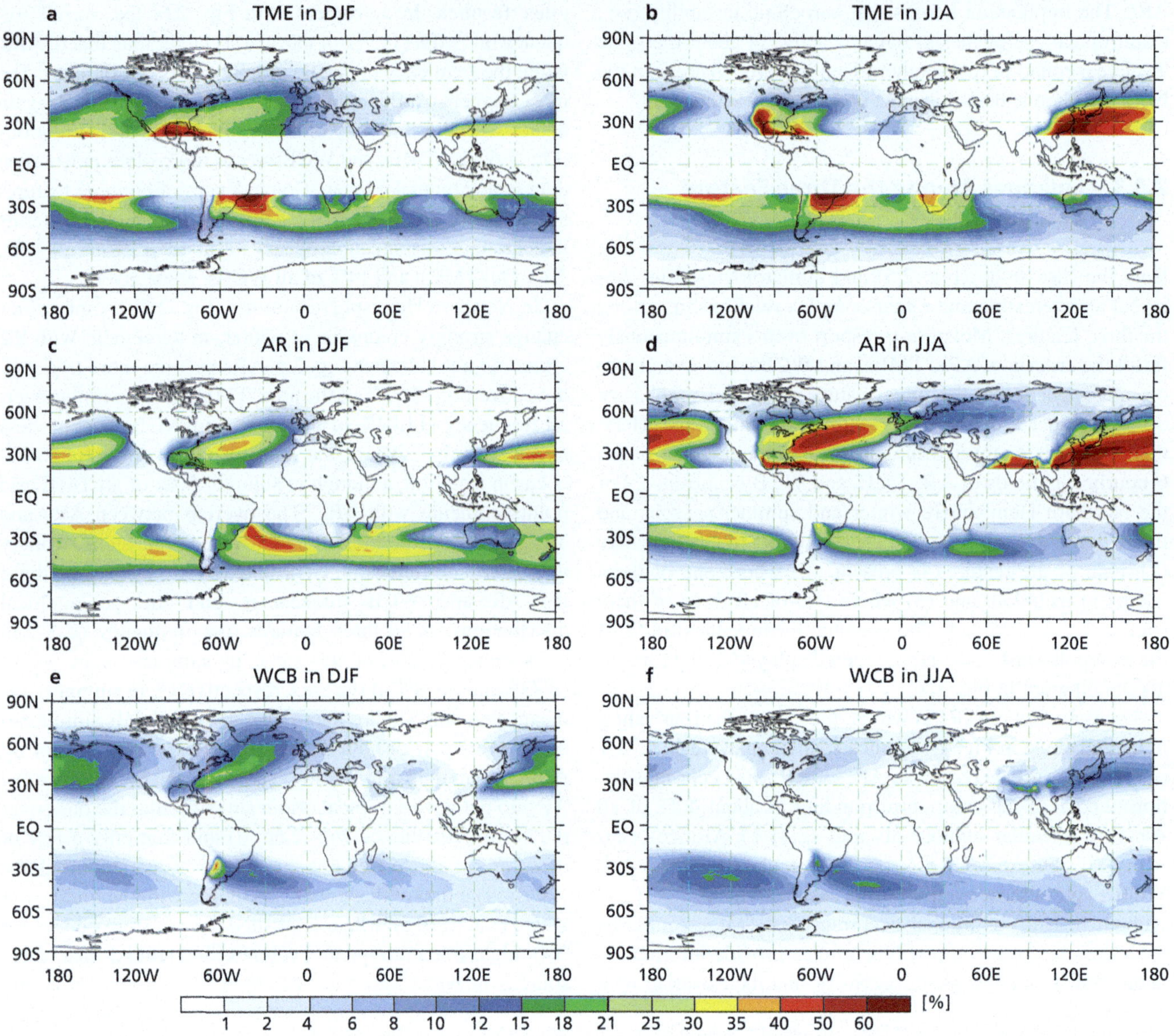

Fig. 2.6 Climatological frequencies (in %) of TMEs (**a, b**), ARs (**c, d**), and WCBs (**e, f**), derived from ERA-Interim data for the time-period 1979–2014. Left-hand panels show boreal winter (DJF); right-hand panels show boreal summer (JJA). See text for exact definitions of the three features

maintenance of the large IWV values in the cyclone's warm sector, which are characteristic of ARs. Similar conclusions were reached in the detailed case studies by Bao et al. (2006), and Sodemann and Stohl (2013)—see also Sect. 2.4.

2.3.4 Summary

For the first time, climatologies of all three features (ARs, TMEs, and WCBs) have been derived from the same global reanalysis data set, enabling comparison of the seasonality of their geographical frequency patterns, and the statistics of their overlap. The main results can be summarized as follows:

- The three features reveal a different and asynchronous seasonality: ARs are more frequent in summer, WCBs in winter, and TMEs have a weak seasonality with a peak in summer (considering hemispherically integrated frequencies). This reflects the generally enhanced vertically integrated humidity values in the warm season (which is beneficial for the formation of ARs) and the stronger baroclinicity and large-scale forcing for ascent in the cold season (which is beneficial for the formation of WCBs). For TMEs, it appears that the enhanced moisture values in summer and the stronger low-level winds in winter lead to the fairly flat seasonal cycle. For specific regions, seasonality can be pronounced, e.g., the very strong frequency

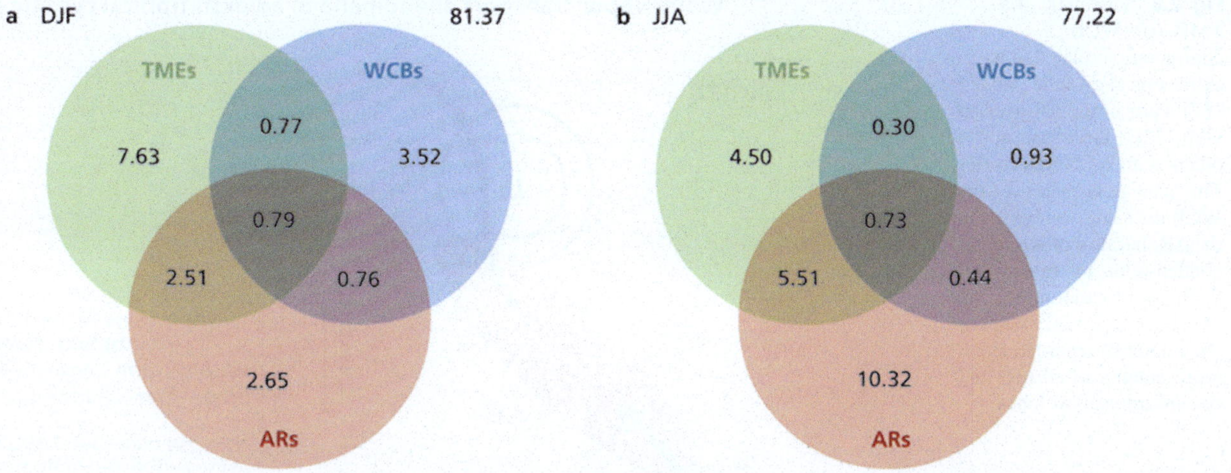

Fig. 2.7 Venn diagrams for (**a**) boreal winter (DJF) and (**b**) boreal summer (JJA), showing the percentage of area in the Northern Hemisphere extratropics (north of 20°N) that is, on average, covered by ARs, TMEs, and WCBs only and simultaneously by two or three of the features. The value in the *top right* of each panel indicates that, on average, about 80% of the area is not covered by any of the three features

maximum of TMEs in summer over the US Great Plains and the western North Pacific, and a maximum of AR frequencies near the US West Coast in winter. WCBs typically occur further poleward than TMEs and ARs, in agreement with their ascent in the main mid-latitude baroclinic zone. It is important to critically comment on the seasonality of ARs in this section. The use of geographically and temporally uniform AR selection criteria (fixed thresholds for IWV, IVT, and length) together with the higher humidity content in mid-latitudes in summer leads to a clear summer maximum in the frequency of ARs in most parts of the globe (cf. Fig. 2.6c, d) and very large frequencies of more than 60% south of Japan. (Note also that in this region the frequency of cyclones is enhanced in summer compared to winter; cf. Fig. 4 in Wernli and Schwierz 2006). In contrast, the recent climatology by Guan and Waliser (2015) uses a combination of relative and absolute criteria (IVT larger than the local 85th percentile in the considered season, and IVT larger than 100 kg m^{-1} s^{-1}) and, as a consequence, their frequencies are limited to a maximum of 15% and show less geographical and seasonal variability. Further research is required to better understand whether many summer ARs identified with the fixed-threshold approach are meaningful, or whether seasonally varying selection criteria are essential when identifying ARs year-round.

- Generally, the spatial overlap of objectively identified TME, AR, and WCB features is rather limited. Most grid points that belong to one of the three features do not meet the criteria of the other features. This is an important result, because it highlights the difference of the three concepts and, therefore, justifies a separate treatment of these three concepts. The most frequent overlap occurs between TMEs and ARs, i.e., between the two features that highlight regions of strong horizontal moisture transport. That TMEs and ARs do not agree more closely is mainly because most ARs do not originate in the tropics. There is little overlap of these features with WCBs, reflecting the strong contrast between horizontal moisture transport (TMEs and ARs) and vertical ascent of moist air masses (WCBs), which usually reduces the former because of precipitation.

Together, these findings clearly show that TMEs, ARs, and WCBs are related but by no means identical concepts of atmospheric flow. They focus on different aspects of moisture transport and must not be used as synonyms. In particular, the strong ascent in WCBs and the associated condensation and rain-out are detrimental to the strong horizontal moisture transport required for ARs, and, therefore, WCBs limit the meridional extension of ARs. ARs most likely occur in regions with weak baroclinicity and weak large-scale ascent (where strong horizontal moisture transport can easily occur), in contrast to WCBs, which require strong baroclinicity, i.e., strongly sloping isentropic surfaces, because of their strong ascent. In some cases, however, WCBs and ARs overlap almost perfectly (e.g., Fig. 2.5a), pointing to a balance between rain-out because of ascent in a WCB and strong surface evaporation and low-level moisture convergence. Next, Sect. 2.4 discusses the detailed moisture budget along ARs, and then Sect. 2.5 summarizes basic dynamical aspects of the generation and depletion of IWV along ARs. The large case-to-case variability must be highlighted: individual cyclones can be associated with strongly different patterns of TMEs, ARs, and WCBs, and this pattern can change along the life cycle of a cyclone (as shown in Fig. 2.5). Therefore, no prototype configuration of the three features fits most situations.

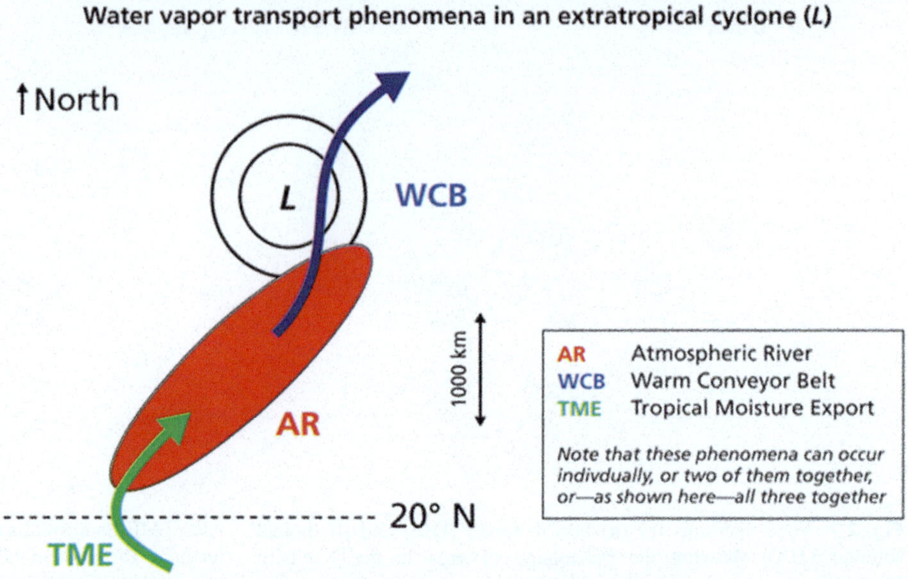

Fig. 2.8 Schematic of a TME–AR–WCB configuration related to an extratropical cyclone (label "*L*"). Parts of the AR overlap with TME trajectories, and others with WCB trajectories. The main ascent phase of the WCB does not overlap with the AR, because of strong condensation and rain-out. Such a configuration is observed (e.g., Fig. 2.12) but should not be regarded as representative of all ARs and/or extratropical cyclones

That said, Fig. 2.8 illustrates a schematic depiction of a specific flow configuration that shows a common interaction. In the equatorward part of the AR, partial overlap with a TME occurs. Further poleward partial overlap of the AR with WCB occurs, but, as the WCB rises, its moisture is strongly reduced because of rain-out and, therefore, most parts do not fulfill the AR criteria. Note that, consistent with the climatological results presented in this section, this situation does not include areas where all three phenomena coincide.

Finally, the presented climatologies and feature overlap values are, of course, quantitatively sensitive to the specific thresholds used to define the features. However, the main conclusions related to the different seasonality, geographical pattern, and the limited overlap of the features likely remain valid for moderate modifications of thresholds. The results contribute to a comprehensive portrayal of the fascinating complexity of the atmosphere's water cycle and, ideally, will stimulate further research into TMEs, ARs, and WCBs.

2.4 Water Vapor Transport in ARs

The term "atmospheric river" creates powerful associations: imagine a stream of water (vapor) in the atmosphere that matches in volume the Earth's largest actual rivers on land (Sect. 2.2). ARs are most frequent over ocean areas, where moisture is abundant; cyclones can intensify substantially because of the lower surface friction over water than over land, and thus can form extensive frontal structures. When the air masses that carry these massive vapor masses are forced upward (e.g., by topography as ARs make landfall, or when transitioning to a WCB, see Sect. 2.5), they cool, and the water vapor condenses and precipitates. Precipitation can sometimes persist for extended periods in one region, which can cause enormous disasters, such as flash floods and landslides (see Chap. 5, "Effects of ARs"). A question that needs to be answered to understand the causes of such hydrological extremes is, therefore, how water is transported in an AR. In addition, understanding where the moisture sources are located could provide insight into what controls AR intensity.

The moisture contained and transported within ARs is their defining characteristic. Images of IWV and IVT sometimes show ARs as narrow filaments that extend from a tropical reservoir of high IWV (Sect. 2.2). This spatial connection suggests the tropics as the principal source of the water vapor transported poleward within ARs. Using different methods, several studies of the air mass and moisture transport within ARs have shown that the true picture is more complex (Bao et al. 2006; Ralph et al. 2011; Sodemann and Stohl 2013; Cordeira et al. 2013). In particular, these studies indicate that although some tropical water vapor may indeed reach higher latitudes within ARs, the bulk of the transported moisture originates from large-scale horizontal moisture convergence in the subtropics and mid-latitudes (Sect. 2.3), implying shorter average transport distances of the water vapor than the AR length. Unlike the "channel flow" perspective suggested in a vertically integrated view, ARs are time-evolving, dynamic, three-dimensional (3-D) features embedded in the large-scale atmospheric flow (Sect. 2.5). Considering all dimensions naturally adds complexity to understanding AR moisture transport.

This section summarizes the current state of knowledge on how moisture is transported within ARs, including where moisture sources are located, and how they are connected to the formation, maintenance, and shape of ARs, and to their decay. The moisture budget of ARs and its changes during their life cycle are considered first. Next, the spatial dimensions of how water vapor is transported horizontally and vertically within ARs are explored. A discussion follows of the advantages and drawbacks of different available methods from which these results were obtained. This section concludes with how moisture transport in ARs could affect dynamical processes, and research areas where knowledge on AR moisture transport should advance.

2.4.1 Moisture Budget During the AR Life Cycle

An AR life cycle can be divided into three stages: development, evolution, and decay (Cordeira et al. 2013). During an AR's life cycle, moisture continuously enters and leaves the moisture band that constitutes the AR. Moisture is added to an AR by processes such as southerly inflow, lateral convergence, and surface evaporation. Moisture is removed by precipitation, outflow, and moisture divergence. So far, the various contributions of these terms in their entire life cycle to the moisture budget of an AR have been only partly evaluated.

The traditional analysis of an AR's moisture budget often focuses on the mature stage of its life cycle. Maps of IWV and IVT typically suggest a connection between a tropical reservoir and a landfall region in the extratropics (Zhu and Newell 1998). Early studies of ARs noted their close relationship with cyclones and their life cycles (Newell et al. 1992). These studies also found that there appears to be a relationship between ARs and explosive cyclones ("bombs") (Zhu and Newell 1994), and that tropical cyclones can induce a strong poleward transport of tropical moisture that feeds into ARs (Stohl et al. 2008). While that instantaneous perspective allowed the transient nature of these phenomena and their association with specific synoptic patterns to be identified, only limited insight was gained into moisture sources and the actual moisture transport in ARs. This has probably led to the misleading perception that the moisture is sourced upwind of the AR's entrance region, transported in the atmosphere all the way along the AR, and lost where the AR terminates. In reality, the picture is more complicated, because an AR is not a static feature that "hoses" water from a reservoir to a target area. ARs evolve over time and change both their shape and position. Furthermore, a vertically integrated perspective masks the fact that air masses ascend as they move poleward, water condenses, and thus moisture is replenished rather than materially conserved during poleward transport.

Following the air masses of precipitation extremes backward in time to their origin provides insight into the leading transport processes. Using trajectory analyses, Bao et al. (2006) found that although poleward transport of tropical moisture may occur along an IWV band (i.e., an AR), in many cases, local moisture convergence is primarily responsible for forming the bands of enhanced IWV. For the case of an AR making landfall on the US West Coast during 1–5 January 1997, 4-day backward trajectories released in the region of high IWV show air mass origins that cover a wide range of latitudes of 10–40°N (Fig. 2.9a). In addition, moisture continuously converges laterally into the AR along the cold front (Fig. 2.9b, contours). The convergence enhances the IWV anomaly and replenishes moisture lost by precipitation. Without that replenishment, the IWV anomaly of an AR would be dissipated quickly. A climatological study by Ryoo et al. (2015) confirmed that these characteristics are indeed typical for ARs.

Considering trajectories alone indicates air flow, but does not delineate the moisture's evaporation sources. The spatial extent of an AR's moisture sources becomes apparent when Lagrangian moisture source diagnostics are applied on the backward trajectories (see Sect. 2.4.3 for details). Using such a diagnostic on backward trajectories, the source regions of moisture for a precipitation extreme in western Norway from an AR were shown to cover vast areas of the North Atlantic (Fig. 2.10a). Moisture converged from both north and south of the AR, supplemented by transport of moisture from the subtropics within two former hurricanes, more than a week before the AR impinged on the Norwegian coast at 60°N (Stohl et al. 2008). Such Lagrangian backward tracing of air masses is the basis for several methods that can be used to identify AR moisture sources. The analysis in Fig. 2.10a identifies where either evaporation or precipitation dominated within air masses along the 12-day transport path toward the target region in western Norway. A detailed accounting of increases and decreases of moisture in air parcels that move along the trajectories enables a clearer delineation of how convergence and the tropics contribute to the final target region (Fig. 2.10b). In combination, both methods show a large footprint of the AR moisture sources and underline the role of both local convergence and long-range transport in an AR's moisture budget (again, see Sect. 2.4.3).

A less well-established part of an AR's moisture budget concerns the evaporation within the AR band itself from the underlying surface. Observations of evaporation during cold-frontal passages over the ocean show that, despite high wind speeds, evaporation is small within the warm sector, ahead of the cold front (Persson et al. 2005). The reason is that air

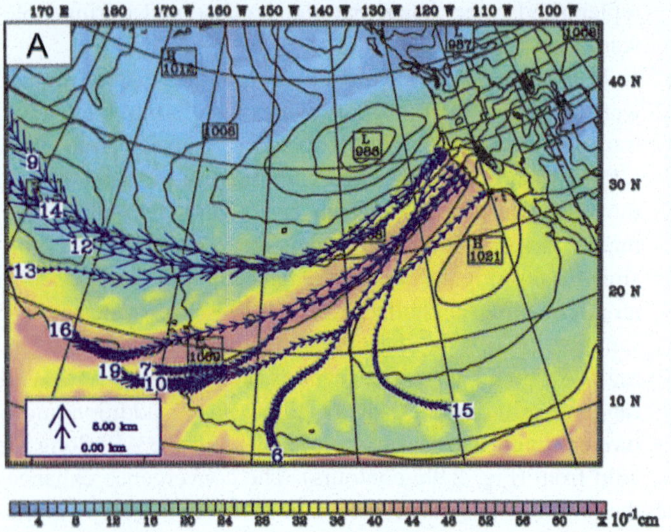

Fig. 2.9 Air mass trajectories (*thick black lines*) calculated for 96 h backward for an AR making landfall at the US West Coast. *Contours* in (**a**) are sea-level pressure (hPa), and contours in (**b**) are convergence at 0 and -1.0×10^{-5} s^{-1}. (Bao et al. 2006, Fig. 7b, c)

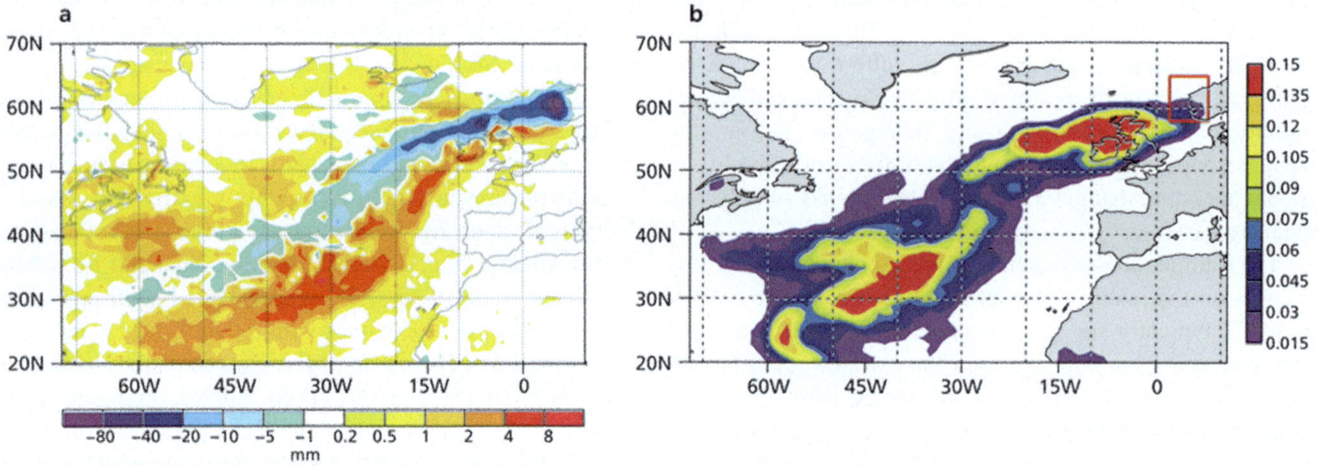

Fig. 2.10 (**a**) Time-integrated Lagrangian $e - p$ (evaporation minus precipitation; mm) budget over 12 days before arrival of intense precipitation at the west coast of Norway during a landfalling AR on 13–14 September 2005. Negative values indicate that precipitation dominates moisture removal from the air mass; positive values indicate net evaporation into air masses on the way to the target area (2–8°E and 58–65°N) for the same period in southern Norway. Note the asymmetric color scale. (Stohl et al. 2008, Figure 9d). (**b**) Moisture sources for precipitation in the target area (*red box*) identified from the Lagrangian moisture accounting method of Sodemann et al. (2008a). Units in (**b**) are kg m^{-2} day^{-1} of evaporation contributing to precipitation in the target area

masses advected in the warm sector are relatively moist, maybe close to saturation, and therefore do not lead to substantial evaporation from the surface. Thus, within an AR characterized by poleward advection of moist air, evaporative contributions from the underlying surface may also be expected to be negligible. On the other hand, evaporation from falling hydrometeors may also moisten the lower troposphere and thus reduce the moisture loss within the AR. Further studies on these aspects of the AR moisture budget are clearly needed.

In summary, horizontal convergence in the extratropics and long-range transport from subtropical and tropical areas appear to make relevant contributions to the moisture budget of an AR. The joint role of these source processes during the life cycle of an AR is exemplified in Fig. 2.11. In a case study of two intense AR events in October 2010, tropical source

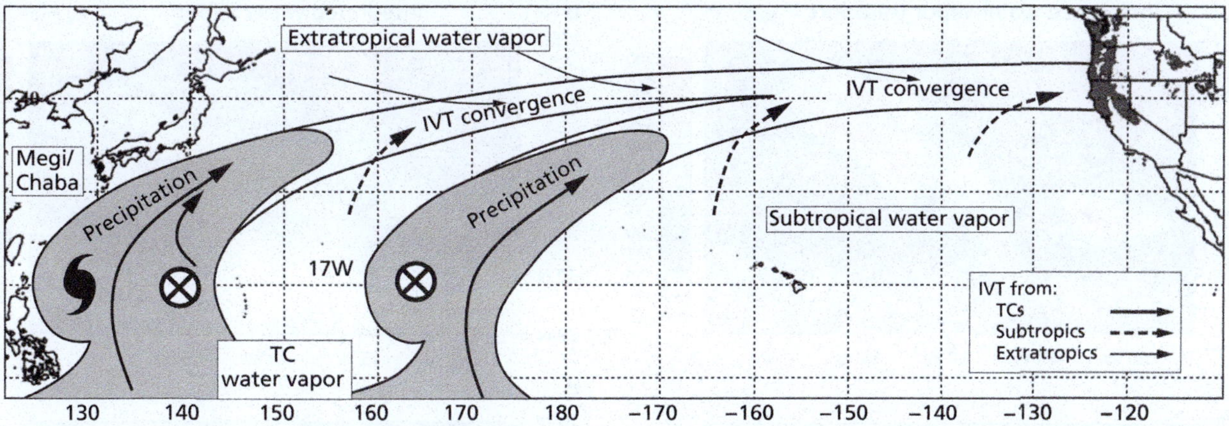

Fig. 2.11 Schematic overview of water vapor source regions during AR development (i.e., tropical source regions) and during AR evolution (i.e., subtropical and extratropical source regions) of two ARs making landfall in October 2010 in the western USA. The 48-h accumulated precipitation ending at 1200 UTC 25 Oct 2010 is *contoured* and *shaded in gray* above 50 mm. *Thick solid*, *dashed*, and *thin solid arrows* denote moisture contributed by tropical, subtropical, and extratropical regions at different stages of the AR life cycle. (Cordeira et al. 2013, Fig. 11e)

regions related to former tropical cyclones were more influential during the development phase. Subtropical and extratropical contributions took over during the later AR evolution phase, as indicated by the solid and dashed arrows in Fig. 2.11.

Toward the end of an AR's life cycle, moisture-removal processes become increasingly important. Eckhardt et al. (2004) highlighted the co-occurrence of ARs and the so-called WCB of extratropical cyclones. WCBs source their moisture in the warm sector of the cyclone, and sometimes within ARs (Sect. 2.3). A key characteristic for separating ARs and WCBs is that the first are defined by high water vapor transport, whereas the latter are characterized by rapid and strong ascent, leading to condensation, deep cloud formation, and precipitation (Sects. 2.3 and 2.5). This is illustrated by comparing an infrared (IR) satellite image and a picture of IWV for the same AR event and at the same time. Where the IWV imagery shows the familiar AR signature of an elongated extension of high IWV from a subtropical reservoir to mid-latitudes (Fig. 2.12b), the IR image shows no or mostly warm low-level clouds in the southwestern part of the AR (Fig. 2.12d), as indicated by the dark blue colors. This is a region with little vertical motion. As the AR transitions into the WCB further to the northeast, vertical motion increases and more widespread cloud formation sets in, eventually forming the broad band of intense, high-reaching clouds typical of a WCB (Fig. 2.12d, red areas near the UK). Note that removal of moisture from precipitation along the way is also expected within the area of the AR that is dominated by shallow clouds. Such removal along the way has not yet been quantified in detail.

2.4.2 Horizontal and Vertical Moisture Transport and AR Maintenance

AR bands can extend over thousands of kilometers (Ralph et al. 2004). This can be identified from IWV or IVT that exceed some threshold value (Sect. 2.2). However, as just discussed in Sect. 2.4.1, most of the moisture is not transported along the whole AR. Rather, moisture convergence along the AR continuously replenishes moisture lost by precipitation within the AR itself. The distance over which moisture is actually transported seems to depend primarily on the size of the cyclone with which the AR is associated. Examples shown in Dacre et al. (2015) suggest that in smaller cyclones at the poleward or downstream end of the North Atlantic storm track, AR moisture is mainly of local origin. However, ARs associated with deep and large troughs can indeed transport moisture over thousands of kilometers, including northward transport by about 20° of latitude and more (Bonne et al. 2015).

Sodemann and Stohl (2013) found that ARs can also be associated with more than one cyclone. This requires that a pronounced wave pattern is present in the upper troposphere, resembling an anticyclonic life cycle (a so-called LC1; see Sect. 2.4 and Fig. 2.15) in Thorncroft et al.'s (1993) cyclone life cycle classification. At the western edge of the AR, small cyclones can develop from frontal waves. The cold fronts of these small cyclones add moisture to the AR by convergence with their warm-sector air masses ahead, while their WCB air stream taps the moisture from the AR. Thus, such frontal wave cyclones each contribute moisture to the AR at one end, and feed on it for their spin-up and development at the other. Subsequent cyclones thus profit from the moisture

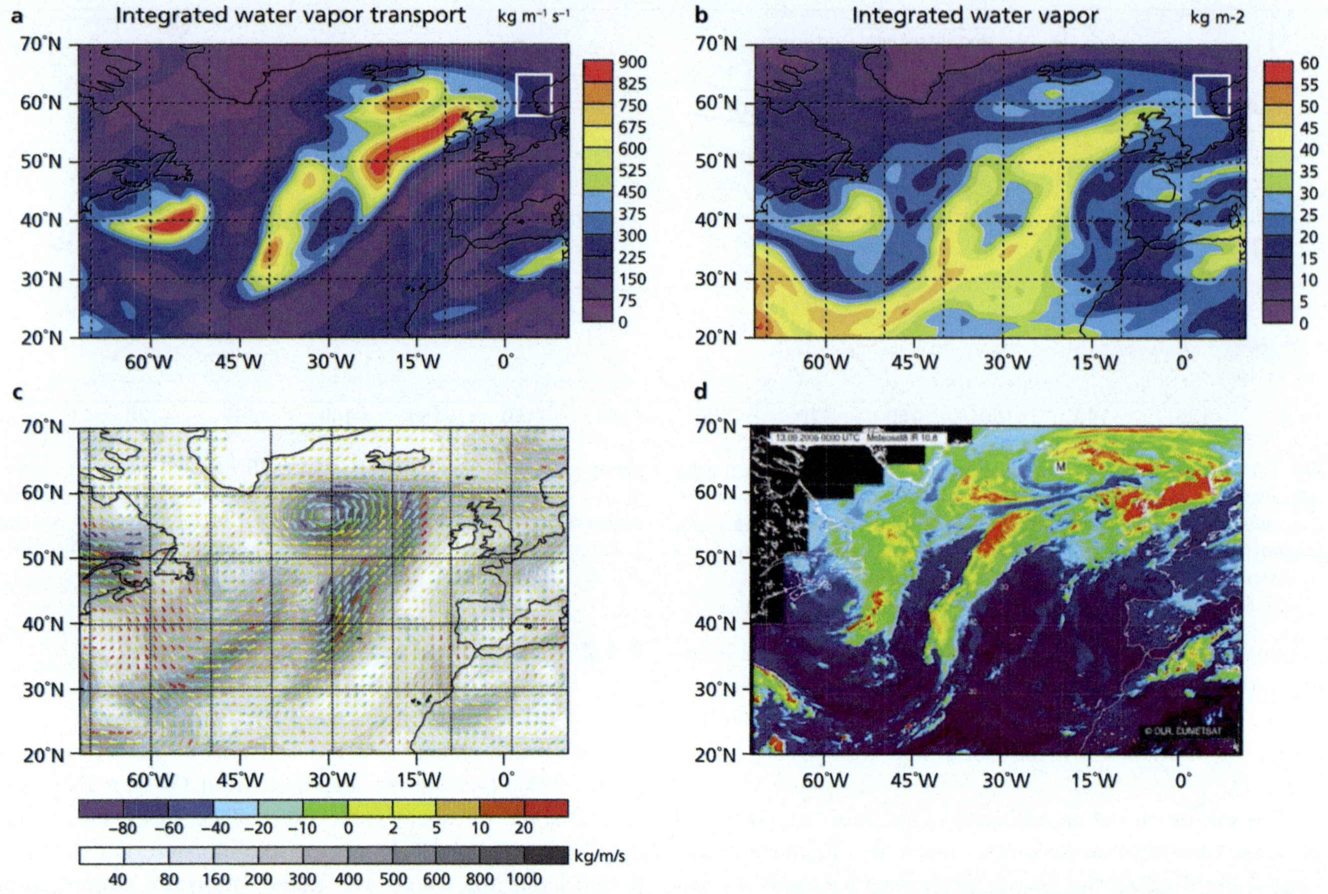

Fig. 2.12 Case study of an AR event that caused flooding on the western coast of Norway during 13–14 Sep 2005. The panels show (**a**) integrated vapor transport (IVT, kg m^{-1} s^{-1}) and (**b**) integrated water vapor (IWV, kg m^{-2}) at 00 UTC on 13 Sep 2005 from European Centre for Medium-Range Weather Forecasts (ECMWF) ERA-Interim reanalyses, (**c**) vertically integrated water vapor transport (*arrows*) and surface freshwater flux (*shading*, kg m^{-1} s^{-1}) at 00 UTC 12 Sep 2005 from ECMWF analyses, (**d**) Meteosat-8 infrared brightness temperatures in 10.8 m channel at 00 UTC on 13 Sep 2005. *Red colors* indicate high (cool) cloud-top temperatures. (Panels (**c**, **d**) from Stohl et al. 2008, Figs. 6c and 4a)

transported poleward by previous cyclones, leading to a "hand-over" of moisture between subsequent short-lived cyclones (Sodemann and Stohl 2013; see also Fig. 2.13). According to this picture, ARs can be longer-lived than the individual cyclones. Following Dacre et al. (2015), the AR can therefore be considered as the footprint which a decaying cyclone leaves behind on its way toward higher latitudes. Although detailed studies do not exist on how a moisture "hand-over" process between cyclones influences the moisture sources that contribute to the AR, it is clear that the warm sectors of the individual cyclones each contribute moisture to the AR, thus generating a large and complex source region for the moisture present in the AR. It is also evident that the moisture source regions will be very different for different stages of the AR evolution, with lower-latitude source regions being initially most important, but with higher-latitude sources becoming more important with time (as just discussed in Sect. 2.4.1). Yet, although the moisture of low-latitude origin may diminish as the AR reaches higher latitudes, it may be important for spawning daughter cyclones and thereby prolonging the AR's development. Furthermore, even small amounts of low-latitude moisture can be important for extreme events at high latitudes, because they might cause critical precipitation thresholds for flooding to be exceeded.

Another aspect of ARs is their 3-D structure. Although most moisture is concentrated in the lower troposphere (Ralph et al. 2004), the upper troposphere can guide their formation dynamically (Sodemann and Stohl 2013; Cordeira et al. 2013), and, at later development stages, AR air masses can rise into the upper troposphere within the WCB (Eckhardt et al. 2004). Thereby, the moisture source regions are also very different for different altitudes. As air masses ascend in a (moist) isentropic fashion in a WCB, a vertical cross-section through an AR shows moisture that originates from lower-latitude sources stacked over moisture that originates from higher latitudes (Fig. 2.14a, d). Thus, at low altitudes, moisture is expected to be mainly of local origin, whereas at

Fig. 2.13 Schematic view of the moisture transport during anticyclonic (LC1-like) wave breaking with a meridional upper-level jet. *Dashed white line* shows the orientation of the upper-level jet, *solid white lines* show sea-level pressure (SLP), and *shaded colors* indicate oceanic moisture from source regions at different latitudes (*red* most southerly, *blue* most northerly). The right-hand side of the figure shows a vertical cross-section of moisture originating from different latitudes lifted to different altitudes. (Sodemann and Stohl 2013, Fig. 10a)

higher altitudes, moisture arrives from distant sources, as also shown schematically in Fig. 2.13 (right margin). Even though this elevated moisture of distant origin contributes relatively little to the total IWV, it could be responsible for precipitation enhancement over topography according to the seeder–feeder mechanism (Dacre et al. 2015), and high-altitude latent heat release is also important for the dynamics of cyclone evolution, and for atmospheric stability.

One feature that complicates the analysis of water source regions is the evaporation of falling precipitation below the cloud base, as evidenced by stable water isotope studies (Yoshimura et al. 2010). Thus, trajectory studies of the origin of water vapor in the lower troposphere may not always correctly identify the source region of the water vapor itself, which may be found at lower latitudes (i.e., typical of higher-altitude moisture sources) than the trajectories indicate. The future potential of stable isotopes to study processes and moisture sources that contribute to an AR is discussed further in Sect. 2.4.4.

2.4.3 Methods for Obtaining an AR Water Budget

Which methodology is used to trace and identify AR water budgets turns out to be central to the information obtained on moisture transport in ARs. The initial, most straightforward approach to quantify moisture transport in ARs was to consider time-averages or snapshots of the vertically integrated Eulerian IVT (Fig. 2.12a). When displayed as vectors, as in the original work by Newell et al. (1992), implicit streamlines of the IVT seem to indicate source regions in the tropics (Fig. 2.12c). As discussed above, however, this perspective is misleading because ARs are 3-D, time-evolving, dynamical features with a complex moisture budget, and their moisture sources and sinks are not faithfully represented by static snapshots of vertical moisture integrals (see Sects. 2.2 and 2.5).

The time-evolving nature of ARs substantially complicates budget studies. For example, evaluating Eulerian moisture budgets from inflow, outflow, evaporation, and precipitation within ARs requires a calculation domain that adjusts to the shape and position of the evolving AR feature. Lagrangian forward and backward trajectories provide an intuitive perspective that follows the flow. However, trajectories "belonging" to the AR must first be identified. Air-parcel trajectories have been used in several studies to investigate the air-mass transport within ARs. Bao et al. (2006) and Ryoo et al. (2015), for example, investigated the air-mass transport with 2- to 4-day backward trajectories for different landfalling ARs, and inferred the moisture origin from the transport path of the air masses. Cordeira et al. (2013) traced quantities such as water vapor mixing ratio along 3-day forward and backward trajectories over the Pacific that indicated moisture uptake and loss during the air-mass transport. The specific humidity budget along trajectories has been used in many studies to reveal information about sources and sinks of water vapor during transport (Stohl and James 2004; Stohl et al. 2008). When such a Lagrangian $e - p$ (evaporation minus precipitation) diagnostic is combined with a weighting algorithm that takes care of moisture lost from or added to an air parcel over several days, spatial maps can be obtained of the evaporative moisture sources of either precipitation events or water vapor in an air mass (Sodemann et al. 2008a; Bonne et al. 2015). Figure 2.10b shows the moisture sources of the AR event from 13–14 Sep 2005 using the quantitative moisture source diagnostic of Sodemann et al. (2008a), with the data set of Läderach and Sodemann (2016) based on the ECMWF's ERA-Interim reanalysis data. The moisture source map can be interpreted as part of total evaporation that contributes to precipitation falling in the target domain at the west coast of Norway (2–5°E, 58–65°N). The moisture sources cover a wide area and extend into the subtropical North Atlantic. In comparison to the $e - p$ map (Fig. 2.10a), more local sources are identified in the moisture accounting map (Fig. 2.10b). These local sources appear to have been masked by rain-out of air masses on the way toward the target area in the $e - p$ map.

Both trajectory-based methods, however, have common and individual limitations. An important limitation of common off-line calculation of Lagrangian trajectories is the temporal and spatial interpolation of the meteorological fields done during trajectory calculations. In particular, in the

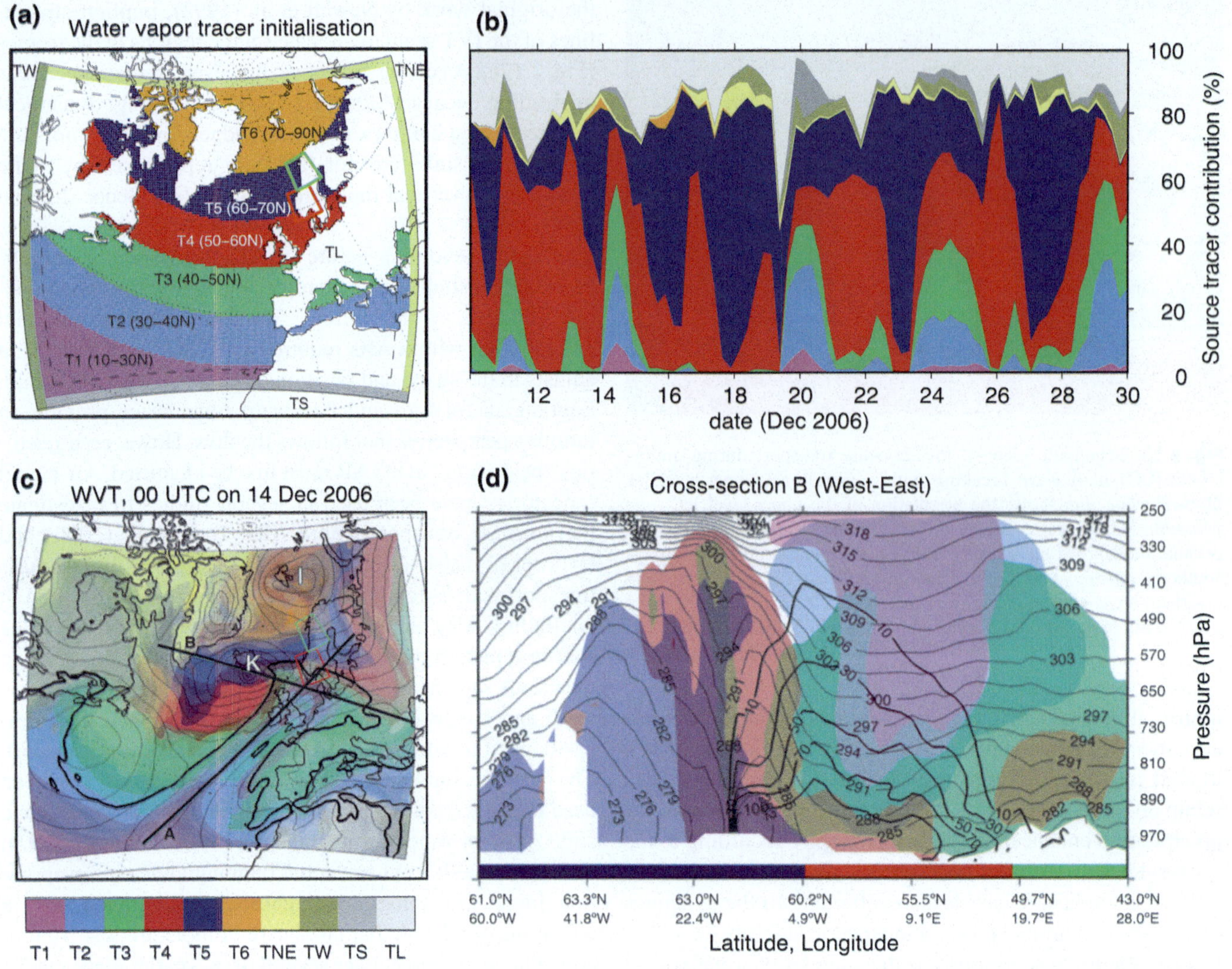

Fig. 2.14 (**a**) Initial water tracer evaporation source regions in a regional model simulation with latitude band initialization. (**b**) Time-series of Eulerian water-vapor tracer contributions over southern Norway (*red box* in Panel **a**). (**c**) Snapshot of water-vapor tracers during an AR making landfall in Dec 2006. *Colors* show contributions from moisture with sources at different latitudes to IWV. *Black contour* at 14 mm IWV outlines AR region. (**d**) Vertical cross-section in West–East direction (letter *b* in panel **c**) of tracer fraction >30% of specific humidity (*shading*) and meridional moisture flux (*black contours*)

light of continuously increasing spatial resolution of the wind field data, it is important to increase the temporal resolution of input fields as well (Bowman et al. 2013). One promising solution to such aspects of the temporal and spatial interpolation during off-line calculations may be on-line trajectory calculations (Miltenberger et al. 2013), but they have not been applied in this context so far. Because the origin of water vapor is not easily derived from common water-vapor measurements, new observational approaches are needed. In particular, the inclusion of stable water isotopes in AR field measurements could help to constrain the contribution of different moisture sources to their budget (Coplen et al. 2008; Yoshimura et al. 2010).

A second, alternative on-line calculation method to study moisture sources and transport in atmospheric models are computational water-vapor tracers. Implemented as a secondary hydrological cycle in a model, the water tracers passively take part in all resolved and parameterized processes that affect the model's water vapor (Sodemann and Stohl 2013). In other words, water released from a specific oceanic source region can be tagged with a "color" tag, evaporated from the surface, and then be traced as water vapor of the same color through the atmosphere (see Fig. 2.13). Tracer water released from different pre-defined source regions—for example, from specific latitude bands in the North Atlantic Ocean—can be tagged with different colors, which provides detailed insight into the vertical and horizontal co-location of moisture that evaporates from different regions and time-periods (Fig. 2.14a, c). The tagged water vapor will also produce precipitation with the same color tag. For a winter season with frequent AR events in southern Norway, the water-tracer method allowed the different source contri-

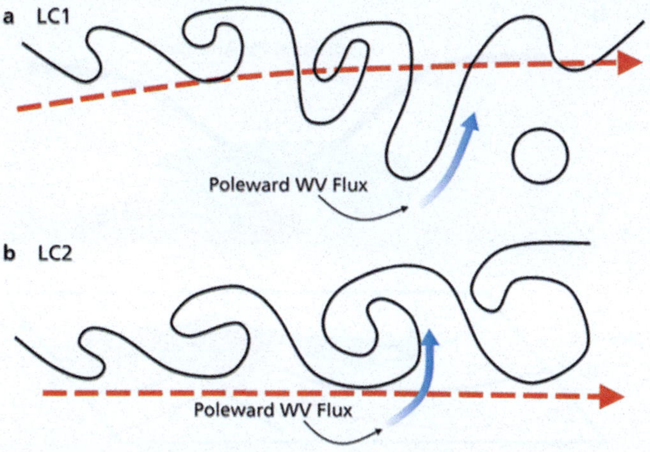

Fig. 2.15 Schematic representations of mid-latitude storm track evolution typifying (**a**) an LC1-type (anticyclonic wave breaking) life cycle and (**b**) an LC2-type (cyclone wave breaking) life cycle. The *black contour* represents a characteristic potential temperature contour on the 2-potential vorticity unit (PVU) surface. The *dashed black line* identifies the approximate position of the mean jet stream axis at each stage. The *gray-to-black arrow* indicates the potential region of poleward water vapor (WV) flux. (Adapted from Thorncroft et al. 1993)

butions for each event to be quantified (Fig. 2.14b). As ARs impinge on the target domain (Fig. 2.14a, red box), the contributions of southerly tracers to precipitation in the area increase, as shown by the spikes in green, light blue, and purple tracers, at the expense of local tracers, colored dark red and dark blue (Fig. 2.14b). On average, during the studied period, remote sources contributed only about one-third of the total rainfall (Sodemann and Stohl 2013). Although achieving budget studies for fixed target domains is straightforward, considering the budget of ARs again requires adoption of a budget calculation that moves with the AR during its life cycle.

It is important, thereby, to point out that both Lagrangian and Eulerian tracer methods, at least to first order, provide consistent results. Although the contribution of remote water vapor is generally larger for the Lagrangian approach, both methods show a consistent general pattern of a blend of southerly moisture sources that have converged horizontally within the AR zone. This information can be obtained explicitly from the Lagrangian moisture diagnostic as a quantitative map of moisture source contributions that shows the evaporation sources during the study period as a meridionally-extending core zone that is supplemented by moisture supplied from the entire North Atlantic Ocean basin. Ultimately, Eulerian and Lagrangian methods are complementary and can work hand-in-hand to provide a comprehensive view of the moisture cycle in ARs: The water-vapor tracers show the entire moisture-transport process of the AR and the surrounding atmosphere; the Lagrangian method provides a detailed mapping of the moisture source regions.

2.4.4 Conclusions, Implications, and Future Directions

Moisture transport within ARs is clearly more complex than suggested from 2-D, vertically integrated snapshots of water vapor and water vapor transport. During the entire life cycle of an AR, moisture converges horizontally into ARs before being transported poleward, ascending, partially condensing, and, finally, precipitating. The moisture delivered by ARs as they make landfall is, therefore, an intricate vertically and horizontally variable combination of water from different source regions. Quantitative model studies are still very limited, both in the Eulerian and Lagrangian model frameworks. The available studies commonly suggest that although long-range moisture transport from the tropics is important for ARs making landfall at rather low latitudes (e.g., in California), it is the exception rather than the rule for ARs impinging on higher-latitude regions (e.g., Northern Europe). More quantitative studies in other regions are clearly needed to more firmly establish the AR moisture budget. These should also include the role of surface evaporation within the AR.

Knowledge about AR moisture transport—and the moisture source regions—has important implications for atmospheric energy transport and radiative effects. Condensational latent heat release couples oceanic sources to far-away atmospheric regions, transferring energy meridionally and vertically. Injecting high amounts of water vapor in a cold and dry polar environment substantially affects the atmosphere's radiative budget. Cloud formation from this water vapor is a further aspect of significance for the climate system. Such aspects of ARs have, so far, received little attention (with the exception of Woods et al. 2013 and related work). Clearly, in the framework of a changing climate, the connection between moisture transport in ARs and energy transfer to high latitudes calls for further analysis.

So far, several conclusions in this chapter are based on only a few or even individual case studies. In addition, current analyses are predominantly based on an atmospheric model's representation of reality, rather than direct observations. The currently used numerical methods are powerful tools for further investigations of how ARs modulate meridional water vapor transport. Eulerian and Lagrangian methods can thereby be seen as complementary. At the same time, model studies can benefit from further improvements such as on-line trajectory calculations for the Lagrangian approach (Miltenberger et al. 2013) or cloud-resolving modeling for Eulerian water-vapor tagging to reduce the uncertainty in model simulations (Winschall et al. 2014).

To more closely tie the water-vapor transport as simulated by atmospheric models to reality, an observational tracer of water-vapor origin and transport would be required. Stable water isotopes are potentially such a tool, because they inte-

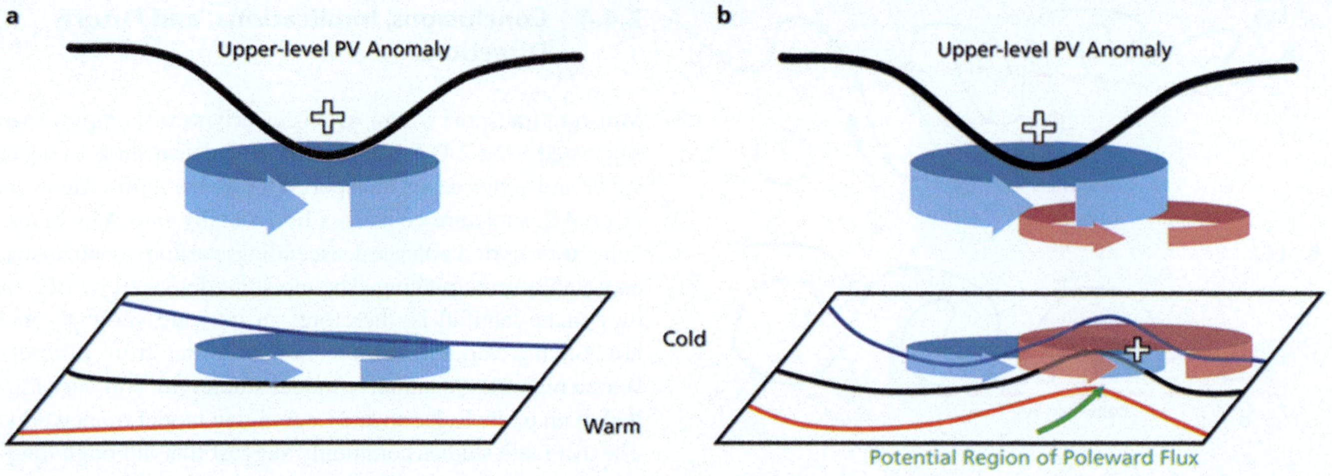

Fig. 2.16 A schematic representation of cyclogenesis with the approach of an upper-level PV anomaly over a low-level baroclinic zone. In (**a**) the cyclonic circulation associated with the upper-level PV anomaly (indicated by *blue upper-level arrow around the "+" symbol*) induces a weak cyclonic circulation (given by *arrow thickness*) to the near surface. The sense of the low-level cyclonic circulation will induce temperature advections ahead of and behind the upper-level PV anomaly. In (**b**) the warm temperature anomaly that has developed can be represented by a low-level positive PV anomaly (represented by the *low-level* "+"). The cyclonic circulation associated with the low-level PV anomaly will induce a weak upper-level cyclonic circulation, given by the *red arrows*, thus reinforcing the upper-level PV anomaly and slowing down its eastward progression. The *green arrow* indicates a potential region of poleward water vapor (WV) flux. (Adapted from Hoskins et al. 1985)

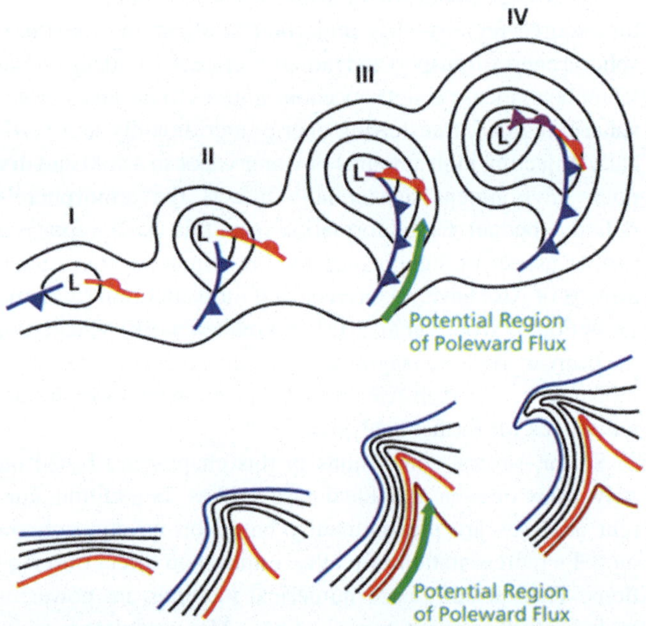

Fig. 2.17 Conceptual model for cyclone evolution following the Norwegian cyclone model that shows idealized lower-tropospheric geopotential height and fronts in the *top panel* and lower-tropospheric potential temperature in the *bottom panel*. The *green arrow* indicates a potential region of poleward water vapor flux. (Adapted [*colorized*] from Schultz et al. (1998) and based on Bjerknes and Solberg (1922))

grate the entire atmospheric transport history during phase changes from source to sink (e.g., Jouzel et al. 2013; Sodemann et al. 2008b). So far, this information has been used in only a few AR case studies (Coplen et al. 2008, 2015; Yoshimura et al. 2010; Bonne et al. 2015). Future field campaigns will, ideally, be able to cover the large spatial extent of moisture sources during AR events, including tracers of moisture origin, such as stable isotopes. This can provide the information necessary to tie model simulations to reality for the location of moisture sources, the spatial extent and duration of moisture transport, and the contribution of different components to the AR moisture budget.

2.5 ARs and Extratropical Dynamics

The Earth's atmosphere is a fluid that primarily comprises dry air with comparatively small concentrations of water vapor. This atmospheric fluid is bound to Earth via gravity and is constrained by the laws of fluid dynamics that describe relationships among fluid velocity, pressure, density, and temperature. The atmosphere is in a perpetual state of motion from spatio-temporal variations in solar radiation on a rotating planet covered mostly in water. These combine to give rise to preferential regions of enhanced atmospheric water vapor concentrations and preferential flow patterns that can transport, aggregate, or deplete—because of cloud formation and precipitation—large quantities of atmospheric water vapor.

Atmospheric water vapor is produced within the hydrologic cycle in large quantities via evaporation of liquid water, sublimation of ice, and transpiration in vegetated areas, among other less common processes. An estimated 420 × 103 km^3 of water vapor is added to the atmosphere via evaporation each year, whereas an additional 70 × 10^3 km^3 of water vapor is added to the atmosphere via transpiration each year (Anderson and Strahler 2008, p. 125). The amount of water vapor per unit mass of air is constrained by the relationship between partial pressure of water vapor and temperature. For example, the saturation [water] vapor pressure, or the partial pressure of water vapor at saturation, increases in a roughly exponential nature from ~6 hPa at 0°C to ~16 hPa at 15°C (see Fig. 4.8 of McElroy 2002, p. 38) following the closed form of the Clausius–Clapeyron equation (see Miller 2015 [Eq. 7.45]). A large majority of global atmospheric water vapor content is contained within the troposphere at higher barometric pressures (following from the relationship between temperature and pressure from the Ideal Gas Law) and at more tropical latitudes at warmer temperature (see Fig. 5.4 of Anderson and Strahler 2008). This relationship is readily observed via sub-daily Special Sensor Microwave Imager (SSM/I) satellite imagery of IWV (see Sect. 2.2), with the maximum in IWV content across the tropics in the so-called IWV reservoir and the minimum in IWV across polar regions.

Further analysis of the sub-daily global IWV from SSM/I satellite imagery highlights the filamentary structures of poleward extrusions of enhanced regions of IWV that have been defined in previous literature as tropospheric rivers (Newell et al. 1992) or in previous sections of this book as ARs. As shown in Sect. 2.2, these filamentary structures are found in the warm-sector region of cyclones ahead of surface cold fronts; herein, their existence is attributed to high-frequency synoptic-scale circulations embedded within the mid-latitude storm track and the development of mid-latitude cyclones. This Sect. 2.5 also investigates the high-frequency synoptic-scale processes that comprise the mid-latitude storm track and their relationship to ARs. Additional information on many of the mid-latitude cyclone dynamics discussed next in Sects. 2.5.1 and 2.5.2 is available in traditional synoptic-scale textbooks (e.g., Bluestein 1992; Carlson 1998; Holton 2004; Martin 2006, among others). See Lau et al. (2011) for a review of intra-seasonal variability in the atmosphere–ocean system that link weather and climate processes, and Newman et al. (2012) for a review of the relative contributions of synoptic- and low-frequency eddies to time-mean atmospheric moisture transport. The subsequent review provides evidence for the formation and presence of ARs in the warm sector of canonical mid-latitude cyclones; however, with notable case-to-case variability in cyclone structures (and their evolution), notable case-to-case variability also exists in the processes that lead to AR formation.

2.5.1 Mid-Latitude Storm Track and Cyclogenesis

The general circulation of the atmosphere is maintained through the generation and depletion of atmospheric available potential and kinetic energy (Lorenz 1955). The mid-latitude storm track is, consequently, the manifestation of the depletion of atmospheric available potential energy and the generation of atmospheric kinetic energy observed via the release of baroclinic and barotropic instability in conjunction with development of large-scale eddies across preferential regions of the globe. The synoptic-scale eddies comprising the mid-latitude storm track largely take on the form of two paradigms associated with baroclinic Rossby wave life cycle behavior (Fig. 2.15; Thorncroft et al. 1993). The two life cycles, identified as Life Cycle 1 (LC1) and Life Cycle 2 (LC2), apply to the development and advection of mid-latitude upper-tropospheric troughs and their upstream (or downstream) ridges. During LC1 (Fig. 2.15a), or "anticyclonic" wave breaking (AWB), troughs tend to tilt from southwest-to-northeast, and zonally contract as they propagate along an anticyclonic arc toward the equator. In contrast, during LC2 (Fig. 2.15b), or "cyclonic" wave breaking (CWB), troughs tend to tilt from southeast-to-northwest, and zonally expand as they propagate along a more cyclonic arc toward the pole. Rossby wave breaking (RWB) in the exit regions of mid-latitude jet streams, particularly associated with LC1 events, can result in penetration of mid-latitude troughs into subtropical latitudes (Webster and Holton 1982; Matthews and Kiladis 1999). Such low-latitude RWB events have been implicated in the development of tropical cloud plumes (e.g., McGuirk et al. 1987, 1988), enhanced convection downstream of the trough (e.g., Moore et al. 2010), and the initiation of poleward water vapor transport out of the tropics (Knippertz 2007) along possible ARs that follow a pathway similar to a TME (e.g., Knippertz and Wernli 2010; Knippertz et al. 2013; see Sects. 1.1.4 and 1.1.5, as well as Chaps. 5 and 7. Although water vapor transport is the signature of ARs in the lower troposphere, recent work suggests that RWB is a characteristic of ARs in the upper troposphere (Payne and Magnusdottir 2014).

The genesis of individual surface cyclones, and their circulations that are able to mobilize greater amounts of water vapor in the lower troposphere (e.g., Bao et al. 2006; Sodemann and Stohl 2013) in the above-mentioned LC paradigms, can be described to first order by quasi-geostrophic (QG) theory. The QG theory explains atmospheric momentum in conjunction with a balance between the Coriolis force and pressure gradient force (i.e., geostrophic balance), and any small displacements or accelerations away from geostrophy that inertia affords (Holton 2004). By definition, the geostrophic wind is non-divergent such that any divergence is the result of fluid expansion (or contraction) generated by fluid acceleration that follows the ageostrophic wind. Furthermore, atmospheric layers bound by regions of divergence or convergence must contain vertical motion according to the QG continuity equation. Thus, the total horizontal wind, V, comprises both the geostrophic wind, V_g, and the ageostrophic wind, V_a, and vertical motion is driven by convergence or divergence of V_a. Following QG theory, cyclogenesis therefore relates to upper-tropospheric divergence downstream of an upper-tropospheric trough. This creates a region of enhanced tropospheric upward vertical motion and influences lower-tropospheric convergence and the generation of cyclonic vorticity (Sutcliffe 1939, 1947; Sutcliffe and Forsdyke 1950) in the vicinity of a lower-tropospheric thermal gradient. The aforementioned cyclogenesis process has been named "Type B" (Petterssen and Smebye 1971) or "Stream II" (Davies 1997) cyclogenesis and is the result of cyclonic vorticity advection by the geostrophic wind in the upper troposphere that over-spreads an area of geostrophic warm air advection in the lower troposphere, or simply cyclonic vorticity advection by the thermal wind (Trenberth 1978).

An alternative way to visualize the cyclogenesis processes is through "potential vorticity (PV) thinking," detailed in Hoskins et al. (1985). For adiabatic and frictionless flow, PV, as defined by Ertel (1942), is the product of earth's vorticity and the static stability for a given isentropic surface. Maxima in tropospheric PV are typically found near the poles, where strong static stability and high absolute vorticity exist from the poleward increase of the Coriolis parameter. PV maxima extend equatorward from higher latitudes in the presence of strong static stability and troughs in the geopotential height field known as PV streamers (Appenzeller and Davies 1992), whereas minima in PV extend poleward from lower latitudes and are associated with weak static stability and ridges in the geopotential height field.

Figure 2.16 illustrates how an upper-tropospheric region of enhanced PV (e.g., a positive PV anomaly or PV streamer) can induce extratropical cyclogenesis: a positive PV anomaly will induce a cyclonic circulation through its Rossby penetration depth that extends to the near surface. Note that the southerly flow on the east side of the cyclonic circulation may support poleward water vapor flux and AR formation if proximate to the IWV reservoir (AR formation may also occur separate from the IWV reservoir; see Sect. 2.5.2). If the underlying surface is characterized by lower-tropospheric baroclinicity, the cyclonic circulation will also support the formation of a warm temperature anomaly at the surface downstream of the upper-tropospheric PV anomaly. The warm temperature anomaly is, consequently, also a lower-tropospheric positive PV anomaly and will, in turn, act to induce its own cyclonic circulation aloft. A positive feedback mechanism is established as the modified upper-tropospheric circulation advects enhanced PV air aloft equatorward and, by the same circulation, advects weak PV air aloft poleward, which reinforces the pre-existing upper-tropospheric PV anomaly. According to QG theory, the requirements for cyclogenesis and cyclone intensification are also met with an upshear-tilted disturbance, which ensures maximum cyclonic (or potential) vorticity advection by the thermal wind over the surface cyclone (Martin 2006).

The above-mentioned storm track dynamics related to RWB and QG/PV frameworks for cyclogenesis provide evidence for synoptic-scale processes that may influence the poleward transport of water vapor from the IWV reservoir into mid-latitudes in association with RWB. These synoptic-scale processes can, consequently, also influence the development and evolution of mid-latitude cyclones and in turn also influence the formation and maintenance of atmospheric water vapor transport along ARs. Section 2.5.2 explains this concept in more detail.

2.5.2 Mid-Latitude Cyclones and ARs

The first modern theory for cyclone evolution originated at the Bergen School of Meteorology in the early twentieth century and is known as the Norwegian Cyclone Model (NCM; Bjerknes 1919; Bjerknes and Solberg 1922). The NCM was developed for cyclones typically located over the Northeast Atlantic, but is applicable to cyclones also located over the Northeast Pacific, and generally in regions where ARs are commonly observed (see Chap. 4). More recent studies have shown that a variety of cyclone and frontal evolutions, beyond the scope of the NCM, are possible (e.g., Browning 1990; Shapiro and Keyser 1990; Evans et al. 1994; Smigielski and Mogil 1995; Young 1995; Bosart et al. 1998; Schultz et al. 1998). The NCM (Fig. 2.17) begins with a small-scale cyclonic disturbance along the polar front that results in the advection of cold air equatorward to the west of the cyclone center, and advection of warm air poleward to the east of the cyclone center, as just described in Sect. 2.5.1. Contraction of the baroclinic zone by the temperature advection and deformation patterns results in the development of the cold and warm fronts. The cold front is observed to rotate faster than the warm front and propagates eastward, eventually catching up to the warm front. Schultz et al. (1998) showed that frontal features favor a meridional elongation and zonal contraction because the background flow is likely highly amplified and diffluent in exit region of the North Atlantic (or North Pacific) jet stream. The NCM thus describes a cyclone evolution that favors the meridional elongation of the cyclone cold front, a contraction of the thermal gradient along the cyclone cold front (i.e., frontogenesis), enhanced southerly flow in the cyclone warm sector along a LLJ stream, and a developing corridor of enhanced

IWV and poleward water vapor flux (Lackmann 2002) that may develop into an AR. The NCM concludes with the formation of an occluded front in the region of weak baroclinicity associated with the thermal ridge that extends from the warm sector toward the cyclone center.

Observational and numerical studies show that cyclones evolve within different combinations of planetary and synoptic-scale flow configurations, variations in lower-tropospheric baroclinicity and lower-tropospheric moisture, and static stability (e.g. Bosart 1981; Reinhold and Pierrehumbert 1982; Gyakum 1983; Reed and Albright 1986; Lackmann et al. 1996; Carlson 1998). For example, the introduction of moisture and an environment with weak static stability can greatly accelerate QG forcing for (saturated) vertical motion and the cyclogenesis processes just described in Sect. 2.5.1 (see Bluestein 1992 [Eq. 5.9.28]). Cyclones may then contain large and rapidly-expanding cloud masses that illuminate likely regions of strong diabatic heating and the poleward advection of lower-tropospheric warm and moist air (e.g., Browning and Pardoe 1973; Browning 1986; Reed and Albright 1986; Wernli and Davies 1997; Schultz 2001; Martin 2006) in the possible location of an AR. This region of strong diabatic heating and latent heat release in the cyclone warm sector can have a large effect on the dynamical evolution of both the synoptic-scale and mesoscale characteristics of the developing cyclone. PV will be redistributed above and below a mid-tropospheric region of latent heat release because of the imposed vertical gradients of diabatic heating (see Martin 2006 [Eq. 9.23]). PV will therefore increase below the diabatic heating maximum and strengthen lower-tropospheric frontal zones and cyclonic circulations, whereas PV will decrease above the diabatic heating maximum, strengthening anticyclonic circulations (i.e., a shortwave ridge) aloft and enhancing the upstream upper-level horizontal PV gradient. The strength of the frontal circulations (i.e., the cross-front ageostrophic circulations; e.g., Sawyer 1956 and Eliassen 1959) consequently increases as a result of mid-tropospheric latent heat release in the presence of weak static stability and warm and moist air (e.g., Hoskins and Bretherton 1972). This cross-front ageostrophic circulation, which may transect an AR located on the warm side of a baroclinic zone that is undergoing frontogenesis, is schematically illustrated in Fig. 2.18 by Cordeira et al. (2013). The end result of this process is a "complex interdependence" among frontal-induced precipitation, lower-tropospheric PV maxima, the LLJ, and warm-sector water vapor transport (Lackmann 2002) that can influence the structure and evolution of ARs (e.g., Sodemann and Stohl 2013; Cordeira et al. 2013).

The development of poleward water vapor transport that may form into ARs in the warm sector of the mid-latitude cyclone can, therefore, occur in conjunction with (1) processes that lead to cyclogenesis and (2) processes that lead to a modification of the ensuing cyclogenesis process. Consequently, processes that lead to modification of the cyclogenesis process in association with saturated ascent, precipitation, and latent heat release may also play a detrimental role in the maintenance and evolution of an AR (e.g., the AR's life cycle). Maintenance of AR conditions within an Eulerian framework (i.e., local conservation of IWV or increases in the magnitude of IWV) relates a net positive or zero sum of IVT convergence, evaporation, and precipitation in the IWV budget. In the presence of precipitation, IWV is conserved in conjunction with evaporation and IVT convergence. Cordeira et al. (2013) paradoxically demonstrate that the initial formation of an AR during cyclogenesis might occur during a period with a net negative sum of the IWV budget related to a decrease in IWV via precipitation. The

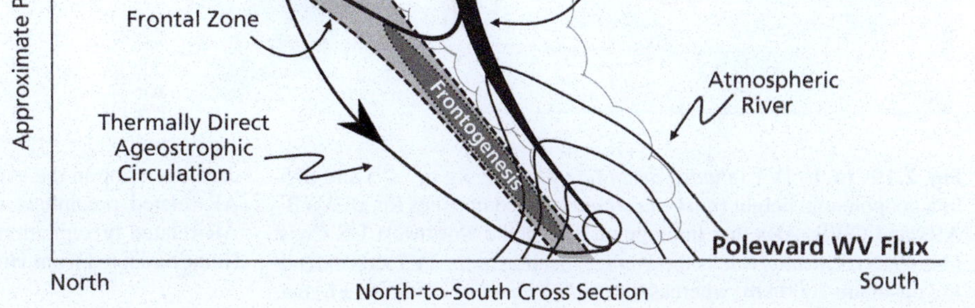

Fig. 2.18 Schematic cross-section representation of the vertical structure of a tropospheric frontal zone (*dashed lines*) with poleward water vapor (WV) flux (*thin lower-tropospheric contours*) along an AR that contains frontogenesis (*shaded*) and a strong thermally-direct ageostrophic circulation (*counter-clockwise rotating arrow*) within the equatorward entrance region of an intense tropopause-level jet stream (*thick contours labeled 50, 70, and 90 m s*$^{-1}$). (Originally modeled after Shapiro (1982) and has been adapted from Cordeira et al. 2013)

associated latent heat release, however, can profoundly affect the subsequent evolution of the synoptic-scale flow that results in an enhanced upper-tropospheric PV gradient, a stronger upper-tropospheric jet stream, and a thermally-direct (frontogenetic) ageostrophic cross-front circulation, which may subsequently aggregate IWV along the AR via IVT convergence, and offset the deleterious effect of precipitation in the IWV budget. A more robust review of water vapor budgets and water vapor source regions for ARs appeared in Sect. 2.4.

2.5.3 Linking Extratropical Dynamics to Hydrometeorological Effects

As discussed above (Sect. 2.5.1), RWB in the exit regions of mid-latitude jet streams—whether anticyclonic or cyclonic—can result in penetration of mid-latitude troughs into subtropical latitudes, and, in turn, enable poleward transport of water vapor along an AR in the warm sector of the mid-latitude cyclone. The type of RWB will determine the impingement angle of the water vapor transport vector along ARs on the coastal topography, modulating the intensity of orographic precipitation and the possibility of flooding over the region. The ensuing discussion focuses on the US West Coast; however, the physical mechanisms of RWB modulation of ARs are globally applicable.

Approximately two-thirds of all detected ARs that made landfall on the US West Coast during 1979–2009 in the NASA MERRA reanalysis data set related directly to RWB over the eastern North Pacific (i.e., RWB–ARs; Hu et al. 2017). As water vapor transport along these ARs arrives at the coast, it encounters the topography of the region: the north-to-south-oriented Cascade Range of the Pacific Northwest or the northwest-to-southeast-oriented Sierra Nevada or Coast Range of California. The impinging angle of the water vapor transport vector on topography can critically determine the intensity of orographic precipitation and the flooding of the affected watersheds (Ralph et al. 2003; Neiman et al. 2011). As expected, the impinging angles of the water vapor transport vector along ARs associated with anticyclonic wave breaking (i.e., AWB–ARs) and cyclonic wave breaking (i.e., CWB–ARs) are very different, as exemplified for the Pacific Northwest US (Fig. 2.19b). Landfalling CWB–ARs over the Pacific Northwest USA predominantly contain water vapor transport that impinges on topography at angles between 10° and 70°, with a median of 36° (i.e., associated with a south-southwesterly IVT direction; Fig. 2.19b),

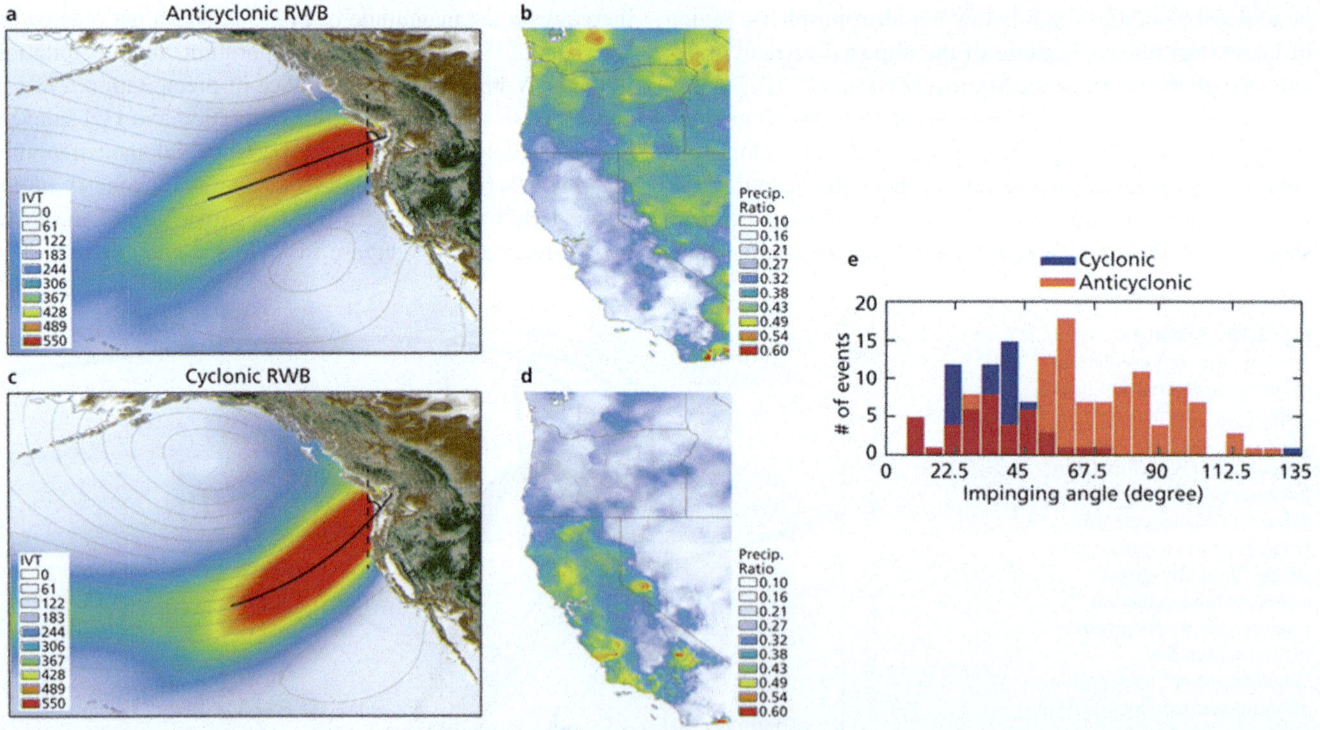

Fig. 2.19 (**a**, **b**) IVT (*shaded according to scale:* kg m^{-1} s^{-1}) and 875-hPa geopotential height (*dashed contours*; *m*) composites for all AWB–ARs and CWB–ARs that impinge on the Pacific Northwest US Coast (44–49°N). The *blue line* is the average location of the IVT axis extending upstream 2000 km, whereas the *dashed blue lines* indicate ±1 standard deviation in the average location for the IVT axis. (**c**, **d**) Ratio of AR-related precipitation from all AWB–ARs and CWB–ARs to all AR-related precipitation for all US West Coast locations (36–49°N). (Image adapted from Hu et al. 2017)

whereas AWB–ARs contain a broader range and a median angle of 62° (i.e., associated with a west-southwesterly IVT direction; Fig. 2.19a). These different impinging angles influence the topographically-modulated AR-related precipitation.

Along the Pacific Northwest US Coast (e.g., Washington and Oregon) and some mountainous regions of the Pacific Southwest US Coast (e.g., California), more than 70% of total AR-related precipitation results from RWB–ARs (not shown; see Hu et al. 2017). Interestingly, when the precipitation related to AWB–ARs or CWB–ARs is isolated, a clear pattern emerges. The Pacific Northwest US Coast is predominantly affected by AWB–ARs, which account for more than 40% of AR-related precipitation (Fig. 2.19c). In contrast, a larger portion of AR-related precipitation in the Pacific Southwest US Coast is associated with CWB–ARs (Fig. 2.19d). Clearly, from these results, intense precipitation preferentially occurs over mountains ranges where the water vapor transport vector along an AR is oriented parallel to the upslope terrain vector. These sharp differences in AR-related precipitation associated with AWB and CWB result in significant differences in streamflow. Hu et al. (2017) analyzed the 20 most intense AR-related streamflow events in the Chehalis River basin in western Washington, and the Russian River basin in northern California. The analysis found that AWB–ARs play a dominant role in explaining the top streamflow events measured within the Chehalis, while CWB–ARs were responsible for most of the top streamflow events in the Russian River basin (Fig. 2.20). It is important to emphasize that many other factors—including antecedent soil moisture conditions, the phase of precipitation (liquid or solid), and the response of the watershed—will all influence the streamflow response. In the case of the Chehalis River basin, AWB–ARs affect the basin more frequently and are also associated with more precipitation falling in liquid rather than snow form.

2.5.4 Summary

The mid-latitude storm track comprises synoptic-scale eddies that are characterized by Rossby waves. These Rossby waves can sometimes penetrate into the subtropics where they may initiate poleward water vapor transport that could serve as the impetus for AR formation. Alternatively, these Rossby waves may also initiate cyclogenesis in the presence of favorable synoptic-scale conditions that relate to the vertical structure of the atmosphere, moisture content, and static stability. The horizontal circulations associated with mid-latitude cyclogenesis may further contribute to poleward water vapor transport and the initial formation of an AR in the warm sector of a cyclone. This warm sector of the cyclone is characterized by enhanced water vapor and weak static stability that can play a crucial role in cyclone intensification related to saturated ascent, precipitation, and diabatic heating. These can, in turn, influence the evolution and strength of frontal circulations and ARs. As such, cyclones can simultaneously deplete IWV along the AR via precipitation in the presence of saturated ascent and generate IWV along the AR via IVT convergence in the presence of frontogenesis. This generation and depletion of IWV along an AR modulates the AR life cycle, which is intimately related to the life cycle of the mid-latitude cyclone, the evolution of the large-scale flow, and the dynamical processes related to precipitation. In fact, mid-latitude dynamics, as characterized by the different

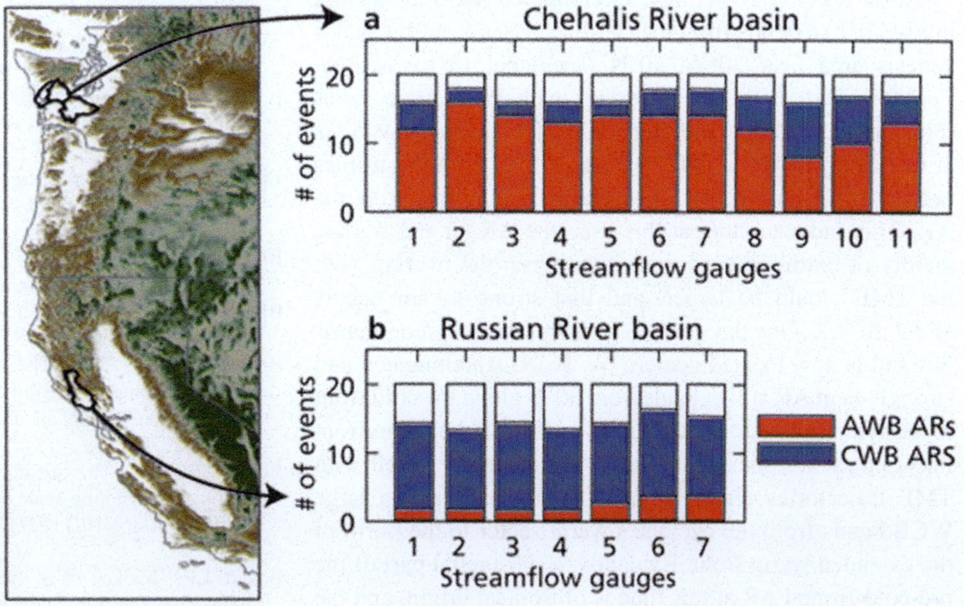

Fig. 2.20 Number of top 20 streamflow events to AWB–ARs (*red*) and CWB–ARs (*blue*) for each gauge within the (**a**) Chehalis River basin and (**b**) the Russian River basin (Hu et al. 2017)

types of RWB, can modulate the spatial pattern and intensity of precipitation and streamflow on the US West Coast.

Improved understanding of AR-related large-scale dynamics and associated hydrologic effects could improve prediction skill. The inter-annual variability of RWB over the Northeast Pacific is modulated in part by the El Niño–Southern Oscillation (ENSO), because the number of AWB events is higher during La Niña years, while CWB is higher during El Niño (Ryoo et al. 2013). Furthermore, climate change could modify RWB dynamics. In fact, Barnes and Hartmann (2012) found a poleward shift of RWB events and a significant decrease of CWB events as a result of poleward shift of mid-latitude jets under increased greenhouse gas forcing. How these changes could affect both large-scale atmospheric dynamics and local hydrology is an important topic for future research.

2.6 A Case Study Example

As an illustration of the three features and their linkage, Fig. 2.5 shows a North Atlantic cyclone in November 1992. This cyclone formed as a frontal wave at 0000 UTC 22 November on a meridionally elongated cold front, downstream of a pronounced, narrow, and elongated upper-level trough, and then experienced rapid intensification from the strong vertical coupling of the upper-level trough and the low-level disturbance. Eventually, at 0000 UTC 24 November, the system became a mature, deep, and stationary Icelandic Low (see the detailed discussions in Wernli (1997) and Rossa et al. (2000)). Figure 2.5 illustrates the development of this cyclone in terms of the sea-level pressure field and the superimposed, objectively identified TMEs, ARs, and WCBs.

At 00 UTC 22 November, an elongated AR (red shading in Fig. 2.5a) extends from the Caribbean Sea over the cyclogenesis area near 30°W, 40°N (see label *L*) to Western Europe. Most of this AR overlaps with WCB trajectories (blue contours mark the outline of WCB trajectory positions at this time), while a TME feature exiting from the tropics between 30° and 40°W only partially overlaps with the AR. This indicates that at this time the AR air masses are mainly of extratropical origin (otherwise, the overlap with the TME would be larger) and that strong ascent occurs along the AR. One day later (Fig. 2.5b), the cyclone intensifies and is now located near 16°W, 56°N. An elongated and strongly bent AR still extends from 30°N along the cold front to the cyclone center and then eastward along the warm front into Central Europe. At this time, the AR agrees well with TME trajectories along the cold-frontal part, and a large WCB bends from the cyclone's warm sector to the north of the extended warm front. Evidently, a substantial part of the pre-cold-frontal AR at this time is of tropical origin, and the almost parallel and only weakly overlapping arrangement of the three features along the warm front points to the non-trivial relative motion of their associated air parcels. One day later (Fig. 2.5c), the cyclone has attained a central sea-level pressure of about 960 hPa near Iceland. Along its cold front, extending from 35°N to Scandinavia, a narrow AR coincides almost perfectly with a TME which extends further poleward. The WCB related to this cyclone is now fully detached from the AR and TME features and extends from Greenland to Finland near 70°N.

This brief case study thus illustrates the complex relationship of the three flow features, in this case with an AR that is first well aligned with a WCB and later with a TME. The WCB mainly highlights the region where moist air ascends in the frontal regions of the cyclone, the AR emphasizes the strong filamentary poleward moisture transport mainly in the cyclone's warm sector, and the TME reveals that some of the air that contributes to the strong water vapor transport in the AR and the rain-out in the WCB is actually of tropical origin.

Acknowledgements We are most grateful to Hanin Binder, Maxi Boettcher, Hanna Joos, Erica Madonna, Gregor Pante, and Michael Sprenger for their support in establishing the ERA-Interim climatologies of WCBs and TMEs. We also thank MeteoSwiss for granting access to ECMWF data, and Paul Neiman and Duane Waliser for their very supportive and constructive reviews.

References

Anderson BT, Strahler AH (2008) Visualizing weather and climate. Wiley, Hoboken

Appenzeller C, Davies HC (1992) Structure of stratospheric intrusions into the troposphere. Nature 358:570–572. https://doi.org/10.1038/358570a0

Bao J-W, Michelson SA, Neiman FM et al (2006) Interpretation of enhanced integrated water vapor bands associated with extratropical cyclones: their formation and connection to tropical moisture. Mon Weather Rev 134:1063–1080. https://doi.org/10.1175/mwr3123.1

Barnes EA, Hartmann DL (2012) Detection of Rossby wave breaking and its response to shifts of the midlatitude jet with climate change. J Geophys Res 117:D09117. https://doi.org/10.1029/2012jd017469

Bjerknes J (1919) on the structure of moving cyclones. Mon Weather Rev 47:95–99. https://doi.org/10.1175/1520-0493(1919)47<95:otsomc>2.0.co;2

Bjerknes J, Solberg H (1922) Life cycle of cyclones and the polar front theory of atmospheric circulation. Geophy Publ 3:3–18

Bluestein H (1992) Synoptic-dynamic meteorology in midlatitudes. Oxford University Press, Oxford

Bonne J-L et al (2015) The summer 2012 Greenland heat wave: in situ and remote sensing observations of water vapor isotopic composition during an atmospheric river event. J Geophys Res 120(7):2970–2989

Bosart LF (1981) The presidents' day snowstorm of 18–19 February 1979: a subsynoptic-scale event. Mon Weather Rev 109:1542–1566. https://doi.org/10.1175/1520-0493(1981)109<1542:tpdsof>2.0.co;2

Bosart LF, Bracken WE, Seimon A (1998) A Study of cyclone mesoscale structure with emphasis on a large-amplitude inertia–

gravity wave. Mon Weather Rev 126:1497–1527. https://doi.org/10.1175/1520-0493(1998)126<1497:asocms>2.0.co;2

Bowman KP, Lin JC, Stohl A et al (2013) Input data requirements for Lagrangian trajectory models. Bull Am Meteorol Soc 94:1051–1058. https://doi.org/10.1175/BAMS-D-12-00076.1

Browning KA (1971) Radar measurements of air motion near fronts. Part two: some categories of frontal air motion. Weather 26:320–340

Browning KA (1986) Conceptual models of precipitation systems. Wea Forecast 1:23–41. https://doi.org/10.1175/1520-0434(1986)001<0023:cmops>2.0.co;2

Browning KA (1990) Organization of clouds and precipitation in extratropical cyclones. In: Newton CW, Holopainen EO (eds) Extratropical cyclones: the Erik Palmén memorial volume. American Meteorological Society, Boston, pp 129–153

Browning KA, Harrold TW (1969) Air motion and precipitation growth in a wave depression. Q J R Meteorol Soc 95:288–309

Browning KA, Pardoe CW (1973) Structure of low-level jet streams ahead of mid-latitude cold fronts. Q J R Meteorol Soc 099:619–638. https://doi.org/10.1256/smsqj.42203

Carlson TN (1980) Airflow through midlatitude cyclones and the comma cloud pattern. Mon Weather Rev 108:1498–1509

Carlson TN (1998) Mid-latitude weather systems. American Meteorological Society, Boston, p 507

Coplen TB, Neiman PJ, White AB et al (2008) Extreme changes in stable hydrogen isotopes and precipitation characteristics identified in a land-falling Pacific storm using a novel precipitation collector and meteorological observatory. Geophys Res Lett 35:L21808. https://doi.org/10.1029/2008GL035481

Coplen TB, Neiman PJ, White AB et al (2015) Categorization of northern California rainfall for periods with and without a radar brightband using stable isotopes and a novel automated precipitation collector. Tellus B 67:48. https://doi.org/10.3402/tellusb.v67.28574

Cordeira JM, Ralph FM, Moore BJ (2013) The development and evolution of two atmospheric rivers in proximity to Western North Pacific tropical cyclones in October 2010. Mon Weather Rev 141:4234–4255. https://doi.org/10.1175/mwr-d-13-00019.1

Dacre HF, Clark PA, Martinez-Alvarado O et al (2015) How do atmospheric rivers form? Bull Am Meteorol Soc 96:1243–1255

Davies HC (1997) Emergence of the mainstream cyclogenesis theories. Meteor Z 6:261–274

Dee DP et al (2011) The ERA-interim reanalysis: configuration and performance of the data assimilation system. Q J R Meteorol Soc 137:553–597

Eckhardt S, Stohl A, Wernli H et al (2004) A 15-year climatology of warm conveyor belts. J Clim 17:218–237

Eliassen A (1959) On the formation of fronts in the atmosphere, The Rossby Memorial Volume. Rockefeller Institute Press, New York

Ertel H (1942) Ein neuer hydrodynamischer Erhaltungssatz. Die Naturwissenschaften 30:543–544. https://doi.org/10.1007/bf01475602

Evans MS, Keyser D, Bosart LF et al (1994) A satellite-derived classification scheme for rapid maritime cyclogenesis. Mon Weather Rev 122:1381–1416. https://doi.org/10.1175/1520-0493(1994)122<1381:asdcsf>2.0.co;2

Fröhlich L, Knippertz P, Fink AH et al (2013) An objective climatology of tropical plumes. J Clim 26:5044–5060. https://doi.org/10.1175/JCLI-D-12-00351.1

Grams CM, Wernli H, Boettcher M et al (2011) The key role of diabatic processes in modifying the upper-tropospheric wave guide: a North Atlantic case-study. Q J R Meteorol Soc 137:2174–2193

Green JSA, Ludlam FH, McIlveen JFR (1966) Isentropic relative-flow analysis and the parcel theory. Q J R Meteorol Soc 92:210–219

Guan B, Waliser DE (2015) Detection of atmospheric rivers: evaluation and application of an algorithm for global studies. J Geophys Res Atmos 120:12514–12535. https://doi.org/10.1002/2015JD024257

Guan B, Waliser D, Ralph FM (2018) An Inter-comparison between reanalysis and dropsonde observations of the total water vapor transport in individual atmospheric rivers. J Hydrometeorol 19(2):321–337. https://doi.org/10.1175/JHM-D-17-0114.1

Gyakum JR (1983) On the evolution of the QE II storm. II: dynamic and thermodynamic structure. Mon Weather Rev 111:1156–1173. https://doi.org/10.1175/1520-0493(1983)111<1156:oteoti>2.0.co;2

Holton JR (2004) An introduction to dynamic meteorology. Elsevier, Amsterdam

Hoskins BJ, Bretherton FP (1972) Atmospheric frontogenesis models: mathematical formulation and solution. J Atmos Sci 29:11–37. https://doi.org/10.1175/1520-0469(1972)029<0011:afmmfa>2.0.co;2

Hoskins BJ, Mcintyre ME, Robertson AW (1985) On the use and significance of isentropic potential vorticity maps. Q J R Meteorol Soc 111:877–946. https://doi.org/10.1256/smsqj.47001

Hu H, Dominguez F, Wang Z et al (2017) Linking atmospheric river hydrological impacts on the U.S. west coast to Rossby wave breaking. J Clim 30:3381–3399. https://doi.org/10.1175/JCLI-D-16-0386.1

Joos H, Wernli H (2012) Influence of microphysical processes on the potential vorticity development in a warm conveyor belt: a case study with the limited area model COSMO. Q J R Meteorol Soc 138:407–418

Jouzel, J., G. Delaygue, A. Landais, V. Masson-Delmotte, C. Risi, and F. Vimeux (2013), Water isotopes as tools to document oceanic sources of precipitation, Water Resour. Res., 49, 7469–7486, https://doi.org/10.1002/2013WR013508

Knippertz P (2007) Tropical–extratropical interactions related to upper-level troughs at low latitudes. Dyn Atmos Oceans 43:36–62. https://doi.org/10.1016/j.dynatmoce.2006.06.003

Knippertz P, Martin JE (2005) Tropical plumes and extreme precipitation in subtropical and tropical West Africa. Q J R Meteorol Soc 131:2337–2365. https://doi.org/10.1256/qj.04.148

Knippertz P, Martin JE (2007) A Pacific moisture conveyor belt and its relationship to a significant precipitation event in the semi-arid southwestern United States. Wea Forecast 22:125–144. https://doi.org/10.1175/WAF963.1

Knippertz P, Wernli H (2010) A Lagrangian climatology of tropical moisture exports to the Northern Hemispheric extratropics. J Clim 23:987–1003. https://doi.org/10.1175/2009JCLI3333.1

Knippertz P, Wernli H, Gläser G (2013) A global climatology of tropical moisture exports. J Clim 26:3031–3045. https://doi.org/10.1175/JCLI-D-12-00401.1

Lackmann GM (2002) Cold-frontal potential vorticity maxima, the low-level jet, and moisture transport in extratropical cyclones. Mon Weather Rev 130:59–74. https://doi.org/10.1175/1520-0493(2002)130<0059:cfpvmt>2.0.co;2

Lackmann GM, Bosart LF, Keyser D (1996) Planetary- and synoptic-scale characteristics of explosive wintertime cyclogenesis over the Western North Atlantic ocean. Mon Weather Rev 124:2672–2702. https://doi.org/10.1175/1520-0493(1996)124<2672:passco>2.0.co;2

Läderach A, Sodemann H (2016) A revised picture of the atmospheric moisture residence time. Geophys Res Lett 43:924–933. https://doi.org/10.1002/2015GL067449

Lau WKM, Waliser DE, Waliser D (2011) Predictability and forecasting. In: Intraseasonal variability in the atmosphere-ocean climate system. Springer, Berlin, pp 433–476. https://doi.org/10.1007/978-3-642-13914-7_12

Lorenz EN (1955) Available potential energy and the maintenance of the general circulation. Tellus 7:157–167. https://doi.org/10.1111/j.2153-3490.1955.tb01148.x

Lu M, Lall U, Schwartz A et al (2013) Precipitation predictability associated with tropical moisture exports and circulation patterns for a major flood in France in 1995. Water Resour Res 49:6381–6392. https://doi.org/10.1002/wrcr.20512

Madonna E, Wernli H, Joos H et al (2014) Warm conveyor belts in the ERA-Interim data set (1979-2010). Part I: Climatology and potential vorticity evolution. J Clim 27:3–26

Martin JE (2006) Mid-latitude atmospheric dynamics: a first course. Wiley, Chichester

Matthews AJ, Kiladis GN (1999) The tropical–extratropical interaction between high-frequency transients and the Madden–Julian oscillation. Mon Weather Rev 127:661–677. https://doi.org/10.1175/1520-0493(1999)127<0661:tteibh>2.0.co;2

McElroy MB (2002) The atmospheric environment: effects of human activity. Princeton University Press, Princeton

McGuirk JP, Thompson AH, Smith NR (1987) Moisture bursts over the tropical pacific ocean. Mon Weather Rev 115:787–798. https://doi.org/10.1175/1520-0493(1987)115<0787:mbottp>2.0.co;2

McGuirk JP, Thompson AH, Schaefer JR (1988) An Eastern pacific tropical plume. Mon Weather Rev 116:2505–2521. https://doi.org/10.1175/1520-0493(1988)116<2505:aeptp>2.0.co;2

Miller ST (2015) Applied thermodynamics for meteorologists. Cambridge University Press, Cambridge

Miltenberger AK, Pfahl S, Wernli H (2013) An online trajectory module (version 1.0) for the nonhydrostatic numerical weather prediction model COSMO. Geosci Model Dev 6:1989–2004. https://doi.org/10.5194/gmd-6-1989-2013

Moore RW, Martius O, Spengler T (2010) The Modulation of the subtropical and extratropical atmosphere in the pacific basin in response to the Madden–Julian oscillation. Mon Weather Rev 138:2761–2779. https://doi.org/10.1175/2010mwr3194.1

Neiman PJ, Schick LJ, Ralph FM et al (2011) Flooding in Western Washington: the connection to atmospheric rivers. J Hydrometeorol 12:1337–1358. https://doi.org/10.1175/2011jhm1358.1

Newell RE, Newell NE, Zhu Y et al (1992) Tropospheric rivers? - A pilot study. Geophys Res Lett 19:2401–2404. https://doi.org/10.1029/92gl02916

Newman M, Kiladis GN, Weickmann KM et al (2012) Relative contributions of synoptic and low-frequency eddies to time-mean atmospheric moisture transport, including the role of atmospheric rivers. J Clim 25:7341–7361. https://doi.org/10.1175/jcli-d-11-00665.1

Payne AE, Magnusdottir G (2014) Dynamics of landfalling atmospheric rivers over the north pacific in 30 years of MERRA reanalysis. J Clim 27:7133–7150. https://doi.org/10.1175/jcli-d-14-00034.1

Persson POG, Neiman, PJ, Walter B et al (2005) Contributions from California coastal-zone surface fluxes to heavy coastal precipitation: A CALJET case study During the Strong El Niño of 1998. Mon Wea Rev 133:1175–1198

Petterssen S, Smebye SJ (1971) On the development of extratropical cyclones. Q J Roy Met Soc 97:457–482. https://doi.org/10.1002/qj.49709741407

Pfahl S, Madonna E, Boettcher M et al (2014) Warm conveyor belts in the ERA-Interim data set (1979-2010). Part II: Moisture origin and relevance for precipitation. J Clim 27:27–40

Pfahl S, Schwierz C, Croci-Maspoli M et al (2015) Importance of latent heat release in ascending air streams for atmospheric blocking. Nat Geosci 8:610–615

Ralph FM, Neiman PJ, Kingsmill DE et al (2003) The impact of a prominent rain shadow on flooding in California's Santa Cruz mountains: A CALJET Case Study and sensitivity to the ENSO cycle. J Hydrometeorol 4:1243–1264. https://doi.org/10.1175/1525-7541(2003)004<1243:tioapr>2.0.co;2

Ralph MF, Neiman PJ, Wick GA (2004) Satellite and CALJET aircraft observations of atmospheric rivers over the eastern North pacific ocean during the winter of 1997/98. Mon Weather Rev 132:1721–1745

Ralph FM, Neiman PJ, Kiladis GN et al (2011) A multiscale observational case study of a Pacific atmospheric river ex hibiting tropical–extratropical connections and a mesoscale frontal wave. Mon Weather Rev 139:1169–1189

Ralph FM, Iacobellus SF, Neiman PJ et al (2017a) Dropsonde observations of total water vapor transport within North Pacific atmospheric rivers. J Hydrometeorol 18:2577–2596

Ralph FM, Dettinger M, Lavers D et al (2017b) Atmospheric rivers emerge as a global science and applications focus. Bull Am Meteorol Soc 98:1969–1973. https://doi.org/10.1175/BAMS-D-16-0262.1

Ralph FM, Dettinger MD, Cairns MM et al (2018) Defining "atmospheric river": how the glossary of meteorology helped resolve a debate. Bull Am Meteorol Soc 99:837–839. https://doi.org/10.1175/BAMS-D-17-0157.1

Reed RJ, Albright MD (1986) A case study of explosive cyclogenesis in the Eastern Pacific. Mon Weather Rev 114:2297–2319. https://doi.org/10.1175/1520-0493(1986)114<2297:acsoec>2.0.co;2

Reinhold BB, Pierrehumbert RT (1982) Dynamics of weather regimes: quasi-stationary waves and blocking. Mon Weather Rev 110:1105–1145. https://doi.org/10.1175/1520-0493(1982)110<1105:dowrqs>2.0.co;2

Rossa AM, Wernli H, Davies HC (2000) Growth and decay of an extra-tropical cyclone's PV-tower. Meteorog Atmos Phys 73:139–156

Rossby CG et al (1937) Isentropic analysis. Bull Am Meteorol Soc 18:201–209

Rutz JJ, Steenburgh WJ, Ralph FM (2014) Climatological characteristics of atmospheric rivers and their inland penetration over the western United States. Mon Weather Rev 142:905–921

Rutz JJ, Steenburgh WJ, Ralph FM (2015) The inland penetration of atmospheric rivers over western North America: a Lagrangian analysis. Mon Weather Rev 143:1924–1944

Ryoo J-M, Kaspi Y, Waugh DW et al (2013) Impact of Rossby wave breaking on U.S. west coast winter precipitation during ENSO events. J Clim 26:6360–6382. https://doi.org/10.1175/jcli-d-12-00297.1

Ryoo J-M, Waliser DE, Waugh DW et al (2015) Classification of atmospheric river events on the U.S. West Coast using a trajectory model. J Geophys Res 120:3007–3028. https://doi.org/10.1002/2014JD022023

Sawyer JS (1956) The vertical circulation at meteorological fronts and its relation to frontogenesis. Proc Roy Soc London A234:346–362

Schultz DM (2001) Reexamining the cold conveyor belt. Mon Weather Rev 129:2205–2225. https://doi.org/10.1175/1520-0493(2001)129<2205:RTCCB>2.0.CO;2

Schultz DM, Keyser D, Bosart LF (1998) The effect of large-scale flow on low-level frontal structure and evolution in midlatitude cyclones. Mon Weather Rev 126:1767–1791. https://doi.org/10.1175/1520-0493(1998)126<1767:teolsf>2.0.co;2

Shapiro MA (1982) Mesoscale weather systems of the central United States. University of Colorado, Boulder, p 78

Shapiro MA, Keyser D (1990) Fronts, jet streams and the tropopause. In: Newton CW, Holopainen EO (eds) Extratropical cyclones: the Erik Palmén memorial volume. Amer Meteor Soc, Boston, pp 167–191

Shaw WN, Lempfert RGK (1906) The life history of surface air currents. A study of the surface trajectories of moving air. Meteor. Office Memoir No. 174, reprinted in: Selected Meteorological Papers of Sir Napier Shaw, Macdonald, 1955, pp 15–131

Smigielski FJ, Mogil HM (1995) A systematic satellite approach for estimating central surface pressures of mid-latitude cold season oceanic cyclones. Tellus A. https://doi.org/10.3402/tellusa.v47i5.11581

Sodemann H, Stohl A (2013) Moisture origin and meridional transport in atmospheric rivers and their association with multiple cyclones. Mon Weather Rev 141:2850–2868. https://doi.org/10.1175/mwr-d-12-00256.1

Sodemann H, Schwierz C, Wernli H (2008a) Interannual variability of Greenland winter precipitation sources: Lagrangian moisture diagnostic and North Atlantic Oscillation influence. J Geophys Res 113:D03107

Sodemann H, Masson-Delmotte M, Schwierz C et al (2008b) Interannual variability of Greenland winter precipitation sources. 2. Effects of North Atlantic Oscillation variability on stable isotopes in precipitation. J Geophys Res 113:D12111

Steinschneider S, Lall U (2016) Spatiotemporal structure of precipitation related to tropical moisture exports over the eastern United States and its relation to climate teleconnections. *J Hydrometeorol* 17:897–913. https://doi.org/10.1175/JHM-D-15-0120.1

Stohl A (2001) A 1-year Lagrangian "climatology" of airstreams in the Northern Hemisphere troposphere and lowermost stratosphere. J Geophys Res 106(D7):7263–7279

Stohl A, James P (2004) A Lagrangian analysis of the atmospheric branch of the global water cycle. Part I: method description, validation, and demonstration for the August 2002 flooding in Central Europe. J Hydrometeorol 5(8):656–678

Stohl A, Forster C, Sodemann H (2008) Remote sources of water vapor forming precipitation on the Norwegian west coast at 60°N – a tale of hurricanes and an atmospheric river. J Geophys Res 113:D05102. https://doi.org/10.1029/2007JD009006

Sutcliffe RC (1939) Cyclonic and anticyclonic development. Q J R Meteorol Soc 65:518–524. https://doi.org/10.1002/qj.49706528208

Sutcliffe RC (1947) A contribution to the problem of development. Q J R Meteorol Soc 73:370–383. https://doi.org/10.1002/qj.49707331710

Sutcliffe RC, Forsdyke AG (1950) The theory and use of upper air thickness patterns in forecasting. Q J R Meteorol Soc 76:189–217. https://doi.org/10.1002/qj.49707632809

Thorncroft C, Hoskins B, Mcintyre M (1993) Two paradigms of baroclinic-wave life-cycle behaviour. Q J R Meteorol Soc 119:17–55. https://doi.org/10.1256/smsqj.50902

Trenberth KE (1978) On the interpretation of the diagnostic quasi-geostrophic omega equation. Mon Weather Rev 106:131–137. https://doi.org/10.1175/1520-0493(1978)106<0131:OTIOTD>2.0.CO;2

Webster PJ, Holton JR (1982) Cross-equatorial response to middle-latitude forcing in a zonally varying basic state. J Atmos Sci 39:722–733. https://doi.org/10.1175/1520-0469(1982)039<0722:CERTML>2.0.CO;2

Wernli H (1997) A Lagrangian-based analysis of extratropical cyclones. II: a detailed case study. Q J R Meteorol Soc 123:1677–1706

Wernli H, Davies HC (1997) A Lagrangian-based analysis of extratropical cyclones. I: the method and some applications. Q J R Meteorol Soc 123:467–489. https://doi.org/10.1002/qj.49712353811

Wernli H, Schwierz C (2006) Surface cyclones in the ERA40 data set (1958- 2001). Part I: novel identification method and global climatology. J Atmos Sci 63:2486–2507

Winschall A, Pfahl S, Sodemann H et al (2014) Comparison of Eulerian and Lagrangian moisture source diagnostics – the flood event in eastern Europe in May 2010. Atmos Chem Phys 14:6605–6619. https://doi.org/10.5194/acp-14-6605-2014

Woods C, Caballero R, Svensson G (2013) Large-scale circulation associated with moisture intrusions into the Arctic during winter. Geophys Res Lett 40:4717–4721. https://doi.org/10.1002/grl.50912

Yoshimura K, Kanamitsu M, Dettinger M (2010) Regional downscaling for stable water isotopes: a case study of an atmospheric river event. J Geophys Res 115:D18114. https://doi.org/10.1029/2010JD014032

Young MV (1995) Types of cyclogenesis. In: Bader MJ, Forbes GS, Grant JR, Lilley RBE, Waters AJ (eds) Images in weather forecasting. Cambridge University Press, Cambridge, pp 213–286

Zhu Y, Newell RE (1994) Atmospheric rivers and bombs. Geophys Res Lett 21(18):1999–2002

Zhu Y, Newell RE (1998) A proposed algorithm for moisture fluxes from atmospheric rivers. Mon Weather Rev 126:725–735. https://doi.org/10.1175/1520-0493(1998)126<0725:APAFMF>2.0.CO;2

Observing and Detecting Atmospheric Rivers

3

From Satellites to Aircraft, Radars, AR Observatories, Regional Mesonets, Reanalyses, and AR Detection Methods

F. Martin Ralph, Allen B. White, Gary A. Wick, Michael L. Anderson, and Jonathan J. Rutz

3.1 Introduction

Now that Chap. 2 has described the structure, processes, and mechanisms associated with atmospheric rivers (ARs), it is relevant to discuss the methods used to observe and detect ARs during the last 15 years, and new observing methods that are emerging. This involves observing systems to provide ground truth of the main AR ingredients, such as water vapor and wind, and process information on AR mechanisms and effects, as well as gridded atmospheric reanalysis products that enable ARs to be studied from historical, regional, and global perspectives. Detection methods (ARDMs) are also critical to identifying ARs in regional and global models, to evaluate their performance in simulating and predicting ARs, as well as to determine how ARs will behave in the future, based on global climate projections.

The first three sections of this chapter describe some of the observing technologies used to monitor ARs, followed by pertinent research experiments and methods to identify ARs. Section 3.2 focuses on satellites, which help to provide a global context for ARs but leave gaps in observing the full structure and evolution of ARs. Section 3.3 introduces the concept of an AR observatory, a land-based collection of instruments that help track ARs as they make landfall. These AR observations are important because some of the satellite ARDMs that work so well over the oceans have limitations over land because of the surface's widely varying emissivity. Section 3.4 describes other ground-based observations used to track the effects of ARs as they penetrate inland. The example given here is the network of observations currently available in California, where ARs are the primary source of annual precipitation and water supply. Section 3.5 describes research field experiments devoted to improving our understanding of ARs, with special emphasis on aircraft-based studies. These episodic campaigns often involve additional specialized observing capabilities located on air- and/or ship-borne platforms that are not part of the routine observing systems operated by weather agencies worldwide. Section 3.6 describes how ARs are represented in reanalyses. Section 3.7 provides an evolution of the various methodologies used to identify ARs, and describes some of the AR climatologies that have been created using these techniques.

Woven through this chapter are comments on where important observational gaps remain, including those associated with predictions and with physical process understanding. F.M. ("Marty") Ralph, Michael D. Dettinger, and Daniel R. Cayan identified many gaps in a proposal to California's Department of Water Resources (DWR) to create a mesoscale observing network. To support enhanced flood response and emergency preparedness, this network would be tailored to California's needs for meteorological information about extreme precipitation and the ARs that produce them. This vision, created between about 2006 and 2008, became the roadmap for future investments and observing system enhancements (White et al. 2013). Figure 3.1 summarizes this vision, including an update on what has been implemented in the 10 years since its creation. Based largely on this vision—and on additional demonstrations of the value of

F. M. Ralph
Center for Western Weather and Water Extremes, Scripps Institution of Oceanography, University of California–San Diego, La Jolla, CA, USA
e-mail: mralph@ucsd.edu

A. B. White (✉) · G. A. Wick
National Oceanic and Atmospheric Administration (NOAA), Earth System Research Laboratory, Boulder, CO, USA
e-mail: allen.b.white@noaa.gov; gary.a.wick@noaa.gov

M. L. Anderson
State of California Department of Water Resources, Sacramento, CA, USA
e-mail: Michael.L.Anderson@water.ca.gov

J. J. Rutz
Science and Technology Infusion Division, National Weather Service, Salt Lake City, UT, USA
e-mail: jonathan.rutz@noaa.gov

F. M. Ralph et al. (eds.), *Atmospheric Rivers*, https://doi.org/10.1007/978-3-030-28906-5_3

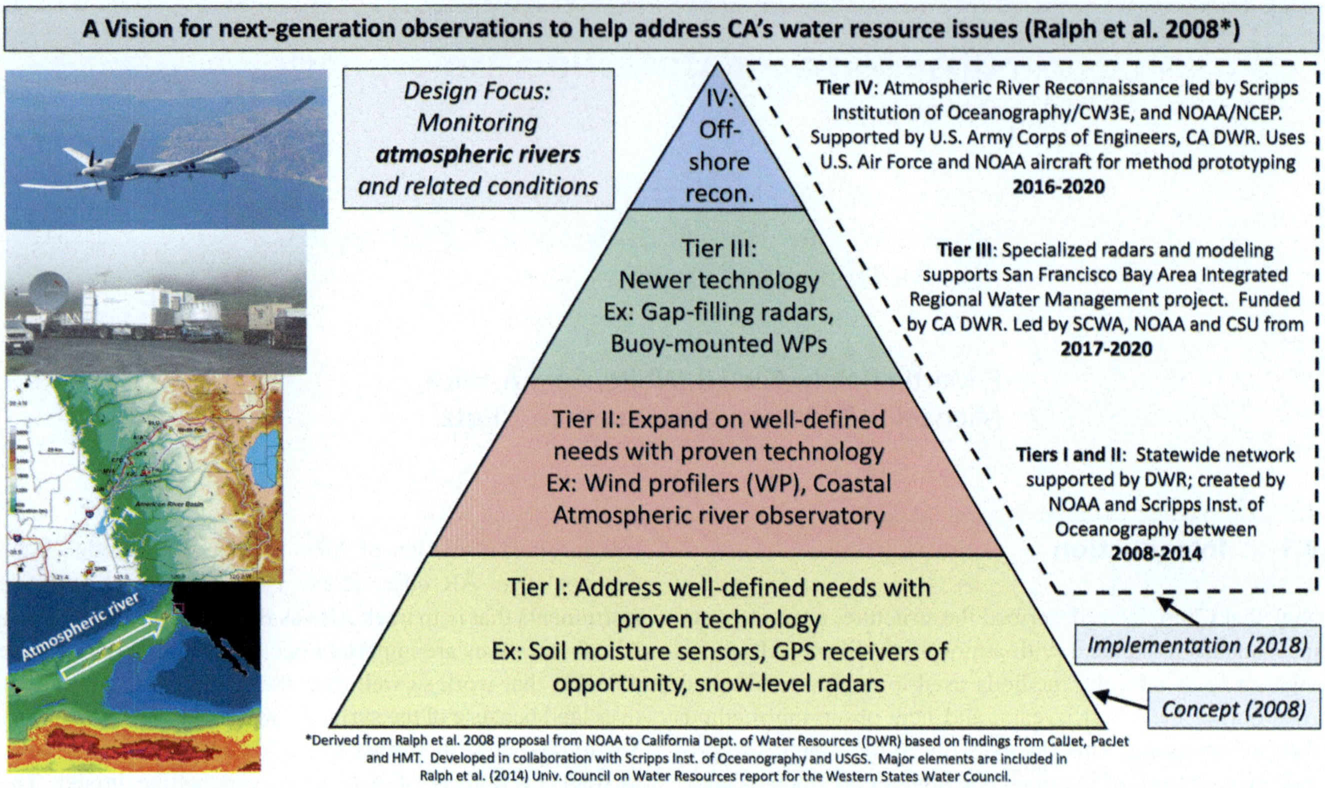

Fig. 3.1 Vision from 2008, and implementation as of 2018, of specialized observations designed largely to monitor AR conditions offshore and over California, including a statewide mesonet of roughly 100 observing sites installed across the state (Tiers 1 and 2). Tiers 3 and 4 are under development, with significant efforts underway starting in 2016–2017. (Note that manned aircraft are being used to prototype AR Recon)

such information through the National Atmospheric Administration's (NOAA's) Hydrometeorological Testbed (HMT; Ralph et al. 2013a; https://hmt.noaa.gov)—the Western States Water Council (WSWC) requested a report on the associated needs and gaps for the entire western US. This led to the organization of a cross-disciplinary, interagency team that surveyed information and observation gaps, and produced a report recommending what hydrometeorological observational capabilities would be needed to support western US water management, flood control, and climate adaptation. This report (Ralph et al. 2014), and its associated lengthy appendix (https://www.fs.fed.us/psw/cirmount/meetings/mtnclim/2012/.../2011_Ralph.pdf), included contributions from 20 federal, state, local, university, and other organizations. This report identified offshore measurements of ARs, boundary layer measurements over land (largely about water vapor transport by ARs and the monsoon), scanning radar estimates of precipitation, and better snow measurements in the mountains (snow density and albedo) as examples of observaional gaps related to ARs.

Key elements of the original 2008 vision and recommendations from the 2014 report to WSWC have been implemented, or are under prototyping, as Fig. 3.2 notes. These observational capabilities have largely been motivated by issues that surrounded monitoring and predicting ARs, and have grown over the years along with the scientific understanding and operational experiences gained since then, which are described throughout this book. The networks and methods that are now in place, or are being prototyped, are globally unique. Combined with the development of new forecast performance metrics (Wick et al. 2013b), an AR scale (Ralph et al. 2019), a weather model tailored to AR prediction (Martin et al. 2018), a community-developed technical definition of what an AR is (Ralph et al. 2018a), a center focused on the relevant problems (Center for Western Weather and Water Extremes; CW3E.ucsd.edu), these observations provide a strong foundation for emerging monitoring and prediction methods for the western US. They can also serve as a model for consideration in other regions affected by landfalling ARs, such as Western Europe (e.g., Lavers et al. 2012), including the Iberian Peninsula (e.g., Ramos et al. 2015); South America (e.g. Viale et al. 2018); South Africa (Blamey et al. 2018); New Zealand (Kingston et al. 2016); and Australia.

3.2 Satellite Observations of ARs

Satellite observations have been critical to research and applications associated with ARs. Satellite sampling provides a regular, global view of the Earth's atmosphere and clouds, and the surface below. Satellite observations,

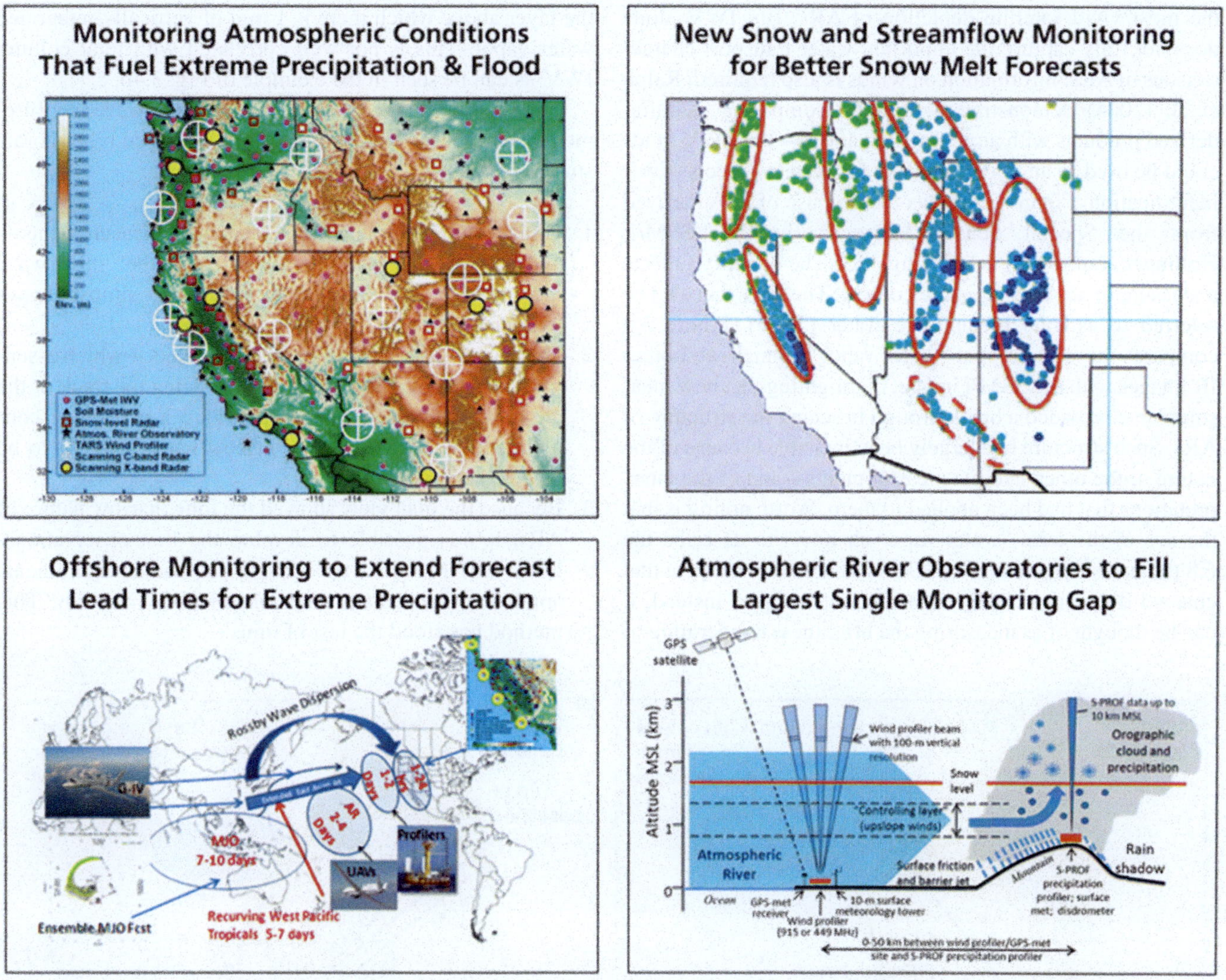

Fig. 3.2 Four broad conceptual elements of the vision for the twenty-first-century monitoring in the western US derived from a cross-disciplinary, multi-agency report, "A Vision for Future Observations for Western US Extreme Precipitation and Flooding" (Ralph et al. 2014)

obtained multiple times per day, offer important insight into the presence, spatial extent, and strength of AR events not possible from any other existing observing system (as distinguished, say, from numerical model analyses). These satellite data also provide a key source of input data to numerical reanalyses, enhancing the role of reanalyses as an additional resource to characterize ARs. This section focuses on two specific types of satellite measurements that have contributed in fundamental ways to this subject: (1) passive microwave radiometric imagery, which, in combination with research aircraft data, enabled the first offshore measurements of ARs presented in Ralph et al. (2004); and (2) radio occultation measurements, which were found to provide uniquely valuable water vapor profile information over the ocean (Neiman et al. 2008a). Additional knowledge has been gained from cloud and precipitation radars on National Aeronautics and Space Administration (NASA) satellites: e.g., Matrosov (2013), Cannon et al. (2018). Regarding infrared (IR)-based satellite measurements, Ralph et al. (2004) showed that high clouds blocked Geostationary Operational Environmental Satellite (GOES)-based measurements of integrated water vapor (IWV), because mid- and high-altitude clouds were often present over an AR, and interfered with retrievals. Although many additional types of satellite measurements have in some way helped with AR studies, they have not been as vital as microwave radiometry, radio occultation, and radar, and thus are not addressed here.

3.2.1 Microwave Radiometry: SSM/I

Much of the growth of research and interest in ARs in the early 2000s was fueled by the compelling imagery derived from passive microwave satellite measurements, as first documented by Ralph et al. (2004). Retrievals of the total column IWV from passive microwave radiometers provide perhaps the most vivid satellite depiction of ARs. But IWV alone does not fully capture the important water-transport characteristics of ARs: information on winds is also required. Ralph et al. (2004) demonstrated—through comparing satellite-derived products with aircraft observations—that IWV fields could be used as an effective proxy for certain aspects of AR identification. Figure 3.3 shows the first use of IWV derived from the Special Sensor Microwave/Imager (SSM/I; Hollinger et al. 1990) in studying ARs. The quantity reflects a measure of an AR's moisture content. The IWV can also be referred to as total precipitable water (TPW). Figure 3.4 compares more traditional "water vapor channel" and other IR images with the SSM/I image, highlighting that new measurements provided a breakthrough in seeing the structure of ARs. Such structure had largely been invisible because of the nature of the other satellite measurements—and of the interpretations that had been applied to them. Worth noting is that the use of the term "water vapor channel" to describe the 6.8 μm measurements from GOES did not help, because that channel does not actually measure water vapor; instead, it can be thought of as measuring the brightness temperature of the layer above which there is 1 mm of vertically integrated water vapor. This is not well correlated with total column IWV, as can be seen in the example of Fig. 3.4b, c.

The satellite-derived SSM/I passive microwave measurements had four invaluable attributes that helped reinvigorate studies of ARs:

1. Unlike GOES "water vapor channel" measurements—and other IR-based measurements—the microwave method could see through the clouds that almost always overlaid an AR, and still measure IWV.
2. Unlike the research aircraft measurements—which essentially measured along a single line tracing the track of the aircraft—the SSM/I images provided a plan-view context. This allowed the long, narrow nature of ARs to be detected and studied.
3. Because the plan view showed the long, narrow nature of ARs, it was possible to develop the first observations-based ARDM, which used length and width requirements applied to a constant threshold of IWV = 2 cm. This method has stood the test of time.

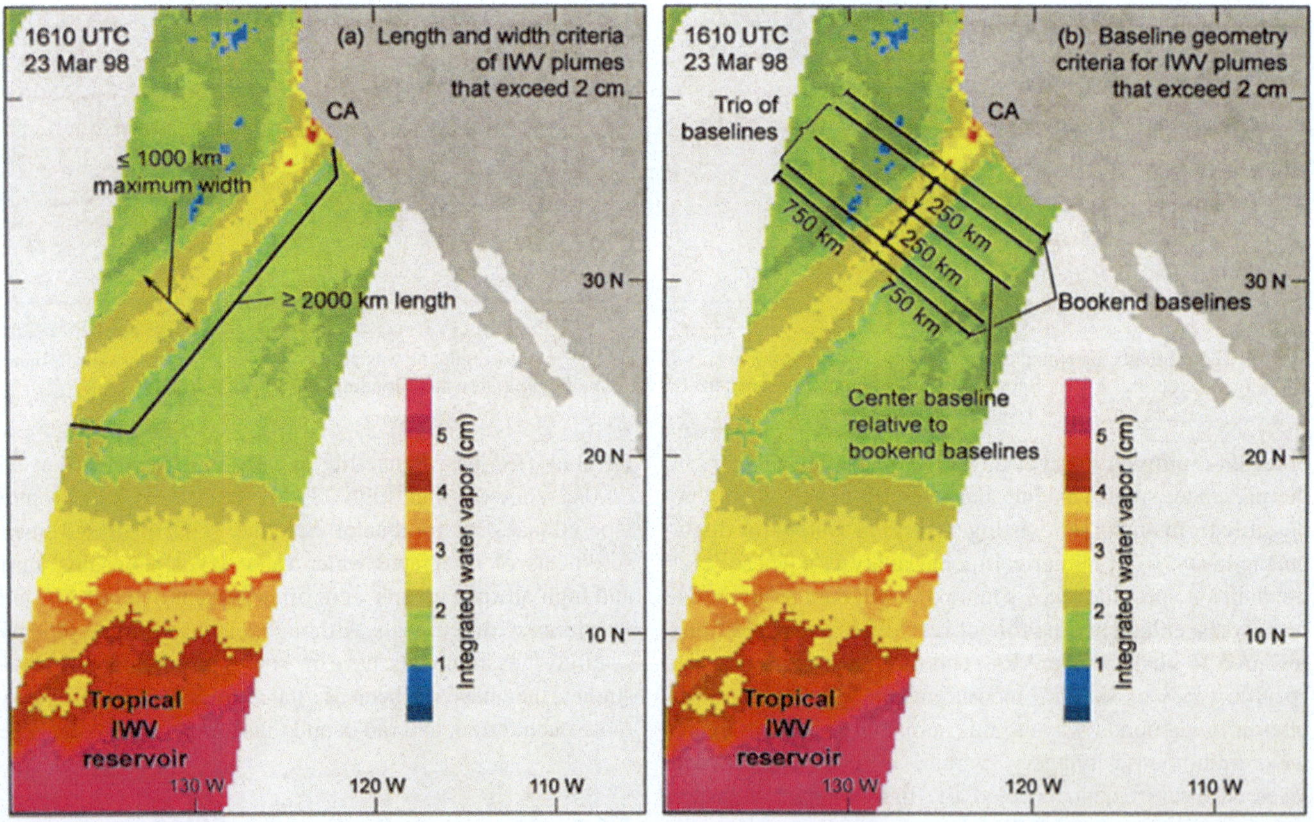

Fig. 3.3 First use of satellite-based Special Sensor Microwave/Imager (SSM/I) observations ("retrievals") of integrated water vapor (IWV) to document AR conditions, with a graphical portrayal of how the satellite measurements were used in the first observations-based AR detection method (ARDM). Graphical depiction of the methodology used to generate composite 1500-km-wide baselines of SSM/I-derived IWV, cloud liquid water, rain rate, and surface wind speed across moisture plumes measured by SSM/I over the eastern Pacific during the CALJET winter of 1997–1998: (**a**) length and width criteria of IWV plumes that exceeded 2 cm. (**b**) baseline geometry criteria relative to the SSM/I swaths for IWV plumes that exceeded 2 cm. (From Ralph et al. 2004)

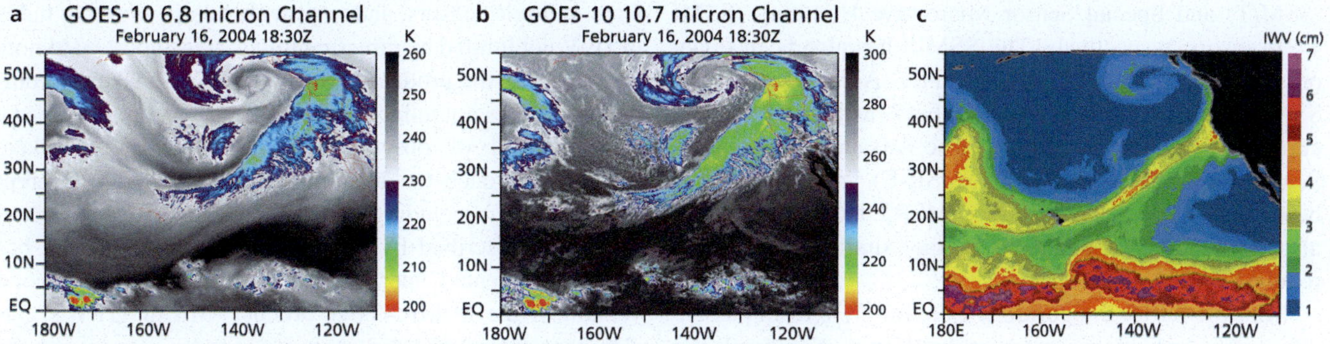

Fig. 3.4 Comparison of the representation of ARs provided by satellite-based infrared (IR) and passive microwave imagery. Panels (**a**, **b**) correspond to IR observations from the Geostationary Operational Environmental Satellite (GOES)-10 satellite at 6.8 μm ((**a**), a water vapor channel) and 10.7 μm ((**b**), a thermal IR channel); panel (**c**) shows a retrieval of integrated water vapor (IWV) from passive microwave channels from the Special Sensor Microwave/Imager (SSM/I) on the Defense Meteorological Satellite Program (DMSP) F-13, F-14, and F-15 satellites. All images correspond to 16 Feb 2004. The GOES images are single scenes sampled at 1830 UTC; the SSM/I IWV image is a composite of retrievals between 1200 and 2400 UTC. (This case was documented by Ralph et al. 2006)

4. The availability of daily SSM/I measurement swaths enabled many ARs to be sampled, compared to very few with the aircraft.

In the 10 years before the Ralph et al.'s (2004) paper introduced to the science community the potential of SSM/I data to study ARs, ten papers at most—and none in the previous few years—had touched on the AR subject. By 12 years after that 2004 paper, there had been hundreds of papers (Ralph et al. 2017). It is not an overstatement that the SSM/I measurements—and the methods developed to use them to detect and study ARs—transformed research on this subject. In no small way, this discovery paved the way to a topic that is the focus of this entire book.

Satellite-derived IWV imagery has been extensively applied to AR science. In Ralph et al. (2004), characteristics of IWV imagery were used to define criteria to identify ARs in the northwest Pacific Ocean adjacent to North America. Using these criteria as well as SSM/I-derived IWV products, Neiman et al. (2008a) developed a multi-year climatology of landfalling ARs along the west coast of the US, which Dettinger et al. (2011) further extended. The climatology and associated IWV imagery proved instrumental in documenting the hydrologic effect of ARs in the US West Coast states (e.g., Ralph et al. 2005a, 2006, 2011, 2013b; Neiman et al. 2008a, 2011; Guan et al. 2010).

Examples of applying satellite-derived IWV products to AR studies in other regions include Lavers et al. (2012), Garreaud (2013), Gorodetskaya et al. (2014), and Neff et al. (2014). The imagery has also been used to validate the representation of ARs in numerical weather prediction (NWP) forecast models (Wick et al. 2013b) and reanalyses (Jackson et al. 2016). Beyond purely scientific applications, the satellite products further supported many broad public outreach activities, including reports by various media outlets. The IWV imagery is still the source of many of the fundamental graphics on AR informational web pages such as https://www.esrl.noaa.gov/psd/atmrivers/ and monitoring web pages such as cw3e.ucsd.edu and http://www.remss.com/about/projects/atmospheric-river-watch.

Visible and IR imagery from geostationary and polar-orbiting satellites—which is avaialable with higher resolution and more frequently from other satellite platforms—does not as clearly indicate the location and extent of ARs. Visible and most thermal IR frequencies primarily indicate the location and height of cloud tops, which are typically not well correlated with the primary moisture transport corridors of ARs. IR measurements at wavelengths near 6.8 μm available from geostationary satellites provide a measure of atmospheric water vapor content, but only from the upper and middle levels of the atmosphere—not where the bulk of the water vapor in an AR is located. Comparing near-contemporaneous IR and passive microwave imagery of an AR in Fig. 3.4 illustrates how the IWV retrievals best reflect the position of the AR core. The cloud extent is clearly not aligned with the maximum atmospheric moisture content, and provides less indication of its magnitude.

A long record of passive microwave retrievals of IWV extending back to 1987 is available via measurements from SSM/I and follow-on instruments. The SSM/I sensor—providing conically scanned, polarized measurements at frequencies of 19.35, 22.235, 37.0, and 85.5 GHz—flew on the Defense Meteorological Satellite Program (DMSP) F-10, F-11, F-13, F-14, and F-15 polar-orbiting satellites. Starting with the DMSP F-16 satellite launched in 2003, the SSM/I sensor was replaced with the Special Sensor Microwave Imager Sounder (SSMIS; Kunkee et al. 2008). The SSMIS combines the imaging channels from the SSM/I with the temperature and humidity sounding capabilities previously provided by the Special Sensor Microwave/Temperature

(SSM/T) and Special Sensor Microwave Humidity (SSM/T-2) sensors, respectively. The SSMIS has also been flown on the F-17, F-18, and F-19 satellites. Although the F-19 SSMIS has since failed, the F-16, F-17, and F-18 satellites currently provide observations. The Advanced Microwave Scanning Radiometer for EOS (AMSR-E), which functioned between 2002 and 2011, and the AMSR2, launched in 2012, also provide measurements at the similar frequencies required for IWV retrieval.

Multiple retrieval algorithms exist for IWV from passive microwave radiometric measurements (e.g., Schlüssel and Emery 1990; Alishouse et al. 1990; Lojou et al. 1994; Petty 1994; Wentz 1995, 1997), and products are available through various sources (two examples include http://www.remss.com/ and https://www.esrl.noaa.gov/psd/psd2/coastal/satres/realtime.html). The retrieval of IWV from passive microwave radiometers is quite accurate, and comparison of cross-satellite retrievals of IWV has been a valuable tool in ensuring the long-term stability of microwave-based satellite products (Wentz et al. 2007). The retrievals are based primarily on measurements near the 22-GHz water vapor absorption line, and require a low and minimally variable surface background emissivity. Although this is satisfied by the ocean, land and ice surfaces are highly emissive and variable at microwave frequencies. As a result, retrievals are only possible over ice-free oceans. The effect of land emission on side lobes in the microwave measurements further restricts the retrieval of IWV within ~100 km of land. The spatial resolution of the microwave measurements is a function of frequency, and ranges between 43 × 69 km at 19.35 GHz, and 28 × 37 km at 37.0 GHz for the SSM/I. The resolution of a single corresponding IWV retrieval can be roughly approximated as ~40 km. Because of over-sampling along the scans, the data are often gridded at a higher resolution, commonly 25 km for the products noted above. Although the swath widths (1394 km from SSM/I, 1707 km from SSMIS) are too narrow to enable full global coverage daily from a single satellite, the operational DMSP has had multiple satellites in closely spaced orbits at a given time (from about 1996 through the time of writing), and the combination of data from two to three satellites enabled 12-hourly coverage with small or no gaps between swaths.

IWV imagery from a single SSM/I or SSMIS sensor can provide only twice-daily sampling at most latitudes because of the constraints of polar satellite orbits. Unique approaches have been used to provide more frequent depictions of the global evolution of IWV. One such approach is the Morphed Integrated Microwave Imagery at the Cooperative Institute for Satellite Studies (CIMMS) TPW (MIMIC-TPW; Wimmers and Velden 2011), a product developed at the University of Wisconsin. In this product, a time-weighted interpolation procedure is used along with vertically averaged NWP model wind outputs to advect IWV observations from multiple sensors, and to provide hourly evolving fields of IWV not limited by direct sampling. It is important to note that although the product is highly useful and visually compelling, it does not reflect a direct observable, but is instead a careful blend of observations and model-derived data. Another approach to blending observations from different sensor types, including satellite sounders (more detail below), is described by Kidder and Jones (2007) and has been implemented operationally within NOAA. A second version of the MIMIC-TPW product (MIMIC-TPW2) has also been introduced that includes sounder data to provide sampling over land regions.

Despite the long historical record, the future of traditional IWV fields from passive microwave imagers is somewhat in doubt. After the failed effort to merge civilian and military environmental satellite programs under the National Polar Operational Environmental Satellite System (NPOESS), only one more DMSP satellite with an SSMIS sensor (F-20) exists, and there are no present plans for future launches. The Japanese currently operate the AMSR2 on their Global Change Observation Mission–Water (GCOM-W) satellite, but plans are uncertain for any future instruments on later satellites. New concepts for potential future passive microwave sensors are being developed by the US Air Force and in Europe, but there were no firm commitments or proposed timelines at the time of writing.

Retrievals of IWV are also derived from passive microwave and IR satellite sounders. Measurements that represent several relatively broad vertical levels of the atmosphere enable total water vapor content to be estimated. Weng et al. (2003) demonstrated an approach using microwave sounding channels such as are available on the Advanced Microwave Sounding Unit (AMSU), SSMIS, and the Advanced Technology Microwave Sounder (ATMS). The Microwave Integrated Retrieval System (MiRS; Boukabara et al. 2011) is now being used to operationally retrieve IWV and other associated atmospheric and precipitation parameters from multiple microwave sounders. A key advantage of the MiRS variational retrieval technique is its ability to retrieve IWV over land as well as water. Although these microwave-derived sounder products have not been used as explicitly in past AR studies, future use of IWV imagery may likely be more closely tied to sounder rather than imager data. Given the large current operational forecast effect of microwave sounders (e.g., Gelaro et al. 2010), future launches are more certain. Although clouds can obscure infrared soundings, Waliser et al. (2012) used retrievals of IWV generated from the Atmospheric Infrared Sounder (AIRS) to develop a 2-year global AR climatology.

However, because current satellites cannot provide direct estimates of integrated vapor transport (IVT)—which characterizes ARs more physically than IWV—they are significantly limited in monitoring ARs solely on their own. The problem results primarily from an inability for current sensors to derive accurate, complete vertical profiles of wind speed and direction. Vertical water vapor profiles can be derived approximately from passive microwave sounders. Wind measurements, however, are constrained to surface measurements from scatterometers and scattered measurements at a few discrete levels derived from the motion (drift) of observed cloud tops (e.g., Nieman et al. 1997). Neither approach has been found to consistently, reliably measure the winds at the level of an AR core—a measurement needed to accurately quantify IVT using satellite data alone. Future goals of deploying a space-based wind Light Detection and Ranging (LIDAR; e.g., Baker et al. 2014) would support much improved global wind profiling, but clouds in the AR environment still pose significant challenges for accurate retrieval of IVT.

Absent these additional capabilities, a critical contribution of satellite observations is in informing numerical model reanalyses which provide an important, otherwise-missing representation of the IVT and other AR characteristics over the open oceans. To more comprehensively describe the meteorological state than existing observations alone, atmospheric reanalyses (e.g., Kalnay et al. 1996; Dee et al. 2011; Gelaro et al. 2017) are generated by assimilation of available satellite and in situ observations into a static numerical weather-prediction system. While not a direct observation, the reanalyses combine the strengths of the observations and numerical models to produce meteorological fields consistent with available observations and incorporated model physics. Current reanalyses are highly mature and extensively utilized in the detection and analysis of ARs as described below.

3.2.2 Radio Occultation: COSMIC

Satellite observations of moisture in the presence of clouds have been available from microwave imagers such as SSM/I that typically provide only the vertically integrated amount of water vapor. Microwave sounders such as AMSU, the Microwave Humidity Sounder (MHS), and the new ATMS instrument provide retrieved moisture profiles with limited vertical resolution that depend on the "first guess" profile (Boukabara and Garrett 2018). More precise vertical moisture information is available, thanks to a novel use of Global Positioning System (GPS) navigation satellites, through the creation of a secondary small-sat constellation called COSMIC (Constellation Observing System for Meteorology, Ionosphere, and Climate) that observes GPS signals just before they are occulted by the earth (Anthes et al. 2008). The refraction of the GPS-satellite-transmitted signals by the atmosphere can be measured precisely. Because the refraction is influenced measurably by the vertical gradients of water vapor mixing ratio and temperature, these data contain useful information. The retrieval methods are sophisticated and can have errors, partly because the ray paths through the atmosphere integrate information over a horizontal distance >250 km in length. Nonetheless, the method has proven useful in AR studies.

After Wick et al.'s (2008) careful comparison between SSM/I and COSMIC water vapor information, Neiman et al. (2008b) showed that "soundings" derived from about a dozen COSMIC samples across a strong AR could be used to create a detailed cross-section of water vapor mixing ratio in the AR and its broader environment. Additionally, a modeling study by Ma et al. (2011) used a regional model and modern data-assimilation methods to show that assimilation of the COSMIC data affected the AR's predicted landfall position.

These early studies of ARs using radio occultation data provide a foundation for future efforts to explore use of COSMIC data in AR prediction. A recent example of this is the use of airborne GPS sensors deployed with CalWater (Ralph et al. 2016) and during AR Recon 2018. The airborne radio occultation system provides high vertical-resolution profiles of atmospheric refractivity, and derived temperature and moisture, from observed measurements of propagation delays of Global Navigation Satellite System (GNSS) signals (Haase et al. 2014). The measurements are complementary to high-altitude aircraft dropsonde observations because they sample the surrounding environment up to 350 km from the flight track. The system has been deployed on the National Science Foundation (NSF) G-V aircraft to improve forecasting hurricanes (Chen et al. 2018) and on the National Oceanic and Atmospheric Administration's (NOAA's) G-IV aircraft in AR studies.

3.2.3 Satellite-Based Cloud and Precipitation Radars: CloudSat and GPM

The use of radar-based measurements from low-orbit NASA satellites has provided another valuable window into AR structure in terms of clouds and precipitation. Two key studies are guides to the subject: Matrosov (2013) and Cannon et al. (2017).

- Matrosov (2013) used Dual-Frequency Precipitation Radar (DPR) measurements from CloudSat to examine the occurrence of cold (signs of ice) and warm precipitation over the ocean within 265 satellite crossings of ARs. The fractions of cold and warm precipitation were similar, but the cold precipitation had, on average, higher rain rates. The ratio of the rain-band width in an AR to the IWV-based AR width was smaller for higher values of maximum IWV in the AR.

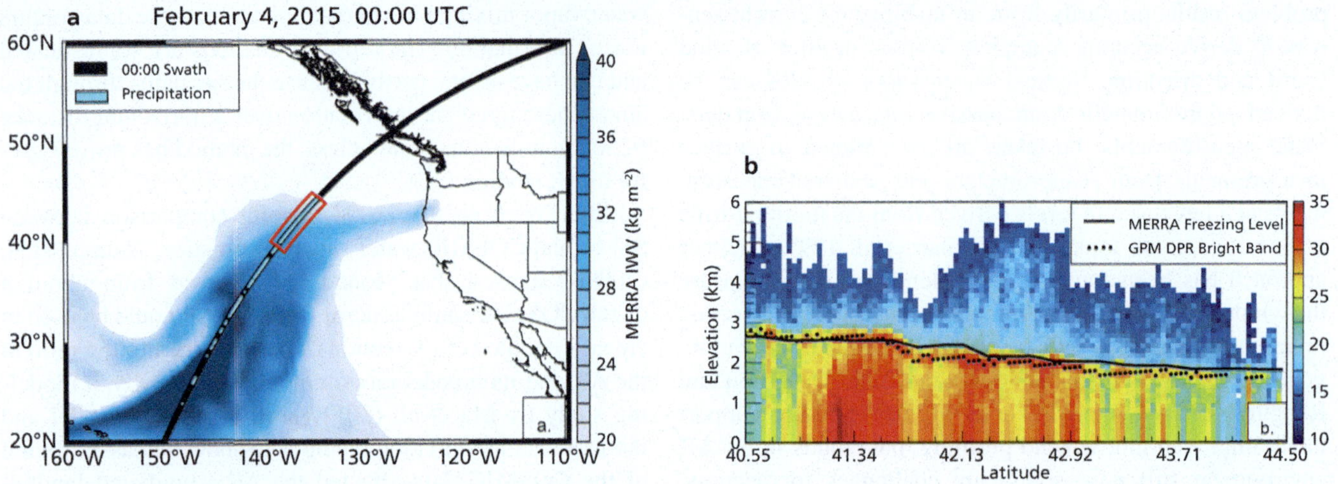

Fig. 3.5 (**a**) Dual-Frequency Precipitation Radar–Global Precipitation Measurement (DPR–GPM) swath through AR conditions on 4 Feb 2015 at 000 UTC and (**b**) the vertical profile of reflectivity from the Ka band satellite-borne radar along the center of the DPR–GPM swath subset within the *red box* in (**a**). (From Cannon et al. 2017)

- To more carefully explore variations in snow-level in ARs, and to compare the estimated rain rate from the Global Precipitation Measurement (GPM) Mission Core Observatory with reanalysis estimates of precipitation rate, Cannon et al. (2017) used an independent catalog of ARs over the northeast Pacific to identify GPM overpasses of ARs (Fig. 3.5). The locations of the overpasses were then used as a basis for extracting the associated parameters from the Modern-Era Retrospective analysis for Research and Applications (MERRA) reanalyses. This revealed important differences between satellite-derived and reanalysis-derived precipitation rates. Reanalysis indicated a much larger fraction of total rainfall was within very light rain rates compared to the satellite-based measurement. The difference could have important implications in terms of reanalysis that properly represents AR water vapor budgets (see Chap. 8), for example.

3.3 AR Observatories

This section describes a collection of commercially available instruments that, together, form an AR observatory (ARO; Fig. 3.6). An ARO provides researchers, forecasters, water and emergency managers, and the general public with continuous surface and upper-air meteorological conditions associated with landfalling ARs. The ARO concept was developed based on years of research on the atmospheric processes associated with ARs. AROs have been used episodically in research field experiments on the east and west coasts of the US and in Mexico for the past decade. More recently, a "picket fence" of semi-permanent AROs was established on the US West Coast to monitor AR conditions and quantitatively measure their impacts.

3.3.1 AR Characteristics Not Readily Observed Using Traditional Meteorological Methods

Meteorology has long recognized the importance of regularly observing phenomena (and/or their origins) using either in situ or remotely sensed observations. The need to detect thunderstorms and precipitation gave rise to weather radar. Detecting tornadoes continues to push weather radar technology. The need to see hurricanes and extratropical cyclones more readily over the oceans contributed to the rise of offshore airborne weather reconnaissance and weather satellites. With the emerging awareness of the key roles of ARs in producing extreme precipitation and flooding—as well as high-impact weather, more generally—it became clear that measuring key attributes of ARs is difficult using sensors designed for other purposes. This section summarizes key gaps, including the winds in the "controlling layer" (defined below) and the amount of IVT. Although radiosondes measure these parameters, their horizontal and temporal spacing are wholly inadequate for measuring ARs. This inadequacy likely slowed recognition of the importance of ARs well beyond when it might have occurred had radiosondes been spaced much more closely together in time and space.

The Low-Level Jet and the "Controlling Layer"

Browning and Pardoe (1973) summarized the structure of the low-level jet (LLJ) in extratropical cyclones, and Browning (1980) showed how critical these were in creating

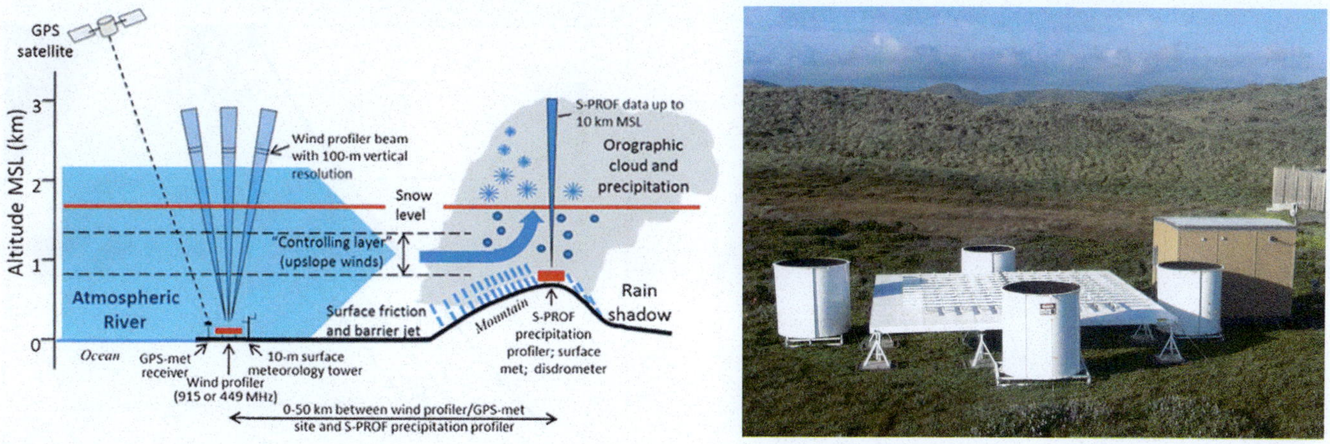

Fig. 3.6 *Left* Schematic summary of an AR observatory (ARO) and *right* photo of part of the ARO installed and operating at Bodega Bay, California, since 2014. (From White et al. 2013)

orographically enhanced rain in Britain. These results were extended using research aircraft measurements offshore during CalJet and PacJet in 1998 and 2001 (Ralph et al. 1999, 2004) and coastal measurements of precipitation and conditions aloft, and took advantage of new technology—wind-profiling radar—created since the earlier work in the UK. These wind profilers provide wind speed and direction profiles hourly (or faster) at roughly 100 m vertical resolution in all weather conditions. The offshore aircraft measurements—plus the remarkable SSM/I satellite measurements of IWV, combined with the AR concept described convincingly by Zhu and Newell (1998)—led to the seminal observational papers that documented AR offshore (Ralph et al. 2004, 2005a). Just before these two papers, though, two studies examined conditions onshore, specifically orographic precipitation (Neiman et al. 2002; Ralph et al. 2003). Schematic summaries of two of these key papers are shown here in Fig. 3.7. They both used the hourly wind profiler data and hourly precipitation data onshore to study processes that control orographic precipitation rates. Neiman et al. (2002) found that winds at roughly 1 km mean sea level (MSL) most strongly controlled hourly rain rates in the coastal mountains, and described this as the "controlling layer." In many cases, the surface winds were relatively decoupled from the controlling layer, especially when a coastal barrier jet was present. Ralph et al. (2003) used a case study of a record flood to diagnose the role of the controlling level in that case, including especially the role of wind direction in modulating the lateral edge of a rain-shadowed area as it struck a coastal mountain range downwind of another coastal range farther south. By combining vertical profiles from 17 ARs measured offshore by aircraft, Ralph et al. (2005a) found that in ARs this controlling layer was moist neutral—meaning it would not resist vertical lift, once saturated. Thus, ARs are characterized by conditions ideal for orographic precipitation because ARs combine the strongest low-altitude horizontal winds with the largest water vapor contents anywhere in the broader extratropical cyclone, with a vertical thermodynamic structure that does not resist being forced upward when the AR hits coastal mountains.

Temporal and Horizontal Spatial Scales of ARs Relative to the Operational Radiosonde Network

These studies revealed the key role of the LLJ, and how it is often decoupled from the surface, thus pointing to the inadequacy of surface observations for inferring AR conditions aloft. Another facet of these storms revealed that ARs are relatively narrow, and that the core AR conditions often last over a specific location for less than 12 h. Ralph et al. (2013b) also found that the storm-total upslope water vapor flux over a coastal site plays a key role in determining the storm-total precipitation and associated hydrological effects. In fact, the storm-total runoff was seven times greater for the 10% longest-duration ARs studied, relative to all storms on average. Because the water vapor flux varied substantially during these events, even rising and falling multiple times, the hourly time resolution of the wind profiler and related data was key to knowing the storm-total water vapor flux. Neiman et al. (2009) and Ralph et al. (2013b) took advantage of Global Positioning System/Meterology (GPS/MET) technology to add water vapor information to wind information, and calculated a proxy for IVT by multiplying IWV measured by GPS/MET with the upslope component of the wind in the controlling layer, as observed by the wind-profiling radar. The addition of water vapor in this way, and the development of a storm-total upslope flux (which required knowing the start and end times of AR conditions), pushed the correlations between forcing from about 50% of the variance in hourly precipitation being explained by variations in

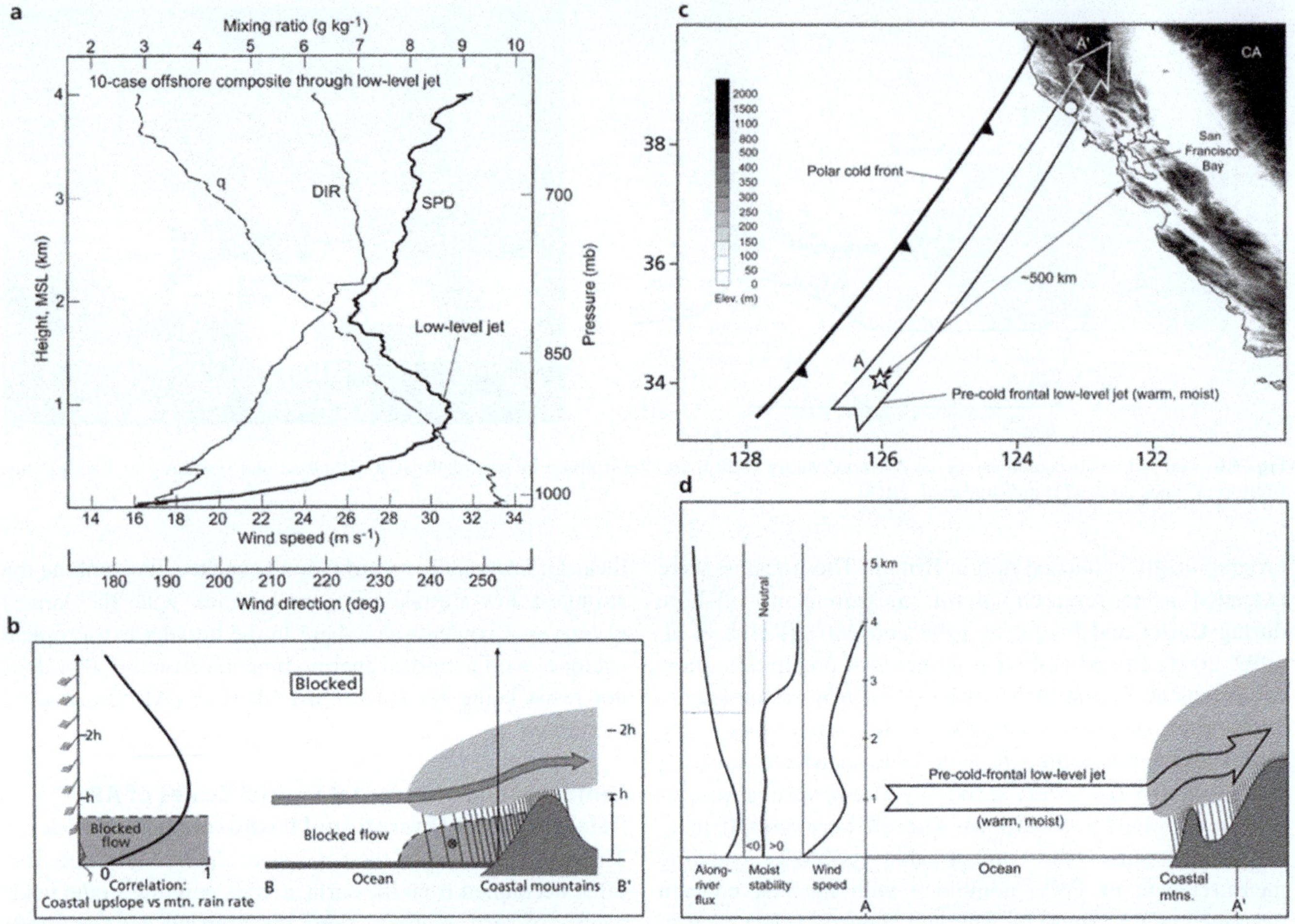

Fig. 3.7 Schematics from Neiman et al. (2002) and Ralph et al. (2005a) highlighting the role of winds aloft (near 1 km MSL) in controlling orographic rainfall downwind. (**a, b**) Conceptual representation of orographic rainfall distribution in California's coastal mountains, and the impact of terrain-blocked flow on this distribution: (**a**) plan view, and (**b**) cross-section perspective, with representative coastal profiles of wind velocity (*flags* and *barbs* as in Fig. 3.3) and correlation coef (based on the magnitude of the upslope flow at the coast vs the rain rate in the coastal mountains) shown on the *left*. The variable h in (**b**) is the scale height of the mountain barrier. The spacing between the rain streaks in (**b**) is proportional to rain intensity. The symbol "⊗" within the blocked flow in (**b**) portrays a terrain-parallel barrier jet (from Fig. 19 of Neiman et al. 2002) (**c, d**) Conceptual representation focusing on conditions in the pre-cold-frontal LLJ region of a landfalling extratropical cyclone over the northeastern Pacific Ocean. (**c**) Plan-view schematic showing the relative positions of an LLJ and trailing polar cold front. The average position of the 17 dropsondes used in this study is shown with a *star* (~500 km offshore of San Francisco), and the Cazadero microphysics site is marked with a *bold white dot*. The points A and A' along the LLJ provide the approximate endpoints for the cross-section in (**d**). (**d**) Cross-section schematic along the pre-cold-frontal LLJ [i.e., along A–A' in (**c**)] highlighting the offshore vertical structure of wind speed, moist static stability, and along-river moisture flux at the location of the altitude scale. Schematic orographic clouds and precipitation are shown, with the spacing between the rain streaks proportional to rain intensity (from Fig. 13 Ralph et al. 2005a)

upslope winds alone in the controlling layer to 75% for storm-total precipitation in ARs (Ralph et al. 2013b). Although it is not laid out here, studies have also found that the snow level aloft can also change rapidly during the passage of an AR, and that these changes have major effects on ensuing streamflow (e.g., White et al. 2002). Based on the facts that normal temporal spacing of operational radiosondes is 12 h, and that their spatial separations are usually hundreds of km on the US West Coast, it is clear that traditional radiosonde systems do not monitor ARs well.

Summary of the Gaps

Although the traditional land-based observing network is not able to adequately detect ARs, consider the possibility that satellites could provide the necessary measurements of the controlling layer and IVT offshore. As Ralph et al. (2004) showed, ARs are often blanketed by mid-level and high clouds that interfere with frequent determination of IWV at least from GOES, and Neiman et al. (2002) and Ralph et al. (2005a) showed how highly variable the vertical shear can be between the ocean surface and the LLJ aloft. With the advent of much-improved cloud-tracked wind capabilities, the amount of such data has grown dramatically. However, these measurements are also stymied by the prevalence of high clouds over exactly the region where observations are needed most.

In summary, the inherent characteristics of ARs—combined with their important role in precipitation, water supply, and flooding—revealed the need to create a new type of observing approach to monitor AR conditions with necessary detail. This set the stage for the development of the ARO array along the coast (described next in Sect. 3.3.2), and for development of an airborne dropsonde-based strategy for offshore (see Sect. 3.5).

Specifically, for an ARO and associated network to be effective at monitoring ARs, it must observe winds aloft in the controlling layer; measure water vapor aloft; make these measurements on roughly hourly time-scales in clear, cloudy, and precipitating conditions; and operate unattended.

3.3.2 ARO Instrumentation

Doppler Wind Profilers

Starting in the 1980s, scientists from the NOAA Wave Propagation Laboratory began using research-developed Doppler wind-profiling radars (Carter et al. 1995)—hereafter referred to as wind profilers—to probe the vertical wind structure of the atmosphere, from the planetary boundary layer through the free troposphere and into the lower stratosphere. To monitor ARs, height coverage for winds in the lowest 4 km of the atmosphere is sufficient, with higher height coverage desirable, especially as ARs penetrate inland and approach higher terrain (e.g., the Sierra Nevada). Resolving winds with better than 200-m vertical resolution is desirable to fully resolve AR features such as the LLJ, as well as warm-frontal- and cold-frontal-induced wind shear.

Wind profilers are ground-based radars that transmit pulses of electromagnetic radiation upward, resulting in scatter from turbulence or other atmospheric constituents—most notably clouds and precipitation. Biologic targets in the atmosphere (birds, insects, trees) and man-made objects (airplanes, buildings, towers) also scatter the transmitted signals from wind profilers. The key in the signal processing of the backscatter returns is to filter out signals from non-atmospheric targets to allow the atmospheric tracers to accurately retrieve the horizontal and vertical wind. To resolve the three-dimensional (3-D) wind vector, wind profilers are configured to transmit one vertical and at least two off-vertical, nearly orthogonal beams. Wind profilers generally transmit at a fixed frequency between 40 and 3000 MHz.

With the CALJET project, and subsequently, researchers began using NOAA's 915-MHz wind profilers to examine the vertical structure of winter storms that affect the US West Coast, including those storms that include embedded ARs (e.g., Ault et al. 2011; Kingsmill et al. 2013; Neiman et al. 2002, 2004, 2013a, b, 2014, 2016; Ralph et al. 2003, 2006, 2011, 2013b; White et al. 2003, 2015a). Researchers at NOAA's Environmental Technology Laboratory (ETL) also developed a 449-MHz wind profiler (Fig. 3.6) to provide deeper height coverage than the smaller 915-MHz wind profilers. Wind profilers operating at other radar frequencies exist throughout the world.

Because each of the wind-profiler technologies has its own advantages and disadvantages for monitoring ARs and other weather phenomena that are most important to coastal and marine weather, researchers at NOAA/ETL conducted a wind profiler technology evaluation study in California during 2005 and 2006. This study was funded by NOAA's Integrated Ocean Observing System (IOOS), and led to a final report with recommendations for which technology was better suited to coastal weather applications (Earth System Research Laboratory [ESRL] Physical Sciences Division 2007). Table 3.1 highlights some of the technological and sampling differences between the different types of wind profilers discussed here. Based in part on findings from the IOOS technology evaluation, ESRL's Physical Sciences Division has built and deployed a semi-permanent network of ¼-scale 449-MHz wind profilers as part of a coastal ARO network (see Sect. 3.3.4). The 449-MHz wind profiler was selected for this application because it provides increased height coverage compared to the smaller 915-MHz wind profilers for both winds and temperatures retrieved from radio acoustic sounding systems (RASS; a corollary instrument often used with wind profilers). Also, NOAA has both experimental and operational licenses to operate atmospheric monitoring equipment using the 449-MHz frequency from the National Telecommunications and Information Administration (of the US Department of Commerce), whereas many users share the 915-MHz frequency.

Surface Meteorology Towers

Because wind profilers are unable to provide wind measurements down to the surface—nor a more complete set of meteorological conditions at the surface—it is customary to install a surface meteorology tower along with the wind profiler. At the ARO sites, NOAA typically installs a walk-up tower with wind speed and direction measurements at the 10-m level; pressure, temperature, and relative humidity measurements at the ~2-m level; and precipitation measurement at the ~1-m level. Often, surface radiation sensors (solar and net) are included with the tower's other sensors. NOAA uses research-grade sensors from commercial manufacturers for all tower measurements. Meteorological sensor signals are acquired through a data logger, and 2-min averages are transmitted once every hour to a data hub in Boulder, Colorado, through one of three communication methods: telephone, satellite, or cellular services. Figure 3.8 shows a

Table 3.1 Physical, operating, and sampling characteristics of wind profilers

	404-MHz (NPN)	915-MHz (Boundary layer)	449-MHz (¼-scale)
Antenna type	Coaxial-collinear phased array	Flat rectangular microstrip patch	Coaxial-collinear phased array
Antenna diameter (m)	13	2	6
Beamwidth (deg.)	4	10	10
Peak transmit power (W)	6000	500	2000
Transmit pulse width (μs)	3.3[a], 20[a]	0.417[b], 0.708[a,b]	0.708[b], 2.833
Height coverage (m)	500[c]–16,000	120–4000	180[d]–8000
Vertical resolution (m)	320[e], 900[e]	63, 106[b]	106, 212[b,e]
Temporal resolution (min)	60	60	60
Advantages for ARs	Full troposphere coverage	Best vertical resolution in the lower atmosphere	Vertical resolution resolves AR features. Operating frequency is secure
Disadvantages for ARs	Network not located in coastal region. Operating frequency no longer available for NOAA's use	Operating frequency not secure (crowded band). Temperature retrievals from RASS can be noisy	Phased-array antenna beam steering unit switches have finite life cycle (hardware issue for maintenance)

[a]Pulse-coding was used in selected operating modes to boost signal power and increase altitude coverage. (For more information on pulse coding, see Ghebrebrhan 1990)
[b]These settings reflect how the profilers were operated during the IOOS technology evaluation. Other degraded transmit and sampling resolutions are possible
[c]Signal attenuators prevent accurate radar reflectivity data below 1 km
[d]This minimum detectable range has been achieved with the ¼-scale 449-MHz profilers using a 0.7-μs pulse
[e]Over-sampling increased vertical resolution (compared to the transmit pulse length)

Fig. 3.8 The 10-m meteorological tower deployed at Bodega Bay, California. The tower is instrumented with an anemometer at a height of 10 m, as well as with pressure, temperature, relative humidity, solar, and net radiation sensors at a height of ~2 m. A rain gauge is mounted on the post to the left of the tower. Midway up the tower is a solar panel that powers the sensors. (Photo Credit: C. King)

surface meteorological tower deployment. Figure 3.9 shows sample time-series of the surface meteorological data collected on a 10-m tower.

Part of the motivation for an ARO is to investigate how an AR affects the precipitation at and downwind of the coast. For the US West Coast states, the downwind focus is often the precipitation that falls in the coastal and inland mountain ranges. Therefore, it is common to deploy or take advantage of existing precipitation gauges along the windward slopes of the terrain downwind (~10s of km for the coast ranges or ~100 km for the Sierra Nevada) from the ARO to monitor the full impacts of a landfalling AR.

Global Positioning System/Meterology (GPS/ MET)

Water vapor is the fuel that generates precipitation, and Global Navigation Satellite Systems (GNSS) such as the Global Positioning System (GPS) offer a robust and reliable method of calculating IWV (Bevis et al. 1992; Duan et al. 1996) with high temporal resolution under all weather conditions (Gutman et al. 2004). Here, following the proverbial expression of "one person's trash is another person's treasure," the atmosphere-caused delay in GPS signal transmission time—which is "noise" for geodetic scientists who rely on GPS for very accurate position information—is used.

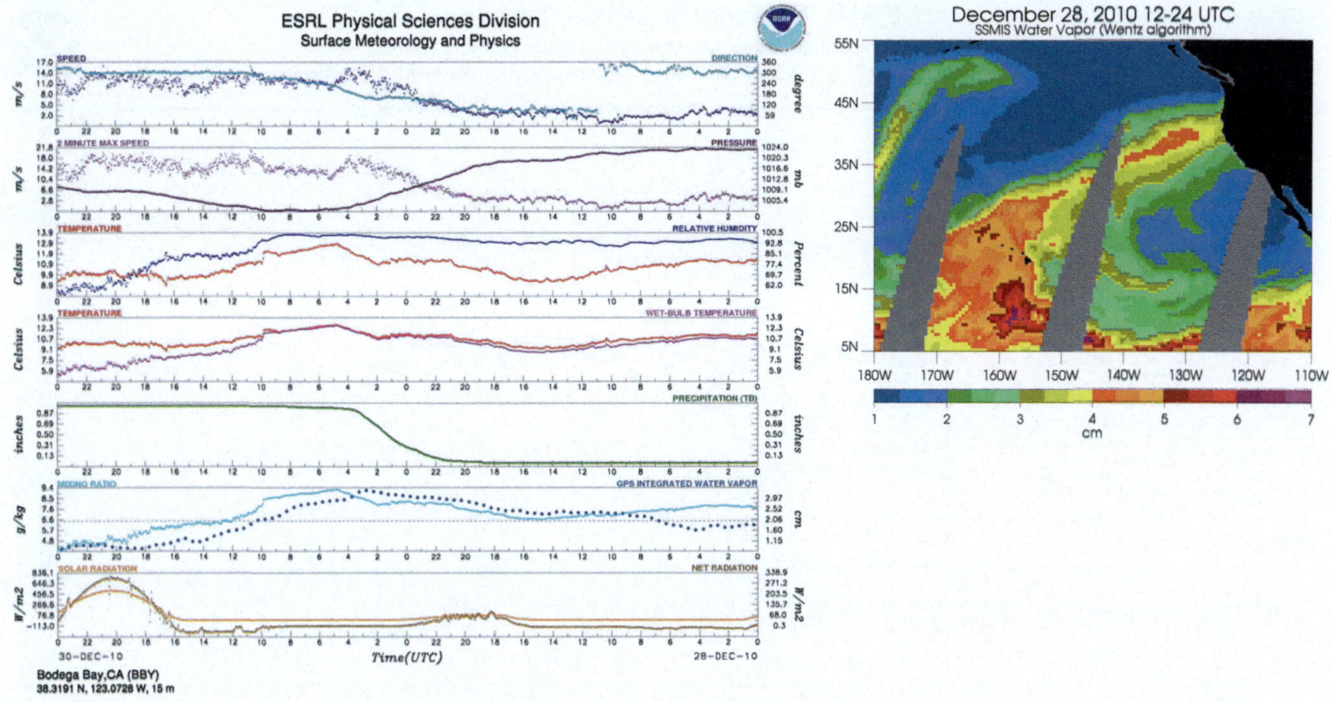

Fig. 3.9 *Left* An example of the surface meteorology time-series plot generated from the 10-m meteorology tower at Bodega Bay, California for the period 28–29 December 2010. Time proceeds from *right to left* along the horizontal axes. Atmospheric surface variables plotted in each panel from *top to bottom* are wind speed and wind direction, 2-min. Maximum wind speed and surface pressure, temperature and relative humidity, temperature and wet-bulb temperature, accumulated precipitation, integrated water vapor (IWV) and mixing ratio, and solar and net radiation. Units for these variables are given *along the vertical axes*. The *horizontal dashed line* on the IWV panel is drawn at 2 cm—the minimum IWV threshold used to identify AR conditions. *Right* Corresponding composite passive microwave satellite image of IWV (see Sect. 3.2) showing the AR present during the afternoon satellite overpasses on 28 December 2010

Also, unlike microwave satellite retrievals, GPS can provide accurate water vapor estimates over land. Therefore, a GPS receiver is an integral part of an ARO. An example time-series of GPS-generated IWV is shown in the second-to-last row of the meteorology time-series shown in Fig. 3.9.

3.3.3 The ARO Water Vapor Flux Tool

NOAA's water vapor flux tool (WVFT; Neiman et al. 2009; White et al. 2013) combines observations from the ARO with operational output from NOAA's high-spatial-resolution (3–13 km), rapidly updated (hourly), weather-prediction models to provide end users with a quantitative means to determine how well the forecast models simulate meteorological conditions in an AR as it passes over the ARO site. The WVFT incorporates upslope wind and IWV thresholds developed by Neiman et al. (2009) to signify when AR conditions are present (i.e., upslope winds $\geq$12.5 m s^{-1}; IWV $\geq$2 cm). When available, the WVFT also incorporates downwind rain gauge information to allow end users to see how well the models simulate how the AR affects precipitation in downwind terrain, for example.

Figure 3.10 provides a typical example of the WVFT from the ARO installed at Bodega Bay, California, for an AR that made landfall in early February 2015. In this case, the forecast model used in the WVFT is the operational "Rapid Refresh" (RAP) model run by NOAA's Environmental Modeling Center (EMC; http://www.emc.ncep.noaa.gov/). Other available models to use in WVFT are the operational High-Resolution Rapid Refresh (HRRR) model run by EMC, or the research versions of each of these operational models. Focusing on the lower two panels of Fig. 3.10, there is generally good agreement among the RAP model 3-h forecasts of the upslope wind in the controlling layer, the IWV, and the upslope IWV flux (i.e., the product of these two variables), which are the so-called orographic forcings for precipitation associated with ARs.

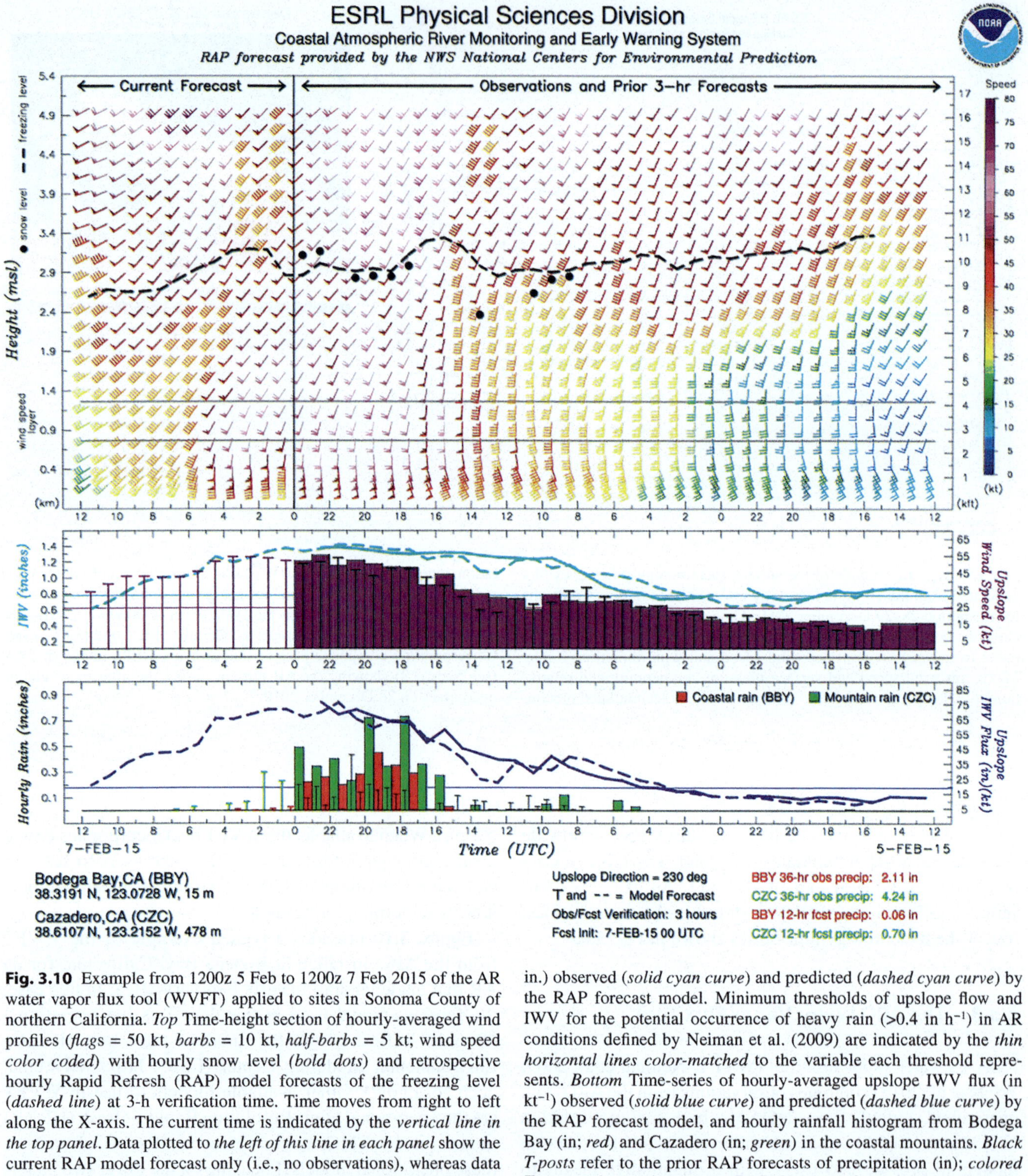

Fig. 3.10 Example from 1200z 5 Feb to 1200z 7 Feb 2015 of the AR water vapor flux tool (WVFT) applied to sites in Sonoma County of northern California. *Top* Time-height section of hourly-averaged wind profiles (*flags* = 50 kt, *barbs* = 10 kt, *half-barbs* = 5 kt; wind speed *color coded*) with hourly snow level (*bold dots*) and retrospective hourly Rapid Refresh (RAP) model forecasts of the freezing level (*dashed line*) at 3-h verification time. Time moves from right to left along the X-axis. The current time is indicated by the *vertical line in the top panel*. Data plotted to *the left of this line in each panel* show the current RAP model forecast only (i.e., no observations), whereas data plotted to *the right of the line in each panel* are a combination of observations and model output. *Middle* Time-series of hourly-averaged upslope flow (kt; from 230°) observed (*histogram*), and predicted (*T posts*) in the layer between 750 and 1250 m MSL (bounded by the *thin horizontal lines in the top panel*), and integrated water vapor (IWV; in.) observed (*solid cyan curve*) and predicted (*dashed cyan curve*) by the RAP forecast model. Minimum thresholds of upslope flow and IWV for the potential occurrence of heavy rain (>0.4 in h^{-1}) in AR conditions defined by Neiman et al. (2009) are indicated by the *thin horizontal lines color-matched* to the variable each threshold represents. *Bottom* Time-series of hourly-averaged upslope IWV flux (in kt^{-1}) observed (*solid blue curve*) and predicted (*dashed blue curve*) by the RAP forecast model, and hourly rainfall histogram from Bodega Bay (in; *red*) and Cazadero (in; *green*) in the coastal mountains. *Black T-posts* refer to the prior RAP forecasts of precipitation (in); *colored T-posts* refer to the current RAP forecast of precipitation (in.) for Bodega Bay (*red*) and Cazadero (*green*). Minimum threshold of upslope IWV flux for the potential of heavy rain, calculated by multiplying the thresholds for upslope flow and IWV, is indicated by the *horizontal blue line*

(Neiman et al. 2002 defined the controlling layer as the layer where upslope wind speed and precipitation intensity are most highly correlated in the coastal mountains; at 1 km MSL for this site.) However, for this particular case, the RAP model tends to under-predict the precipitation observed both at the coastal ARO site in Bodega Bay, California (BBY) and the inland coastal mountain site in Cazadero, California (CZD). Current research focuses on understanding these differences, which likely result from how precipitation is parameterized in numerical weather prediction.

During the winter of 2007–2008, NOAA operated three prototype AROs along the California coast at part of NOAA's Hydrometeorological Testbed (HMT). On 4–5 January 2008, a strong AR impacted each ARO as the AR (and parent storm) translated southeastward. Figure 3.11 illustrates the bottom portion of the WVFT (without model output) and how it could have been used to provide forecast lead time for the heavy orographic rainfall that occurred in the coastal mountains. In each panel, the upslope IWV flux was calculated based on a wind direction (230°, 225°, and 195° from top to bottom, respectively) normal to the local terrain orientation. First, it is clear that the maximum IWV flux is associated with the periods of maximum orographic rainfall enhancement. Second, in this case, the AR translated southward at an average speed of ~12 m s^{-1}, which could have given southern California about a ½-day warning for the AR conditions that were leading to heavy rainfall in northern California.

3.3.4 The US West Coast ARO "Picket Fence"

By 2006, a major proposal concept, consisting of 4 "Tiers" (as was described in Sect. 3.1 and summarized in Fig. 3.1) was developed to create a meso-scale network (mesonet) of

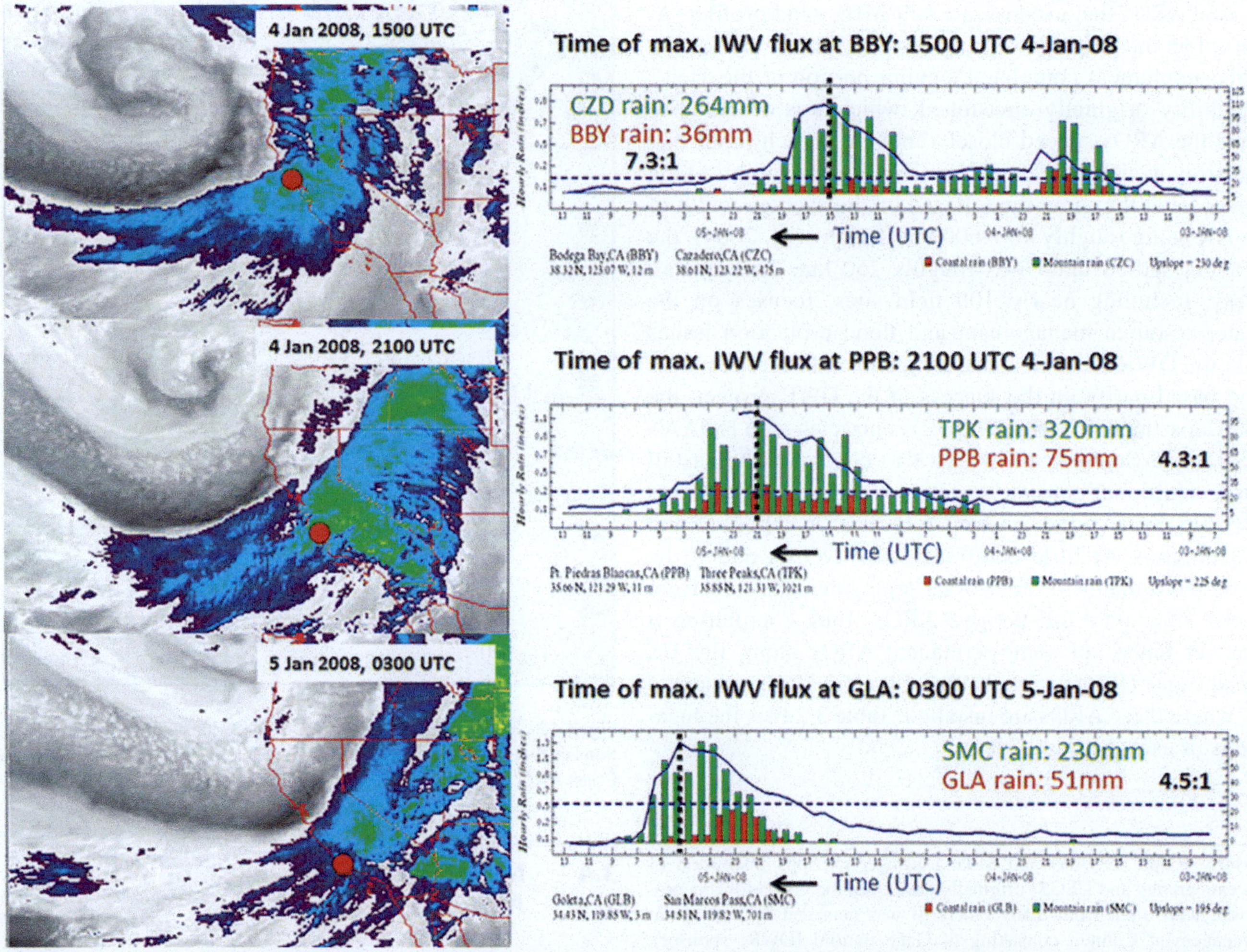

Fig. 3.11 *Left*—Enhanced infrared satellite imagery for the dates and times shown in the *upper right* for an AR making landfall on the California coast. *Right*—*Bottom panel* of the water vapor flux tool (WVFT) (as in Fig. 3.10 except without numerical model forecasts) highlights the relationship between upslope integrated water vapor (IWV) flux (based on upslope wind directions of 230°, 225°, and 195° from *top panel to bottom panel*, respectively) and the orographically enhanced coastal mountain rainfall (orographic ratios shown in *bold black text*)

observations across California based on the HMT findings on the role of ARs in creating extreme precipitation and flooding in the region. NOAA submitted to the California DWR the roughly $10 million, 5-year proposal[1] focused on Tier 1 of the observing system concept (described next in Sect. 3.4) including three sensor types and involving nearly 100 field sites. Work started in 2008. Support for this major project was based a variety of factors, including HMT's success in demonstrating the importance of ARs to water supply in California, and on the strong partnership that developed among NOAA (F.M. "Marty" Ralph and Allen B. White), Scripps scientists (David R. Cayan and Michael D. Dettinger), and the California DWR (Michael L. Anderson and Arthur Hinojosa, Jr.).

About a year after this initial effort started, a proposal to add four AROs along the coast as part of Tier 2 was submitted and soon awarded. This $4-milion-dollar addition focused on a contract addendum from the DWR with NOAA's ESRL in 2010 to construct and install four coastal AROs that use ¼-scale 449-MHz wind profilers as part of an integrated observing system (White et al. 2013). This supplement completed a major portion of the Tier-2 capability originally envisioned, which was designed to have the AROs spaced closely enough to each other that the array would typical measure an AR at two points along the coast at the same time. Based on the average width of an AR being roughly 400–600 km (Ralph et al. 2004), the spacing settled upon was roughly 250 km. This overall array, including nearly 100 field sites, focuses on the water resource management and flood protection issues that the DWR faces in a changing climate. More recently, and based partly on the success of the DWR project, the US Department of Energy (DOE) contracted with NOAA's ESRL and Scintec, AG in 2013 to construct and install three additional coastal ¼-scale 449-MHz wind profilers to help with wind energy forecasts in Oregon and Washington. At little additional cost, NOAA added the extra instruments to these DOE deployments to transform these sites into full-fledged AROs, thus completing a "picket fence" of semi-permanent AROs along the US West Coast (White et al. 2015a). Figure 3.12 shows a map of where these AROs are installed. Table 3.2 lists the locations of each ARO.

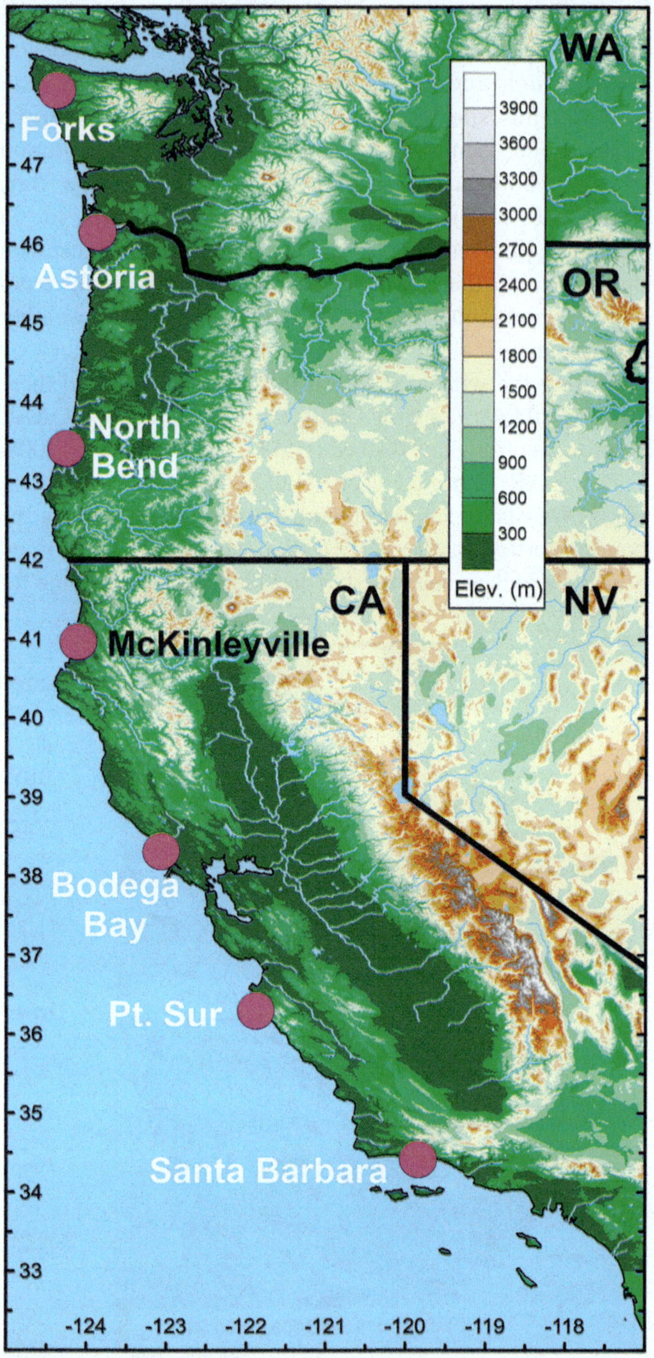

Fig. 3.12 Terrain base-map of the US West Coast states with the locations of the seven AR observatories (AROs) that constitute the US West Coast ARO "picket fence." (Adapted from White et al. 2015a)

[1]Marty Ralph (NOAA) and Mike Dettinger (Scripps Institution of Oceanography and USGS) originally developed the vision for this proposal, with input from many others. It was managed by a Program Management Council consisting of Gary Bardini (DWR; sponsor), Marty Ralph (NOAA; Proposal/Contract PI), and David Cayan (Scripps and USGS; Partner). The Program Management Council oversaw execution by the Project Management Team, which was led by Allen White (NOAA), and included Michael Anderson (DWR) and Mike Dettinger (Scripps and USGS).

3.4 Network Observations: Monitoring ARs over California

In 2005, California approved financing to help improve flood management and response. At the same time, NOAA's HMT (https://hmt.noaa.gov) study in northern California was testing new monitoring methods for extreme precipitation from

Table 3.2 Locations of the seven AROs that comprise the US West Coast ARO "picket fence"

Location	Station ID	LAT (O)	LON (O)	ELEV (M)	Installation date
Forks, WA	FKS	47.975	−124.398	95	07/21/2015
Astoria, OR	AST	46.157	−123.883	3	09/03/2015
North Bend, OR	OTH	43.420	−124.240	5	10/15/2015
McKinleyville, CA	ACV	40.972	−124.110	58	11/18/2015
Bodega Bay, CA	BBY	38.319	−123.073	16	03/21/2013
Point Sur, CA	PTS	36.304	−121.888	13	01/06/2017
Santa Barbara, CA	SBA	34.429	−119.847	3	07/01/2016

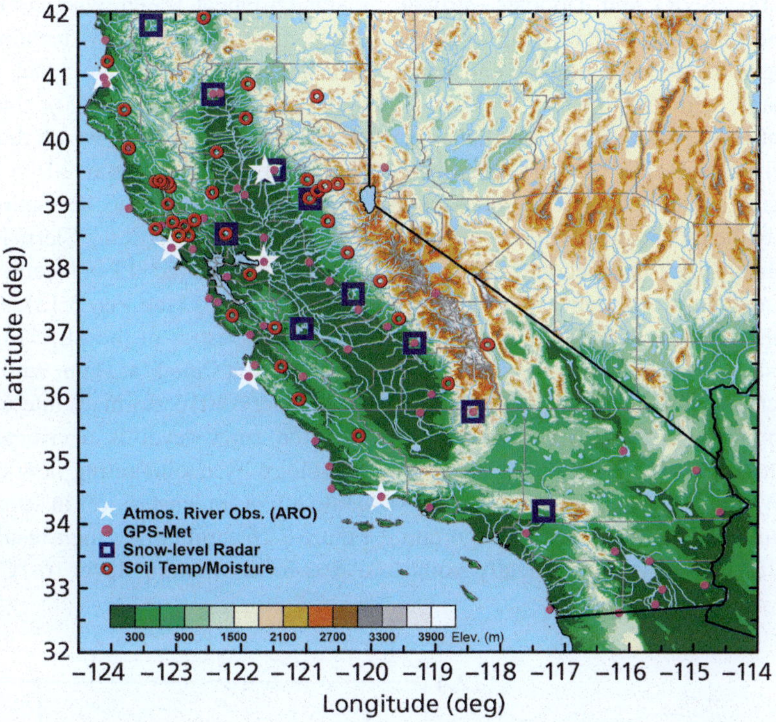

Fig. 3.13 Base-map of California indicating the locations of the AR monitoring network consisting of six AR observatories (AROs; *white stars*), 58 Global Positioning System/Meteorology (GPS/MET) sites (*pink dots*), ten snow-level radars (*open blue squares*; see Sect. 3.4.2), and 39 HMT sites where soil moisture is measured (*red circles*; see Sect. 3.5.3). These complement pre-existing soil moisture networks operated by the Scripps Institution of Oceanography, the Natural Resources Conservation Service, and the National Centers for Environmental Information. NRCS Snow Telemetry (SNOTEL) sites measure snow depth and snow-water equivalent. (Adapted from White et al. 2013)

ARs. The combination of these events led to the installation of an AR observing network for California (Sect. 3.3.4 described how this was developed; see White et al. 2013 for a description of the network). The network consists of over 100 sensors that help monitor the freezing level, onshore water vapor flux, spatial distribution of integrated water vapor, precipitation, and soil moisture conditions. Data from the network are transferred to a data hub in Boulder, Colorado, where near-real-time data products are generated and displayed on the Internet (http://www.esrl.noaa.gov/psd/data/obs/datadisplay/). Data are also transmitted to NOAA's National Weather Service (NWS) Weather Forecast Offices, and are available at the California Data Exchange Center (http://cdec.water.ca.gov/). Figure 3.13 shows a terrain base-map with the locations of these sensors. The following paragraphs describe the different sensors and offer sample observations.

3.4.1 AR Observatories (AROs)

The AR observatories (AROs) and their associated data products were described in Sect. 3.3. The four AROs in California are spaced approximately 270 km apart, on average, enabling some observation of almost all landfalling ARs in California.

3.4.2 Snow-Level Radar

The freezing level is a critical characteristic of ARs that can differentiate a beneficial storm from a flood-producing storm. The freezing level is located at the top of the melting layer, where precipitation begins to melt from snow or ice into rain in the atmosphere. Figure 3.14 shows an example of a melting layer depicted by vertical profiles of radar reflectivity (backscatter intensity in the form of range-corrected signal-to-noise ratio; SNR) and Doppler vertical velocity (DVV) measured with a 915-MHz wind profiler. The so-called radar "bright band" or region of enhanced radar reflectivity associated with the melting process is clearly evident. The bright band results when snowflakes or snowflake aggregates begin to melt. The wet surface reflects more than the frozen surface, which increases the backscatter intensity, and ultimately the radar reflectivity. As the precipitating particles completely melt and become all liquid, their density increases and so does their fall velocity. Hydrodynamically unstable raindrops break into smaller raindrops. These smaller drops reflect less to the radar than the larger drops. This decreases the radar reflectivity in the bottom part of the melting layer (see Fig. 3.14). The snow level was defined as the altitude of maximum radar reflectivity in the bright band by White et al. (2002), who also developed an automated algorithm to detect this important variable during precipitation events using vertically pointing radar. Based on comparisons with temperature profiles measured by rawinsondes, the snow level exists at a temperature of about 1 °C (White et al. 2010).

Engineers from NOAA's ESRL found a way to decrease the cost and footprint of the vertically pointing radar used to identify the snow level by inventing a snow-level radar (SLR; Johnston et al. 2017). The SLR is a frequency-modulated continuous-wave (FM-CW) radar that operates in the S-band frequency range. The radar uses two small (2-m-diameter) antennas—one to transmit and one to receive—mounted on a standard boat trailer with commercially available radar electronics. The FM-CW technology allows the SLR to achieve sensitivity nearly equal to similar pulsed S-band radars (White et al. 2000), but with a much-reduced power-aperture product, which dramatically reduces cost by a factor of 5 (Johnston et al. 2017). This development enabled seven of these SLRs to be deployed in major watersheds along the west slope of the Sierra Nevada, with an additional three SLRs installed at key locations in the coastal and transverse mountains (see Fig. 3.13). Table 3.3 lists some of the key characteristics of the SLR. Figure 3.15 shows a picture of the SLR at Pine Flat Dam, with a sample data plot from a late January 2016 storm. White et al. (2002) demonstrated why the snow level is a critical parameter in mountain hydrology by documenting how changes in the snow level could affect forecast runoff in several California watersheds. Figure 3.16 summarizes their results, where it is evident that a rise in the melting level (0 °C isotherm) of as little as

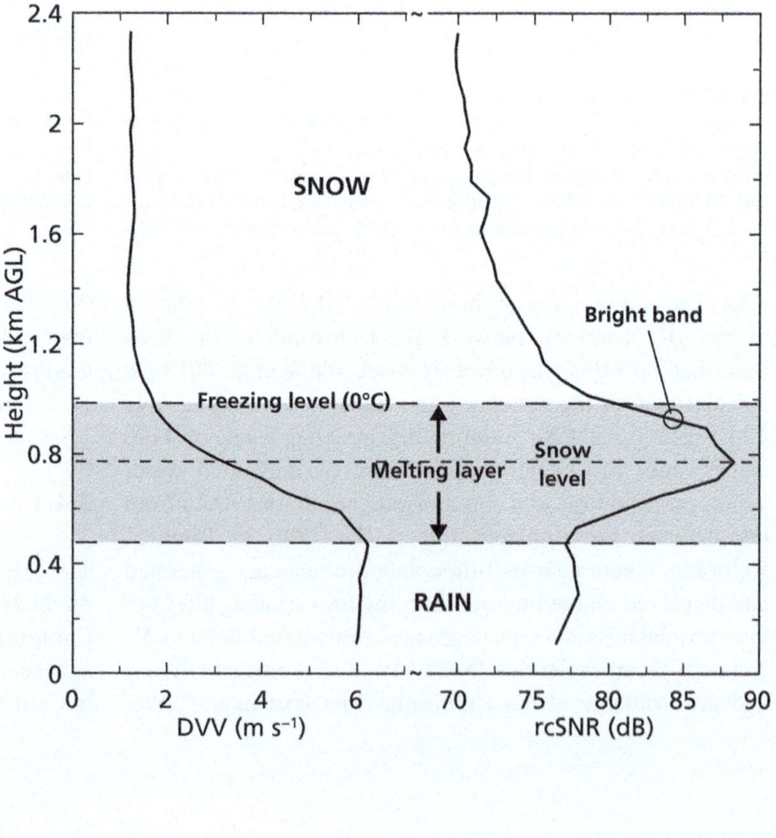

Fig. 3.14 Hourly median profiles of signal-to-noise ratio (SNR) and Doppler vertical velocity (DVV; positive downward) measured with the vertical beam of the 915-MHz wind profiler at Bodega Bay, California, between 1100 and 1200 UTC on 24 February 2001. The snow level is indicated by the *bold dashed line* at 0.772 km above ground level (AGL). The freezing level measured by a rawinsonde launched from Bodega Bay at 1126 UTC is shown by the *dashed line* at 0.994 km AGL. For illustration, the bottom of the melting layer is estimated to be at the bottom of the bright band, which is also where DVV is largest. The profiles were measured in stratiform rain. (Adapted from White et al. 2002)

Table 3.3 Physical characteristics and operating parameters of the SLR (White et al. 2013)

Parameter	Typical configuration	Minimum value	Maximum value
Frequency (GHz)	2.835	2.835	2.835
Antenna diameter (m)	1.2	1.2	1.2
Average transmit power (W)	0.7	0.6	12
Beamwidth (deg)	5.7	5.7	5.7
Range resolution (−3 dB response) (m)	46.4	15.1	116
Range gate spacing (m)	40	13	100
Snow-level determination period (minutes)	10	5	60
Nyquist velocity[1] (m/s)	21.5	10.0	24.0
Number of spectral points	256	64	1024
Number of heights	252	1	512
Lowest height (m above radar)	20	10	100
Highest observed height (m above radar)	10,060	6000	51,000

2000 ft. can more than triple the anticipated runoff in the watershed for a prescribed quantitative precipitation forecast (QPF) of 4 in. in 24 h.

3.4.3 Integrated Water Vapor (GPS/MET)

The technique for using GPS receiver signals to measure IWV via the satellite occultation technique was described in Sect. 3.2.2. Aside from being an integral part of the coastal AROs, an additional 32 GPS/MET stations across California are part of the statewide AR observing network to monitor the inland transport of moisture in ARs. Many of these sites are part of the Plate Boundary Observatory (PBO) operated by UNAVCO (originally *U*niversity *NAV*STAR *Co*nsortium), a non-profit university-governed consortium that uses geodesy to facilitate geoscience research and education. White et al. (2015b) used IWV data from the GPS/MET stations to study how the San Francisco Bay Area gap affected precipitation observed downstream in the Sierra Nevada during an AR event. Figure 3.17 illustrates the main hypothesis behind this study, and demonstrates how the GPS/MET network was used to help test the hypothesis.

3.4.4 Soil Moisture

The timing of storms with ARs that occur during the winter wet season can also determine whether a flood will ensue. For early-season storms or after a drought, the antecedent soil conditions are normally dry, such that the ground absorbs much of the precipitation, thereby minimizing runoff. Later in the wet season, the time between storms determines whether the soils dry out sufficiently to absorb some or all of the rainfall from the next storm (Zamora et al. 2011). Therefore, part of the statewide observing network includes soil moisture and surface meteorology stations. Each measurement site has two to five levels that range from 10 to 100 cm of soil temperature and soil moisture probes. Surface meteorology consists of a tipping-bucket rain gauge, a temperature/relative humidity sensor, and a data logger. All instruments are commercially available. Figure 3.18 shows the HMT-developed design for the soil moisture network deployed across California. Figure 3.19 shows an example of a soil measurement site, along with observed data to illustrate stream flow response to soil moisture conditions in the Russian River watershed in Sonoma County, California. Three separate precipitation events struck the watershed within a 5-day period from late November through early December 2012. Peaks in Russian River stream flow were observed each time the recorded precipitation kept the 10-cm soil at field capacity (the water content left in the soil after excess water has drained off, and downward movement of water has ceased) for a period longer than 3 h. Soil moisture expressed as volumetric water content at field capacity can vary from 15% to 55%, depending on the type of soil. The 15,000-cubic-feet-per-second (cfs) flow peak occurred early on 3 December after the soil at 15-cm depth exceeded the field capacity by 14% volumetric water content, as a result of the saturation–excess runoff (Dunne and Black 1970). The maximum flow stage (the height of water in a river above a reference height) that corresponded to this peak stream flow was 5.98 m, which is 0.42 m below flood stage for this particular location on the Russian River.

Table 3.4 lists the instruments used as part of the soil moisture monitoring stations. NOAA partnered with the California Department of Forestry and Fire Protection (CAL FIRE) for most of the installations shown in Fig. 3.13 because the numerous CAL FIRE station locations across California offered a variety of soil and meteorological conditions, and the data collected at the sites are useful to station operators. Other soil moisture networks in California include the Soil Climate Analysis Network (SCAN) operated by the Natural Resources Conservation Service (NRCS), and the US Climate Reference Network (USCRN) operated by the National Centers for Environmental Information (NCEI).

3.5 Field Campaigns and Experiments

Over the past two decades (1997–2018), several field campaigns and experiments have been devoted to ARs. These projects have led to improved scientific understanding of the structure and dynamics of ARs, better forecasts and warnings for the hazardous effects of ARs, increased awareness and visibility of ARs in the refereed literature and public

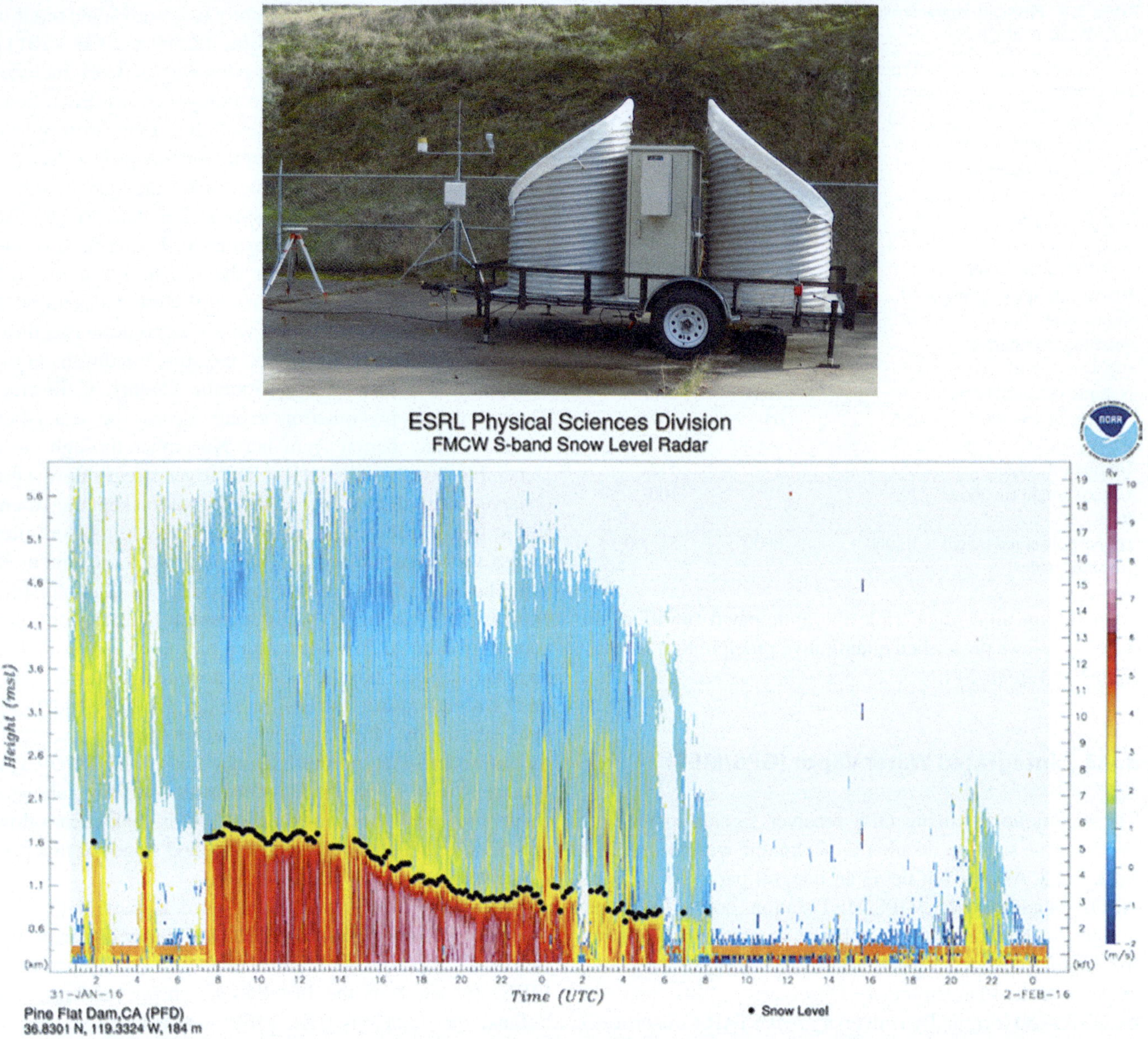

Time (UTC)	0115	0215	0315	0415	0515	0615	0715	0815	0915	1015	1115	1215	1315	1415	1515	1615	1715	1815	1915	2015	2115	2215	2315	0015
Snow Level (m)	none	1595	none	1448	none	none	1642	1687	1734	1639	1640	1687	1592	1537	1553	1398	1315	1155	1117	1003	940	940	1040	936
Snow Level (ft)	none	5231	none	4749	none	none	5385	5535	5689	5377	5379	5533	5221	5041	5093	4585	4314	3788	3663	3289	3084	3084	3411	3070
Sfc Temp (C)	10.75	10.21	9.79	9.61	9.38	9.27	9.08	9.17	8.91	8.52	8.35	8.19	7.68	7.77	7.79	7.74	7.74	7.68	7.27	6.74	6.46	6.60	6.16	5.44

Time (UTC)	0115	0215	0315	0415	0515	0615	0715	0815	0915	1015	1115	1215	1315	1415	1515	1615	1715	1815	1915	2015	2115	2215	2315	0015
Snow Level (m)	1012	1022	991	774	759	797	775	790	none	none	none	none	none	none	none	none	none	none	none	none	none	none	none	none
Snow Level (ft)	3319	3352	3250	2538	2489	2614	2542	2591	none	none	none	none	none	none	none	none	none	none	none	none	none	none	none	none
Sfc Temp (C)	4.72	4.48	4.64	4.59	3.88	3.84	3.77	3.69	3.74	3.78	3.40	2.96	3.38	3.46	2.92	2.82	3.92	6.20	7.65	7.83	6.76	7.77	7.79	7.74

Fig. 3.15 *Top* The snow-level radar (SLR) at Pine Flat Dam, with a collocated surface meteorology station and a global positioning system (GPS) antenna for measuring integrated water vapor IWV; see Sect. 3.4.2). *bottom* A 48-h time–height display from the SLR that indicates the snow level (*black dots*) at 10-min. Resolution. The *color contours* are of the radial velocity (Rv), which in precipitation closely represents the hydrometeor fall velocities (m s^{-1}) indicated by the *color scale on the right*. Time (UTC) and dates are *listed on the horizontal axis*. The *table below the plot* quantifies the snow level altitude during periods of precipitation, and provides collocated surface temperature observations. (Photo credit: Clark King)

domains, and a formal definition of ARs in the American Meteorological Society's (AMS's) *Glossary of Meteorology*. Below are brief summaries of each project, along with some of the major outcomes that resulted from the work that has been published to date. References provide more detail about each project and its findings. Table 3.5 lists each project by name, location, study period, and key observing and/or modeling platforms used during the project.

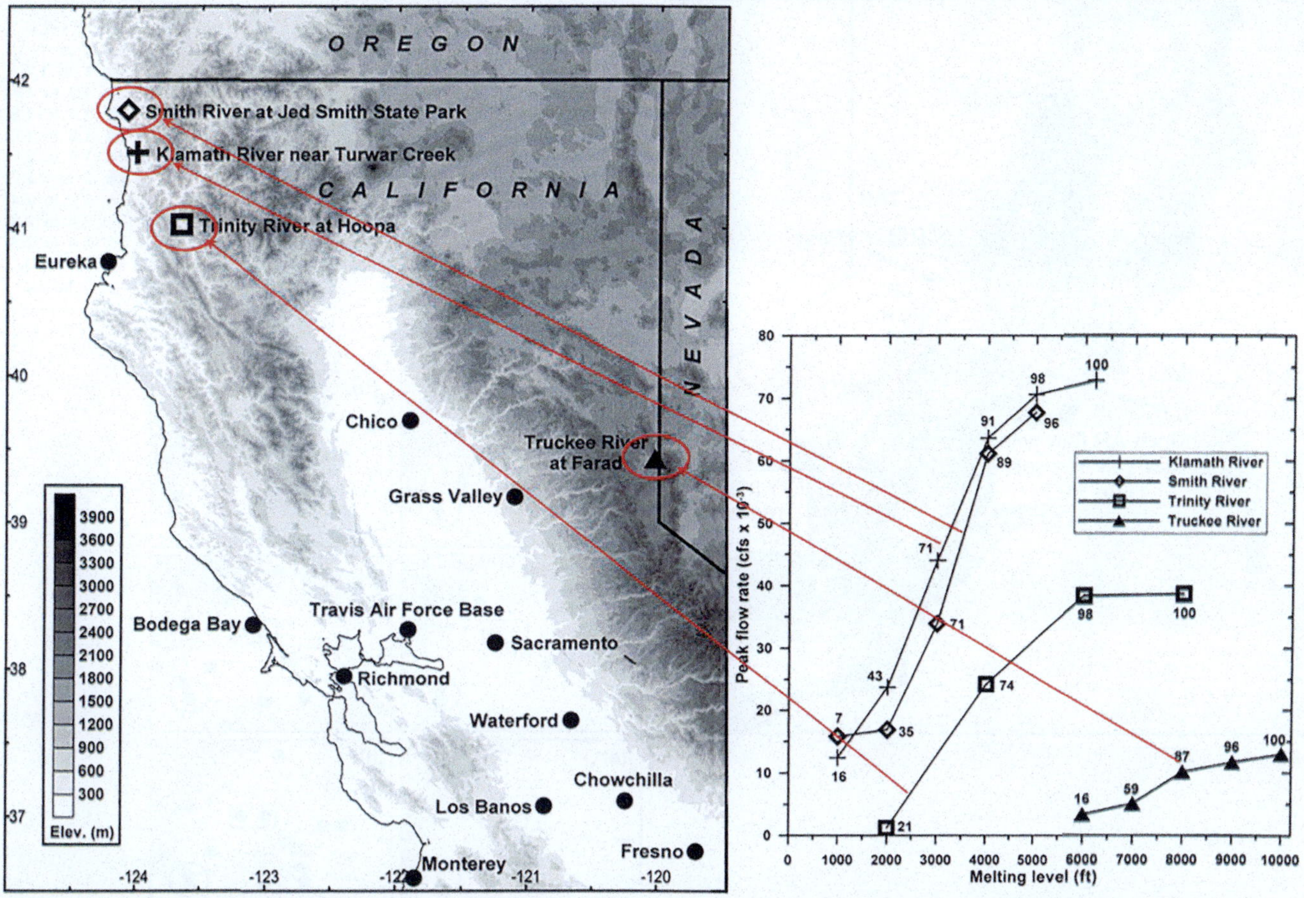

Fig. 3.16 *Left* Terrain base-map of northern California that highlights the locations of four river basins prone to flooding. *Right* River forecast model simulations of the sensitivity of runoff to changes in melting level for these same four river basins. The *posted numbers* give the approximate percentage of basin area below the altitude that corresponds to the melting level. (White et al. 2002)

3.5.1 CALJET

The California Land-falling Jets Experiment (CALJET; Ralph et al. 1999, 2004) was carried out in California and over the eastern Pacific during the winter of 1997–1998, with the overarching goal of helping to improve 0- to 24-h forecasts of damaging weather affecting the western US, with a particular emphasis on California. During CALJET, both ground-based and airborne platforms were used to obtain detailed measurements in winter storm environments across California's coastal domain and over the data-sparse eastern Pacific Ocean. The NOAA WP-3D aircraft measured nearly continuous (1 s) standard navigational and meteorological parameters during 26 CALJET flights in January through March 1998. More than 200 dropsondes were released to obtain high-resolution vertical profiles of wind velocity, temperature, and moisture over the Pacific Ocean offshore of California. A helically scanning, tail-mounted, X-band radar measured radar reflectivity and Doppler velocity within precipitation elements. Ralph et al. (2004) used the NOAA WP-3D measurements to diagnose the horizontal and vertical structure of ARs, which led to the first published two-dimensional (2-D) schematic of an AR (Fig. 3.20). This schematic has been adapted by Cordeira et al. (2013) and extended by Ralph et al. (2017) to represent many more cases in the composite, which has been adopted as the schematic for the AR definition in the AMS *Glossary of Meteorology* (Ralph et al. 2018a). The airborne observations collected were also key in supporting and defining the use of SSM/I and SSMIS satellite observations in characterizing ARs, as presented in Sect. 3.2.1.

The ground-based campaign for CALJET included 12 of NOAA/ETL's 915-MHz wind profilers (Carter et al. 1995) deployed across California. A subset of those were located along the California coast to study orographic precipitation in the Coast Ranges. One of those wind profilers was located at BBY, 35 km upwind of coastal mountain site CZD, where NOAA/ETL deployed for the first time an S-band vertically pointing precipitation profiler (S-PROF; White et al. 2000) and a surface met station with a tipping-bucket rain gauge (see Fig. 5.3 in Chap. 5). This combination of instruments

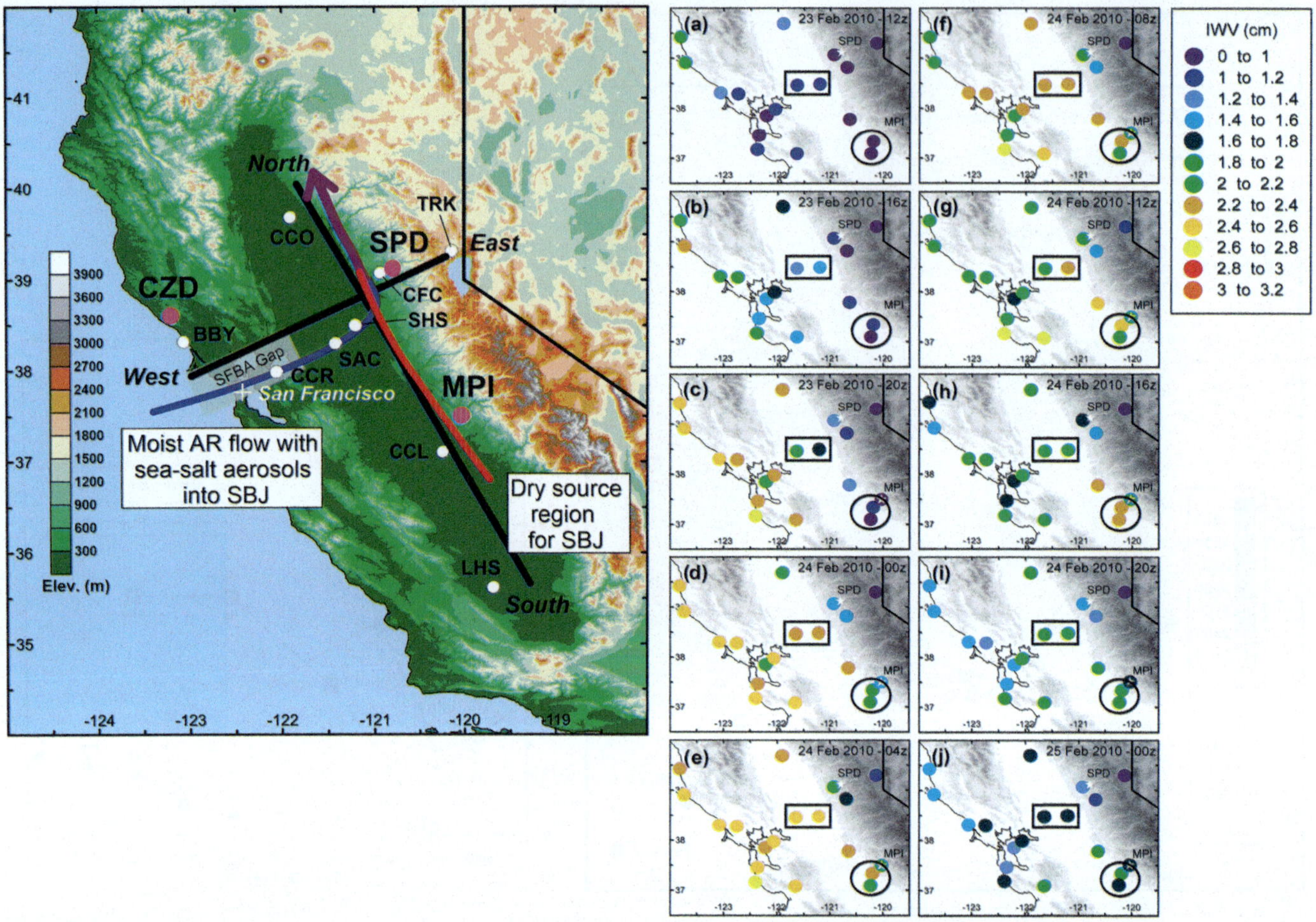

Fig. 3.17 *Left* Terrain base-map of California, with a schematic showing the interaction (*purple curve*) between unimpeded AR flow through the San Francisco Bay Area gap (*blue curve*) with the Sierra Barrier Jet (SBJ; see Sect. 5.2) flowing northward along the eastern side of the Central Valley (*red curve*) during a typical winter storm with an embedded AR. Instrumented sites with Doppler wind profilers in California at Bodega Bay (BBY), Chico (CCO), Chowchilla (CCL) Colfax (CFC), Concord (CCR), Lost Hills (LHS), Sacramento (SAC), Sloughhouse (SHS), and Truckee (TRK) are indicated by *white dots*. Cross-sections (*black lines*) are used to represent AR and SBJ flow characteristics (not shown). Locations of vertically pointing precipitation-profiling radars (part of NOAA's Hydrometeorological Testbed's [HMT's] observing network) at Cazadero (CZD), Sugar Pine Dam (SPD) and Mariposa (MPI) are indicated by *pink dots*. *Right* (**a**–**j**) integrated water vapor (IWV; cm) over central California at 4-h intervals from 1200 UTC 23 Feb to 0000 UTC 25 Feb 2010. The dates and times are shown near the top of panels (**a**–**j**). Two Central Valley Global Positioning System/Meteorology (GPS/MET) sites upwind of SPD are enclosed by a rectangle and two Central Valley GPS/MET sites upwind of MPI are enclosed by an *oval* to illustrate that more water vapor arrives at SPD than at MPI. (White et al. 2015b)

later became known as an ARO (see Sect. 3.3). Neiman et al. (2002) used this observing couplet to study the relationship between upslope flow and mountain precipitation, and found that the correlation between these two variables peaked at 1 km altitude, which also corresponded to the altitude of maximum wind speed in the LLJ measured offshore with the NOAA P-3, as shown in Fig. 3.21. (Sect. 5.2 in Chap. 5 discusses further aspects of orographic precipitation.)

Another major finding from CALJET was the role of warm rain processes in enhancing the orographic precipitation measured at Cazadero (CZD), California. White et al. (2003) used the automated bright band detection algorithm developed by White et al. (2002) to distinguish precipitation falling with a bright band (BB rain)—indicating a contribution from ice processes—from precipitation falling without a bright band (NBB rain), indicating no or little influence from ice processes. Figure 3.22 summarizes the results of White et al. (2003). BB rain has much greater radar reflectivity and greater downward velocity throughout the column than NBB rain because NBB rain consists primarily of tiny drops that fall from a shallower cloud than BB rain (Martner et al. 2008). When the shallow orographic feeder cloud associated with NBB rain and the deeper synoptically driven cloud associated with BB rain exist simultaneously, there is efficient droplet growth through the seeder–feeder process (Bergeron 1965), and this is when the heaviest orographic rainfall enhancement occurs at CZD (Kingsmill et al. 2016).

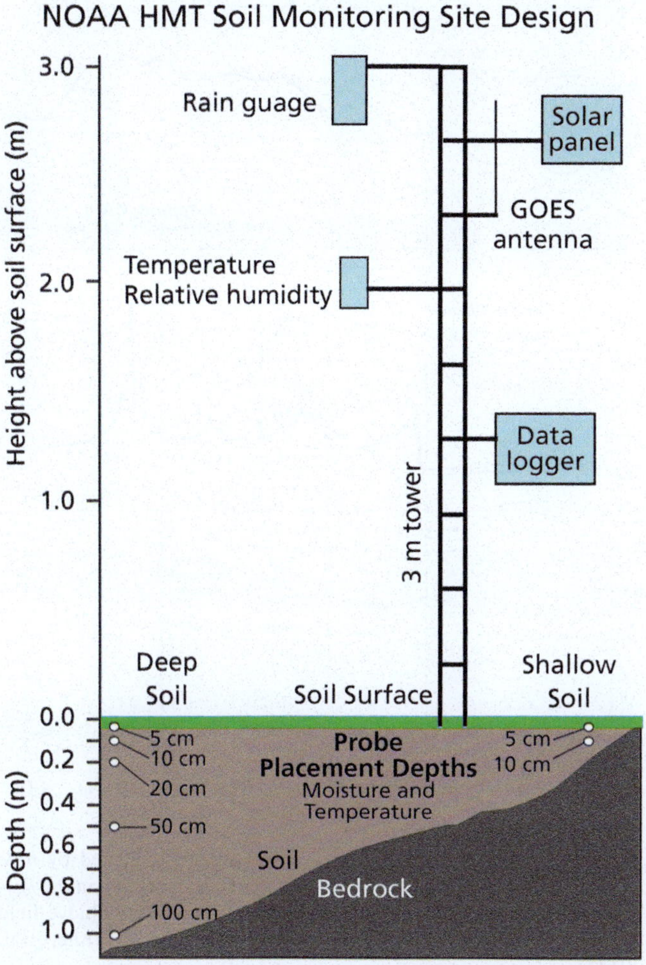

Fig. 3.18 Design of soil moisture measurement system used for the California soil moisture network. (From Zamora et al. 2011)

Other significant results from CALJET have been published by Ralph et al. (2003), Neiman et al. (2004), Persson et al. (2005), and Kingsmill et al. (2006), among others.

3.5.2 PACJET

The Pacific Land-falling Jets Experiment (PACJET) was a follow-on to CALJET and aimed to answer additional questions about US West Coast winter storms that arose after CALJET. PACJET expanded the domain of CALJET to include Oregon and Washington, and involved operational forecaster and emergency manager input from the outset. For example, a PACJET planning meeting was held in Monterey, California, in October 2001, where forecasters and other end users provided input for the upcoming winter's research activities. Like CALJET, the PACJET 2001 and 2002 field phases used the NOAA WP-3D research aircraft deployed from Monterey, California, in 2001 and from Portland, Oregon, in 2002.

PACJET 2003 did not involve a research aircraft but continued to uncover the key atmospheric processes that lead to extreme precipitation and flooding along the US West Coast. PACJET 2003 started to focus its activities on California's flood-prone Russian River watershed. One hypothesis being tested was whether a gap-filling radar could help improve quantitative precipitation estimates (QPE) and quantitative precipitation forecasts (QPF) for the Russian River basin. PACJET researchers tested this hypothesis by deploying NOAA's X-band scanning radar (Matrosov et al. 2007) on the northern California coast at Fort Ross. For example, Lim et al. (2013) developed new QPE algorithms that took advantage of the dual-polarization measurements from this radar. Figure 3.23 shows where the PACJET 2003 observing systems were located, and provides an example of how the X-band radar improved precipitation detection offshore of the Russian River basin.

Morss and Ralph (2007) examined how forecasters and emergency managers made use of the PACJET observations in their daily operations. Other significant results from PACJET have been published by Jorgensen et al. (2003), Neiman et al. (2005), Ralph et al. (2005a), Neiman et al. (2006), Han et al. (2009), and Richiardone and Manfrin (2009), among others.

3.5.3 HMT

Building from the successes of CALJET and PACJET, and based on recommendations from a US Weather Research Program workshop on cool-season QPF held in February 2004 (Ralph et al. 2005b), NOAA's Hydrometeorology Testbed (HMT; Ralph et al. 2013a; hmt.noaa.gov) was born in 2004. The first proposal to create the HMT was presented in a workshop in November 2002, co-led by F.M. Ralph and R. Fulton ("Developing a Testbed Concept for Hydrometeorology in NOAA"; https://hmt.noaa.gov/history/), to join the research and forecasting communities in an effort to improve the forecasting of extreme precipitation. Ralph et al. (2013a) describe the first decade of HMT, which as of 2019 has operated for more than 15 years.

As of 2019, HMT continues to accelerate the prototyping, demonstration, use, and transition of advanced hydrometeorological observations, models, and products to improve forecasts of extreme precipitation that can be used in the hydrologic prediction of stream flow and other surface processes in the Day 1–10 time-period. Currently, HMT is co-managed by the NOAA Office of Oceanic and Atmospheric Research (OAR) Physical Sciences Division (PSD) and the NWS Weather Prediction Center (WPC) in partnership with the NWS Office of Water Prediction.

To help bridge the gap between research and operations, WPC conducts two annual experiments: the Flash Flood and

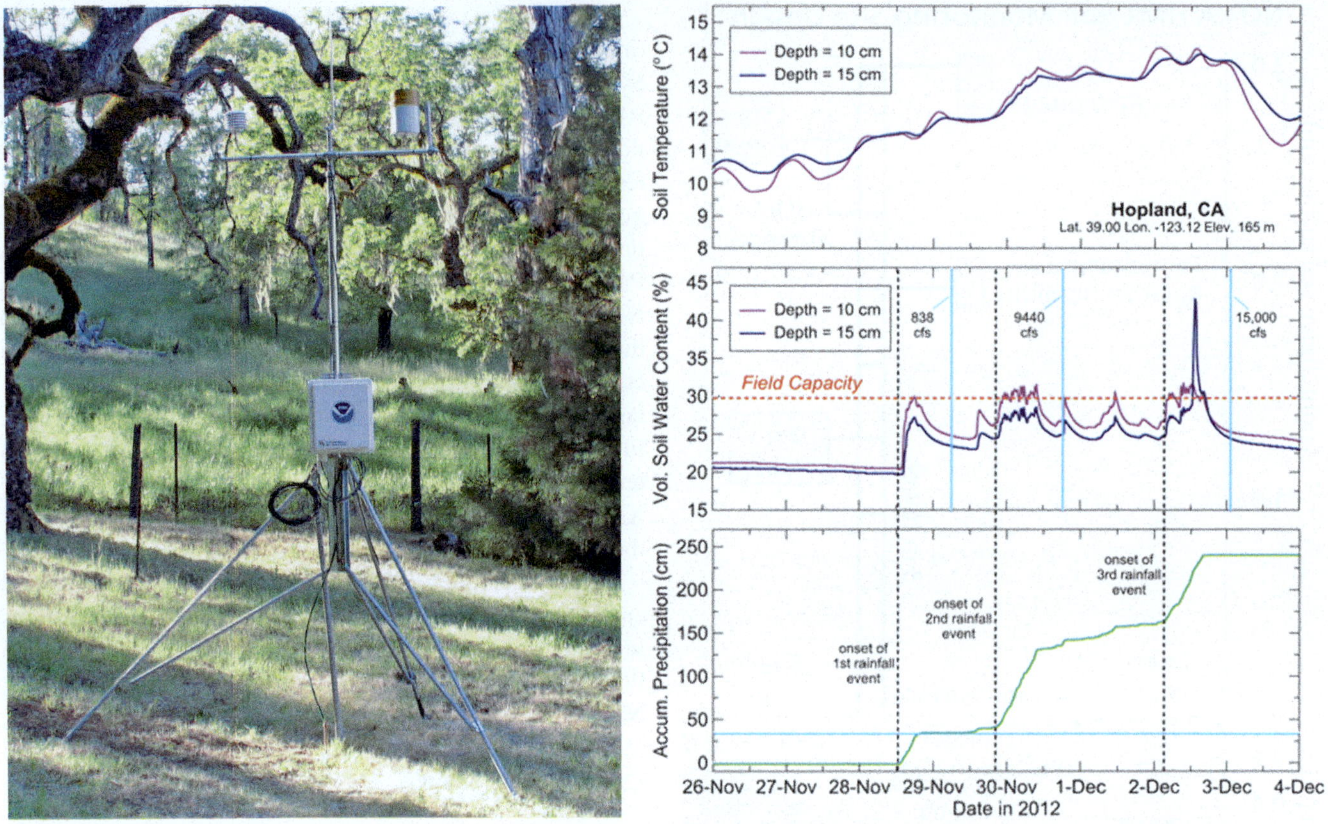

Fig. 3.19 *Left* The soil monitoring station in Hopland, California. Instruments are listed in Table 3.4. *Right* (*top*) Soil temperature (°C), (*middle*) volumetric soil water content (%), and (*bottom*) accumulated precipitation (mm) observed at Hopland, from 0000 UTC 26 Nov. 2012 to 1400 UTC 3 Dec. 2012. Peaks in Russian River stream flow provided by the US Geological Survey are indicated by *blue vertical lines* in the *middle panel*. The *thin horizontal line* in the *bottom panel* indicates the amount of rainfall required to achieve field capacity initially for the 10-cm soil moisture probe. (White et al. 2013)

Table 3.4 Instruments in the soil-probe and surface meteorology stations deployed across California (White et al. 2013)

Variable	Instrument	Type	Accuracy
Air temperature	Campbell scientific CS215	Sensirion SHT 75 Single chip element	±0.3 °C at 25 °C ±0.4 °C over +5 to +40 °C
Relative humidity	Campbell scientific CS215	Sensirion SHT 75 Single chip element	±2% (10–90% RH) ±4% (0–100% RH)
Precipitation	Texas electronics TR-525I	Tipping bucket	±1% (up to 0.254 mm h^{-1}) 0–3% (25.4–50.8 mm h^{-1}) 0–5% (50.8–76.2 mm h^{-1})
Soil temperature	Campbell scientific T107	Thermistor	±0.4 °C (worst case)
Soil wetness	Campbell scientific CS616	Reflectometer	±2%

Intense Rainfall Experiment (FFaIR) held during the summer, and the Winter Weather Experiment (WWE) held during the winter. Both of these testbed activities are designed to enhance and extend forecast skill for high-impact weather, especially precipitation, by facilitating interactions among researchers, operational forecasters, and other end users in a quasi-operational environment. Experiment summaries are available at www.wpc.ncep.noaa.gov/hmt/experimentsummaries.shtml.

HMT has established a long-term observing network in California with funding provided by the California DWR (see Sect. 3.3.4). HMT also conducted a pilot study in the southeast US that focused on the causes of extreme precipitation events in that area. For example, since ARs have been documented to dominate extreme precipitation events along the US West Coast during the winter wet season, Mahoney et al. (2016) examined the role of ARs in extreme precipitation events in the southeastern US. Their results indicated that ARs occur across all seasons in the southeast US (see Fig. 3.24), and match with 52% of the large-scale (> 7000 km^{-2}) heavy (>100 mm day^{-1}) precipitation events.

Table 3.5 Summary of field campaigns and experiments focused on ARs

Experiment name and key publication	Location	Study period	Observing/modeling platform
California Land-falling Jets Experiment (CALJET) Ralph et al. (2004)	California and Eastern Pacific	Winter 1997–98	NOAA WP-3D (dropsondes, tail radar, lower fuselage radar), wind profilers, S-band precipitation profiler, serial rawinsondes
Pacific Land-falling Jets Experiment (PACJET) Ralph et al. (2005a)	Eastern Pacific	Winters 2000–2001 and 2001–2002	NOAA WP-3D (dropsondes, tail radar, lower fuselage radar)
Hydrometeorology Testbed (HMT) Ralph et al. (2013a) White et al. (2013)	California and Pacific Northwest		Ground-based meteorological in situ and remote sensors, research and operational numerical weather prediction models
CalWater Ralph et al. (2016) Creamean et al. (2013)	California and Eastern Pacific	Jan–Feb 2009 Jan–Mar 2010 Winter 2010–2011	Ground-based aerosol and meteorological in-situ and remote sensors
Winter Storms and Pacific Atmospheric Rivers (WISPAR) Neiman et al. (2014)	Eastern Pacific	Feb–Mar 2011	NASA Global Hawk
CalWater2 Ralph et al. (2016)	California and Eastern North Pacific	Jan–Mar 2015	NOAA G-IV, NOAA P-3, DOE G-1, ground-based aerosol and meteorological in situ and remote sensors
El Niño Rapid Response (ENRR) and Sensing Hazards with Operational Unmanned Technology (SHOUT) Dole et al. (2018)	Eastern Tropical Pacific, Kiritimati Is.	Jan–Mar 2016	NOAA G-IV, NASA Global Hawk, U.S.; 4x daily rawinsondes from Kiritimati; Coordinated with AR Recon 2016
Atmospheric River Reconnaissance (AR Recon 2016; leads Ralph and Tallapragada)	Northeast Pacific	Feb 2016	Air Force C-130 s (2); coordinated with ENRR/SHOUT
North Atlantic Waveguide and Downstream Impact Experiment (NAWDEX) Shäfler et al. (2018)	North Atlantic	Sep–Oct 2016	German HALO, DLR Falcon, SAFIRE Falcon, enhanced surface observations
Atmospheric River Reconnaissance (AR Recon 2018, 2019, 2020); Ralph and Tallapragada	Northeast Pacific	Jan–Feb (or Mar) 2018, 2019, 2020	Air Force C-130s (2); all years and NOAA G-IV (2018, 2020)

Researchers and collaborators in HMT have produced a number of publications related to ARs and hydro-meteorology that is too large to list here. Interested readers may find links to all HMT publications on https://hmt.noaa.gov/pubs.

3.5.4 Ghost Nets

Other field campaigns with different primary objectives also yielded observations that have advanced AR-associated research. During the Ghost Nets campaign in 2005 that focused primarily on marine debris, the NOAA WP-3D flew two supplemental flights in late March over an AR in the vicinity of Hawaii. The first flight deployed 44 dropsondes during two transects of the AR; the second flight deployed 23 more in a single transect. The observations, collected near the southern end of the AR, showed that there could be poleward water vapor flux from the tropics into the base of an AR, as suggested by satellite imagery that had suggested a tropical connection (Fig. 3.25; Ralph et al. 2011).

3.5.5 CalWater-1

CalWater is a multi-year program focused on phenomena related to water extremes (drought, flood) along the US West Coast (Ralph et al. 2016). CalWater uses a coupled numerical modeling–observational approach to address questions on the dynamics of extreme precipitation and cloud– aerosol–precipitation interaction (CAPI). CalWater was conducted in two major phases: CalWater-1 collected observations in winters 2009–2011; CalWater-2 collected observations from 2014 to 2016. Primary sponsors were the California Energy Commission, DOE, and NOAA, with additional support for CalWater 2 from NASA and the National Science Foundation (NSF). Here, the field programs that make up CalWater-1, which was a land-focused field campaign to study ARs and CAPI, are emphasized. This included an "early start" experiment in February 2009 at Sugar Pine, California (SPD), an observing site in the Sierra Nevada, where a collection of AR and CAPI observing equipment was collocated. Two AR periods were present during the early start. Ault et al. (2011) contrasted these two AR periods and found differences in the aerosol content of the precipitation samples collected. The first storm showed mostly organic species from biomass burning, whereas the second storm showed a transition from biomass-burning organics to a dominance of long-range transported dust. Since dust can effectively serve as ice nuclei, further research was warranted to see if dust were present in the clouds that induce precipitation. Therefore, the 2011 field campaign included 25 flights of the DOE G-1 aircraft from McClellan

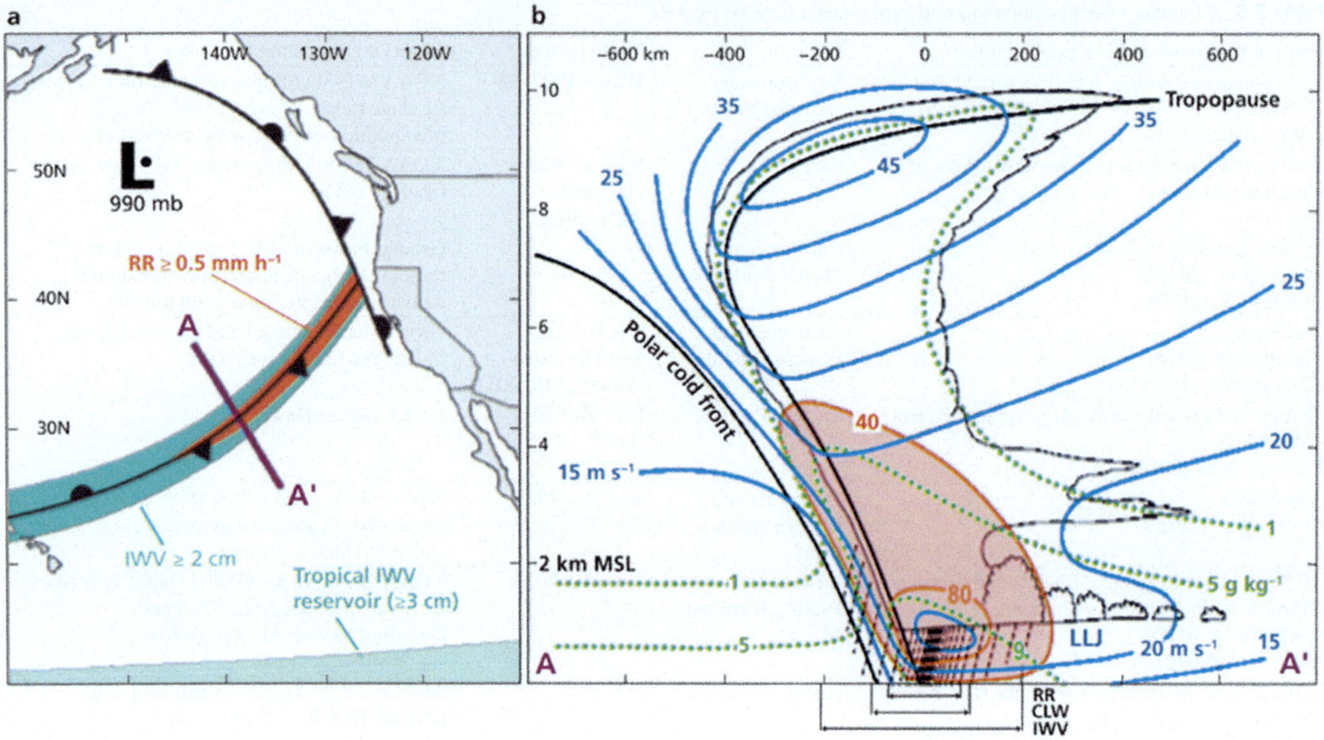

Fig. 3.20 Conceptual representation of an AR over the northeastern Pacific Ocean. (**a**) Plan-view schematic of concentrated integrated water vapor (IWV; IWV ≥2 cm; *dark green*) and associated rain-rate enhancement (RR ≥0.5 mm h^{-1}; *red*) along a polar cold front. The tropical IWV reservoir (>3 cm; *light green*) is also shown. The bold line AA' is a cross-section projection for (**b**). (**b**) Cross-section schematic through an AR (along AA' in **a**), highlighting the vertical structure of the along-front isotachs (*blue contours*; m s^{-1}), water vapor specific humidity (*dotted green contours*; g kg^{-1}), and horizontal along-front moisture flux (*red contours* and shading; ×10^5 kg s^{-1}). Schematic clouds and precipitation are also shown, as are the locations of the mean width scales of the 75% cumulative fraction of perturbation IWV (widest), cloud liquid water (CLW), and RR (narrowest) across the 1500-km cross-section baseline (*bottom*) (Ralph et al. 2004)

Fig. 3.21 Profile of the correlation coefficient between hourly averaged upslope flow measured at Bodega Bay (BBY), California, and hourly rain rate measured downwind in the coastal mountains at Cazadero County, California (CZD) for the 25 CALJET winter-season cases consisting of 468 h of data pairs (*bold curve*). The composite profile of wind speed measured in ten different lower-level jets (LLJs) measured offshore of California near BBY with the National Oceanic and Atmospheric (NOAA) WP-3D (*light curve*). (Adapted from Neiman et al. 2002)

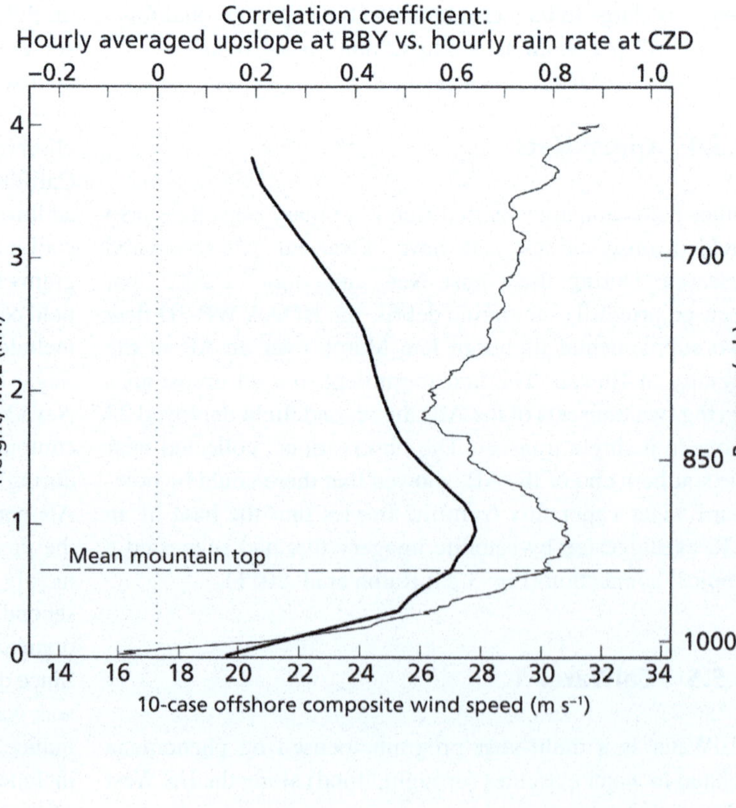

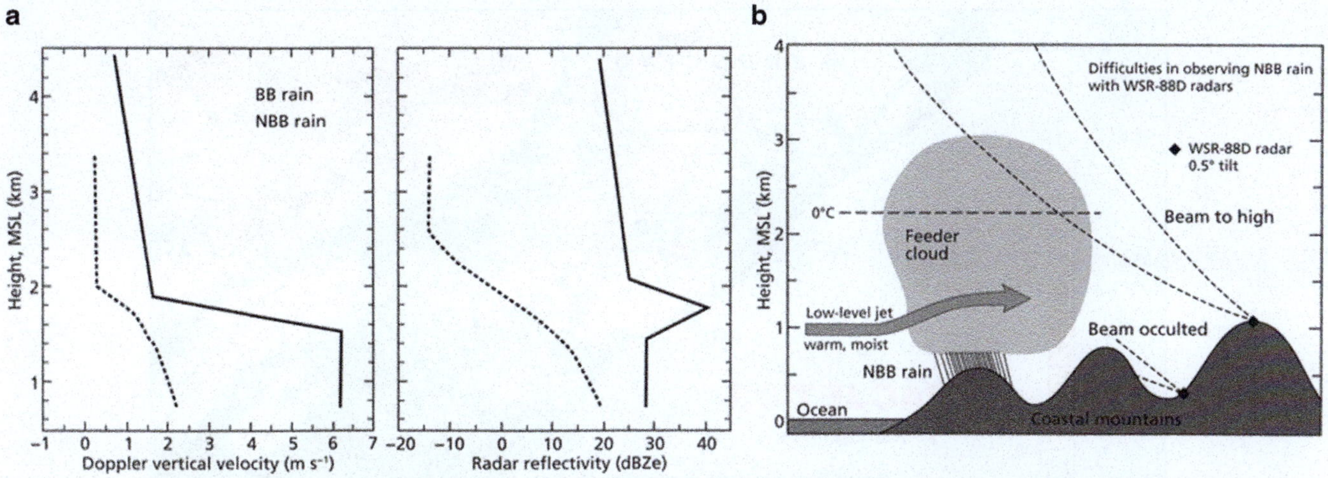

Fig. 3.22 (**a**) Composite winter-season profiles of (*top left*) Doppler vertical velocity (DVV; m s^{-1}; positive downward) and (*top right*) equivalent radar reflectivity factor (dBZ_e) measured by the S-band vertically profiling precipitation profiler (S-PROF) during bright band (BB) rain (*solid*) and non-bright band (NBB) rain (*dashed*). The altitude scale of individual BB profiles was normalized for BB height before the compositing, and the composite BB profiles were then plotted relative to the average BB height. The average rain rate for each rain type is approximately the same (3.95 mm h^{-1}). These profiles were obtained at CZD during winter 1997–1998. (**b**) Conceptual representation of shallow NBB rain in California's coastal mountains, and the inability of the operational Weather Surveillance [Doppler] Radar (WSR)-88D radars to adequately observe it (*bottom*). NBB rain is portrayed falling from a shallow feeder cloud forced by warm and moist onshore flow associated with a land-falling LLJ in an AR (*bold arrow*). (White et al. 2003)

Air Base (Sacramento, California) between 2 February and 6 March 2011 (Fan et al. 2014). The aircraft was instrumented with a suite of aerosol sampling and analysis equipment, in part to investigate the role of aerosols in enhancing or suppressing precipitation along the coast and over the Sierra Nevada.

Another key element explored during CalWater-1 was the interaction of landfalling ARs in California and the Sierra Barrier Jet (SBJ) (see also Chap. 5). Results (Neiman et al. 2013b; Kingsmill et al. 2013) documented that inland-penetrating ARs often ride up and over the SBJ (e.g., see Fig. 5.6 in Chap. 5). Other significant CalWater-1 results were published by Creamean et al. (2013, 2014) and Rosenfeld et al. (2014), among others. (The HMT discussion, Sect. 3.5.3, had additional CalWater publications.)

3.5.6 WISPAR

NASA and NOAA partnered to demonstrate how unmanned aircraft system (UAS) technology can be applied to weather and climate science. Much of this work has been devoted to tropical weather systems, including hurricanes. However, a few of the UAS projects have addressed Pacific weather systems, including a focus on ARs. The first of these was the Winter Storms and Pacific Atmospheric Rivers (WISPAR) project, conducted in February and March 2011, which included three flights with the NASA Global Hawk (GH) UAS.

One of the key observing systems on the GH is the Airborne Vertical Atmospheric Profiling System (AVAPS™) dropsonde system developed by the National Center for Atmospheric Research (NCAR). The AVAPS™ can carry up to 90 dropsondes per GH flight, and each sonde can be deployed remotely during the flight from the GH operations flight control center. In fact, the NOAA UAS Program established the requirement for the number of dropsondes largely based on what would be needed to monitor ARs. The original requirement was 100 sondes. The AVAPS dropsondes provide data similar to rawinsondes (also known as weather balloons), for example, which the NWS routinely launches twice daily—except dropsondes are ejected from the GH and parachuted downward to the surface from GH flight level (approx. 60,000 ft), whereas rawinsondes are carried upward by helium-filled balloons. Data from the GH AVAPS™ are telemetered to the surface, where they undergo data processing and quality control in near real time. Then the data are transmitted through the Global Telecommunications System (GTS) so they can be assimilated into NWP systems worldwide. Additional information about the GH dropsonde system is given by Wick et al. (2018b).

While a primary purpose of WISPAR was to demonstrate the performance of the GH dropsonde system, a key science focus was ARs, and an AR was sampled on each of the three GH flights conducted during the mission.

- The first WISPAR flight in mid-February 2011 observed and characterized an AR with a strong tropical connection near Hawaii (i.e., a so-called "Pineapple Express" event).

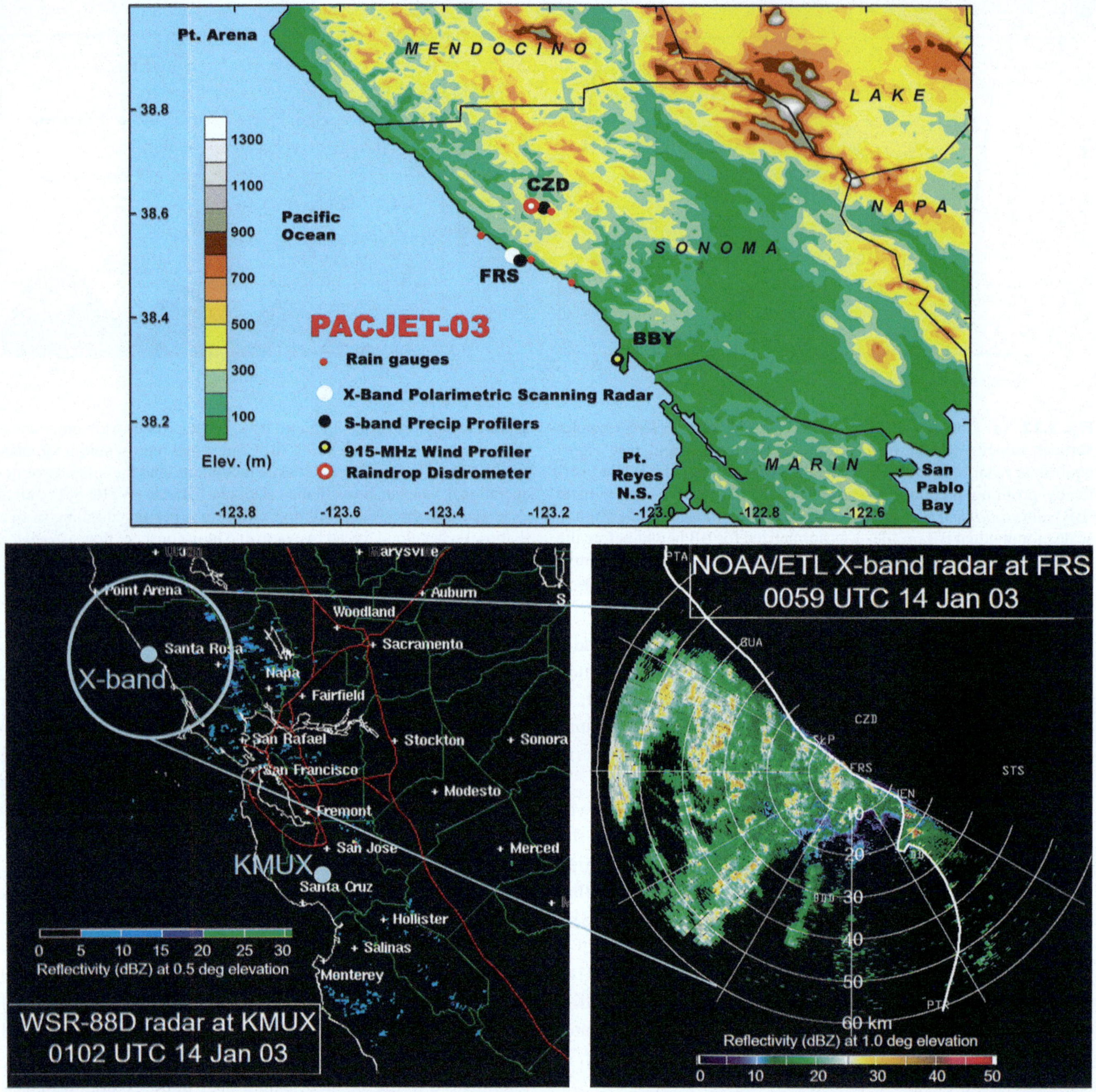

Fig. 3.23 *Top* Base-map indicating the location of the X-band scanning radar at Fort Ross (FRS). Other PACJET 2003 observing equipment was located at Cazadero (CZD) and Bodega Bay (BBY), California, as indicated in the key. *Bottom* Example to illustrate the gap-filling radar concept for precipitation monitoring. The nearest National Weather Service (NWS) operational scanning radar (KMUX) scans too high above the precipitating clouds along the coast north of San Francisco and therefore cannot measure the precipitation echoes detected locally by the X-band radar

- The second flight in early March was designed to obtain targeted observations to support forecasted cyclogenesis in the US Midwest that was verified a couple of days later. A flight of the NOAA G-IV, deployed at the time to support the Winter Storms Reconnaissance program, was coordinated with the GH mission, and deployed 44 dropsondes across the same AR the GH sampled.
- The third flight in mid-March targeted the Arctic region north of Alaska to demonstrate the use of the GH and AVAPS™ in the Arctic, and to compare the GH observations with profiles obtained from Barrow, Alaska. The transit to and from the GH base in southern California, however, enabled two high-density dropsonde transects of a Pacific AR.

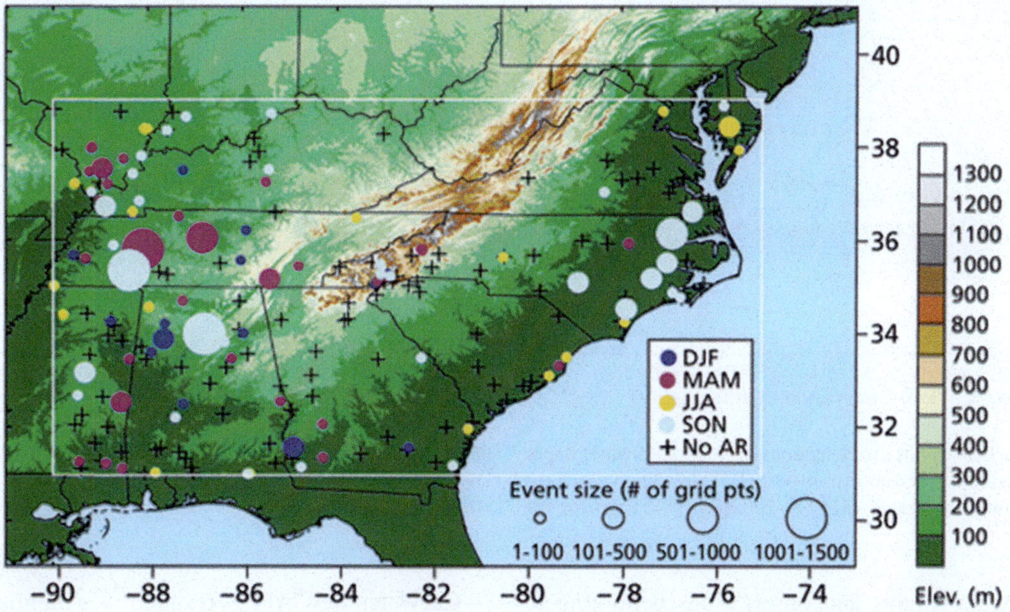

Fig. 3.24 Season of occurrence [winter (DJF) = *dark blue*, spring (MAM) = *pink*, summer (JJA) = *gold*, fall (SON) = *light blue*] of heavy precipitation events matched with ARs within 250 km and 24 h, plotted over terrain (elevation, m; *shaded* as in legend). Location indicated by *circle* is the center point of the heavy precipitation. *Circle size* indicates size (in number of grid points with ~38 km spacing from the National Centers for Environmental Protection–Climate Forecast System Reanalysis (NCEP–NCAR) as shown in legend at bottom right. *Black + signs* indicate heavy precipitation events in which no AR was matched. (Mahoney et al. 2016)

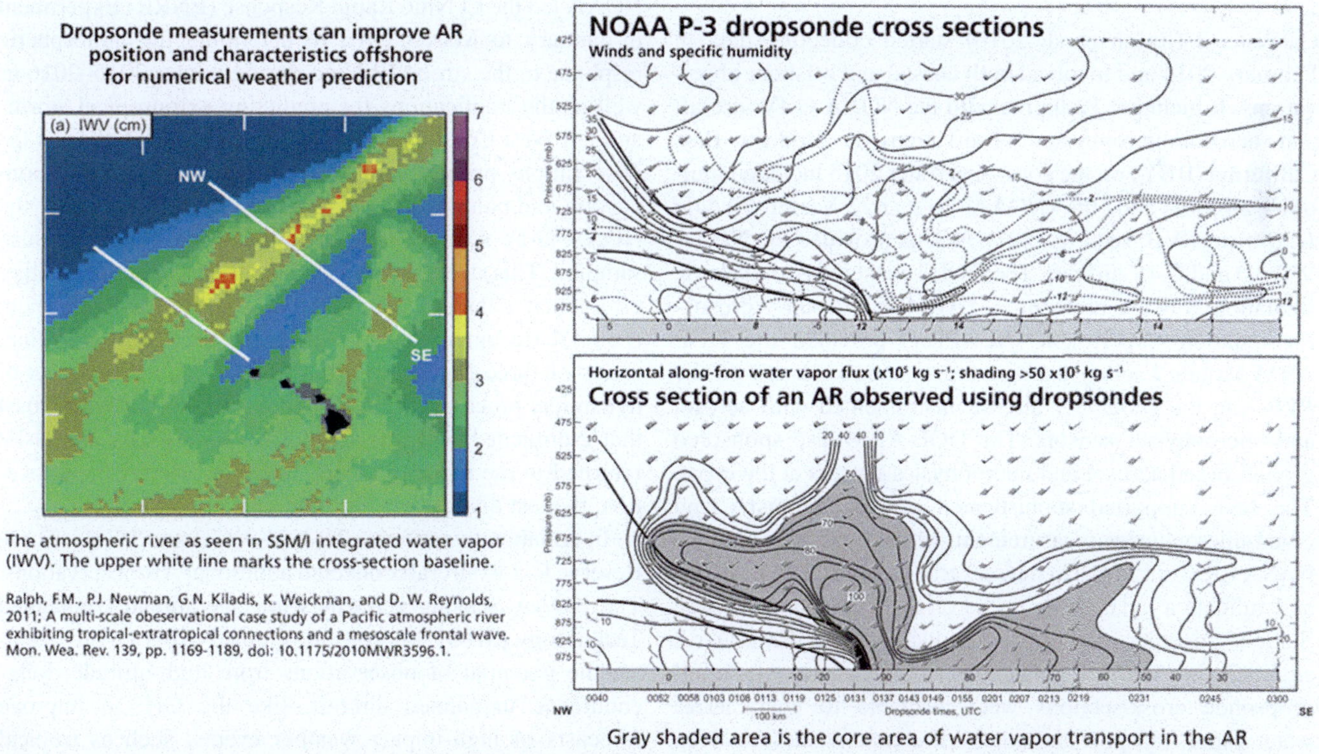

Fig. 3.25 Cross-section derived from dropsonde data during the GhostNets field campaign of 2005. (From Ralph et al. 2011)

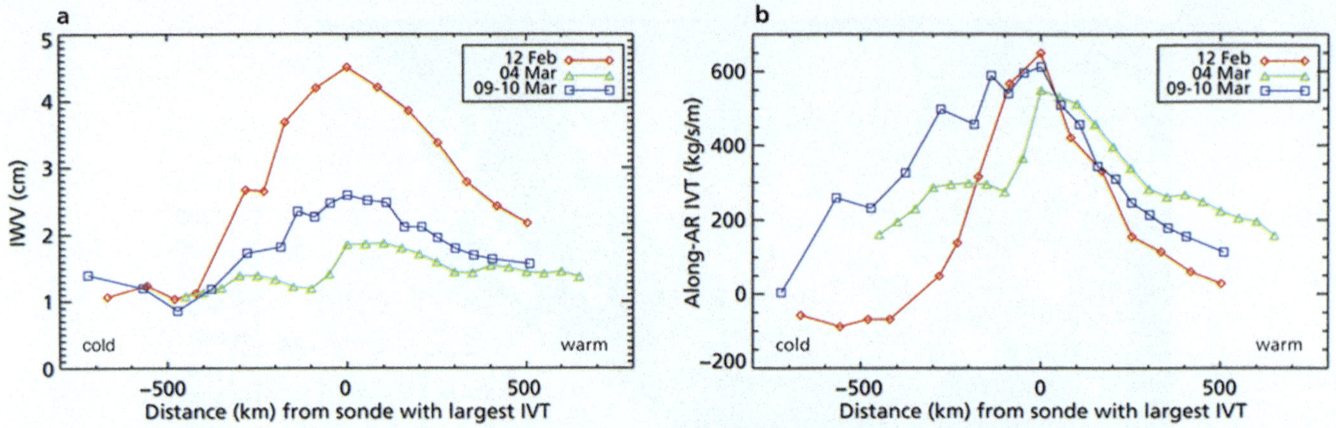

Fig. 3.26 Comparison of AR cross-sections of integrated water vapor (IWV) and integrated water vapor transport (IVT) obtained from the NASA Global Hawk. (**a**) Traces of 1000–200-hPa IWV (cm) for the three cross-sections. The traces are centered on the maximum value of IWV. (**b**) As in (**a**), except for AR-parallel IVT (kg s^{-1} m^{-1}). (Wick et al. 2018b)

Results from the flights and observations pertaining to ARs can be found in Neiman et al. (2014) and Ralph et al. (2017). Figure 3.26 illustrates the differences in the AR cross-sections of IWV and IVT obtained from three AR transects by the GH.

3.5.7 CalWater-2

CalWater-2 (Ralph et al. 2016) started collecting data in February 2014 and involved both coastal and offshore observations. It included 12 flights with the NOAA G-IV aircraft and aerosol/microphysics observations at Bodega Bay, California (BBY) on the coast. CalWater 2015 included multiple aircraft and the NOAA Research Vessel *Ronald H. Brown* (RVB; see Fig. 3.27). NOAA provided the NOAA WP-3D and G-IV aircraft. The DOE-sponsored Atmospheric Radiation Measurement (ARM) Cloud–Aerosol–Precipitation Experiment (ACAPEX) provided the DOE ARM Mobile Facility 2 observing system, mounted on the *RVB*, and the DOE G-1 aircraft instrumented with aerosol and microphysics sensors. The DOE-ARM also sponsored ground-based aerosol and microphysics sensors at the coast. The NSF supported sophisticated aerosol chemistry and cloud microphysics measurements at the coast as well. The NASA ER-2 aircraft also flew several missions, in part to test and validate a radar being tested on the International Space Station. Together, the flights conducted under CalWater-2 significantly expanded the number of AR cases in which full dropsonde cross-sections were available to characterize water vapor transport and other AR characteristics, as Sect. 3.5.11 will summarize (Ralph et al. 2016; Ralph et al. 2017; Guan et al. 2018). Additional airborne activities associated with CalWater-2 in 2016 are included in Sect. 3.5.8 below (ENRR and SHOUT).

CalWater-2/ACAPEX resulted in a number of other significant scientific results published by Cordeira et al. (2017), Creamean et al. (2015), Martin et al. (2017), Neiman et al. (2016, 2017), and White et al. (2015b), among others.

3.5.8 ENRR and SHOUT

NOAA led the El Niño Rapid Response (ENRR) Experiment in January to March 2016 to determine the atmospheric response to the strong El Niño during winter 2015–2016 as well as the implications for predicting extratropical storms and US West Coast rainfall (Dole et al. 2018). Although the observations primarily focused on organized tropical convection and poleward convective outflow, as sampled by the NOAA G-IV near the heart of the El Niño, ARs were also sampled. This sampling was via both the campaign and other efforts that collaborated with the broader ENRR Experiment (i.e., AR Reconnaissance described in Sect. 3.5.10). A final series of three flights from the G-IV examined a series of dynamical processes from the tropics to the US West Coast that culminated with the landfall of an AR. As the G-IV returned to the mainland from its Hawaii deployment location, the last flight made multiple transects of the AR.

In collaborative efforts, the NASA GH and US Air Force Reserve C-130 aircraft collected additional AR observations. The NOAA Sensing Hazards with Operational Unmanned Technology (SHOUT) project used the NASA GH to evaluate the potential of observations from high-altitude, long-endurance unmanned aircraft, like the GH, to improve forecasts of high-impact weather events, such as tropical cyclones and major winter storms. The SHOUT project collaborated with ENRR to examine how observations in the extratropical eastern North Pacific affected forecasts of severe weather on the North American west coast and Alaska.

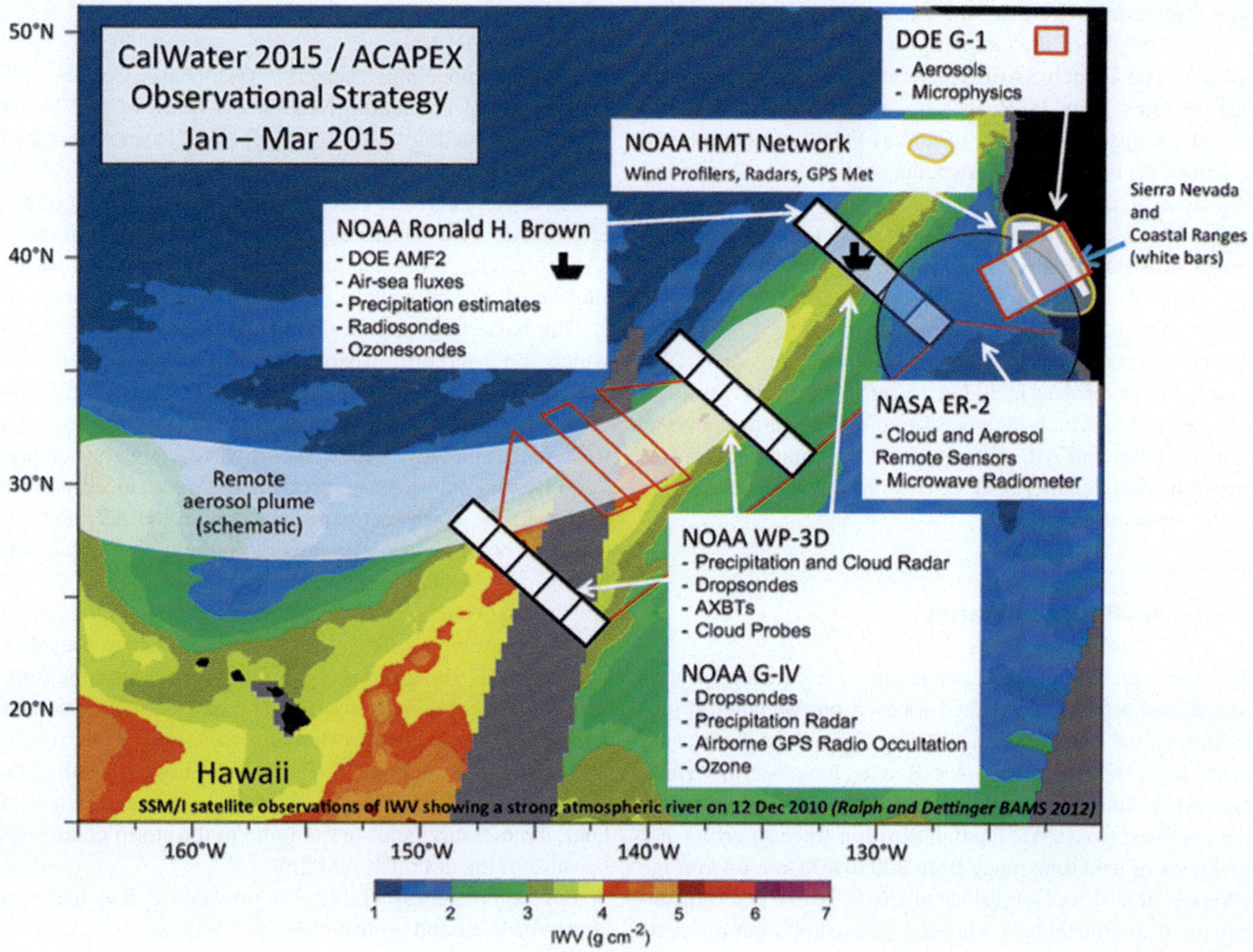

Fig. 3.27 Conceptual design of the CalWater-2/ACAPEX field program. (Ralph et al. 2016)

The observations explicitly targeted regions in which forecast models exhibited sensitivity to initial conditions, but these regions frequently corresponded physically to components of ARs, which was the focus of the AR Recon effort, led by F.M. Ralph and Vijay S. Tallapragada, described below in Sect. 3.5.10. Wick et al. (2018a) and Kren et al. (2018) summarized the effect of the SHOUT observations. Regionally, the observations resulted in small positive effects on multiple forecast fields. Interestingly, the observations from the final GH flight had their largest positive effect on forecasts in the southeastern US associated with a severe weather outbreak that occurred downstream a couple of days later (Kren et al. 2018).

3.5.9 NAWDEX

A European-led campaign directly relevant to AR studies was the international North Atlantic Waveguide and Downstream Impact Experiment (NAWDEX; Shäfler et al. (2018), http://nawdex.ethz.ch) conducted in September–October 2016 with observations over the North Atlantic. NAWDEX's overarching scientific aim was to increase physical understanding and quantify the effects of diabatic processes on disturbances to the jet stream near North America, their influence on downstream propagation across the North Atlantic, and their consequences for high-impact weather in Europe. The experiment comprised several nationally funded projects. Although many components were closely related to ongoing AR research, one project funded through France and Norway explicitly cited ARs as an objective: This project cited goals of studying ARs in the eastern Atlantic, and identification and quantification of errors in model-based representations of ARs.

The experiment incorporated three aircraft as well as enhanced ground-based observations in the UK and France. The German High Altitude and LOng Range Research Aircraft (HALO) and DLR (Deutsches Zentrum für Luft-

und Raumfahrt; German Aerospace Center) Falcon were deployed as part of the German, Swiss, UK, and US contributions. The French SAFIRE (Service des Avions Français Instrumentés pour la Recherche en Environnement—the French facility for airborne research) Falcon was employed as part of the French and Norwegian contributions. A primary objective of the SAFIRE Falcon was to collect measurements along ARs that specifically targeted diabatic processes. Instrumentation included a Doppler cloud radar, high-spectral resolution LIDAR, and in situ microphysical probes. These observations were to be compared directly to AR representations from both satellite platforms and NWP analyses. The project included 15 intensive observing periods, several of which included sampling of warm conveyor belts (WCBs) and ARs. Results from the campaign are still emerging, but Schäfler et al. (2018) present initial highlights of the observational results.

3.5.10 AR Reconnaissance

The Atmospheric River Reconnaissance "AR Recon" project formulated a targeting method focused on AR landfall prediction on the US West Coast. (Chap. 8 provides the broader basis for—and vision of—AR Recon, longer-term.) Here, context is summarized and the existing data described. On the US West Coast, AR landfall position forecast errors at 1 to 4 days of lead time range from 200 to 400 km, on average (Wick et al. 2013b; DeFlorio et al. 2018, 2019), and can contribute to significant errors in extreme precipitation forecasts (e.g., Ralph et al. 2010, 2011). The recent addition of moist processes in an adjoint method enabled a conclusion that errors in the location and characteristics of ARs offshore were the leading source of initial condition error for landfalling storm forecasts on the US West Coast (Doyle et al. 2014). These forecast errors affect water decisions in the west, including those associated with mitigating flood risk and drought, as well as those associated with restoring endangered salmon (http://cw3e-web.ucsd.edu/firo/).

The AR Recon project is a multi-year, inter-agency, cooperative effort to collect unique dropsonde observations in and around ARs off the US West Coast to improve AR-landfall-associated weather forecasts during the cool season. The project collected data with multiple aircraft in three ARs in February 2016 (two Air Force C-130 s; in coordination with the ENRR field campaign). AR Recon executed flights into six ARs in January–February 2018 (involving a mix of two Air Force C-130 s, and NOAA's G-IV). A similar field effort took place in February–March 2019, with two AF C-130 aircraft observing six storms over 5 weeks. AR Recon formally became "operational" in 2020 when the National Winter Season Operations Plan (NWSOP) directed the deployment of the AF and NOAA Weather Recon' aircraft to monitor ARs over the pacific using methods developed in AR Recon 2016, 2018, 2019.

Global modeling centers (National Centers for Environmental Prediction [NCEP], US Navy, ECMWF) and regional modeling efforts (Coupled Ocean/Atmosphere Mesoscale Prediction System [COAMPS®], West-WRF [Weather Research and Forecasting]) are collaborating on this effort. An AR Data Assimilation Steering Committee (ARDASC) has formed to bring together diverse expertise and substantial institutional capacity to carry out the collaboration.

The basic targeting strategy is illustrated in Fig. 3.28a, which is a 3-aircraft mission. Two aircraft focus on the AR itself, according to findings from the moist adjoint method. The third aircraft targets the AR as well, but also the region to its north and west, where a short-wave trough or potential vorticity anomaly may be present. It is recognized that the trough to the northwest can interact with the AR, and that subtle errors in their representation in the model initial conditions can translate into large errors in forecasts of the AR and its associated landfall and precipitation. Figure 3.28b shows an example of a 3-aircraft mission, where the AR is shown via its signature in IVT. The scope of the measurements is geographically substantial, representing the equivalent of a radiosonde network roughly the size of the western 40% of the contiguous US (Fig. 3.29). However, unlike the geographically fixed locations of the radiosonde sites on land, the measurements are tailored to the storm conditions, because of the aircraft's mobility.

For completeness, Table 3.6 provides a few technical characteristics and milestones from experience gained in 2016, 2018, and 2019.

3.5.11 Synthesis of Airborne Cross-Sections Across ARs into a Composite of AR Structure and TIVT

The previous sections have summarized an extensive set of observations by aircraft. Based on the earliest flights in 1998, 2005, and 2011, a total of five AR cross-sections offered enough coverage by dropsondes to sample the full width of each AR, including the one from the Ghost Nets field campaign shown in Fig. 3.25. Preliminary analysis, based on defining the edges of the AR using IWV or IVT thresholds, indicated that an average AR would transport about 3.1×10^8 kg s^{-1} of water vapor. This is referred to as total integrated vapor transport (TIVT), which is a parameter analogous to the flux of water in a terrestrial stream. However, the small sample size raised concerns about its representativeness. This gap was part of the motivation for CalWater 2 to sample far more cases (Ralph et al. 2016), and then to add to these using AR Recon data. By the end of AR Recon 2016, 21 storms had adequate dropsonde coverage

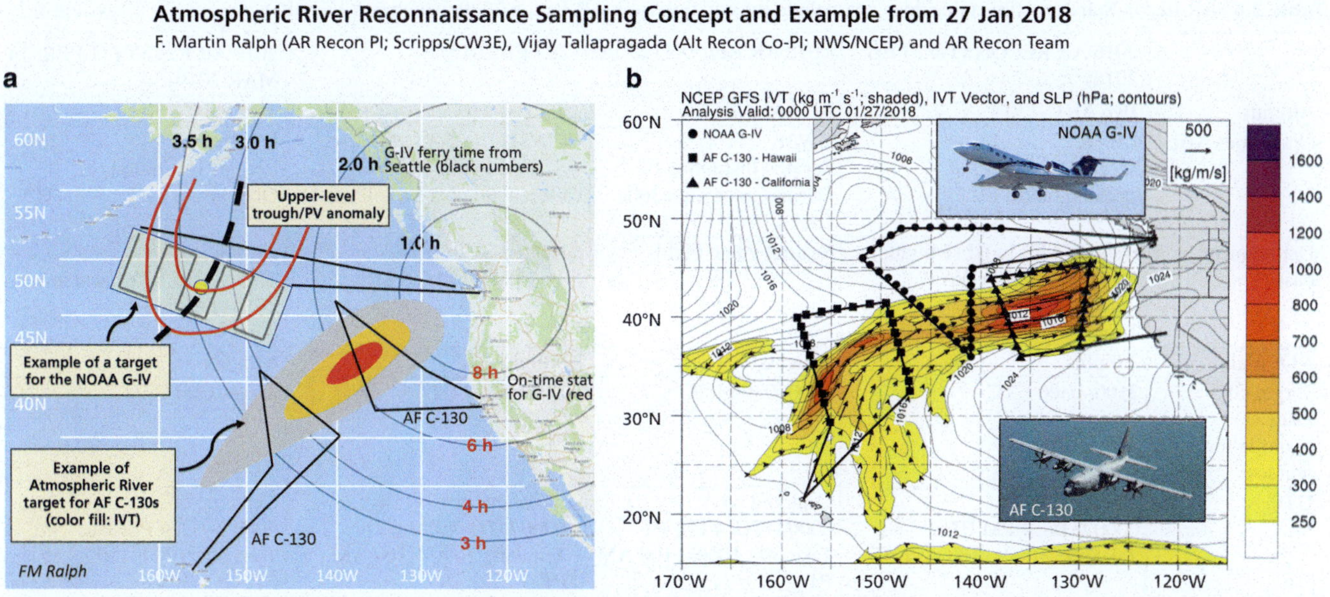

Fig. 3.28 (**a**) *Left* AR Recon targeting concept and example using three aircraft, executed on 27 Jan 2018. (**b**) *Right* In addition, the moist adjoint method is used to identify regions of large initial condition error impacts, which largely match the location of the AR

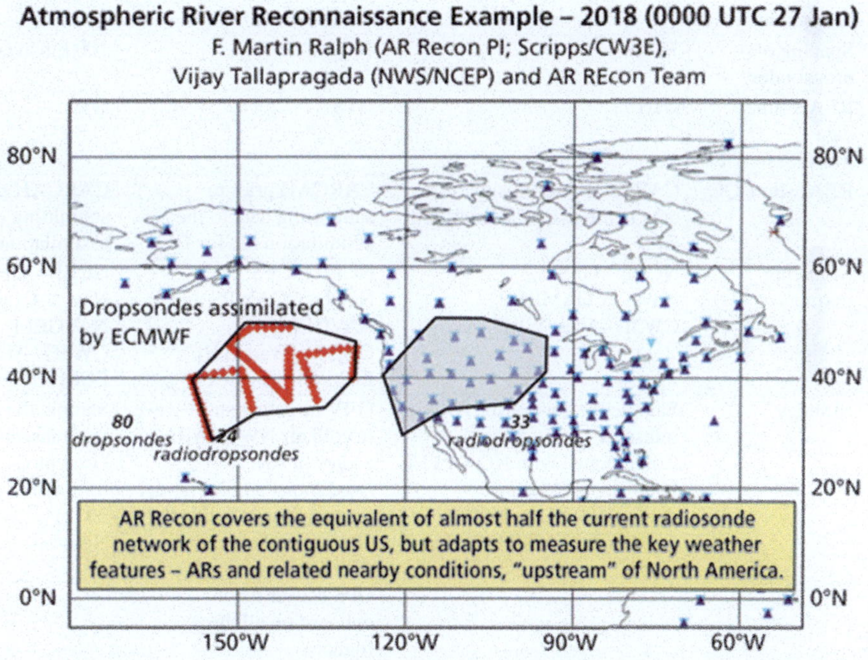

Fig. 3.29 Dropsonde locations for the first three-aircraft AR Recon mission, demonstrating the large geographic area covered

(Fig. 3.30). These data were collected from four types of aircraft, during six airborne field campaigns conducted over nearly 20 years.

Ralph et al. (2017) developed a method to composite these into a single cross-section, as well as to compare characteristics of them from the subtropical cases versus the midlatitude cases. In short, they found that the mean TIVT was more than 4.7×10^8 kg s^{-1}, which was >50% more than the mean of the first five cases before CalWater and AR Recon. The average TIVT is 2.6 times the discharge of the Amazon River into the Atlantic—meaning that ARs are, on average, the largest freshwater "rivers" on earth.

Because these 21 samples might still have suffered from not being representative, a companion study was undertaken to compare them to reanalysis representations of the same ARs, and then to a composite of thousands of ARs available from the broader reanalysis data set. Guan et al. (2018) presented the remarkable result that the TIVT from the reanalyses (NASA's MERRA) of the 21 same cases match the aircraft observations within 3%. Even more remarkably,

Table 3.6 AR Recon Summary (this table does not include information from AR Recon 2020, which had 17 IOPs)

	2016 AR RECON (1–21 FEB 2016)	2018 AR RECON (25 JAN–28 FEB 2018)		2019 AR RECON (1 FEB–14 MAR 2019)
Aircraft	Two AF C-130 s	NOAA G-IV	Two AF C-130 s	Two AF C-130 s
Operations base	CW3E Scripps Institution of Oceanography, La Jolla, CA	CW3E Scripps Institution of Oceanography, La Jolla, CA	CW3E Scripps Institution of Oceanography, La Jolla, CA	CW3E Scripps Institution of Oceanography, La Jolla, CA
Deployment bases	Hickam AFB in HI (3 flights) McChord AFB in WA (2 flights) Travis AFB in CA (1 flight)	Seattle, Paine Field	Hawaii Travis AFB in CA	Hawaii Travis AFB, CA (used in 2016 and 2018) San Diego, CA
# Of flights	3 flights	3 flights	6 flights	6 flights
Target GFS Data assimilation window	2100–0300 UTC	2100–0300 UTC	2100–0300 UTC	2100–0300 UTC
UTC	Centered on 0000 UTC, 4 PM PT Take offs:1000–1100 AM PT	Centered on 0000 UTC, 4 PM PT Take offs:1000–1100 AM PT	Centered on 0000 UTC, 4 PM PT Take offs:1000–1100 AM PT	Centered on 0000 UTC, 4 PM PT Take offs:1000–1100 AM PT
# Of storms	3 storms measured by both aircraft	3 storms with both C130s	4 storms were measured by both aircraft2 storms were measured by one aircraft each in late February	5 storms were measured by both aircraft1 storm was measured by one aircraft
# OF dropsondes	48 per flight on average	40 per flight on average	25 per flight on average	25 per flight on average
Spacing of dropsondes	70 km	100 km average	100 km average	100 km averages
REAL-time data transmission	GTS	GTS	GTS	GTS
REAL-time QC	CARCAH prior to submitting data to the assimilation file for GFS	CARCAH prior to submitting data to the assimilation file for GFS	CARCAH prior to submitting data to the assimilation file for GFS	CARCAH prior to submitting data to the assimilation file for GFS
Modeling partners	NCEP – GFS Navy – COAMPS CW3E – West WRF	NCEP – GFS Navy – COAMPS and NAVGEM CW3E – West WRF ECMWF	NCEP—GFS Navy—COAMPS and NAVGEM CW3E—West WRF6 ECMWF	NCEP—GFS and GFSFV3navy—COAMPS and NAVGEM CW3E—West WRF ECMWF
Strategy	Sample the core of the AR via transects Collaborated with NOAA's El Niño Rapid Response deployment of the NOAA G-IV and the NASA Global Hawk	G-IV targeting short-wave in cold air NW of AR, and/or in AR Used moist adjoint model for targeting (Doyle et al. 2012) in collaboration with NRL GPS sensor flown onboard on all three flights	Sample the core of the AR via transects Used moist adjoint model for targeting (Doyle et al. 2012) in collaboration with NRL	Sample the core of the AR via transects Used moist adjoint model for targeting (Doyle et al. 2012) in collaboration with NRL Used ensemble sensitivity methods for targeting, in collaboration with NCEP and SUNY Albany GPS sensor flown onboard

comparison *of the mean of the 21 cases with the mean of thousands of cases from reanalysis showed that the mean of the 21 cases matched those of all cases within 5%*. This comparison is shown in Sect. 3.6.

Finally, the compositing of the dropsondes themselves yielded a mean cross-section. By using reanalyses it was also possible to calculate the composite synoptic environment and horizontal structure of a mean AR. These methods and results are presented in Ralph et al. (2017) and are shown in Fig. 3.31. The AMS *Glossary of Meteorology* definition of "atmospheric river" uses this figure in the formal definition, which was developed by a committee with substantial community input through three open meetings (Ralph et al. (2018a) describe the process of developing the definition; it is also covered in Sect. 2.2.1 in Chap. 2).

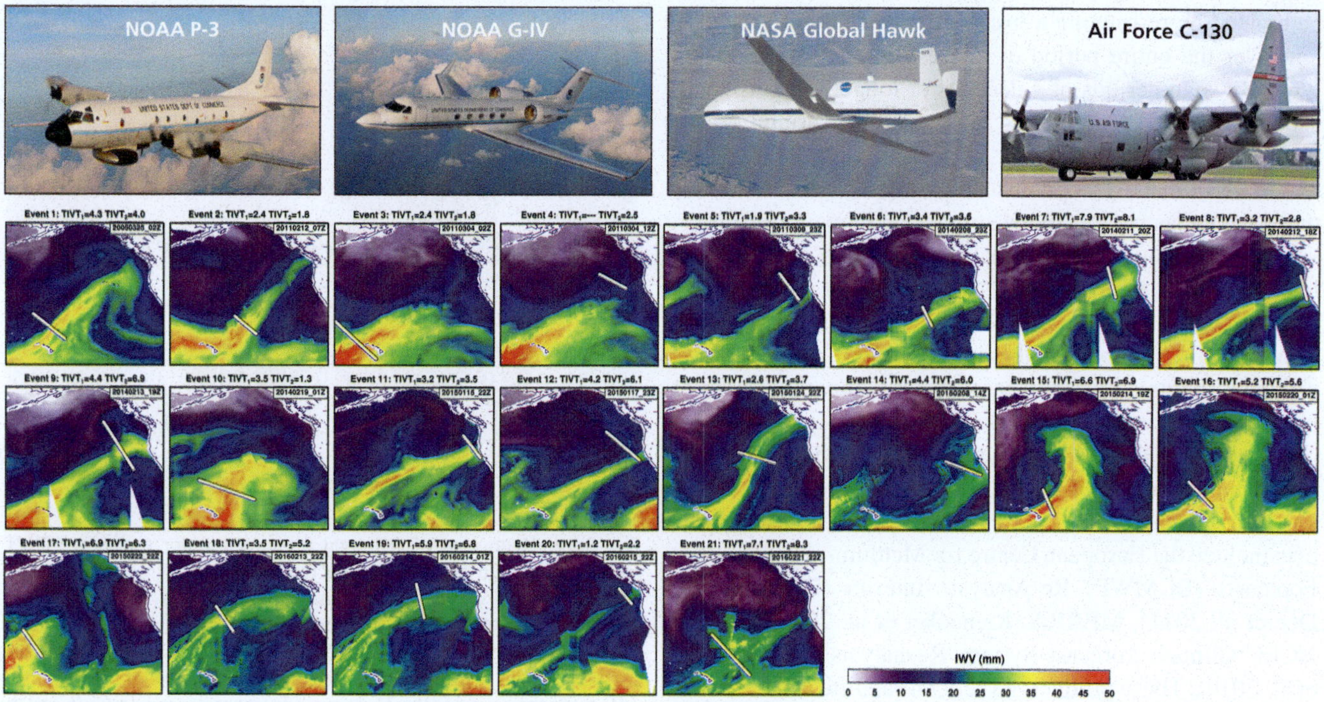

Fig. 3.30 Snapshots of each of the 21 aircraft-observed ARs are shown overlaid on satellite-observed integrated water vapor (IWV), with the baseline (*white line*) marking the location of aircraft track used in the analysis. Each of the four aircraft types used to collect these data over nearly 20 years is shown. (From Ralph et al. 2017)

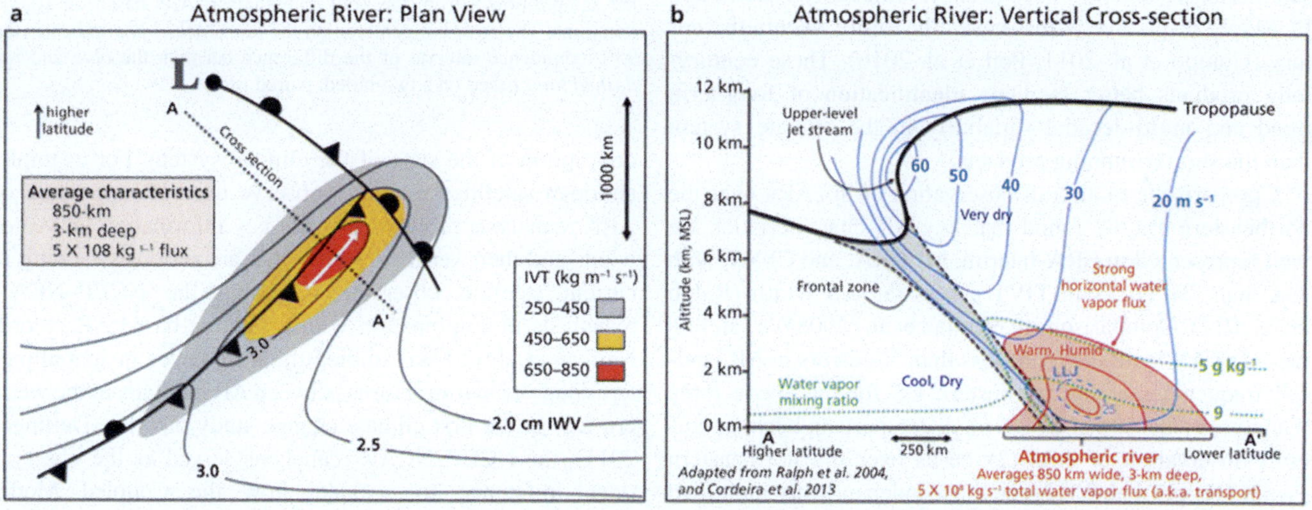

Fig. 3.31 Composite schematic of AR structure based on (**a**) aircraft observations of 21 ARs, and (**b**) used in the AMS *Glossary of Meteorology* definition of ARs. (From Ralph et al. 2017, 2018a)

3.6 ARs in Reanalyses

In weather forecasting, "analysis" is a technique for creating a gridded representation of the current state of the earth system (e.g., atmosphere, ocean, land surface, etc.) based on judiciously combining information from the forecast and observations. The analysis can then be used as initial conditions for producing the next forecast. The more complete spatio-temporal coverage of the analysis products provides an opportunity for the study of large-scale weather phenomena (such as ARs) that are difficult to be resolved by observations alone, especially before satellite products become routinely available. Since the forecast system constantly changes and improves over time, which may lead to spurious changes in the associated analysis products, a special version of analysis, namely, retrospective analysis (reanalysis), can be created by re-running a fixed version of the forecast system over the entire period of interest. Reanalysis products typically span periods of 4 decades or longer, making them especially useful for studying long-term AR variability and trends.

Different reanalysis products differ in their data assimilation systems as well as in the sources and amount of observational data injected into that system. The original and longest (1948 onward) reanalysis product that makes use of rawinsonde data is the NCEP–NCAR reanalysis (Kalnay et al. 1996). It is one of the most widely used reanalysis products, making it a useful baseline for comparing different types of observations and later generations of reanalysis products. More modern reanalysis products generally feature improved treatment of the global water cycle—one of the key shortcomings in the NCEP–NCAR product. These products include the European Centre for Medium-range Weather Forecasts (ECMWF) Re-Analysis Interim (ERA-Interim; Dee et al. 2011), MERRA (Rienecker et al. 2011), and the NCEP Climate Forecast System Reanalysis (CFSR; Saha et al. 2010). These products all rely on satellite observations, and the period of availability is therefore limited to the satellite era (1979 onward). To extend the coverage to earlier periods, reanalysis products that depend only on surface observations—which are longer-running than upper-air and satellite observations—have been developed by NOAA and ECMWF, respectively, that cover the entire twentieth century (Compo et al. 2011; Poli et al. 2016). These century-long products better facilitate identification of long-term trend and multi-decadal variability in the climate system than the shorter-running products.

Comparisons to aircraft observation of six ARs over the northeastern Pacific found that key AR characteristics are well represented in ERA-Interim, MERRA, and CFSR, with less than 7% errors in TIVT across the AR width (Ralph et al. 2012). Comparisons to Neiman et al. (2008a)'s satellite record of AR landfalls found less than 5% errors in AR landfall frequencies along the western US for the above three reanalyses (Jackson et al. 2016). A comparison between aircraft observation of ARs (21 cases over the northeastern Pacific; Ralph et al. 2017) and those independently identified and measured by a reanalysis-based detection algorithm (about 6000 cases over the same domain during the winters of 1979–2016) was conducted by Guan et al. (2018). They found the mean TIVT of the roughly 6000 (21) ARs identified in the ERA-Interim reanalysis to be within 5% (3%) of the 21 aircraft observations (Fig. 3.32). These studies indicate the key elements of ARs are generally well captured by the reanalysis products, although errors of different magnitudes still remain, depending on the particular product.

Compared to other types of data, the reanalysis products have played key roles in AR studies, given their more complete spatio-temporal coverage and more comprehensive description of the state of the climate system. For example, although satellites routinely observe the IWV signature of ARs, reanalysis products provide key information on vector winds and their vertical profiles that are seriously lacking in current satellite observations. Using the NCEP–NCAR reanalysis in combination with a satellite-based AR record, Neiman et al. (2008a) studied the differences in spatial patterns and vertical structures between AR landfalls in the western US. In the first climate change study on ARs (Dettinger 2011), the NCEP–NCAR reanalysis served as the observational reference to evaluate how the Coupled Model Intercomparison Project (CMIP) Phase 3 models represented historical AR activity in central California. Using a simple ARDM based on IWV and 925-hPa wind at a single model grid cell, they showed varied representation of winter AR frequency and probability distributions of AR IWV and upslope wind by seven CMIP3 models. Lavers et al. (2012) developed a more complex method based on IVT to detect ARs along a pre-determined cross-section (e.g., the west coast of Europe or North America) and applied it to five reanalysis products. Generally good agreement in winter AR frequency in Britain was found between the different reanalysis products. Later studies more fully used the spatial IVT patterns provided by

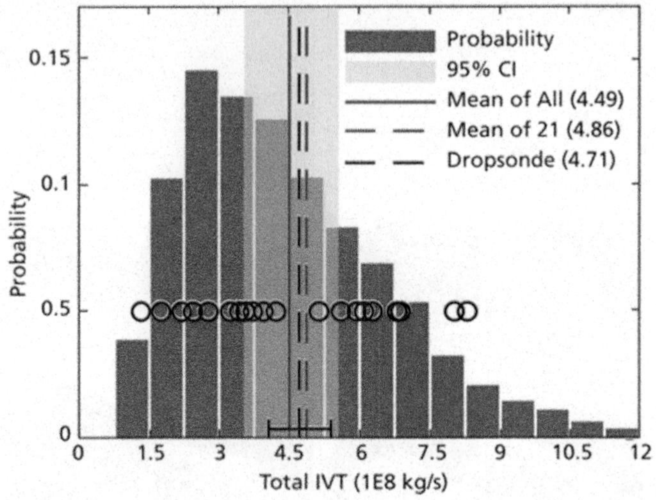

Fig. 3.32 Histogram of AR total integrated vapor transport (TIVT) (10^8 kg s^{-1}) based on all ARs detected in ERA-Interim over the northeastern Pacific (AR centroids within 163.4–124.6°W, 23–46.4°N) during 15 January to 25 March of 1979–2016 (*gray bars*). From Guan et al. (2018). Also shown are the mean AR TIVT based on all reanalysis ARs that contributed to the histogram (*red solid*), the subset of the reanalysis ARs that correspond to the 21 dropsonde transects (*red dashed*), and the observed value based on the 21 dropsonde transects as reported in Ralph et al. (2017) (*blue dashed* for the mean, and *blue circles* for individual transects). The mean AR TIVT value is also indicated in the figure legend for each sample. *Red shading* indicates the 95% confidence interval of the mean reanalysis AR TIVT for a random 21-member sample drawn from the pool of reanalysis ARs based on 10,000 iterations. The error bar centered on the *blue dashed line* indicates the 95% confidence interval of the difference between the *blue* and *red dashed lines* based on a two-tailed, paired t-test

reanalyses in their development of AR detection algorithms, and provided regional- (Rutz et al. 2014) to global-scale (Guan and Waliser 2015) climatological description of AR frequency, geometry, intensity, and other characteristics. The AR climatology from three reanalysis products (including their differences as a rough measure of reanalysis uncertainty) provided the observational reference in Guan and Waliser (2017) to evaluate a suite of 24 global weather and climate models in terms of 17 AR metrics. In these studies, reanalysis products provided IVT or other key information on ARs not available from other data products, or not available on spatio-temporal scales similar to reanalyses. Besides the above uses in AR identification, characterization, and model evaluation, the reanalysis products—via their assimilation of satellite and other types of observations—provide a way for these observations to collectively contribute to AR studies.

3.7 AR Identification

A majority of peer-reviewed publications that discuss AR science begin their introductions by first defining ARs, and then acknowledging the seminal work of Zhu and Newell (1998). These authors were the first to use the term "atmospheric rivers" to describe filamentary corridors of intense, vertically integrated water vapor transport. In their analysis, "AR fluxes" occur whenever IVT exceeds a certain value relative to the zonal mean. They showed that ARs, previously referred to as "tropospheric rivers" by Newell et al. (1992), are responsible for ~90% of poleward water vapor transport in the mid-latitudes, despite occupying <10% of earth's longitudinal circumference. In the decades since, as the study of ARs has expanded and matured, authors seeking to improve understanding of these features and their role in weather and climate have developed numerous and differing but largely complementary methods to identify ARs. Indeed, many sections in Chap. 4—which focuses on describing AR characteristics over specific regions—use different methods, based on which is best suited to a particular region.

These methods differ in a wide variety of ways, exemplified in the following list:

- First, authors choose a characteristic variable of ARs on which to perform thresholding: typically, this is either vertically-integrated water vapor (IWV) or vapor transport (IVT).
- Next, the authors decide on whether the threshold will be held constant, or if it will be relative: changing as a function of location, season, and/or climatology.
- Then, they must choose the actual thresholding value (e.g., IVT ≥250 kg m^{-1} s^{-1} for a constant value, or IVT ≥85th percentile for a seasonally/climatologically varying value).
- They must then decide on the geometric criteria (if any) for AR identification: length, width, aspect ratio, consideration of feature curvature, etc.
- Some authors take further steps, such as identifying the centroid and directional orientation of the AR.
- The next decision is how to track the ARs: in a Eulerian sense (looking at single time slices of spatial fields) or a Lagrangian sense (following the AR "object")?
- Some authors include a temporal criterion, where either the AR "object" or conditions over a point must be consistently identified over a number of time-steps.
- Some authors use machine learning techniques to identify ARs without requiring a specific exact threshold or geometric requirements.
- Some algorithms end in filters where non-AR phenomena—which still may have similar features to ARs (e.g., tropical cyclone remnants or cutoff lows)—are filtered out.
- Finally, they must choose whether to identify ARs subjectively by analysis, or objectively using an algorithm.
- Not surprisingly, this process produces a tremendous number of permutations, and the resulting literature attests to the diversity of choices authors have made.

This section outlines the diversity of AR identification methods in use: their defining characteristics, how they differ, and the rationale for why certain approaches are chosen. Some minor historical background is included, where useful. Selected studies are highlighted as examples of certain methods, but each method is not described in detail; this is by no means an exhaustive list of AR identification methods in use. The end of this section briefly discusses current community efforts to provide publicly distributed inter-comparisons of all available algorithms, applied to common data sets, so the larger community has guidance and appropriate context when choosing or designing a detection algorithm for a given science question or region of interest. These efforts also address the physical reasons behind quantifiable differences in relevant characteristics within AR categories produced by different algorithms, such as their duration, intensity, and frequency.

The first set of criteria used to identify ARs as spatial objects (as opposed to the fluxes Zhu and Newell identified) was developed in a study by Ralph et al. (2004), based on satellite data and aircraft observations taken during the CALJET field campaign. These authors defined ARs as features ≥2000 km in length and ≤1000 km in width, with IWV >20 mm throughout. These criteria served as the basis for many studies that followed, including the 8- and 11-year climatologies of AR landfalls along the North American west coast produced by Neiman et al. (2008a) and Dettinger (2011), respectively. These studies subjectively examined elongated plumes of IWV over the northeastern Pacific, and identified events fitting the Ralph et al. (2004) criteria as ARs. The reason for using IWV—as opposed to IVT—is that it was readily available from satellite observations, whereas IVT, which relies on wind data, was not. These climatologies subsequently served as the foundation for beginning to assess AR contributions to precipitation, on the basis of those making landfall along the US West Coast (Ralph et al. 2005a, 2006; Neiman et al. 2008a, 2009); Guan et al. 2010; Dettinger 2011; Rutz and Steenburgh 2012). Shortly after, Wick et al. (2013a) developed an objective, algorithmic technique for determining whether such plumes of IWV fit the Ralph et al. (2004) criteria and should be identified as ARs. This method proved useful in validating the model skill associated with landfalling ARs (Wick et al. 2013b) (see Chap. 6).

Up until this point, most studies examining ARs had focused on ARs either over the Pacific Ocean or that affected the immediate US West Coast. The availability of SSM/I IWV data over the oceans and the concentration of population along the US West Coast made this a logical step. However, researchers have recently been shifting their focus toward IVT as the defining characteristic of ARs, and using reanalysis products to search for ARs over the entire globe. As discussed in Chap. 8, reanalyses combine state-of-the-art observations with modeling techniques to have consistent data sets available at regular time-steps and spatial grid-boxes. This has been made possible, in part, by data collected at AROs (previously described in Sect. 3.3), a number of which are situated along the US West Coast. Ralph et al. (2013b) used data from an ARO located at Bodega Bay (BBY), California, to show remarkable correlations between the upslope component of water vapor flux and precipitation at the nearby observational site of Cazadero (CZD). Correlations between upslope water vapor flux and storm-total runoff were shown to be equally impressive.

Next, Rutz et al. (2014) showed that IVT is better correlated than IWV with cool-season, western US precipitation. Wanting to build upon previous studies that used an IWV threshold of 20 mm along the US West Coast, they also showed that an IVT threshold of 250 kg m^{-1} s^{-1} identified similar AR events along the immediate coast. This allows the analysis to be extended inland where IWV >20 mm is climatologically very rare. Based on these findings, Rutz et al. (2014) identified features ≥2000 km in length, with IVT >250 kg m^{-1} s^{-1} throughout as ARs (unlike Ralph et al. (2004), they did not set a width requirement, since features identified using IVT tend to be more diverse in their shapes). After using reanalysis data to calculate IVT and objectively identify ARs, they calculated the mean duration of AR events at BBY. These results (~20 h) matched those of the observational study by Ralph et al. (2013b), increasing confidence in the use of reanalysis data to investigate AR climatology.

Gershunov et al. (2017) use thresholds for IWV (15 mm) and IVT (250 kg m^{-1} s^{-1}), as well as length and width criteria, to identify ARs. Their mixed approach to AR identification is unique, and produces results for various AR characteristics that (not surprisingly) fall between those of methods that use either IWV or IVT alone.

The AR identification methods discussed above are based on choosing a thresholding variable (e.g., IVT) and a *constant* value (e.g., 250 kg kg m^{-1} s^{-1}) that must be exceeded through an area that meets certain geometric requirements (e.g., 2000 km in length). A number of recent studies have used the exceedance of *relative*, location- and/or seasonally dependent, values (i.e., 85th percentile of climatological IVT) to identify ARs.

A study by Lavers et al. (2012) was one of the first to use a percentile-based AR identification method in analyzing AR impacts on Great Britain (GB). These researchers used the ~85th percentile of IVT along a transect west of GB to identify periods when ARs contributed to floods over GB. A study by Payne and Magnusdottir (2014) follows these criteria to examine the dynamic characteristics of ARs making landfall along the US West Coast. A recent study by Mundhenk et al. (2016) follows similarly, but allows for dynamic adjustment of an anomaly-based threshold on the basis of the study domain. Furthermore, recent studies have noted the effects of ARs on polar regions such as Greenland (Neff et al. 2014) and Antarctica (Gorodetskaya et al. 2014). To do so, the latter of these identifies ARs using modified IWV criteria, which is similar to a percentile-based method in that it accounts for the typically cold and dry air masses over these areas.

There are advantages and disadvantages to using a percentile-based method to identify ARs. One advantage is that it allows for the diagnosis of coherent features that are characterized by anomalous water vapor transport in areas where a fixed threshold might be too high (i.e., restrictive), such as continental interiors or polar regions. One disadvantage is that these features, although anomalous relative to local and/or seasonal climatology, can sometimes be disconnected from the mid-latitude synoptic/dynamic processes typically thought of as being associated with ARs (and described in more detail in Chap. 2).

Aware of these issues, and intending to provide the first global AR detection algorithm and databases, Guan and Waliser (2015) combined a percentile-based method (85th percentile of location- and seasonally-dependent climatological IVT), with a minimum IVT threshold (100 kg m^{-1} s^{-1}) to identify ARs. This minimum threshold acts to exclude regions of very weak IVT, even if they are nonetheless quite anomalous at a certain time and place. Their method requires that an AR be ≥2000 km in length and have an aspect ratio (length/width) ≥2. This aspect ratio criterion is a useful compromise between methods that require ARs to be <1000 km in width (e.g., Ralph et al. 2004) and methods with no width requirement (e.g., Rutz et al. 2014), thus allowing some flexibility in identifying elongated features of various sizes as ARs. Their method facilitated a consistent examination of the basic AR characteristics on a global scale, and has been the basis for global AR studies involving AR climatology and variability, global model evaluation (Guan and Waliser 2017), and examination of AR-related extreme events across the globe (Waliser and Guan 2017), as well as for studying the global impacts of climate change on ARs (Espinoza et al. 2018).

Recently, the research community has taken important steps toward greater agreement on the definition of ARs, and the methods best suited to identify them. The first step was the creation of an AR definition for the AMS's *Glossary of Meteorology*, which provides guidance on which features characterize an AR (Ralph et al. 2018a). The second step was the initiation of the Atmospheric River Tracking Method Intercomparison Project (ARTMIP), which is an ad hoc, international effort that aims to quantify the uncertainties in AR climatology that arise solely from using different methods, as well as describe the advantages and disadvantages these methods present for given science questions and regions of interest (Shields et al. 2018). Within the description of the experimental methodology provided in Shields et al. (2018), a 1-month proof-of-concept study in February 2017 showed differences in AR counts between different methodologies (Fig. 3 from that paper). An ARTMIP Early-Start study (Ralph et al. 2018b) had begun before ARTMIP's creation, but then adopted some of its strategies, although it focused on one location and fewer ARDMs. A subset of ARTMIP participants investigated a cross-section of different detection algorithms applied to various reanalysis data sets at the gridbox that encompassed the Bodega Bay ARO (see Sect. 3.3.), and included the observational data set begun in Ralph et al. (2013b). This study illuminated some of the physical reasons behind differences in catalogs output by different methodologies. Notably, the annual average AR count is the characteristic that was most sensitive to detection method, and varied by a factor of 2; this was found to depend on the strictness of the geometric criteria (Table 5 from that paper). There is much less difference (on the order of 10%) in relevant AR characteristics such as duration and peak intensity (Ralph et al. 2018b). Furthermore, the study showed that choice of reanalysis data set mattered much less than the choice of detection method, even with a resolution difference of as much as 2°. This corroborated other studies that found little difference between AR catalogs produced using different reanalyses (Lavers et al. 2012; Guan and Waliser 2017; Guan et al. 2018).

It is important to note that many researchers have also developed methods to attribute precipitation to ARs or AR conditions, and these methods differ significantly, in addition to methods of identifying ARs themselves. This suggests a large amount of uncertainty in the fraction of total precipitation attributable to ARs (cf. Huning et al. 2017)—a quantity that will be very important as AR climatology and the distribution of precipitation alters as a result of climate change.

References

Alishouse JC, Snyder SA, Vongsathorn J et al (1990) Determination of oceanic total precipitable water from the SSM/I. IEEE Trans Geosci Remote Sens 28:811–816

Anthes RA, Bernhardt PA, Chen Y et al (2008) The COSMIC/FORMOSAT-3 Mission: early results. Bull Am Meteorol Soc 89:313–333

Ault AP, Williams CR, White AB et al (2011) Detection of Asian dust in California orographic precipitation. J Geophys Res 116:D16205

Baker WE, Atlas R, Cardinali C et al (2014) Lidar-measured wind profiles. The missing link in the global observing system. Bull Am Meteorol Soc 95:543–564

Bergeron T (1965) On the low-level redistribution of atmospheric water caused by orography. In: Proc Int Conf on Cloud Physics, Tokyo, Japan. IAMAP/WMO, pp 96–100

Bevis M, Businger S, Herring TA et al (1992) GPS meteorology: remote sensing of the atmospheric water vapor using the global positioning system. J Geophys Res 97(D14):15787–15801

Blamey RC, Ramos AM, Trigo RM et al (2018) The influence of atmospheric rivers over the South Atlantic on winter rainfall in South Africa. J Hydrometeorol 19:127–142. https://doi.org/10.1175/JHM-D-17-0111.1

Boukabara S–A, Garrett K (2018) Tropospheric moisture sounding using microwave imaging channels: application to GCOM-W1/AMSR2. IEEE Trans Geosci Remote Sens 56(9):5537–5549

Boukabara S–A, Garrett K, Chen W et al (2011) MiRS: an all-weather 1DVAR satellite data assimilation and retrieval system. IEEE Trans Geosci Remote Sens 49:3249–3883

Browning KA, Pardoe CW (1973) Structure of low-level jet streams ahead of mid-latitude cold fronts. Q J R Meteorol Soc 99(422):619–638. https://doi.org/10.1002/qj.4970994330

Cannon F, Ralph FM, Wilson AM et al (2017) GPM satellite radar measurements of precipitation and freezing level in atmospheric rivers: comparison with ground-based radars and reanalyses. J Geophys Res-Atmos 122:12747–12764. https://doi.org/10.1002/2017JD027355

Cannon F, Hecht CW, Cordeira JM et al (2018) Synoptic and mesoscale forcing of Southern California extreme precipitation. J Geophys Res-Atmos 123(24):13714–13730. https://doi.org/10.1029/2018JD029045

Carter DA, Gage KS, Ecklund WL et al (1995) Developments in UHF lower tropospheric wind profiling at NOAA's aeronomy laboratory. Radio Sci 30:977–1001

Chen XM, Chen SH, Haase JS et al (2018) The impact of airborne radio occultation observations on the simulation of Hurricane Karl (2010). Mon Wea Rev 146:329–350

Compo GP et al (2011) The twentieth century reanalysis project. Q J Roy Meteorol Soc 137:1–28. https://doi.org/10.1002/qj.776

Cordeira JM, Ralph FM, Moore BJ (2013) The development and evolution of two atmospheric rivers in proximity to western North Pacific tropical cyclones in October 2010. Mon Wea Rev 141:4234–4255

Cordeira JM, Ralph FM, Martin A, Natalie G, Ryan Spackman J, Neiman PJ, Rutz JJ, Pierce R (2017) Forecasting atmospheric rivers during CalWater 2015. Bull Am Meteorol Soc 98(3):449–459

Creamean JM, Suski DJ, Rosenfeld D et al (2013) Dust and biological aerosols from the Sahara and Asia influence precipitation in the Western U.S. Science 339:1572–1578

Creamean JM, Lee C, Hill TC et al (2014) Chemical properties of insoluble precipitation residue particles. J Aerosol Sci 76:13–27

Creamean JM, Ault AP, White AB (2015) Impact of interannual variations in sources of insoluble aerosol species on orographic precipitation over California's Central Sierra Nevada. Atmos Chem Phys 15:6535–6548

Dee DP, Uppala SM, Simmons AJ et al (2011) The ERA-interim reanalysis: configuration and performance of the data assimilation system. Q J R Meteorol Soc 137:553–597. https://doi.org/10.1002/qj.828

DeFlorio MJ, Waliser DE, Guan B et al (2018) Global assessment of atmospheric river prediction skill. J Hydrometeorol 19, 409–426. https://doi.org/10.1175/JHM-D-17-0135.1

DeFlorio M, Waliser DE, Guan B et al (2019) Global Evaluation of Atmospheric River Subseasonal Prediction Skill. Clim Dyn 52, 3039–3060. https://doi.org/10.1007/s00382-018-4309-x

Dettinger M (2011) Climate change, atmospheric rivers, and floods in California – a multimodel analysis of storm frequency and magnitude changes. J Am Water Resour Assoc 47(3):514–523

Dettinger MD, Ralph FM, Das T et al (2011) Atmospheric rivers, floods and the water resources of California. Water 3(2):445–478. http://www.mdpi.com/2073-4441/3/2/445/

Dole RM, Spackman JR, Newman M et al (2018) Advancing science and services during the 2015-16 El Niño: the NOAA El Niño rapid response field campaign. Bull Am Meteorol Soc 99:975–1001. https://doi.org/10.1175/BAMS-D-16-0219.1

Doyle JD, Reynolds CA, Amerault C et al (2012) Adjoint Sensitivity and Predictability of Tropical Cyclogenesis. J Atmos Sci 69:3535–3557. https://doi.org/10.1175/JAS-D-12-0110.1

Doyle JD, Amerault C, Reynolds CA et al (2014) Initial condition sensitivity and predictability of a severe extratropical cyclone using a moist adjoint. Mon Wea Rev 142:320–342. https://doi.org/10.1175/MWR-D-13-00201.1

Duan J, Bevis M, Fang P et al (1996) GPS meteorology: direct estimation of the absolute value of precipitable water. J Appl Meteorol 35:830–838

Dunne T, Black RD (1970) An experimental investigation of runoff production in permeable soil. Water Resour Res 6:478–490

Espinoza V, Waliser DE, Guan B et al (2018) Global Analysis of Climate Change Projection Effects on Atmospheric Rivers. Geophys Res Lett. https://doi.org/10.1029/2017GL076968

Fan J, Leung LR, DeMott PJ et al (2014) Aerosol impacts on California winter clouds and precipitation during CalWater 2011: local pollution versus long-range transported dust. Atmos Chem Phys 14(1):81

Garreaud R (2013) Warm winter storms in Central Chile. J Hydrometeorol 14:1515–1534

Gelaro R, Langland RH, Pellerin S et al (2010) The THORPEX observation impact intercomparison experiment. Mon Wea Rev 138:4009–4025

Gelaro R, McCarty W, Suárez MJ et al (2017) The modern-era retrospective analysis for research and applications, version 2 (MERRA-2). J Clim 30:5419–5454

Gershunov A, Shulgina R, Ralph FM et al (2017) Assessing the climate-scale variability of atmospheric rivers affecting western North America. Geophys Res Lett 44:7900–7908. https://doi.org/10.1002/2017GL074175

Ghebrebrhan O (1990) Full decoding of truncated ranges for ST/MST radar applications. IEEE Trans Geosci Remote Sens 28:14–18

Gorodetskaya IV, Tsukernik M, Claes K et al (2014) The role of atmospheric rivers in anomalous snow accumulation in East Antarctica. Geophys Res Lett 41:6199–6206

Guan B, Waliser DE (2015) Detection of atmospheric rivers: evaluation and application of an algorithm for global studies. J Geophys Res Atmos 120:12514–12535. https://doi.org/10.1002/2015JD024257

Guan B, Waliser DE (2017) Atmospheric rivers in 20 year weather and climate simulations: a multimodel, global evaluation. J Geophys Res Atmos 122:5556–5581. https://doi.org/10.1002/2016JD026174

Guan B, Molotch NP, Waliser DE et al (2010) Extreme snowfall events linked to atmospheric rivers and surface air temperature via satellite measurements. Geophys Res Lett 37:L20401. https://doi.org/10.1029/2010GL044696

Guan B, Waliser DE, Ralph FM (2018) An inter-comparison between reanalysis and dropsonde observations of the total water vapor transport in individual atmospheric rivers. J Hydrometeorol 19:321–337. https://doi.org/10.1175/JHM-D-17-0114.1

Guttman SI, Sahm SR, Benjamin SG et al (2004) Rapid retrieval and assimilation of ground based GPS precipitable water observations at the NOAA forecast systems laboratory: impact on weather forecasts. J Meteorol Soc Japan 82:351–360. https://doi.org/10.2151/jmsj.2004.351

Haase JS, Murphy BJ, Muradyan P et al (2014) First results from an airborne GPS radio occultation system for atmospheric profiling. Geophys Res Lett 41:1759–1765

Han M, Braun SA, Persson PO et al (2009) Alongfront variability of precipitation associated with a midlatitude frontal zone: TRMM observations and MM5 simulation. Mon Wea Rev 137:1008–1028

Hollinger JP, Peirce JL, Poe GA (1990) SSM/I instrument evaluation. IEEE Trans Geosci Remote Sens 28:781–790

Huning LS, Margulis SA, Guan B et al (2017) Implications of detection methods on characterizing atmospheric river contribution to seasonal snowfall across Sierra Nevada, USA. Geophysical Res Lett 44:10445–10453. https://doi.org/10.1002/2017GL075201

Jackson DL, Hughes H, Wick GA (2016) Evaluation of landfalling atmospheric rivers along the U.S. West Coast in reanalysis data sets. J Geophys Res Atmos 121:2705–2718. https://doi.org/10.1002/2015JD024412

Johnston PE, Jordan JR, White AB (2017) The NOAA FM-CW snow-level radar. J Atmos Ocean Technol 34:249–267

Jorgensen DP, Pu Z, Persson PO et al (2003) Variations associated with cores and gaps of a Pacific narrow cold frontal rainband. Mon Wea Rev 131:2705–2729

Kalnay E, Kanamitsu M, Kistler R et al (1996) The NMC/NCAR 40-year reanalysis project. Bull Am Meteorol Soc 77:437–471. https://doi.org/10.1175/1520-0477(1996)077<0437:TNYRP>2.0.CO;2

Kidder SQ, Jones AS (2007) A blended satellite total precipitable water product for operational forecasting. J Atmos Ocean Technol 24:74–81

Kingsmill DE, Neiman PJ, Ralph FM et al (2006) Synoptic and topographic variability of northern California precipitation characteristics in landfalling winter storms observed during CALJET. Mon Wea Rev 134:2072–2094

Kingsmill DE, Neiman PJ, Moore BJ et al (2013) Kinematic and thermodynamic structures of Sierra barrier jets and overrunning atmospheric rivers during a landfalling winter storm in northern California. Mon Wea Rev 141:2015–2036. https://doi.org/10.1175/MWR-D-12-00277.1

Kingsmill DE, Neiman PJ, White AB (2016) Microphysics regime impacts on the relationship between orographic rain and orographic forcing in the coastal mountains of northern California. J Hydrometeorol 17:2905–2922

Kingston DG, Lavers DA, Hannah DM (2016) Floods in the Southern Alps of New Zealand: the importance of atmospheric rivers. Hydrol Process 30:5063–5070. https://doi.org/10.1002/hyp.10982

Kren AC, Cucurull L, Wang H (2018) Impact of UAS Global Hawk dropsonde data on tropical and extratropical cyclone forecasts in 2016. Weather Forecast 33:1121–1141

Kunkee DB, Poe GA, Boucher DJ et al (2008) Design and evaluation of the first special sensor microwave imager/sounder. IEEE Trans Geosci Remote Sens 46:863–883

Lavers DA, Villarini G, Allan RP et al (2012) The detection of atmospheric rivers in atmospheric reanalyses and their links to British winter floods and the large-scale climatic circulation. J Geophys Res 117:D20106. https://doi.org/10.1029/2012JD018027

Lim S, Cifelli R, Chandrasekar V et al (2013) Precipitation classification and quantification using X-band dual-polarization weather radar: application in the hydrometeorology Testbed. J Atmos Ocean Technol 30:2108–2120

Lojou J–Y, Benard R, Eymard L (1994) A simple method for testing brightness temperatures from satellite microwave radiometers. J Atmos Ocean Technol 11:387–400

Ma Z, Kuo Y-H, Ralph FM et al (2011) Assimilation of GPS radio occultation data for an intense atmospheric river with the NCEP regional GSI system. Mon Wea Rev 139:2170–2183

Mahoney K, Jackson DL, Neiman P (2016) Understanding the role of atmospheric rivers in heavy precipitation in the Southeast United States. Mon Wea Rev 144:1617–1632

Martin AC, Cornwell GC, Atwood SA et al (2017) Transport of pollution to a remote coastal site during gap flow from California's interior: impacts on aerosol composition, clouds, and radiative balance. Atmos Chem Phys 17:1491–1509

Martin A, Ralph FM, Demirdjian R et al (2018) Evaluation of atmospheric river predictions by the WRF model using aircraft and regional Mesonet observations of orographic precipitation and its forcing. J Hydrometeorol 19:1097–1113. https://doi.org/10.1175/JHM-D-17-0098

Martner BE, Yuter SE, White AB et al (2008) Raindrop size distributions and rain characteristics in California coastal rainfall for periods with and without a radar bright band. J Hydrometeorol 9:408–425

Matrosov SY (2013) Characteristics of landfalling atmospheric rivers inferred from satellite observations over the eastern North Pacific Ocean. Mon Wea Rev 141(11):3757–3768. https://doi.org/10.1175/MWR-D-12-00324.1

Matrosov SY, Clark KA, Kingsmill DE (2007) A polarimetric radar approach to identify rain, melting-layer, and snow regions for applying corrections to vertical profiles of reflectivity. J Appl Meteorol Climatol 46:154–166

Morss RE, Ralph FM (2007) Use of information by National Weather Service forecasters and emergency managers during CALJET and PACJET-2001. Wea Forecast 22:539–555

Mundhenk B, Barnes EA, Maloney ED (2016) All-season climatology and variability of atmospheric rivers over the North Pacific. J Clim. https://doi.org/10.1175/JCLI-D-15-0655.1

Neff W, Compo GP, Ralph FM et al (2014) Continental heat anomalies and the extreme melting of the Greenland ice surface in 2012 and 1889. J Geophys Res Atmos 119:6520–6536

Neiman PJ, Ralph FM, White AB et al (2002) The statistical relationship between upslope flow and rainfall in California's coastal mountains: observations during CALJET. Mon Wea Rev 130:1468–1492

Neiman PJ, Ralph FM, Persson PO et al (2004) Modification of fronts and precipitation by coastal blocking during an intense landfalling winter storm in Southern California: observations during CALJET. Mon Wea Rev 132:242–273

Neiman PJ, Wick GA, Ralph FM et al (2005) Wintertime nonbrightband rain in California and Oregon during CALJET and PACJET: geographic, interannual, and synoptic variability. Mon Wea Rev 133:1199–1223

Neiman PJ, Ralph FM, White AB et al (2006) A multiwinter analysis of channeled flow through a prominent gap along the Northern California coast during CALJET and PACJET. Mon Wea Rev 134:1815–1841

Neiman PJ, Ralph FM, Wick GA et al (2008a) Meteorological characteristics and overland precipitation impacts of atmospheric rivers affecting the West Coast of North America based on eight years of SSM/I satellite observations. J Hydrometeorol 9:22–47

Neiman PJ, Ralph FM, Wick GA et al (2008b) Diagnosis of an intense atmospheric river impacting the Pacific Northwest: storm summary and offshore vertical structure observed with COSMIC satellite retrievals. Mon Weather Rev 136:4398–4420

Neiman PJ, White AB, Ralph FM et al (2009) A water vapour flux tool for precipitation forecasting. Proc Inst Civil Eng Water Manag 162:83–94

Neiman PJ, Schick LJ, Ralph FM et al (2011) Flooding in Western Washington: the connection to atmospheric rivers. J Hydrometeorol 12:1337–1358

Neiman PJ, Ralph FM, Moore BJ et al (2013a) The landfall and inland penetration of a flood-producing atmospheric river in Arizona. Part 1: observed synoptic-scale, orographic, and hydrometeorological characteristics. J Hydrometeorol 14:460–484

Neiman PJ, Hughes M, Moore BJ et al (2013b) Sierra barrier jets, atmospheric rivers, and precipitation characteristics in Northern California: a composite perspective based on a network of wind profilers. Mon Wea Rev 141:4211–4233

Neiman PJ, Wick GA, Moore BJ et al (2014) An airborne study of an atmospheric river over the subtropical Pacific during WISPAR: Dropsonde budget-box diagnostics, and precipitation impacts in Hawaii. Mon Wea Rev 142:3199–3223

Neiman PJ, Moore BJ, White AB et al (2016) An airborne and ground-based study of a long-lived and intense atmospheric river with mesoscale frontal waves impacting California during CalWater-2014. Mon Wea Rev 144:1115–1144

Neiman PJ, Gaggini N, Fairall CW et al (2017) An analysis of coordinated observations from NOAA's Ronald H. Brown ship and G-IV aircraft in a landfalling atmospheric river over the North Pacific during CalWater-2015. Mon Wea Rev 145:3647–3669

Newell RE, Newell NE, Zhu Y et al (1992) Tropospheric rivers? – a pilot study. Geophys Res Lett 19:2401–2404

Nieman SJ, Menzel WP, Hayden CM et al (1997) Fully automated cloud-drift winds in NESDIS operations. Bull Am Meteorol Soc 78:1121–1133

Payne AE, Magnusdottir G (2014) Dynamics of landfalling atmospheric rivers over the North Pacific in 30 years of MERRA reanalysis. J Clim 27(18):7133–7150. https://doi.org/10.1175/JCLI-D-14-00034.1

Persson PO, Neiman PJ, Walter B et al (2005) Contributions from California coastal-zone surface fluxes to heavy coastal precipitation: a CALJET case study during the strong El Niño of 1998. Mon Wea Rev 133:1175–1198

Petty GW (1994) Physical retrievals of over-ocean rain rate from multichannel microwave imaging. Part II: algorithm implementation. Meteorog Atmos Phys 54:101–122

Poli P et al (2016) ERA-20C: an atmospheric reanalysis of the 20th century. J Clim 29(11):4083–4097. https://doi.org/10.1175/JCLI-D-15-0556.1

Ralph FM, Persson POG, Reynolds D et al (1999) The California Land-falling Jets Experiment (CALJET): Objectives and design of a coastal atmosphere–ocean observing system deployed during a strong El Niño. In: Preprints, Third Symp. on Integrated Observing Systems, Dallas, TX. Amer Meteor Soc, pp 78–81

Ralph FM, Neiman PJ, Kingsmill DE et al (2003) The impact of a prominent rain shadow on flooding in California's Santa Cruz Mountains: a CALJET case study and sensitivity to the ENSO cycle. J Hydrometeor 4:1243–1264

Ralph FM, Neiman PJ, Wick GA (2004) Satellite and CALJET aircraft observations of atmospheric rivers over the eastern North Pacific Ocean during the winter of 1997/98. Mon Wea Rev 132:1721–1745

Ralph FM, Neiman PJ, Rotunno R (2005a) Dropsonde observations in low-level jets over the northeastern Pacific Ocean from CALJET-1998 and PACJET-2001: mean vertical-profile and atmospheric-river characteristics. Mon Wea Rev 133:889–910

Ralph FM, Rauber RM, Jewett BF et al (2005b) Improving short-term (0–48 h) cool-season quantitative precipitation forecasting: recommendations from a USWRP Workshop. Bull Am Meteorol Soc 86:1619–1632

Ralph FM, Neiman PJ, Wick GA et al (2006) Flooding on California's Russian River: the role of atmospheric rivers. Geophys Res Lett 33:L13801. https://doi.org/10.1029/2006GL026689

Ralph FM, Sukovich E, Reynolds D et al (2010) Assessment of extreme quantitative precipitation forecasts and development of regional extreme event thresholds using data from HMT-2006 and COOP observers. J Hydrometeorol 11:1288–1306

Ralph FM, Neiman PJ, Kiladis GN et al (2011) A multi-scale observational case study of a Pacific atmospheric river exhibiting tropical-extratropical connections and a mesoscale frontal wave. Mon Wea Rev 139:1169–1189. https://doi.org/10.1175/2010MWR3596.1

Ralph FM, Wick GA, Neiman PJ et al (2012) Atmospheric rivers in reanalysis products: a six-event comparison with aircraft observations of water vapor transport, WCRP Reanalysis Conf., Silver Spring, MD, 1 p. https://www.wcrp-climate.org/ICR4/posters/Hughes_AT-20.pdf

Ralph FM, Intrieri J, Andra D Jr et al (2013a) The emergence of weather-focused testbeds linking research and forecasting operations. Bull Am Meteorol Soc 94:1187–1210

Ralph FM, Coleman T, Neiman PJ et al (2013b) Observed impacts of duration and seasonality of atmospheric-river landfalls on soil moisture and runoff in coastal Northern California. J Hydrometeorol 14:443–459

Ralph FM, Dettinger MD, White A (2014) A vision for future observations for Western U.S. extreme precipitation and flooding. J Contemp Water Resour Res Educ 153:16–32

Ralph FM, Prather KA, Cayan D et al (2016) CalWater field studies designed to quantify the roles of atmospheric rivers and aerosols in modulating U.S. West Coast precipitation in a changing climate. Bull Am Meteorol Soc 97:1209–1228. https://doi.org/10.1175/BAMS-D-14-00043.1

Ralph FM, Iacobellis SF, Neiman PJ et al (2017) Dropsonde observations of total integrated water vapor transport within North Pacific atmospheric rivers. J Hydrometeorol 18(9):2577–2596. https://doi.org/10.1175/JHM-D-17-0036.1

Ralph FM, Dettinger MD, Cairns MM et al (2018a) Defining "atmospheric river" how the glossary of meteorology helped resolve a debate. Bull Am Meteorol Soc 99:837–839

Ralph FM, Wilson AM, Shulgina T et al (2018b) Comparison of Atmospheric River Detection Tools: How Many Atmospheric Rivers Hit Northern California's Russian River Watershed? *Climate Dynamics*. 10.1007/s00382-018-4427-5

Ralph FM, Rutz JJ, Cordeira JM et al (2019) A scale to characterize the strength and impacts of atmospheric rivers. Bull Am Meteorol Soc 100:269–289. https://doi.org/10.1175/BAMS-D-18-0023.1

Ramos AM, Trigo RM, Liberato ML et al (2015) Daily precipitation extreme events in the Iberian Peninsula and Its association with atmospheric rivers. J Hydrometeorol 16:579–597. https://doi.org/10.1175/JHM-D-14-0103.1

Richiardone R, Manfrin M (2009) Neutral saturated lapse rate: an experimental check from CALJET-1998 and PACJET-2001. Mon Wea Rev 137:4382–4385

Rienecker MM et al (2011) MERRA: NASA's modern-era retrospective analysis for research and applications. J Clim 24:3624–3648. https://doi.org/10.1175/JCLI-D-11-00015.1

Rosenfeld D, Chemke R, Prather K et al (2014) Polluting of winter convective clouds upon transition from ocean inland over Central California: contrasting case studies. Atmos Res 135–136:112–127

Rutz JJ, Steenburgh WJ (2012) Quantifying the role of atmospheric rivers in the interior western United States. Atmos Sci Lett 13:257–261

Rutz JJ, Steenburgh WJ, Ralph FM (2014) Climatological characteristics of atmospheric rivers and their inland penetration over the western United States. Mon Wea Rev 142:905–921. https://doi.org/10.1175/MWR-D-13-00168.1

Saha S, Moorthi S, Pan H–L et al (2010) The NCEP climate forecast system reanalysis. Bull Am Meteorol Soc 91:1015–1057. https://doi.org/10.1175/2010BAMS3001.1

Schäfler A, Craig G, Wernli H et al (2018) The North Atlantic Waveguide and Downstream Impact Experiment Bull Amer Meteor Soc 99:1607–1637. https://doi.org/10.1175/BAMS-D-17-0003.1

Schlüssel P, Emery WJ (1990) Atmospheric water vapour over oceans from SSM/I measurements. Int J Remote Sens 11:753–766

Shields CA, Rutz JJ, Leung L–Y et al (2018) Atmospheric River Tracking Method Intercomparison Project (ARTMIP); project goals and experimental design. Geosci. Model Dev 11:2455–2474

Viale M, Valenzuela R, Garreaud RD et al (2018) Impacts of atmospheric rivers on precipitation in southern South America. J Hydrometeorol 19:1671–1687. https://doi.org/10.1175/JHM-D-18-0006.1

Waliser D, Guan B (2017) Extreme winds and precipitation during landfall of atmospheric rivers. Nat Geosci 10:179–183. https://doi.org/10.1038/NGEO2894

Waliser DE, Moncrieff MW, Burridge D et al (2012) The "year" of tropical convection (May 2008 – April 2010): climate variability and weather highlights. Bull Am Meteorol Soc 93:1189–1218

Weng F, Zhao L, Poe G et al (2003) AMSU cloud and precipitation algorithms. Radio Sci 38:8068–8079

Wentz FJ (1995) The intercomparison of 53 SSM/I water vapor algorithms. Technical report, Remote Sens Syst, Santa Rosa, CA. 19 pp

Wentz FJ (1997) A well-calibrated ocean algorithm for special sensor microwave/imager. J Geophys Res 102:8703–8718

Wentz FJ, Ricciardulli L, Hilburn KA et al (2007) How much more rain will global warming bring? Science 317:233–235

White AB, Jordan RJ, Martner BE et al (2000) Extending the dynamic range of an S-band radar for cloud and precipitation studies. J Atmos Ocean Technol 17:1226–1234

White AB, Gottas DJ, Strem ET et al (2002) An automated brightband height detection algorithm for use with Doppler radar spectral moments. J Atmos Ocean Technol 19:687–697

White AB, Neiman PJ, Ralph FM et al (2003) Coastal orographic rainfall processes observed by radar during the California land-falling jets experiment. J Hydrometeorol 4:264–282

White AB, Gottas DJ, Henkel AF et al (2010) Developing a performance measure for snow-level forecasts. J Hydrometeorol 11:739–753

White AB, Anderson ML, Dettinger MD et al (2013) A twenty-first-century California observing network for monitoring extreme weather events. J Atmos Ocean Technol 30:1585–1603

White AB, Mahoney KM, Cifelli R et al (2015a) Wind profilers to aid with monitoring and forecasting of high-impact weather in the Southeastern and Western United States. Bull Am Meteorol Soc 96:2039–2043

White AB, Neiman PJ, Creamean JM et al (2015b) The impacts of California's San Francisco Bay Area gap on precipitation observed in the Sierra Nevada during HMT and CalWater. J Hydrometeorol 16:1048–1069

Wick GA, Kuo Y, Ralph FM et al (2008) Intercomparison of integrated water vapor retrievals from SSM/I and COSMIC. Geophys Res Lett 35:L21805. https://doi.org/10.1029/2008GL035126

Wick GA, Neiman PJ, Ralph FM (2013a) Description and validation of an automated objective technique for identification and characterization of the integrated water vapor signature of atmospheric rivers. IEEE Trans Geosci Remote Sens 51:2166–2176

Wick GA, Neiman PJ, Ralph FM et al (2013b) Evaluation of forecasts of the water vapor signature of atmospheric rivers in operational numerical weather prediction models. Weather Forecast 28:1337–1352

Wick GA, Dunion JP, Walker J (2018a) Sensing Hazards with Operational Unmanned Technology: Impact study of Global Hawk unmanned aircraft system observations for hurricane forecasting, Final Report. NOAA Tech Memo. OAR-UAS-002, 94 pp

Wick GA, Hock TF, Neiman PJ et al (2018b) The NCAR/NOAA Global Hawk dropsonde system. J Atmos Ocean Technol 35:1585–1604. https://doi.org/10.1175/JTECH-D-17-0225.1

Wimmers AJ, Velden CSA (2011) Seamless advective blending of total precipitable water retrievals from polar-orbiting satellites. J Appl Meteorol Climatol 50:1024–1036

Zamora RJ, Ralph FM, Clark E et al (2011) The NOAA hydrometeorology testbed soil moisture observing networks: design, instrumentation, and preliminary results. J Atmos Ocean Technol 28:1129–1140

Zhu Y, Newell RE (1998) A proposed algorithm for moisture fluxes from atmospheric rivers. Mon Wea Rev 126(3):725–735. https://doi.org/10.1175/1520-0493(1998)126<0725:APAFMF>2.0.CO;2

4 Global and Regional Perspectives

Jonathan J. Rutz, Bin Guan, Deniz Bozkurt, Irina V. Gorodetskaya, Alexander Gershunov, David A. Lavers, Kelly M. Mahoney, Benjamin J. Moore, William Neff, Paul J. Neiman, F. Martin Ralph, Alexandre M. Ramos, Hans Christian Steen-Larsen, Maria Tsukernik, Raúl Valenzuela, Maximiliano Viale, and Heini Wernli

4.1 Introduction

Bin Guan and Jonathan J. Rutz

This chapter explores the global and regional footprints of ARs, which are just beginning to be recognized. Recent advancements, along with the need for benchmarking weather/climate model simulations, motivate a characterization of the basic features of ARs on global scales. This requires AR detection methods (ARDMs) that can be applied for different regions. Section 4.2 follows an AR detection method that considers both geometry and intensity thresholds in identifying ARs. ARs represent a cornerstone atmospheric phenomenon for generating significant wintertime precipitation along the US West Coast and other coastlines, globally. Section 4.3 discusses the links between ARs and

J. J. Rutz (✉)
Science and Technology Infusion Division, National Weather Service, Salt Lake City, UT, USA
e-mail: jonathan.rutz@noaa.gov

B. Guan
University of California, Los Angeles, Los Angeles, CA, USA
e-mail: bin.guan@jpl.nasa.gov

D. Bozkurt · R. Valenzuela
Department of Geophysics, Center for Climate and Resilience Research, University of Chile, Santiago, Chile
e-mail: dbozkurt@dgf.uchile.cl; rvalenzuela@dgf.uchile.cl

I. V. Gorodetskaya
Department of Physics, Centre for Environmental and Marine Studies, University of Aveiro, Aveiro, Portugal
e-mail: irina.gorodetskaya@ua.pt

A. Gershunov · F. M. Ralph
Center for Western Weather and Water Extremes, Scripps Institution of Oceanography, University of California–San Diego, La Jolla, CA, USA
e-mail: sasha@ucsd.edu; mralph@ucsd.edu

D. A. Lavers
European Centre for Medium-Range Weather Forecasts (ECMWF), Reading, UK
e-mail: david.lavers@ecmwf.int

K. M. Mahoney
Physical Sciences Laboratory, National Oceanic and Atmospheric Administration (NOAA), Earth System Research Laboratory, Boulder, CO, USA
e-mail: kelly.mahoney@noaa.gov

B. J. Moore · W. Neff
Cooperative Institute for Research in Environmental Sciences, University of Colorado–Boulder, Boulder, CO, USA

Physical Sciences Laboratory, National Oceanic and Atmospheric Administration (NOAA), Earth System Research Laboratory, Boulder, CO, USA
e-mail: benjamin.moore@noaa.gov; william.neff@colorado.edu

P. J. Neiman
National Oceanic and Atmospheric Administration (NOAA), Earth System Research Laboratory, Boulder, CO, USA
e-mail: Paul.J.Neiman@noaa.gov

A. M. Ramos
Instituto Dom Luiz, Faculdade de Ciências, Universidade de Lisboa, Lisboa, Portugal
e-mail: amramos@fc.ul.pt

H. C. Steen-Larsen
Geophysical Institute, University of Bergen, Bergen, Norway

Bjerknes Centre for Climate Research, University of Bergen, Bergen, Norway
e-mail: Hans.Christian.Steen-Larsen@uib.no

M. Tsukernik
Brown University, Providence, RI, USA

M. Viale
Argentinean Institute for Snow, Glaciology and Environmental Science, (IANIGLA - CONICET), Mendoza, Argentina
e-mail: maxiviale@mendoza-conicet.gob.ar

H. Wernli
Institute for Atmospheric and Climate Science, ETH Zurich, Zurich, Switzerland

F. M. Ralph et al. (eds.), *Atmospheric Rivers*, https://doi.org/10.1007/978-3-030-28906-5_4

climate modes in several regions. Section 4.4 describes the climatology of ARs along the US West Coast based on the extensive field campaigns and additional studies made possible by new satellite and reanalysis data sets. Section 4.5 addresses inland-penetrating ARs, first by highlighting key hydrometeorological and dynamical characteristics of an AR that ravaged the semi-arid desert southwest, and then by exploring more generally the climatology of ARs across the western US and the airflow trajectories associated with inland-penetrating ARs across this region. Section 4.6 describes ARs that occur over the southeastern US, and Sect. 4.7 highlights the climatology and characteristics of ARs across Europe. Section 4.7 focuses on ARs interacting with the very high topography of the Andes over southern South America, and Sect. 4.8 describes the unique aspects of ARs that affect polar regions.

4.2 Global Climatology

Bin Guan and Jonathan J. Rutz

The global footprints of ARs are just starting to be recognized. Adding to the volume of studies that focus on the west coasts of North America and Europe are recent and ongoing studies on ARs in polar regions (Gorodetskaya et al. 2014; Neff et al. 2014), eastern continental and coastal regions (Hirota et al. 2016; Mahoney et al. 2016), and other areas (e.g., Australia, New Zealand) where ARs have historically received less scientific attention. These recent advancements, along with the need for benchmarking weather and climate model simulations, motivate a characterization of the basic features of ARs on global scales. Such an undertaking requires the development of objective ARDMs suitable for application over different regions. Here, one such method is used for a consistent accounting of ARs and their climatology across the globe.

4.2.1 AR Detection Method and Justification

This section follows the ARDM introduced in Guan and Waliser (2015), which considers both geometry and intensity thresholds in identifying ARs. For any given vertically integrated water vapor transport (IVT) map, the method first extracts a set of "objects" (i.e., contiguous areas) based on the IVT magnitude threshold—namely, greater than the 85th percentile specific to each grid cell and season (Fig. 4.1) and a fixed lower limit of 100 kg m^{-1} s^{-1}. Additional thresholds based on IVT direction (within 45° of the AR shape orientation), length (>2000 km), and length/width ratio (>2) are then applied to these objects, resulting in a defined set of ARs. These ARs are further checked for landfalls based on whether they intersect the coastline. The method is applied to European Centre for Medium-Range Weather Forecasts (ECMWF) Reanalysis-Interm (ERA-Interim) (Dee et al. 2011) IVT to produce a global AR catalog for the period of 1979–2015. Based on this catalog, 11 ARs on average are over the globe at any given time, of which two to three are making landfall.

Previous studies have proposed alternative ARDMs, based on different IVT thresholds (e.g., total IVT of 250 or 500 kg m^{-1} s^{-1}, anomalous IVT of 250 kg m^{-1} s^{-1} relative to the climatology) and/or geometric requirements (e.g., >2000 km in length with no width requirement, >1500 km in length and <1500 km in width, and >~1400 km and length/width ratio >1.6) (Rutz et al. 2014; Mahoney et al. 2016;

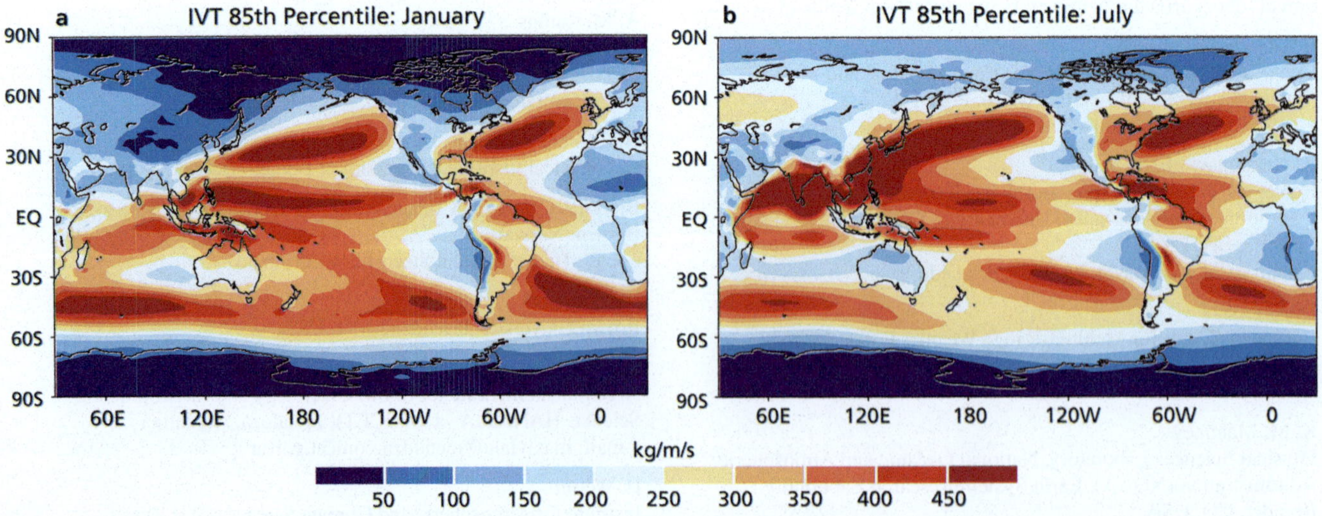

Fig. 4.1 The 85th percentile of integrated water vapor transport (IVT) magnitude (kg m^{-1} s^{-1}) at each grid cell for the months of (**a**) November–March (NDJFM) and (**b**) May–September (MJJAS) over the period of 1979–2015. A total of 12 maps, for 12 overlapping 5-month seasons, are used to threshold 6-hourly IVT in the detection of ARs. Grid cells with IVT magnitude above the greater of the 85th percentile and 100 kg m^{-1} s^{-1} are retained for AR detection. IVT is derived from ERA-Interim reanalysis (Dee et al. 2011). (Updated from Guan and Waliser 2015)

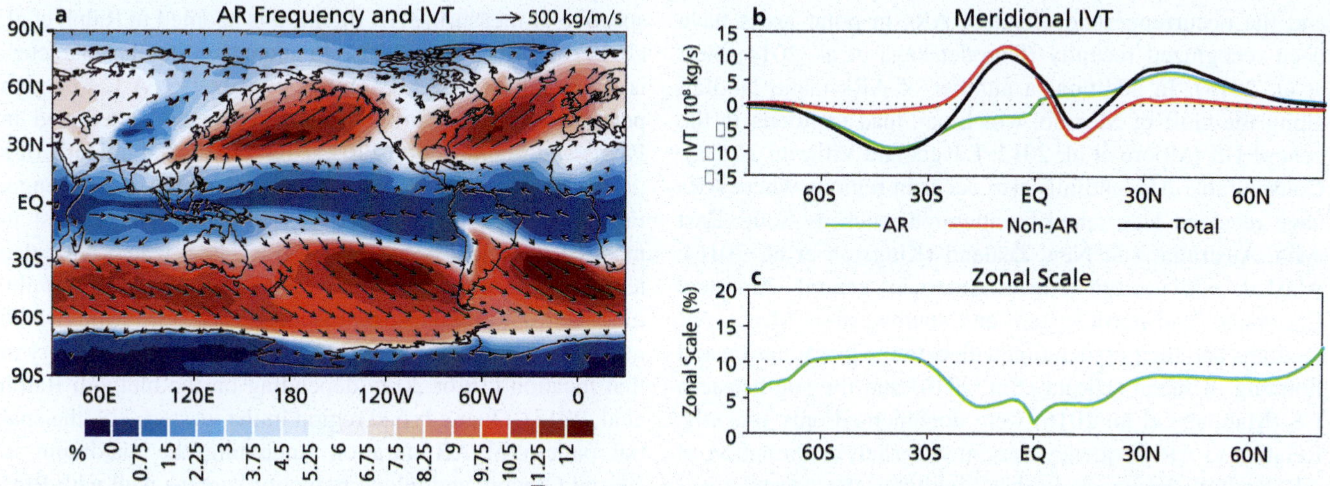

Fig. 4.2 (**a**) Integrated water vapor transport (IVT) AR frequency (percent of time-steps; *shading*) and mean AR IVT (kg m^{-1} s^{-1}; arrows) at each grid cell over the period of 1979–2015. *White* shading in limited areas indicates no AR detected over the analysis period. (**b**) Zonally integrated meridional IVT (kg s^{-1}) associated with AR transport (*green*), non-AR transport (*red*), and their combination (*black*). (**c**) Integrated AR zonal scale expressed as the fraction of the total zonal circumference at given latitudes. (Updated from Guan and Waliser 2015)

Mundhenk et al. 2016). The IVT anomaly-based method in Mahoney et al. (2016) is conceptually similar to the percentile-based method in Guan and Waliser (2015) in allowing regionally- and seasonally-varying values for thresholding total IVT. A more detailed discussion of ARDMs is in Sect. 3.7. The Guan and Waliser (2015) algorithm is used here because of its reasonable performance across different regions and climatologies. Specifically, comparisons with three independent studies conducted for western North America (Neiman et al. 2008a), Britain (Lavers et al. 2011), and East Antarctica (Gorodetskaya et al. 2014) showed over ~90% agreement in detected AR landfall dates. Some notable cases of AR landfalls in Greenland (Neff et al. 2014), Norway (Stohl et al. 2008), the south-central US (Moore et al. 2012), and Japan (Hirota et al. 2016) are also detected by the algorithm used here.

4.2.2 AR Frequency and IVT

AR frequency, calculated at each grid cell as the percentage of reanalysis time-steps when the grid cell is within the boundary of an AR, is shown in Fig. 4.2a. Mean AR IVT, calculated similarly, is also shown. ARs are more frequent in mid-latitude ocean basins than over land and other latitudes. Notable AR frequency maxima are found in the extratropical North Pacific/Atlantic, southeastern Pacific, and South Atlantic.[1] AR frequency minima are found near the equator, over and northeast of the Tibetan Plateau, and in polar areas, reducing to zero in interior Antarctica (white shading on Fig. 4.2a). The north–south gradient in AR frequency and the inland reduction of AR frequency along the west coast of North America are broadly consistent with Rutz et al. (2014) and Mundhenk et al. (2016). Note that AR shapes determined by the 85th percentile IVT often extend further inland than the 250 kg m^{-1} s^{-1} IVT contour (Guan and Waliser 2015), which explains the somewhat higher AR frequency in the interior western US compared to Rutz et al. (2014). The strongest AR IVT, directed eastward and poleward, is located over the extratropical ocean basins, in good agreement with the original calculation of Zhu and Newell (1998) based on 3 years of data and a simple method for isolating filamentary IVT from the large-scale background.

In the Northern Hemisphere, ARs account for 84% of the total meridional IVT, and 8% of the zonal circumference of the earth (i.e., the length of a given latitude circle), between 30 and 50° latitude. These percentages are 88% and 11%, respectively, in the Southern Hemisphere (Fig. 4.2b–c). The slightly lower AR fractional IVT, compared to Zhu and Newell (1998), could be a result of different data sources, analysis periods, and methods. The AR fractional zonal scale, on the other hand, matches their result nicely.

4.2.3 AR Landfall Frequency

AR landfall is defined here as the location with maximum IVT along the coastline intersected by a given AR, and landfall is defined only if the mean AR IVT and the IVT at the landfall location are directed onshore. Most AR landfalls occur along the west coasts of North America, southern South America, and Europe (Fig. 4.3). The effects of AR landfalls on extreme precipitation and floods in these regions have been documented in a number of studies (see Chap. 5). AR landfalls in Greenland and Antarctica are notable as well,

[1]North and south refer to the hemisphere. For example, South Atlantic refers to the sector of the Atlantic Ocean located south of the equator.

and the occurrence and effect of ARs in polar areas have been recognized recently (Gorodetskaya et al. 2014; Neff et al. 2014). In addition, a number of ARs make landfall along the Gulf of Mexico, which can lead to floods in the central US (Moore et al. 2012; Lavers and Villarini 2013a). Considerable AR landfalls also occur in regions where ARs have received less scientific attention, such as South/East Asia, Australia, and New Zealand (Kingston et al. 2016); Alaska; northeastern North America; central America/Caribbean; and northwestern and southwestern Africa. AR landfalls and their association with extreme precipitation and flooding in Japan (Hirota et al. 2016) and the southeastern US (Mahoney et al. 2016) were documented only recently. Relative to AR frequency presented earlier, examination of AR landfall frequency further highlights the coastal locations where the impacts of ARs are expected to be large, i.e., located at the core of the AR where the IVT is strong and directed onshore, as opposed to the edge of the AR where IVT is weak or not directed onshore. Most of the locations with high AR frequencies also have high landfall frequencies (comparing Fig. 4.3 with Fig. 4.2). The most notable exception is along the southeast coast of South America, where AR frequency is as large as that for the west side of the continent but AR landfalls are extremely few, suggesting the notable AR frequency in southeastern South America is contributed by ARs moving *away* from the continent, which may have different overland effects compared to ARs moving *toward* the continent from the open ocean (i.e., landfalling ARs).

4.2.4 AR Duration

Mean duration of ARs is calculated as the average number of hours (based on 6-h time-steps) a given grid cell stays continually within the boundary of an AR. This Eulerian perspective of measuring AR duration is adopted in Ralph et al. (2013) and Rutz et al. (2014) in understanding the characteristics and effects of ARs on the western US. A Lagrangian perspective of measuring AR life cycle was attempted in Payne and Magnusdottir (2014), but automated tracking of a given AR in time and space remains a technical challenge. From the Eulerian perspective, AR duration is longest in most extratropical ocean basins (Fig. 4.4), where ARs also tend to occur more frequently (see Fig. 4.2a). For example, mean AR duration is 17 h at Bodega Bay, California, which matches well with measurements from the AR observatory in that location (16 or 20 h, depending on the method; Ralph et al. 2013). Over a broad region in the western US, the spatial pattern of AR duration (including the maximum in coastal Oregon, and inland reduction) agrees well with Rutz et al. (2014), although the ocean-to-land gradient is somewhat smaller here, which is attributable to the percentile-based IVT threshold used for AR detection.

4.2.5 AR Precipitation Fraction

ARs account for over 30% of the total precipitation in a number of extratropical areas (Fig. 4.5). The notable percentages are associated with enhanced AR activity in these regions (e.g., Fig. 4.2a) and largely consistent with AR precipitation fractions found previously for western, central, and eastern North America and Western Europe (Guan et al. 2010; Dettinger et al. 2011; Rutz and Steenburgh 2012; Rutz et al. 2014; Lavers and Villarini 2015). Notable AR precipitation fractions in the southeastern US and southern Greenland echo recent findings of the roles of ARs in these regions (Neff et al. 2014; Mahoney et al. 2016). Interestingly, large AR precipitation fractions can be seen in the Middle East, centered in Iran. Although the characteristics of ARs in this

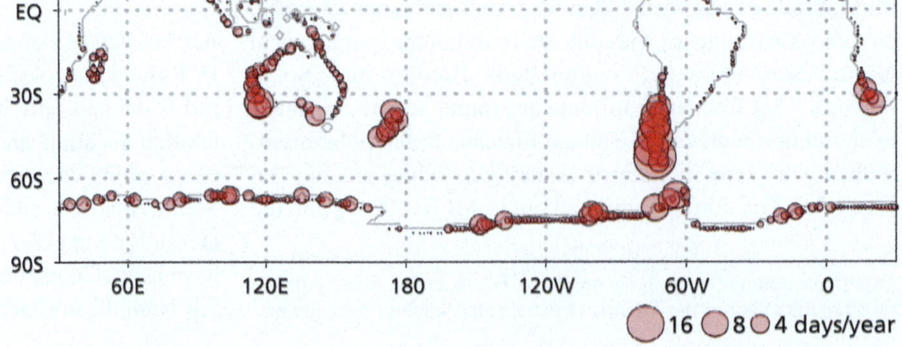

Fig. 4.3 Frequency (days per year) of AR landfalls based on all months of 1979–2015. The frequency values in days per year were obtained by multiplying the fraction of 6-hourly time-steps with AR landfalls (i.e., probability of landfall occurrence) by 365.2425. (Updated from Guan and Waliser 2015)

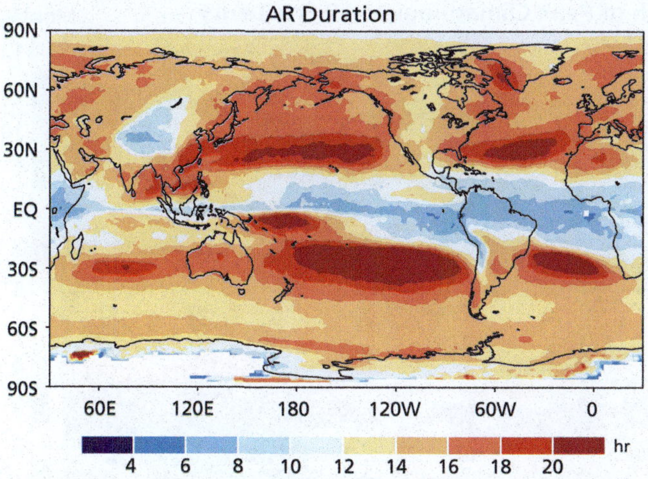

Fig. 4.4 Mean duration (hour) of ARs at each grid cell. Calculations are based on all months of 1979–2015. (Updated from Guan and Waliser 2015)

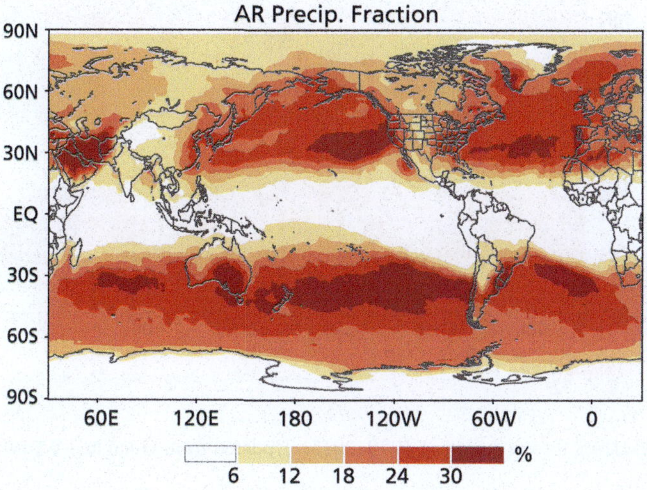

Fig. 4.5 Mean AR fractional contribution to total precipitation over the period of 1997–2015, for which precipitation data from Global Precipitation Climatology Project version 1.2 (Huffman et al. 2001) are available. (Updated from Guan and Waliser 2015)

arid/semi-arid region need further research and understanding, the result here is consistent with the finding that in Iran a small number of extreme events disproportionately contribute a large fraction of the annual precipitation (Alijani et al. 2008), a characteristic of AR precipitation found in many places. Further, it was found that extreme precipitation in Iran is largely governed by the interaction between a mid-tropospheric trough over the Middle East and a low-tropospheric anticyclone around the Arabian Peninsula/Arabian Sea that favors moisture transport from the southern water bodies, with the pattern of the moisture transport reminiscent of ARs (Raziei et al. 2012; their Fig. 5).

4.2.6 Seasonality

The seasonality of ARs can be illustrated by the month of peak AR frequency (Fig. 4.6). Large spatial variations suggest AR activity depends on the particular climatologies in different regions. Southward along the west coast of North America, the month of peak AR frequency is August in Alaska, September/October in western Canada, November in Washington and Oregon, December in northern California, January in southern California, and February in Baja California—consistent with the pattern found in Rutz et al. (2014). Western Europe has a fairly similar seasonal migration. In South America, the month of peak AR frequency varies from March near the southern tip to May/June in the central Andes. In the above regions, the month of peak AR frequency tends to occur sequentially from fall at higher latitudes to wintertime at lower latitudes and is likely connected to the seasonal movement of storm tracks.

The seasonality of AR frequency is further examined by comparing two extended seasons (November–March vs. May–September). The two seasons have contrasting dry and wet conditions in a number of west coast areas of the mid-latitude continents. The largest contrast in AR frequency between the two seasons occurs in the northeastern Pacific–western US, and in southeastern Pacific–central Chile (Fig. 4.7). In both regions, the amplitude of the seasonal contrast is above one-third of the annual mean AR frequency (comparing to Fig. 4.2a), with AR frequencies increased in their respective winter seasons. Interesting seasonal contrasts in AR frequency are also seen in some east coast areas, particularly in East Asia, where the signs of the seasonal anomalies relative to the annual mean AR frequency are opposite to those on the other side of the Pacific, which may have different implications for seasonal precipitation under the monsoonal climatology.

4.2.7 Summary of Sect. 4.2

Advancements in automated ARDMs afford improved characterization and understanding of the global footprints and potential effects of ARs, especially their roles in weather and climate via their shaping of global water cycles. The remarkable agreement between the original estimate by Zhu and Newell (1998) and the results from more sophisticated methods presented herein supports and further highlights the common notion that, in the mid-latitudes, ARs account for more than 90% of the poleward water vapor transport though they occupy only about 10% of the earth's circumference. Landfall occurrences on all continents and major islands mark the role of ARs in global weather and water extremes. The ARDMs described herein and in other studies will complement one another in fur-

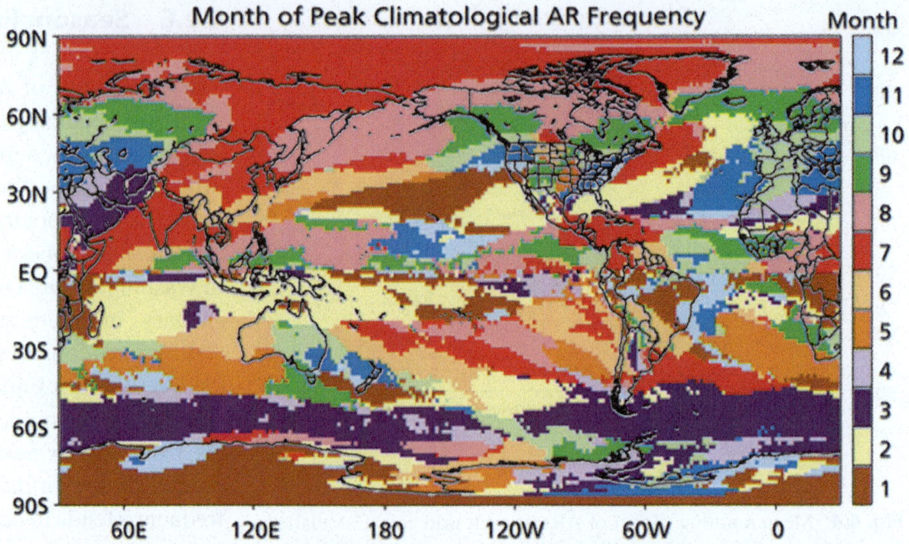

Fig. 4.6 Month of peak climatological AR frequency for the period of 1979–2015. (Updated from Guan and Waliser 2015)

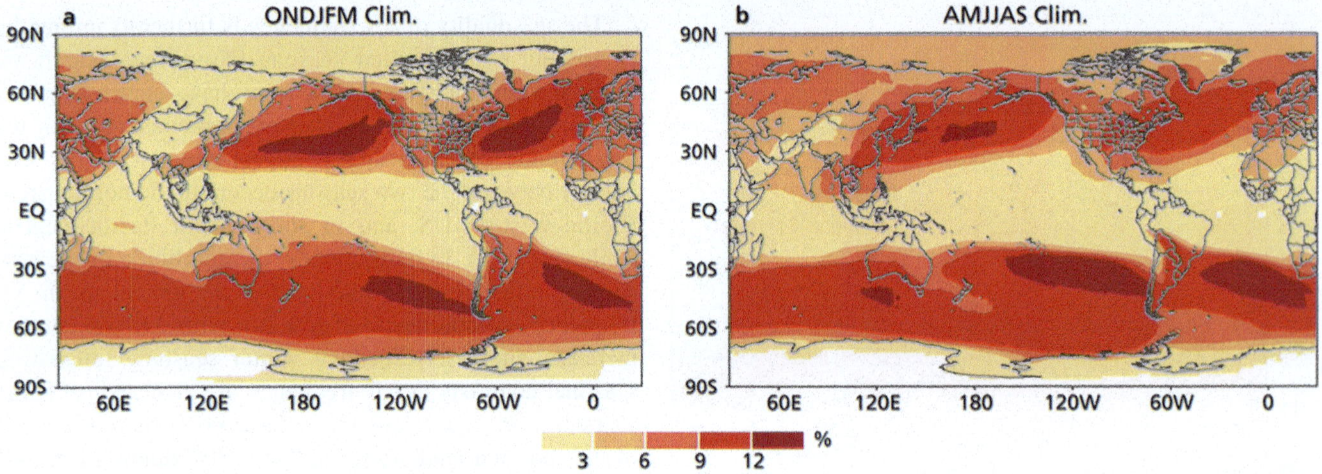

Fig. 4.7 AR frequency (percent of time-steps) in (**a**) ONDJFM and (**b**) AMJJAS for the period of 1979–2015. (Updated from Guan and Waliser 2015)

thering investigations of ARs from global perspectives, such as understanding the representation of ARs in state-of-the-art weather and climate models, and assessing the predictability and predictive skill of ARs in operational forecasting systems.

4.3 Climate Modulation

Bin Guan and Jonathan J. Rutz

Links between ARs and climate modes have been shown to be prominent in several regions, including the west coasts of North America and Europe, where ARs play important roles in regional weather and hydrology. For example, the unusually high frequency of ARs in California's Sierra Nevada during the 2010–2011 winter—a doubling of the normal number of ARs per winter—was associated with the somewhat rare concurrence of the negative phase of two large-scale climate modes over an extended period of that winter—namely, the Arctic Oscillation (AO) and the Pacific/North American (PNA) pattern (Guan et al. 2013). Such influence on extratropical ARs can also come from conditions that originate in the tropics, such as those associated with the Madden–Julian Oscillation (MJO) and El Niño–Southern Oscillation (ENSO) (Guan et al. 2012; Guan et al. 2013; Guan and Waliser 2015; Mundhenk et al. 2016). Studies focused on Europe have also found links between ARs and regional/local climate modes, including the North Atlantic Oscillation (NAO) and the Scandinavian pattern (Lavers et al. 2012; Lavers and Villarini 2013b).

Based on an AR catalog over the period of 1997–2014, Guan and Waliser (2015) presented an accounting of global links between ARs and four climate modes (ENSO, MJO, AO, and PNA). Here, the analysis is extended to the period of 1979–2015, making use of the same global AR catalog

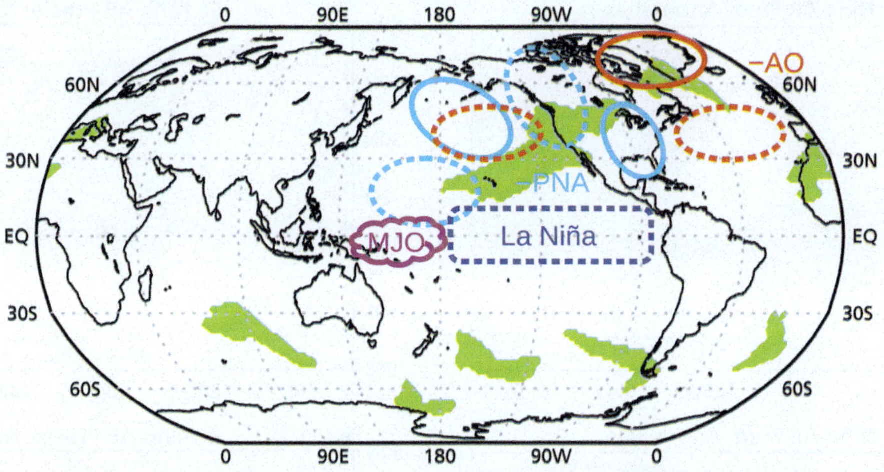

Fig. 4.8 Schematic showing one of the many possible configurations of four climate modes for a given time-period, i.e., the negative phases of the Arctic Oscillation (AO) and Pacific/North American (PNA) pattern, the cold phase of the El Niño–Southern Oscillation(ENSO), and the western Pacific phase of the Madden–Julian Oscillation (MJO). For AO and PNA, the *solid/dashed contours* show the representative locations of the high-/low-pressure anomaly centers associated with the negative phases of the two modes. The *green shading* shows examples of ARs detected on an arbitrary day. The climate modes modulate AR activity through their influence on the large-scale atmospheric circulation

described in Sect. 4.2. The schematic in Fig. 4.8 shows one of the many possible configurations of the four modes examined, to illustrate the patterns and the locations of the "action centers" associated with these modes. Of interest are the influences of these four climate modes on AR frequency and precipitation intensity. Only the extended season of October–March is considered for a focused discussion. This is the season when the four modes considered are the most energetic—and hence potentially more significant to AR activity. It is also the wet season for a number of west coast regions in the Northern Hemisphere with documented AR effects. Composite anomalies of AR frequency and precipitation are calculated for each of the four modes relative to the climatological mean. Negative/positive phases of ENSO, AO, and PNA are defined as a 0.5 standard deviation below and above zero, based on their respective indices provided by the NOAA Climate Prediction Center. The eight phases of MJO are based on the Real-time Multivariate MJO (RMM) index (Wheeler and Hendon 2004), and only strong cases for which the magnitude of the RMM index is >1.0 are considered.

4.3.1 El Niño–Southern Oscillation

A well-known example of climate modes is ENSO, which has a recurring period of about 2–7 years—and has profound effects on weather and climate globally. ARs are less frequent during the cold phase of ENSO in a number of subtropical and extratropical regions. The anomaly patterns are largely reversed during the warm phase of ENSO. In either phase, the anomalies are strongest in the northeastern Pacific, the North Atlantic, and the Gulf of Mexico and the Caribbean. Slightly weaker but coherent anomalies are located in the southern Indian Ocean, the South Pacific convergence zone extending into South America, and the South Atlantic (Fig. 4.9a, b). The magnitudes of the anomalies are coherent and notable in many regions by comparison to the climatological mean AR frequency (Fig. 4.9c). For AR precipitation, the most notable effect is near the US Pacific Northwest and western Canada, and along the eastern US and the Gulf Coast, where AR precipitation is increased during El Niño—albeit with no strong signals during La Niña (Fig. 4.9d, e).

4.3.2 Madden–Julian Oscillation

As the primary component of tropical intra-seasonal variability (Madden and Julian 1971, 1972), the MJO features large-scale, eastward propagation of clouds and precipitation along the equatorial band, with typical periods of 40–50 days (e.g., Matthews 2000). Through interactions with phenomena on multiple time-scales, the MJO plays important roles in regional weather and climate. In the North Pacific, positive anomalies of AR frequency propagate from East Asia to south of the Aleutian Islands during MJO Phases 1–4. This is followed by propagation of negative anomalies during Phases 5–8 in the same region (Fig. 4.10, *left*). Also notable is increased AR frequency offshore the Pacific Northwest of US and Canada during Phase 6. A more pronounced increase in AR frequency is noted south of Alaska and western Canada during Phases 7–8. From tropical North America to Northern Europe, positive anomalies are seen propagating during Phases 1–4. This is followed by somewhat less coherent negative anomalies in this region during Phases 5–8. Similar

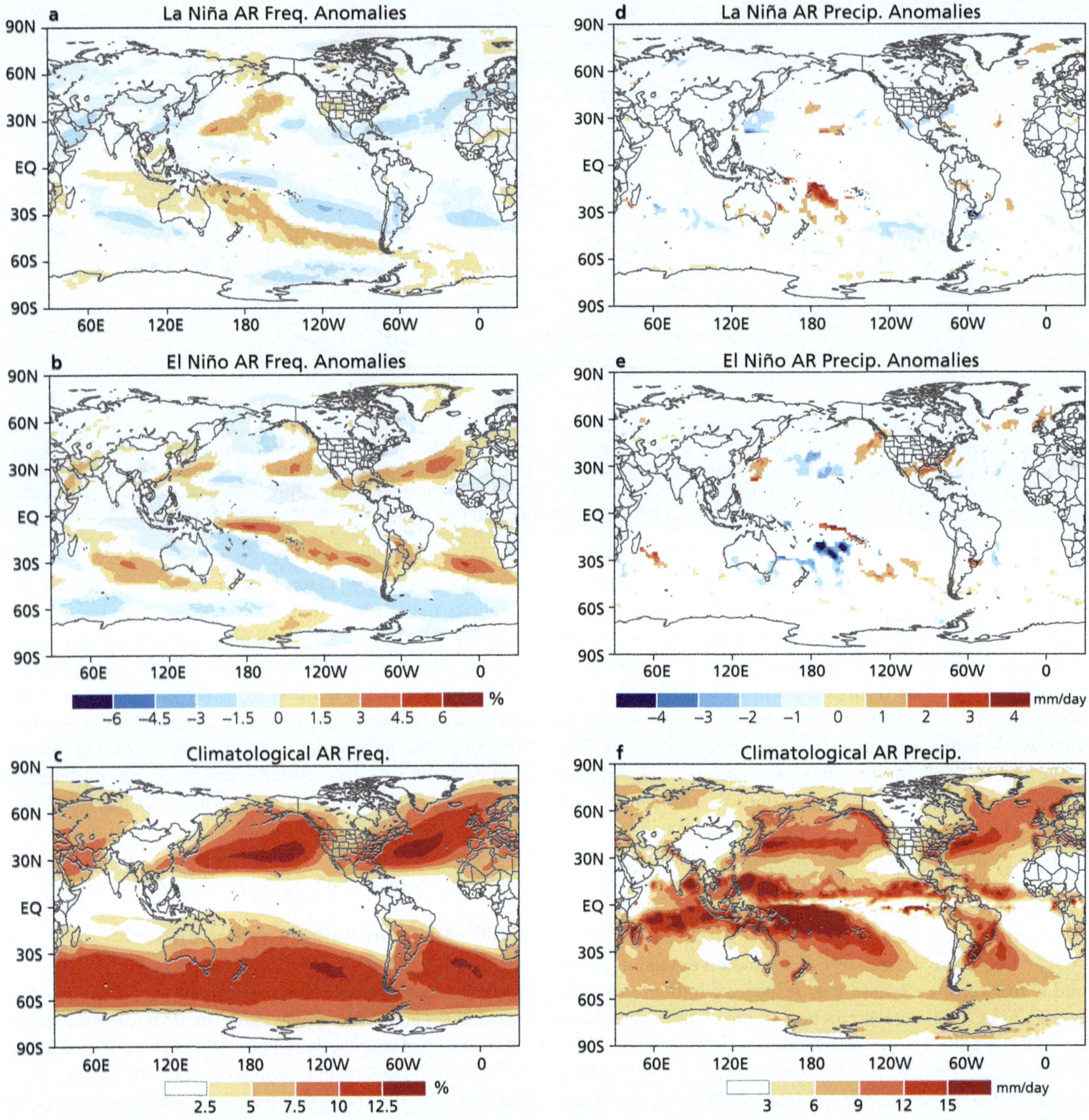

Fig. 4.9 (**a**, **b**) Composite ONDJFM AR frequency anomalies (percent of time-steps) during (**a**) La Niña and (**b**) El Niño conditions. (**c**) ONDJFM climatology of AR frequency, based on which the composite anomalies in (**a**, **b**) are calculated. (**d–f**) as (**a–c**), but for AR precipitation (mm/day). In (**a**, **b**) and (**d**, **e**), values are shown only if they are statistically significant at the 95% level based on 2-tailed z-test and the number of samples contributing to the calculation is >200. AR frequency is based on integrated water vapor transport (IVT) derived from ERA-Interim reanalysis (Dee et al. 2011). Precipitation is from Global Precipitation Climatology Project version 1.2 (Huffman et al. 2001). (Updated from Guan and Waliser 2015)

eastward propagation of AR frequency anomalies is seen in the Southern Hemisphere, with a notable band of anomalies that, when fully developed, extends all the way from the Maritime Continent to east of the Antarctic Peninsula (Fig. 4.10, *right*). AR precipitation is enhanced from East Asia to offshore California during Phases 2–7 and suppressed during Phases 6–3 in this region (Fig. 4.11, *left*; note that MJO Phase 1 follows Phase 8, by definition). AR precipitation anomalies are also seen propagating in the Southern Hemisphere, most notably from Australia to French Polynesia during Phases 4–7 (Fig. 4.11, *right*). Considering the characteristic location of MJO convective anomalies in each phase

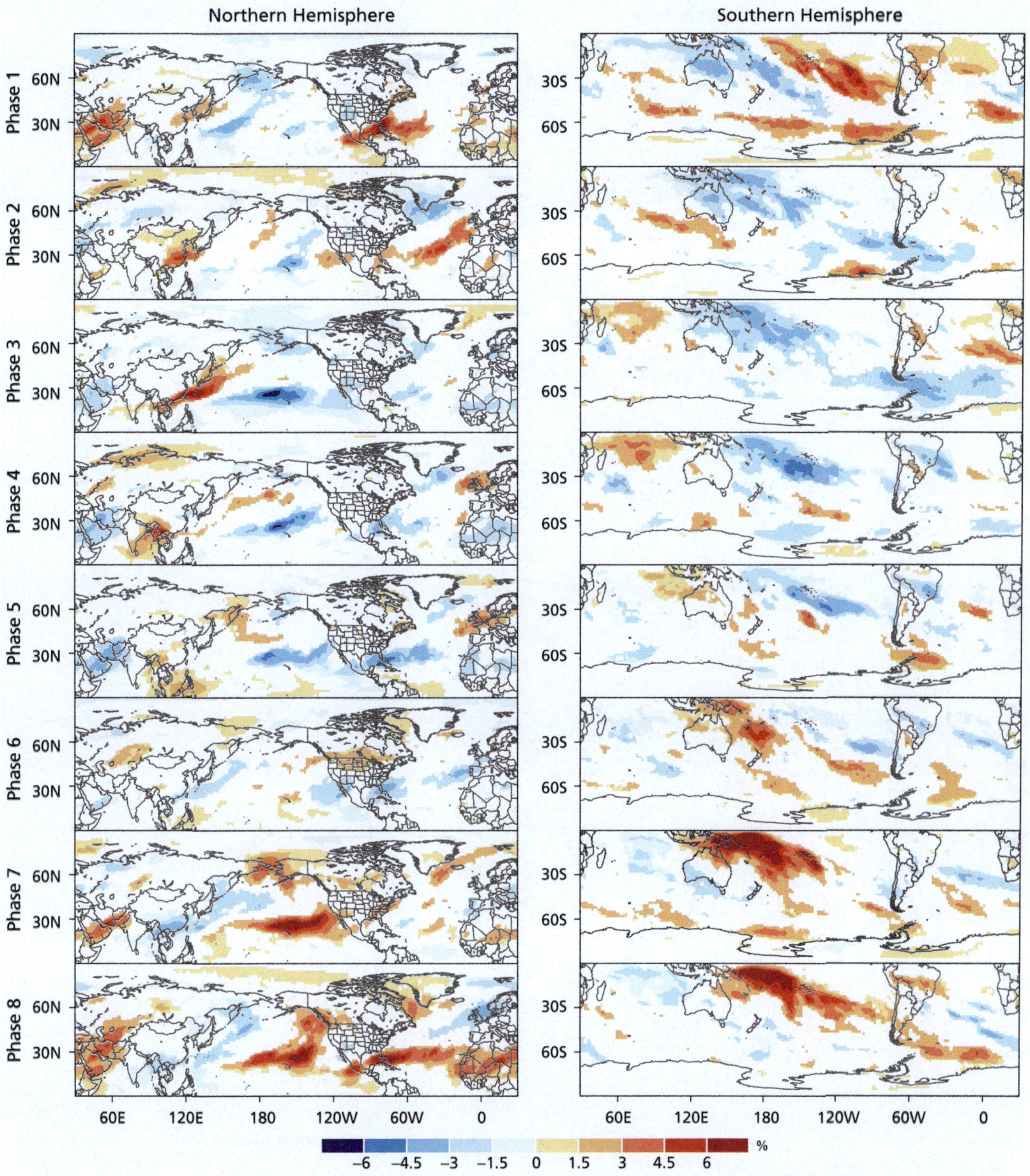

Fig. 4.10 Composite ONDJFM AR frequency anomalies (percent of time-steps) relative to the ONDJFM climatology during each phase of the Madden–Julian Oscillation (MJO). The two hemispheres are shown separately in two columns to improve the visualization. Values are shown only if they are statistically significant at the 95% level based on two-tailed z-test and the number of samples contributing to the calculation is >50. (Updated from Guan and Waliser 2015)

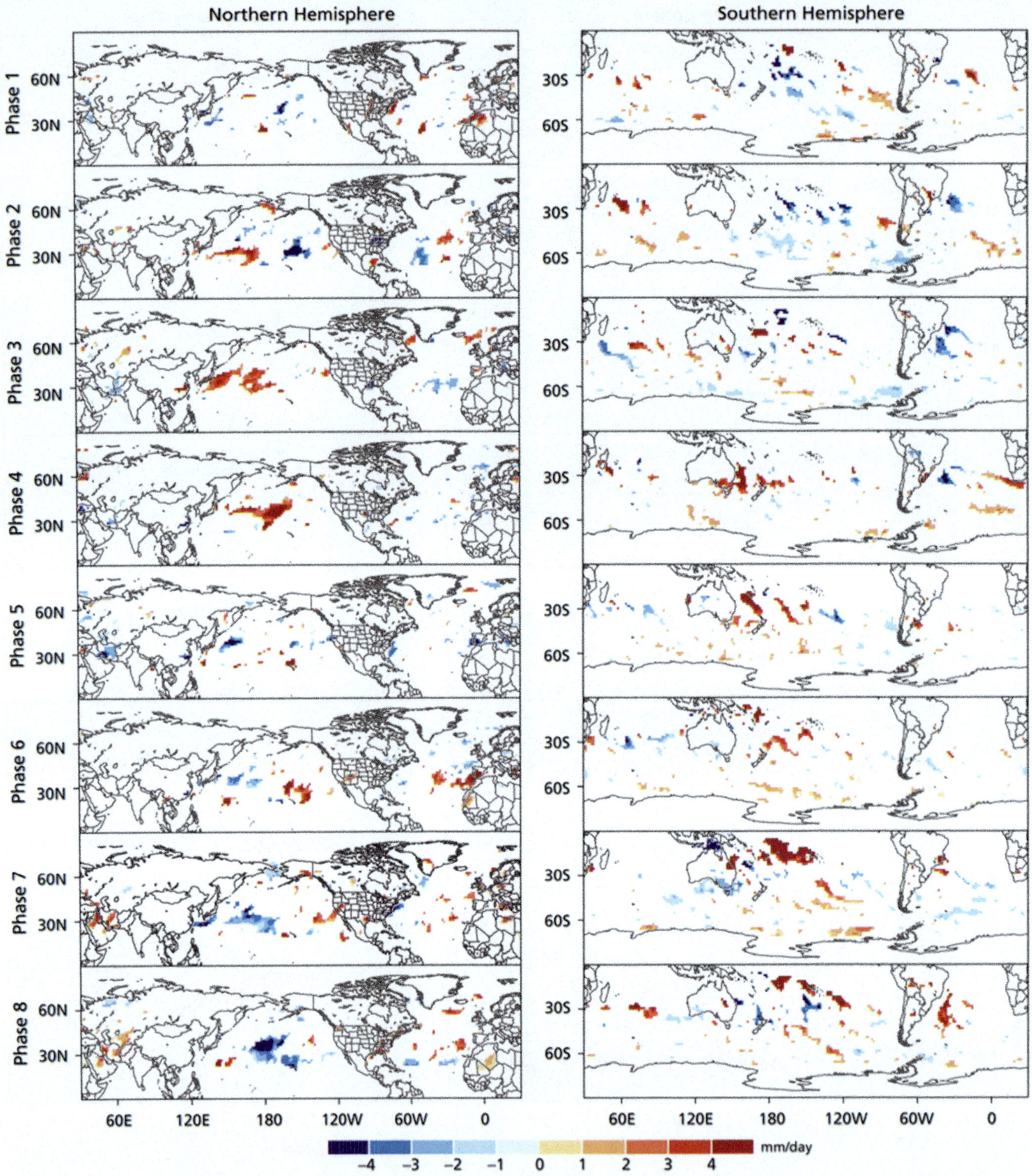

Fig. 4.11 Composite ONDJFM AR precipitation anomalies (mm/day) relative to the ONDJFM climatology during each phase of the Madden–Julian Oscillation (MJO). The two hemispheres are shown separately in two columns to improve the visualization. Values are shown only if they are statistically significant at the 95% level based on two-tailed z-test and the number of samples that contribute to the calculation is >50. (Updated from Guan and Waliser 2015)

(Wheeler and Hendon 2004), the influence of the MJO on AR frequency and precipitation is often downstream of the MJO convection itself.

4.3.3 Arctic Oscillation

The AO is characterized by a seesaw pattern of sea level pressure anomalies between the Arctic area and the surrounding extratropical areas (Thompson and Wallace 1998). The negative phase of the AO, characterized by weakened mid-latitude westerly winds, favors the southward outbreak of cold Arctic air and the associated southward excursion of storminess (Jeong and Ho 2005). Unlike the closely related NAO, the AO has considerable footprints in the Pacific, although its physical nature relative to the NAO raised some questions (Ambaum et al. 2001). The negative phase of the AO is associated with *enhanced* AR frequency in the subtropical Pacific offshore the western US, in the subtropical North Atlantic extending to South Europe, and in the Labrador Sea and western Greenland—and meanwhile *reduced* AR frequency in North Europe (Fig. 4.12a). A largely reversed pattern is seen during the positive phase of AO (Fig. 4.12b). The magnitude and spatial extent of the dipole pattern in the North Atlantic and Western Europe are especially notable. The influence of the AO on AR precipitation is more localized. The most notable anomalies are inshore and offshore northern California, where AR precipitation is increased (decreased) during the negative (positive) phase of the AO (Fig. 4.12c, d).

4.3.4 Pacific/North American Pattern

The PNA features a wave train pattern of four action centers (i.e., pressure systems) with alternating signs that link the Pacific Ocean and the North American continent (Wallace and Gutzler 1981). Consistent with the direction of wind anomaly implicated by the four action centers, AR frequency is increased on the southeastern (northwestern) flank of the low (high) pressure centers (Fig. 4.13a, b). AR precipitation is decreased (increased) during negative (positive) PNA along the eastern US and the Gulf Coast and offshore northern California (Fig. 4.13c, d).

4.3.5 Summary of Sect. 4.3

Links between ARs and modes of climate variability have important implications for subseasonal, seasonal, and long-term AR predictability and predictions, given the different time-scales associated with these climate modes. The general pattern of ENSO and MJO influences on AR frequency in the North Pacific is consistent among studies—based on different ARDMs, analysis periods, definition of winter seasons, and other details—which suggest the robustness in the underlying relationship between ARs and the two modes (Guan and Waliser 2015; Mundhenk et al. 2016). Inshore and offshore California, the influence of AO and PNA on AR precipitation shown here is consistent with their modulation of AR IVT (Guan et al. 2013), based on an independent AR record (Neiman et al. 2008a). Also, the AO's influence on AR frequency in

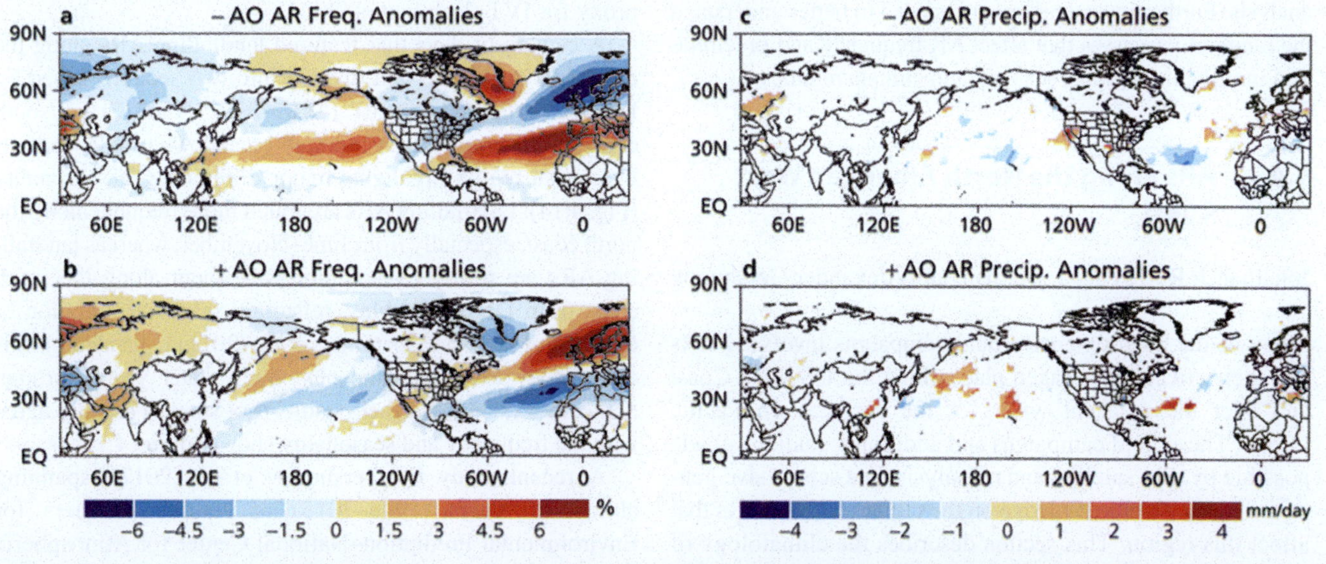

Fig. 4.12 (**a**, **b**) Composite ONDJFM AR frequency anomalies (percent of time-steps) for the Arctic Oscillation. (**c–d**) as (**a–b**), but for AR precipitation (mm/day). In (**a**, **b**) and (**c**, **d**), values are shown only if they are statistically significant at the 95% level based on two-tailed z-test and the number of samples contributing to the calculation is >200. AR frequency is based on integrated water vapor transport (IVT) derived from ERA-Interim reanalysis (Dee et al. 2011). Precipitation is from Global Precipitation Climatology Project version 1.2 (Huffman et al. 2001). (Updated from Guan and Waliser 2015)

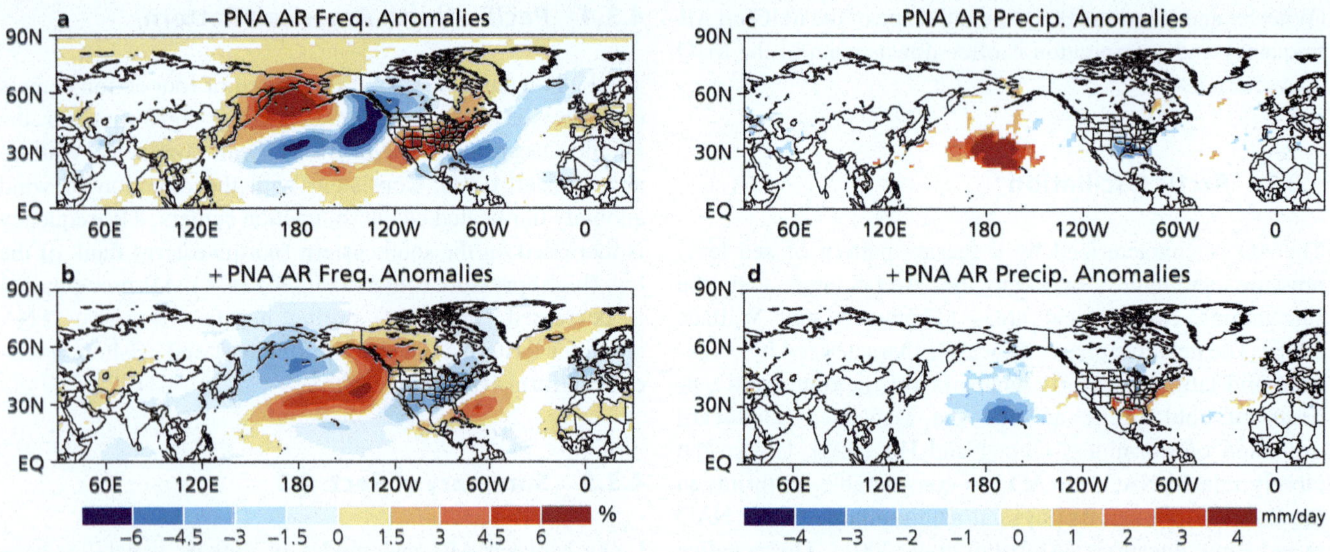

Fig. 4.13 (**a**, **b**) Composite ONDJFM AR frequency anomalies (percent of time-steps) during the Pacific/North American (PNA) pattern. (**c**) ONDJFM climatology of AR frequency, based on which the composite anomalies in (**a**, **b**) are calculated. (**d**, **f**) as (**a**, **c**), but for AR precipitation (mm/day). In (**a**, **b**) and (**d**, **e**), values are shown only if they are statistically significant at the 95% level based on two-tailed z-test and the number of samples that contribute to the calculation is >200. AR frequency is based on integrated water vapor transport (IVT) derived from ERA-Interim reanalysis (Dee et al. 2011). Precipitation is from Global Precipitation Climatology Project version 1.2 (Huffman et al. 2001). (Updated from Guan and Waliser 2015)

Europe is largely consistent with the NAO's result (Lavers and Villarini 2013b), a regional manifestation of the AO.

Compared to how these four modes influence AR frequency, their influence on AR precipitation is highly localized. In this regard, regional- and local-scale modes of variability—in addition to the large-scale modes—may need to be considered. The months of April–September (southern winters) await similar analysis. Further investigations are needed to better understand the range of processes that affect AR frequency and precipitation in different regions, seasons, and climate regimes.

4.4 ARs along the North American West Coast

Jonathan J. Rutz, Paul J. Neiman, and Alexander Gershunov

Many of the most extensive field campaigns involving ARs and their effects have taken place along the US West Coast and over the adjacent waters of the northeastern Pacific Ocean. These field campaigns and additional studies—made possible by new satellite and reanalysis data sets—have generated a wealth of knowledge on the climatology of ARs that affect this region. This section describes the climatology of ARs along the US West Coast and the atmospheric patterns most typically associated with them.

The first climatology of ARs along the North American west coast was produced by Neiman et al. (2008a) and covered an 8-year period from October 1997–September 2005. They defined ARs as narrow plumes of integrated water vapor (IWV) with values >2 cm that were >2000 km long and <1000 km wide, following Ralph et al. (2004). IWV data were retrieved from the Special Sensor Microwave/Imager (SSM/I; Hollinger et al. 1990), and, in the absence of the necessary wind observations, IWV serves as a reasonable proxy for IVT (Ralph et al. 2004).

A catalog of days that featured landfalling ARs along the coast was developed and divided into two sections based on latitude: (1) the "north coast" (41.0°–52.5°N) and (2) the "south coast" (32.5°–41.0°N). A monthly breakdown of AR landfalls by coastal region highlights major differences in seasonality (Fig. 4.14). Landfalling ARs are much more frequent along the north coast, especially from June–November, whereas landfalling ARs are nearly as—or more—frequent along the south coast from December–May, reflecting seasonal migration of the North Pacific jet. Dettinger et al. (2011) expanded on this climatology to include the next three water years (through September 2008), and their results (not shown) broadly agree with the frequency and seasonality shown in Fig. 4.14.

A recent study by Gershunov et al. (2017), spanning almost 70 years (1948–2015) of National Centers for Environmental Prediction–National Center for Atmospheric Research (NCEP–NCAR) reanalysis data, detects landfalling ARs using criteria of IVT >250 kg m^{-1} s^{-1}, IWV >15 mm, and length >1500 km. Their results clearly show the seasonal progression of landfalling ARs and their associated characteristics along North America's west coast (Fig. 4.15). These ARs peak in frequency in September and October along the coast of British Columbia, with the coastal landfall maximum gradually shifting southward along the US West Coast and weakening, reaching the Oregon/California border by December and January, and becoming markedly less frequent through February (Fig. 4.15a). The average duration of landfalling ARs follows a seasonal progression similar to that of frequency, particularly north of ~35°N (Fig. 4.15b). South of there, the average duration does not decrease as rapidly as the frequency, but the longer-lived events are largely confined to January and February. At most latitudes, the mean IVT associated with landfalling ARs is greatest during the cool season (October–March), but south of ~30°N the

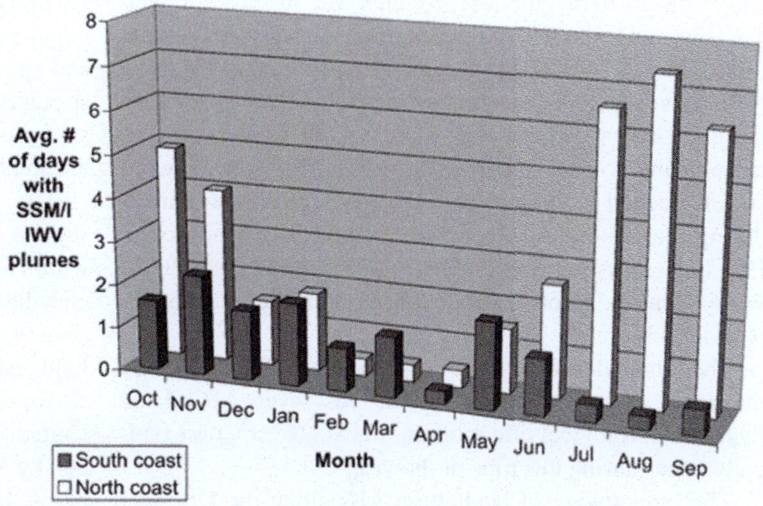

Fig. 4.14 Monthly distribution of the average number of days Special Sensor Microwave Imager (SSM/I)-observed integrated water vapor (IWV) plumes intersected the north-coast and south-coast domains of North America during the water years 1998–2005. (From Neiman et al. 2008a)

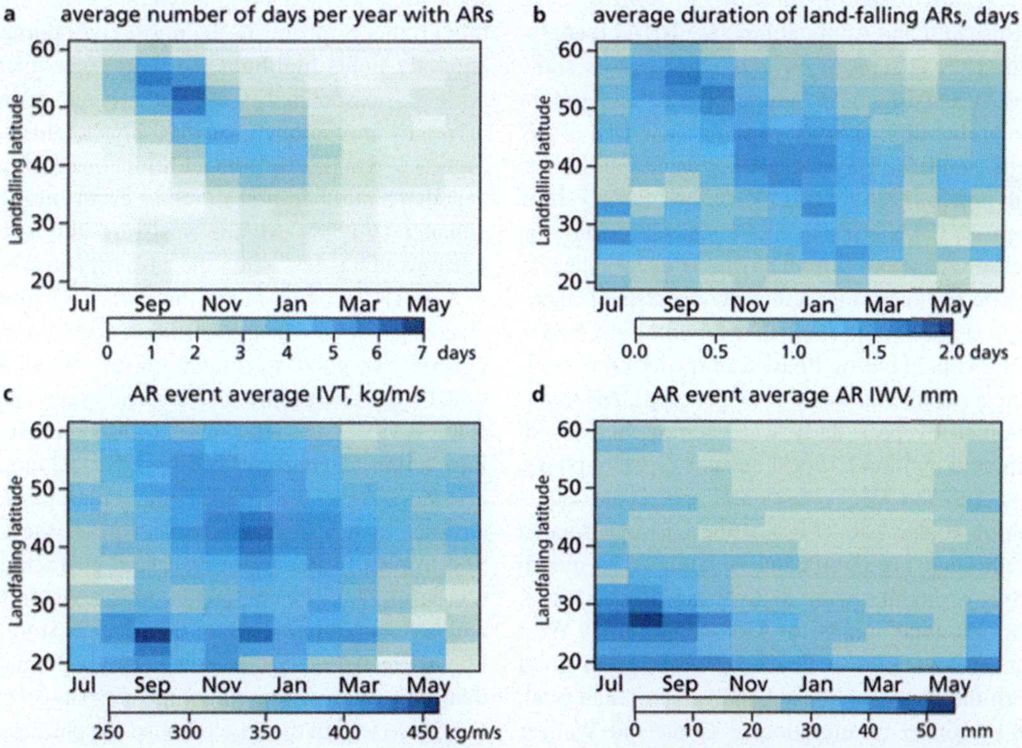

Fig. 4.15 Statistics of landfalling ARs along the west coast of North America as a function of month and landfall latitude, based on National Centers for Environmental Prediction–National Center for Atmospheric Research (NCEP-NCAR) reanalysis data from 1948–2015. (**a**) Number of 6-hourly AR occurrences rounded to days, (**b**) average duration (of consecutive AR occurrences) at landfalling latitude, (**c**) mean integrated water vapor transport (IVT) per AR occurrence, and (**d**) mean integrated water vapor (IWV) per AR occurrence

July–September period also features large IVT events, likely related to decaying tropical storms over the northeastern Pacific (Fig. 4.15c). The mean IWV shows this late summer and autumn phenomenon even more clearly, highlighting the latitudes and seasons most affected by tropical and subtropical moisture (Fig. 4.15d).

There are noteworthy differences in the seasonality of US West Coast AR landfalls found in the climatologies of Dettinger et al. (2011), Gershunov et al. (2017), and Guan and Waliser (2015), the last of which was featured in Sect. 4.2 (Fig. 4.6). These differences are primarily attributable to differences in ARDM. Methods that rely on IWV (e.g., Neiman et al. 2008a; Dettinger et al. 2011), will identify more ARs during the warm season (April–September), as more water vapor is available, while the greater wind speeds needed to support enhanced IVT are less common during these months. In contrast, methods that incorporate IVT (e.g., Gershunov et al. 2017; Guan and Waliser 2015) will identify more ARs during the cool season (October–March).

The remainder of this section focuses on AR climatology during the cool season, given that the most severe AR-related hydrometeorological effects along the US West Coast typically occur during this time of the year.

The climatology of landfalling ARs along the US West Coast was also examined by Payne and Magnusdottir (2014), who presented results based on Modern-Era Retrospective Analysis for Research and Applications (MERRA) reanalysis data during 1979–2011 (Fig. 4.16). Their results show that the latitude of maximum AR landfall frequency (and, in fact, the entire probability distribution function [PDF] of AR landfalls) shifts southward as the cool season progresses, consistent with previous results (Fig. 4.16a). As ENSO shifts from a warm phase (El Niño) to a cool phase (La Niña), the PDF of landfalling ARs goes from being more evenly distributed across latitudes to having a more concentrated maximum near 45°N (Fig. 4.16d). The PDF of landfalling ARs is significantly augmented during Phase 6 of the MJO, becoming more strongly centered near 45°N (Fig. 4.16g). AR intensity, as determined by peak daily water vapor flux, is not strongly modulated by month, ENSO phase, or MJO phase, though lower-numbered MJO phases do seem to produce a larger fraction of weaker ARs (Fig. 4.16b, e, h). AR-related precipitation, although not significantly modulated by month or ENSO phase, is modulated by MJO phase (Fig. 4.16c, f, i). In particular, the total number of ARs along the US West Coast is greatest during MJO Phase 6, which is in good agreement with the California-based analysis in Guan et al. (2012) and the broader examination in Guan and Waliser (2015). An increased number of ARs during El Niño also agrees with Guan and Waliser (2015), although the latter suggested the effect to be weak in and near California compared to further north.

Along with the development of an AR catalog, Neiman et al. (2008a) also presented plan view NCEP–NCAR reanalysis composites of relevant quantities during landfalling ARs along the north-coast and south-coast sections described above. The 500-hPa-height fields during AR landfalls along the north and south coasts (Fig. 4.17a, c) are characterized by greater heights (i.e., a ridge) over parts of the western US, lesser heights (i.e., an eastward extension of a trough) over the Gulf of Alaska, and a strong gradient in heights between, typically oriented from southwest to northeast. The 500-hPa-height anomalies (Fig. 4.17b, d) more clearly show the dipole and gradient in heights, highlighting the importance of a strong ridge over the western US for the northern ARs, and a strong offshore trough (or cut-off low-pressure system) for the southern ARs, as Rutz et al. (2015) also showed.

The composites of 500-hPa-height anomalies clearly show a corridor of enhanced southwesterly flow embedded within an amplified atmospheric pattern, and it is along these corridors that enhanced IVT and ARs are often observed (Fig. 4.18). The north-coast composite (Fig. 4.18a) highlights ARs that extend from the eastern Pacific Ocean northeastward toward southwestern Canada and the Northwestern United States, containing IVT values in excess of 500 kg m^{-1} s^{-1}. The south-coast composite (Fig. 4.18b) highlights ARs that extend toward the southwestern US, with IVT generally <350 kg m^{-1} s^{-1}. Both composites feature ARs somewhat oriented along a synoptic-scale cold frontal zone, though this is clearer in the north-coast composite. The IVT anomaly fields highlight the large-scale cyclonic curvature typically associated with south-coast events, which are driven by anomalously low 500-hPa heights just offshore (cf. Figs. 4.17d and 4.18d). In addition, these anomaly fields suggest that a number of north-coast events may be connected to enhance IVT that extends westward into the central Pacific, whereas this is less often the case for south-coast events.

For AR landfalls along the north and south coasts, composites of IWV constructed from SSM/I and reanalysis are generally in good agreement on the overall spatial distribution of the AR plume and tropical water vapor "reservoir" (Fig. 4.19). The AR plume (northernmost dotted line in Fig. 4.19e, f), characterized by IWV ≥2 cm, extends northeastward from the tropics and subtropics, along and in advance of a cold frontal zone, toward the US West Coast. The tropical water vapor reservoir, characterized as a broad region of IWV >3 cm, covers most regions south of ~20°N and is centered just north of the equator (southernmost dotted line in Fig. 4.19e, f). One noteworthy difference in analyses is that the reanalysis-based composites suggest values of IWV (~2.5 cm) within the core of these AR plumes, which is larger than that observed from SSM/I imagery, further confirming the value of satellite observations over the data-sparse Pacific.

AR landfalls along both coasts are characterized by a strong upward maximum in vertical motion along and just

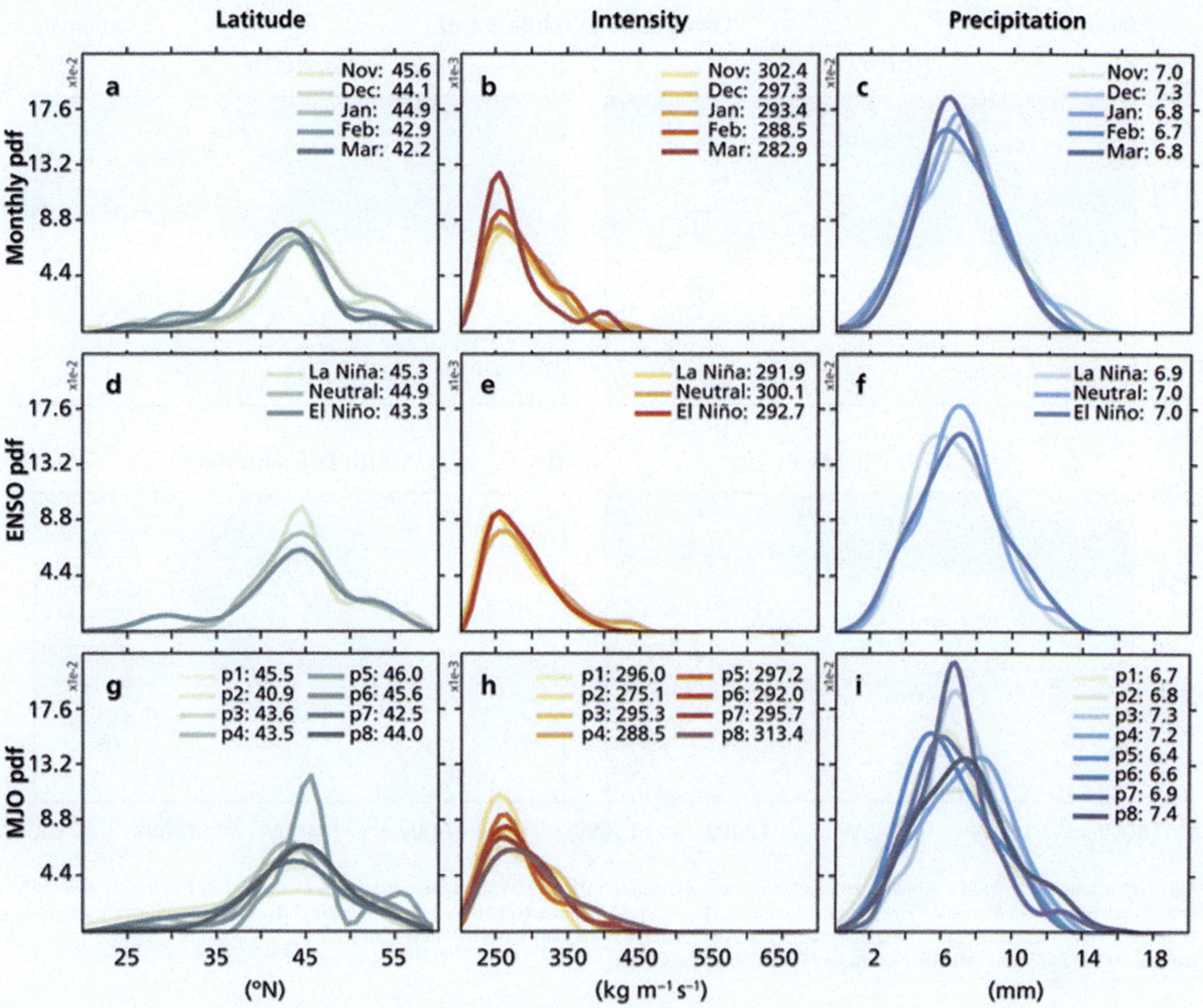

Fig. 4.16 Probability density functions for landfalling ARs over Nov–Mar for the years 1979–2011 sorted according to (**a–c**) month (749 dates), (**d–f**) El Niño–Southern Oscillation (ENSO) phase (749 dates), and (**g–i**) Madden-Julian Oscillation (MJO) phases with amplitudes >1 (469 dates). Each column shows the distribution of (*left*) landfalling latitude, (*center*) landfalling peak daily water vapor flux, and (*right*) landfalling total daily precipitation. The y-axis shows the probability density function for each panel, where the center column is an order of magnitude less than the right and left columns. Averages for each category are shown in the legend in each panel. (From Payne and Magnusdottir 2014)

upstream of where the AR intersects the coast. In these areas, orographic lifting and dynamic processes associated with the cold front combine to produce a region of enhanced rising motion (Fig. 4.20). Not surprisingly, composites of precipitation (not shown) for both north- and south-coast ARs highlight large amounts of precipitation in the vicinity of these rising motions.

The schematic in Fig. 4.21 blends some of the key features identified in the north- and south-coast composites described above into one comprehensive, conceptual representation of ARs that impact the US West Coast. The wintertime plan view schematic depicts high pressure west of the Baja Peninsula and low pressure (i.e., an extratropical cyclone) in the Gulf of Alaska (Fig. 4.21a). Between these two features, a northeastward-directed AR extends along and ahead of the cold frontal zone associated with the cyclone, with strong water vapor flux being located just offshore. Dashed lines indicate precipitation, which is most intense where vertical motion is maximized, in the area where the AR intersects near-coastal topography. Corresponding vertical-profile composites for both winter and summer highlight changes in wind speed, wind direction, mountain-normal water vapor flux, and vertical velocity as a function of height (Fig. 4.21b). In general, wintertime ARs feature stronger wind speeds and a more favorable orientation of flow relative to topography than their more moisture-laden

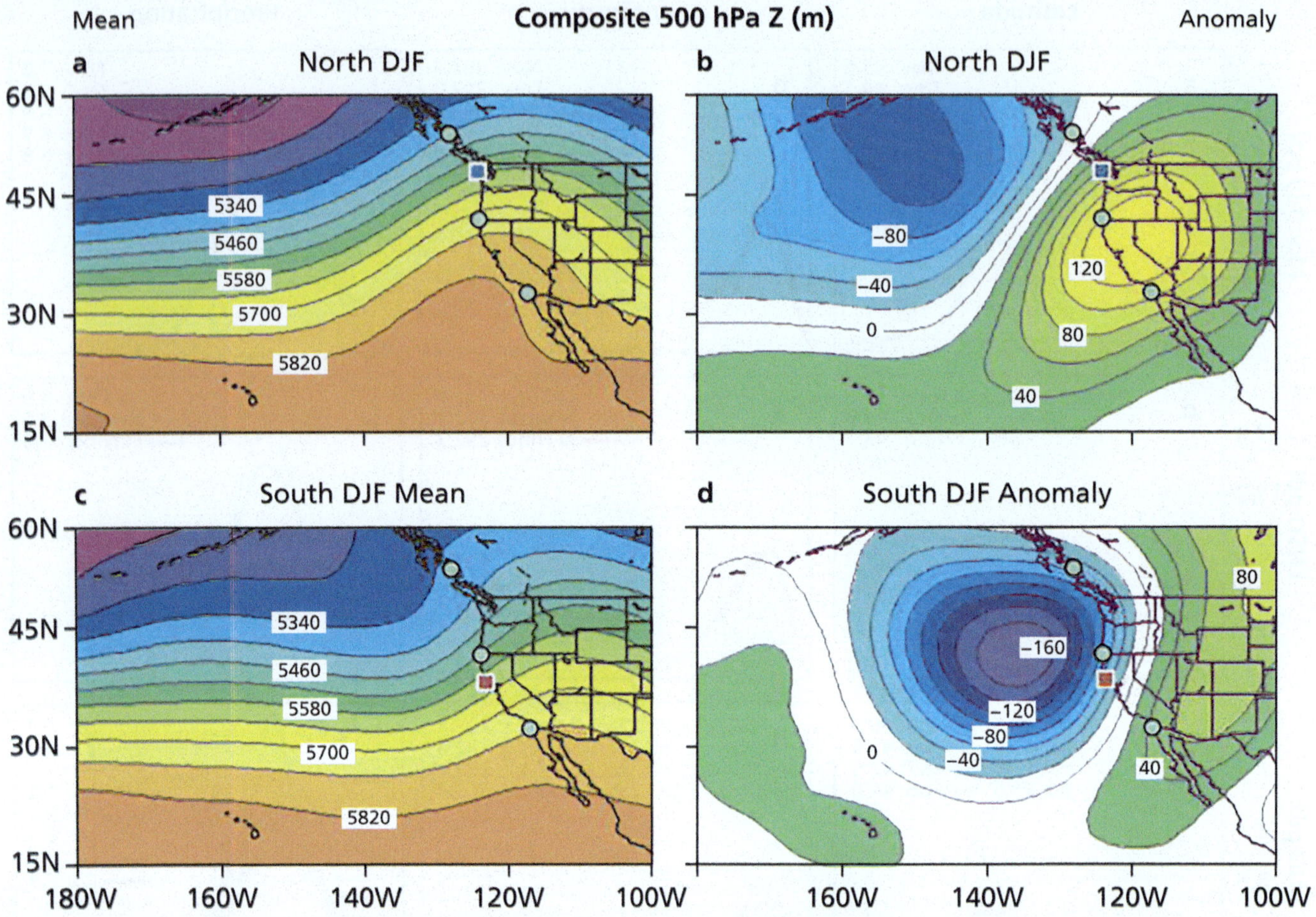

Fig. 4.17 Composite 500-hPa geopotential height (*left*) and anomalies (*right*) derived from the National Centers for Environmental Prediction–National Center for Atmospheric Research (NCEP-NCAR) reanalysis data set for Special Sensor Microwave/Imager (SSM/I) integrated water vapor (IWV) plumes (i.e., ARs as defined in Neiman et al. (2008a)) intersecting the (*top*) north-coast and (*bottom*) south-coast domains on a daily basis in winter (DJF)

summertime counterparts. Hence, upon intersecting topographical barriers along the US West Coast, the wintertime ARs produce more intense upward vertical motions and mountain-normal water vapor fluxes, leading to heavier precipitation and greater impacts.

The fraction of precipitation attributable to ARs, on various time-scales, has been an important topic of research. Dettinger et al. (2011) examined the fraction of cool-season precipitation attributable to landfalling ARs (between 32.5° and 52.5°N) at western US Cooperative Observer Program (COOP) weather stations during water years 1998–2008 and found values ranging from <10% to nearly 50%, with the greatest values across northern California and the smallest values located east of the Rockies and over the southwestern US (Fig. 4.22a). Rutz and Steenburgh (2012) performed a similar analysis using 0.25° gridded data and snow telemetry (SNOTEL) data for comparison. The gridded analysis, shown in Fig. 4.22b, is in broad agreement with the results of Dettinger et al. (2011). Furthermore, the predominantly high-elevation SNOTEL analysis conducted by Rutz and Steenburgh (2012) agrees well with the predominantly low-elevation COOP data. These authors also extended the analysis to consider ARs that make landfall along the coast of Mexico's Baja Peninsula (24°–32.5°N) and found a larger fraction of precipitation attributable to ARs across much of the western US (Fig. 4.22c, d). In particular, the fraction of precipitation attributable to ARs was found to be 10–15% greater over the southwestern US (i.e., up to double the fraction found without including these ARs), highlighting the importance of landfalling ARs along the Baja Peninsula in contributing to western US hydroclimate. Further contributions to precipitation, particularly over the southwestern US, are likely made by cut-off closed lows, which play a major role in the climatology of this region, and are sometimes associated with ARs (Oakley and Redmond 2014; Abatzoglou 2016).

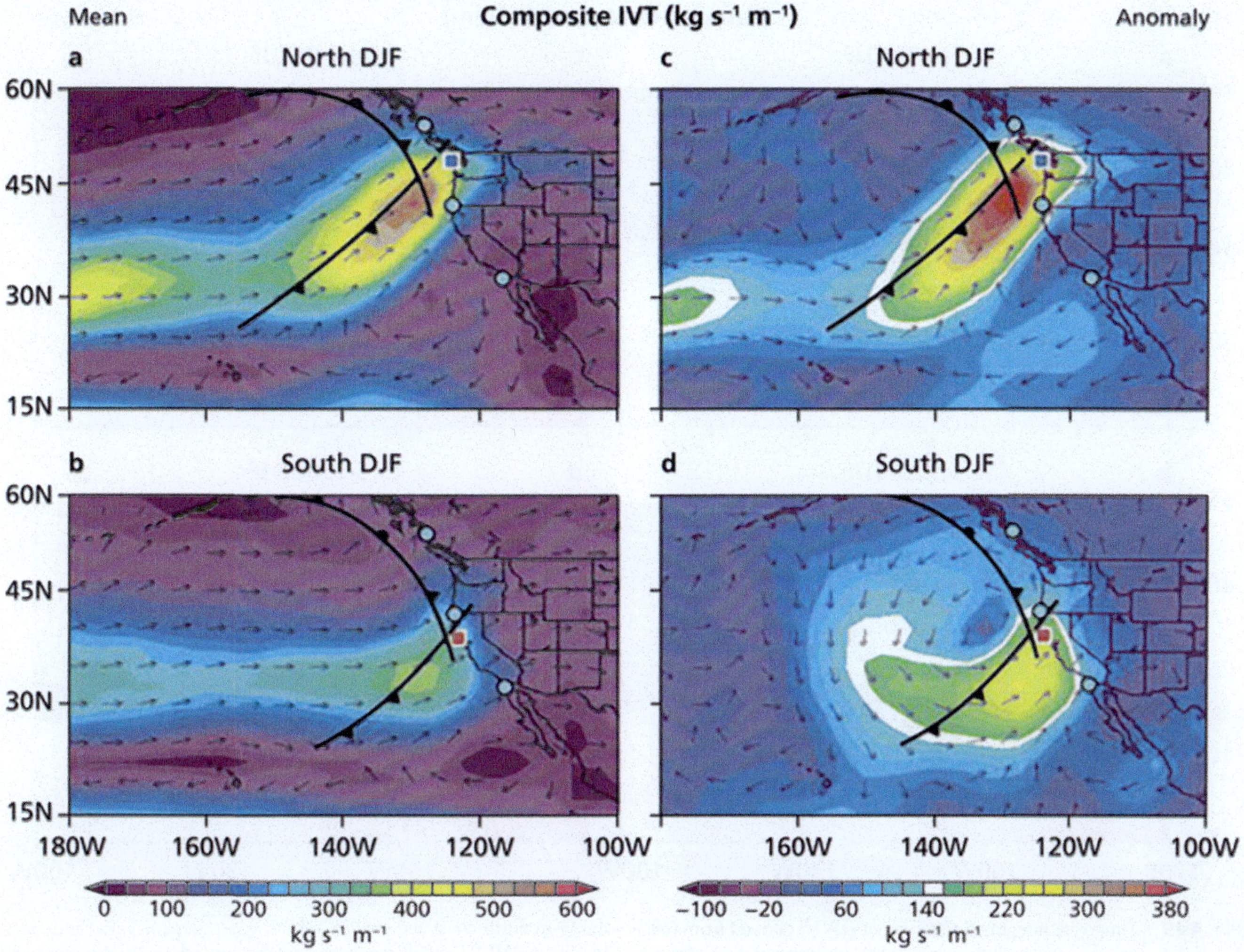

Fig. 4.18 Composite 500-hPa geopotential height (*left*) and anomalies (*right*) derived from the National Centers for Environmental Prediction–National Center for Atmospheric Research (NCEP–NCAR) reanalysis data-set for integrated water vapor transport (IVT; kg s^{-1} m^{-1}). (i.e., ARs as defined in Neiman et al. 2008a) intersecting the (*top*) north-coast and (*bottom*) south-coast domains on a daily basis in winter (DJF)

Gershunov et al. (2017) examined the fraction of precipitation attributable to landfalling ARs by season (Fig. 4.23a), based on 6-km precipitation data from Livneh et al. (2013). In the annual mean, they found this fraction to be >30% along and upstream of the Pacific Ranges and >50% over northern California and southern Oregon, whereas this fraction is <30% over interior basins and <20% over portions of the Great Basin (i.e., eastern Nevada and western Utah). Their seasonal analysis reveals that these fractions are greatest during the October–December period and least during the April–June period, and the spatial distribution locales with higher and lower fractions remain relatively consistent throughout the year. The cool-season results from Gershunov et al. (2017) fall somewhere in between those of Dettinger et al. (2011) and Rutz et al. (2014), which is not surprising given that they use both IWV and IVT to identify ARs, whereas Dettinger et al. (2011) uses only IWV and Rutz et al. (2014) uses only IVT.

4.4.1 Summary of Sect. 4.4

This section has highlighted the climatology of ARs making landfall along the west coast of North America. It has also, to some extent, discussed the climatology of AR-related impacts, albeit from a very broad point of view. The hydrometeorological processes and effects associated with ARs—both individually (e.g., extreme precipitation and flooding) and cumulatively over time (e.g., water resource availability, effects on landscapes and ecosystems)—are discussed more thoroughly in other chapters of this book.

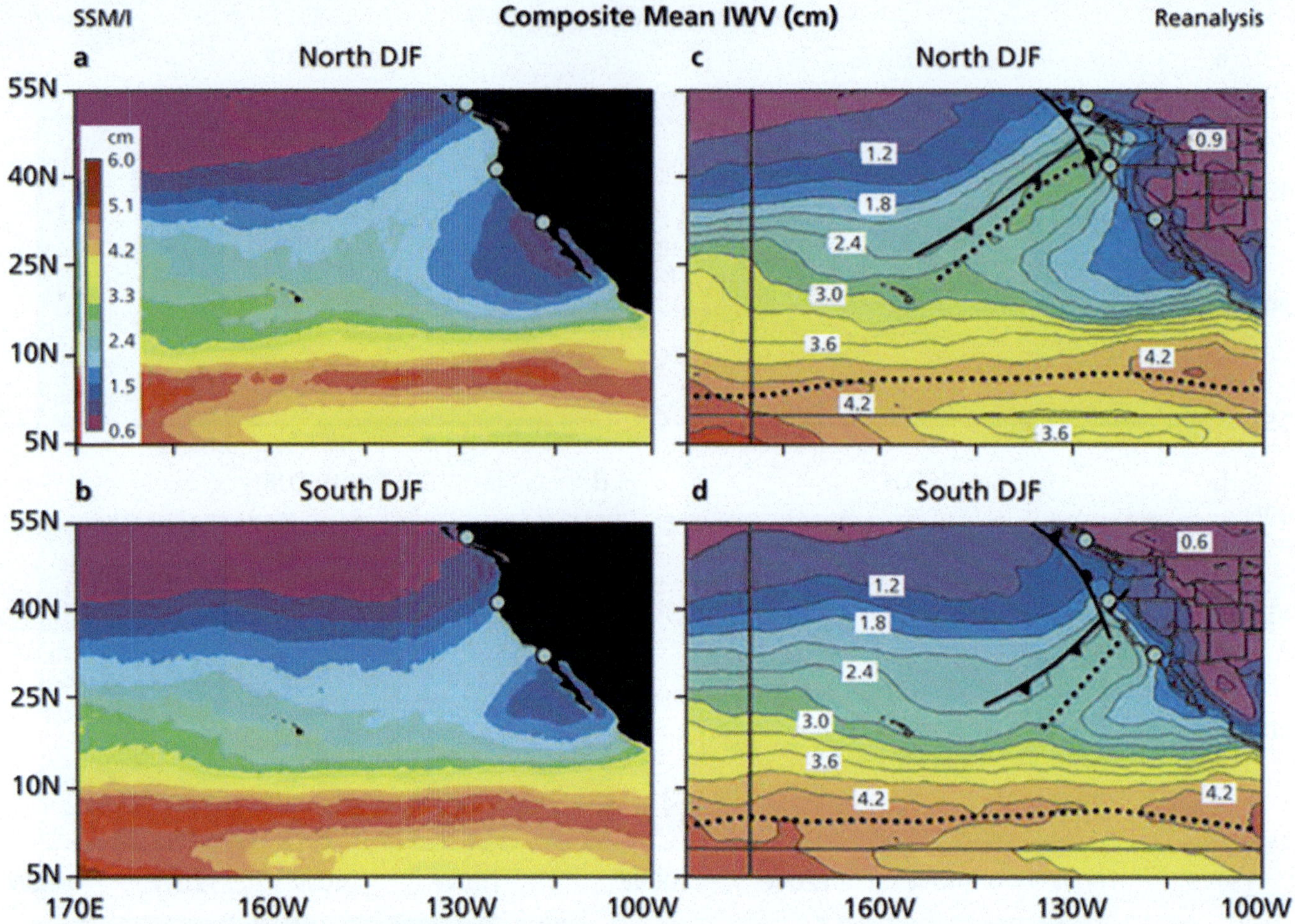

Fig. 4.19 Composite integrated water vapor (IWV) derived from (*left*) Special Sensor Microwave/Imager (SSM/I) imagery and (*right*) the National Centers for Environmental Prediction–National Center for Atmospheric Research (NCEP–NCAR) reanalysis data set for SSM/I IWV plumes (i.e., ARs as defined in Neiman et al. 2008a) intersecting the (*top*) north-coast and (*bottom*) south-coast domains on a daily basis in winter (DJF). Dotted lines represent the core of the IWV plumes and the inter-tropical conversion zone (ITCZ). Standard frontal notation is used to mark approximate positions of relevant synoptic features

4.5 Inland-Penetrating ARs Over the Western United States

Jonathan J. Rutz, Michael Alexander, and Paul J. Neiman

As discussed throughout this chapter, ARs represent a cornerstone atmospheric phenomenon for generating significant wintertime precipitation along the US West Coast and other coastlines globally. ARs and their effects are often thought of as being limited to low-lying coastal regions, such as that found between the west coast of North America and the quasi-continuous topographical barrier that resides 50–250 km inland (hereafter, the Pacific Ranges). However, ARs can at times penetrate eastward of the Pacific Ranges and affect parts of the Intermountain West (IMW). Prime examples include the January 2010 AR event that broke meteorological and hydrological records across the state of Arizona (Neiman et al. 2013; Hughes et al. 2014) and the November 2006 AR that severely impacted Oregon and Washington before causing high-elevation rainfall, road washouts, and extensive flooding in and around Glacier National Park, which lies far from the coast in northwestern Montana (Bernhardt 2006; Neiman et al. 2008b; Mueller et al. 2017). This section addresses inland-penetrating ARs, first by highlighting key hydrometeorological and dynamical characteristics of an AR that ravaged the semi-arid desert southwest, and then by exploring more generally the climatology of ARs across the western US and the airflow trajectories associated with inland-penetrating ARs across this region.

In late January 2010, a series of transient baroclinic waves propagated inland from coastal California and Mexico to the southwestern US. The last, and strongest, of these disturbances was accompanied by an intense AR that contributed to widespread heavy rains, substantial orographic precipitation enhancement along the Mogollon Rim escarpment, record-setting high-elevation snows with blizzard condi-

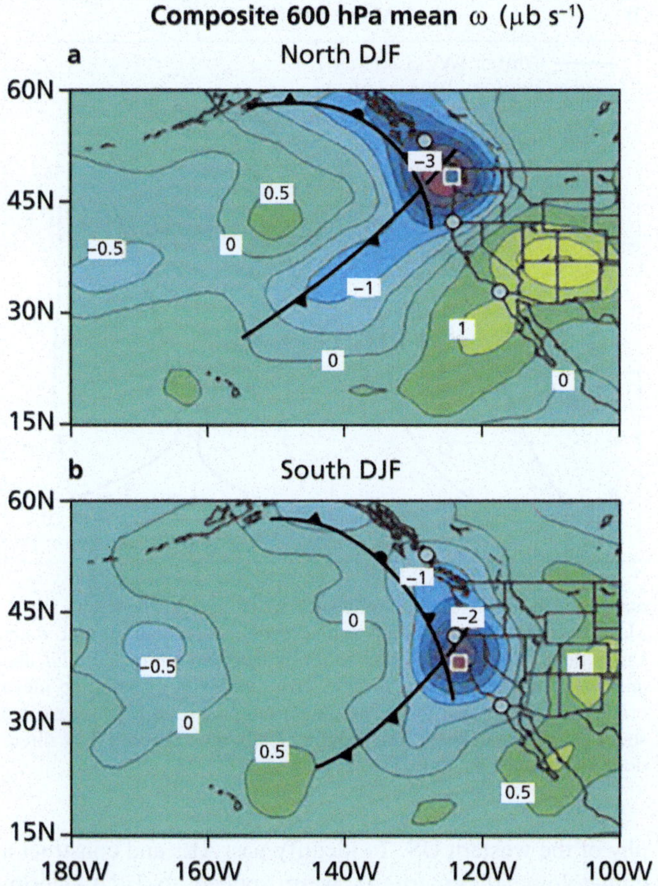

Fig. 4.20 Composite 500-hPa geopotential height (*left*) and anomalies (*right*) derived from the National Centers for Environmental Prediction–National Center for Atmospheric Research (NCEP–NCAR) reanalysis data set for mean vertical velocity ω (μb s^{-1}) (i.e., ARs as defined in Neiman et al. 2008a) intersecting the (*top*) north-coast and (*bottom*) south-coast domains on a daily basis in winter (DJF)

tions, strong thunderstorms, flash flooding, dust storms, power outages, and road closures in Arizona. This storm caused three fatalities and incurred more than $11 million in damages across the state (NCDC 2010). The characteristics and effects of this AR, as documented in the observational study of Neiman et al. (2013), were comparable to those of strong landfalling ARs along the west coasts of mid-latitude continents.

At 0000 UTC 22 January, 2010, a potent AR made landfall along the Baja Peninsula of Mexico and penetrated northeastward through a relatively low-elevation gap in the topography of the peninsula. The AR reached Arizona with IVT values >10 standard deviations above the mean directed perpendicular to the mountainous terrain of the Mogollon Rim. The schematic shown in Fig. 4.24 highlights key dynamical factors that led to the inland penetration and extreme AR effects in Arizona, the latter of which are discussed in more detail in Chap. 5. Back-trajectory analyses (not shown) revealed that the collocated IWV and IVT plumes within the intense AR of 21–22 January strengthened markedly and rapidly during landfall via extratropical dynamical processes, without entraining tropical water vapor. Gridded reanalysis data coupled with regional observing platforms across central Arizona documented strong, moist, northeastward-directed water vapor fluxes centered at ~2 km mean sea level (MSL) and moist-neutral stratification in the lowest 3 km of the atmosphere. The low-level fluxes and moist neutrality in this case matched the mean observed characteristics from 17 dropsondes released into ARs over the Pacific (Ralph et al. 2005). The vertical phasing of the AR with upper-level dynamics is highlighted at 250 hPa, where the left exit of a cyclonically curved polar jet—a favored region of ascent (Beebe and Bates 1955)—was situated across northern Arizona above the downwind terminus of the AR.

Given the presence of weak static stability, strong incoming water vapor fluxes oriented perpendicular to the northwest–southeast escarpment of the Mogollon Rim, and large-scale dynamics, orographically enhanced heavy precipitation was focused in the center of the state and lasted for >24 h. As quantified in a follow-on modeling study by Hughes et al. (2014), the intense character of the incoming vapor fluxes was partly a consequence of the precise positioning of the AR across the relatively low mountains of southern Baja south of 30°N (rather than across the much higher, northern portion of this range) and west of Mexico's Sierra Madre Occidental, because only a fraction of the water vapor within the AR over the eastern Pacific was likely lost to orographic processes upwind of Arizona. This terrain gap in the Baja Peninsula generally serves as a focus for ARs that penetrate further northeastward (Rutz et al. 2015). During the persistent heavy precipitation over central Arizona, unusually high melting levels that exceeded 2.5 km MSL in the AR environment resulted in rain (rather than snow) falling at high elevations over much larger catchment areas than during a typical storm, thus leading to greatly enhanced runoff volumes and flooding. Additional factors that contributed to the runoff included snowmelt from the anomalously large snowpack, previously saturated soils, and relatively impermeable bedrock with overlying thin soils.

This case study—combined with a composite study by Rutz and Steenburgh (2012) that showed substantial contributions of annual precipitation in Arizona that arise from the landfall of ARs across Baja—highlights the fact that cool-season ARs strongly affect Arizona's large storms, floods, and water supplies. Another composite study by Rivera et al. (2014) provided a climatological characterization of AR events that led to cool-season precipitation in Arizona's Verde River basin, demonstrating that the ARs crossed the Baja Peninsula before affecting the Verde basin. This case study provides an example of one AR that penetrated inland over Arizona; the remainder of this section will examine

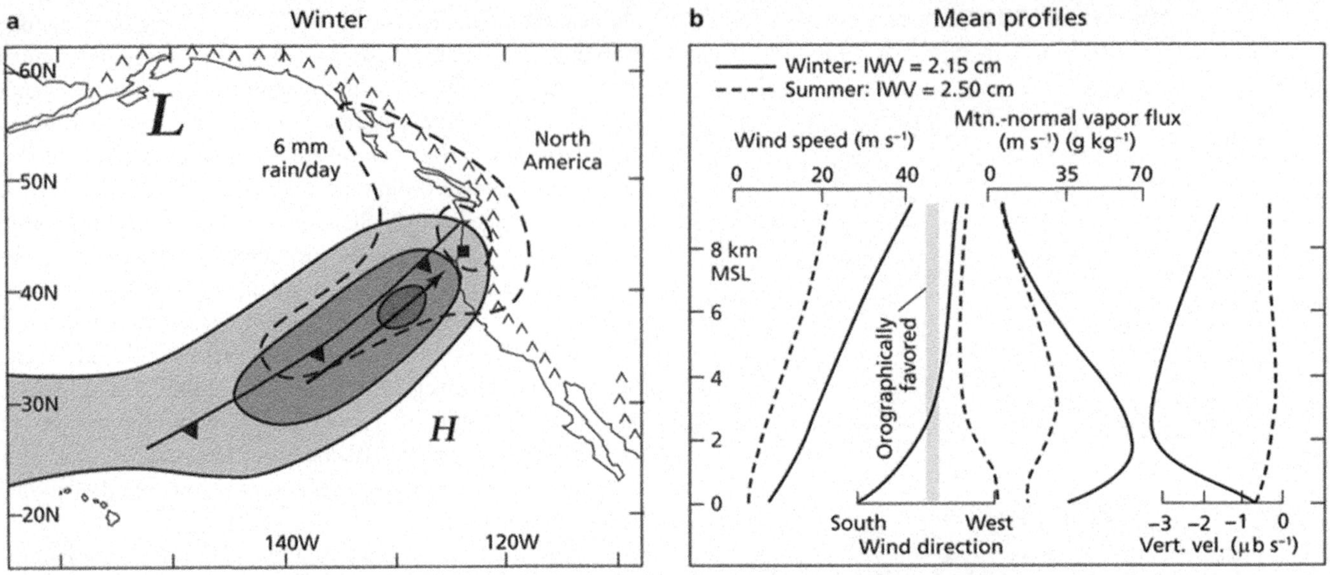

Fig. 4.21 Conceptual representation of synoptic conditions associated with landfalling ARs during DJF, based on an average of the north-coast and south-coast reanalysis composites. *Left:* Plan view of integrated water vapor transport (IVT) (*solid contours*; *light shading*: IVT 250–350 kg m^{-1} s^{-1}, *medium shading*: IVT 350–450 kg m^{-1} s^{-1}, *dark shading*: IVT > 450 kg m^{-1} s^{-1}), daily rainfall (*dashed*; mm d^{-1}), 925-hPa cold front and pre-cold-frontal flow (*bold arrow*). The *black square* marks the position of the composite sounding shown below. *Right:* Mean profiles of wind speed and direction, mountain-normal water vapor flux, and vertical velocity for winter and summer (*solid* and *dashed*, respectively). The *vertical gray-shaded bar* marks the mean orientation orthogonal to the mountain ranges in the north-coast and south-coast domains (i.e., the orographically most favored flow direction). (Neiman et al. 2008a)

more broadly the climatological characteristics of ARs that penetrate into the western US.

ARs are often identified using IWV or IVT, but an IVT-based threshold works best over the IMW for multiple reasons. First, cool-season 24-h precipitation is better correlated with daily-mean IVT than with daily-mean IWV across this region (not shown; see Fig. 2 in Rutz et al. 2014). Furthermore, these greater correlations are most evident along and upstream of topographical barriers, where precipitation tends to be heaviest. For example, Ralph et al. (2013) found that upslope vapor flux explained ~75% of the variability in observed precipitation near the California coast. Second, the general decrease in IWV with increasing surface elevation and further reduction during the cool season results in climatological values over the western US that are well below traditional IWV-based thresholds (i.e., 20 mm) for identifying ARs over the northeastern Pacific. Third, during many AR events, areas of IVT ≥250 kg m^{-1} s^{-1} often penetrate further inland than coinciding areas of IWV ≥ 20 mm. Furthermore, areas of IVT ≥250 kg m^{-1} s^{-1} often overlap with areas of heavy precipitation. An example highlighting both of these aspects is shown in Fig. 4.25.

For these reasons, Rutz et al. (2014) defined ARs as contiguous features with length ≥2000 km and IVT ≥250 kg m^{-1} s^{-1} throughout. In contrast to previous studies (Ralph et al. 2004; Neiman et al. 2008a; Dettinger et al. 2011), a limitation on feature width was not used, because it was found to be inconsequential over the complex topography of the western US. To identify past ARs and construct a climatology, these criteria were applied to ERA-Interim atmospheric reanalysis data during cool-season months beginning in November 1988 and ending in April 2011.

The AR frequency (Fig. 4.26a) is defined as the percentage of reanalysis time-steps in which an AR is identified at a particular grid point. The AR frequency is largest (>15%) along the coasts of Washington, Oregon, and northern California. It decreases rapidly southward along the California coast and eastward across the Cascades and Sierra Nevada—especially the southern "High" Sierra, between about 35° and 38°N—into the IMW, where the AR frequency is smallest (<3%). Relatively low-elevation corridors, such as Oregon's Columbia River basin, serve as conduits for the extension of larger AR frequencies into the interior (Fig. 4.26). Conversely, large decreases in AR frequency occur in the vicinity of high topographical barriers, with smaller AR frequencies extending downstream.

The mean AR duration (Fig. 4.26b) is defined as the mean number of consecutive reanalysis observations in which an AR is identified at a particular grid point, multiplied by the reanalysis time-step (e.g., 6 h). The mean AR duration is characterized by a spatial distribution similar to that of AR frequency, with the largest values (>20 h) along the coasts of Washington, Oregon, and northern California, and the smallest values (<15 h) across portions of the interior, particularly downstream of California's High Sierra. Based on this pattern, the mean AR duration is larger than one might expect

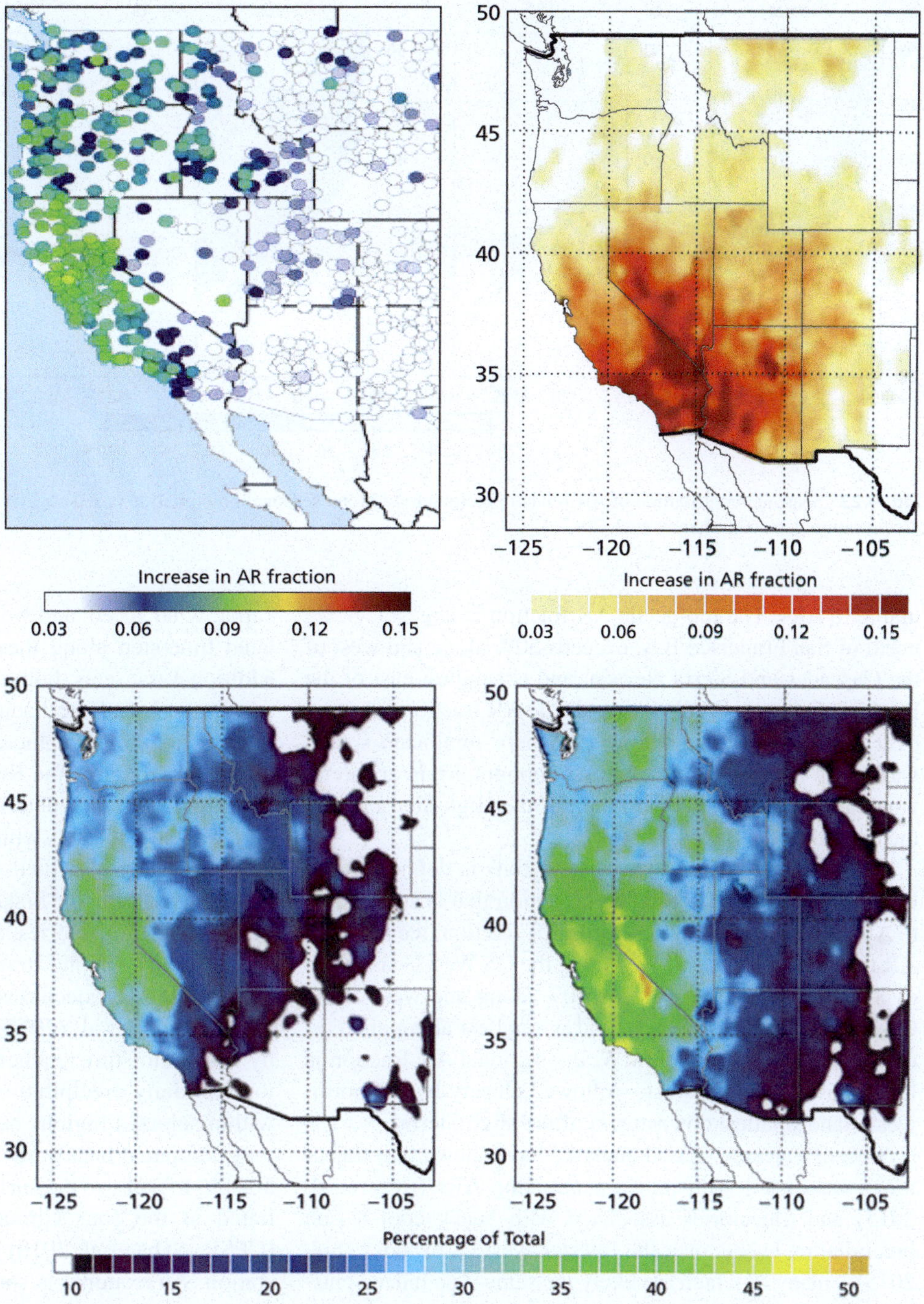

Fig. 4.22 Counter-clockwise from *top left* fraction of cool-season (November–April) precipitation attributable to landfalling ARs between 32.5°–52.5°N at western US cooperative weather stations (Dettinger et al. 2011). *Bottom left:* As previous but using the 0.25° gridded Climate Prediction Center (CPC) Unified Precipitation Data (Rutz and Steenburgh 2012). *Bottom right:* As previous but including landfalling ARs along the Baja Peninsula (24.5°–32.5°N; Rutz and Steenburgh (2012). *Top right:* Difference between the previous two (Rutz and Steenburgh 2012)

(16–18 h) over the southwestern US, given the smaller AR frequencies found there. As with AR frequency, larger mean AR durations extend eastward from Oregon's Columbia River basin through the relatively low-elevation northern US Rockies and into the far northwestern Great Plains, whereas smaller AR durations extend downstream of the High Sierra across the central Great Basin and into the central US Rockies.

ARs are often associated with heavy precipitation events, but are also important for their contributions to long-term hydrology. To assess their contribution to both weather and climate, precipitation must be attributed to ARs. The method of Rutz et al. (2014) considers whether or not an AR was identified in any of the five, 6-hourly reanalysis observations within a 24-h (1200–1200 UTC) accumulation period, and, if so, attributes that 24-h precipitation to an AR. Results shown below are based on a gridded precipitation analysis that agreed very well with point-based observations.

The top-decile AR fraction (Fig. 4.26c) is defined as the fraction of top-decile 24-h precipitation events that are attrib-

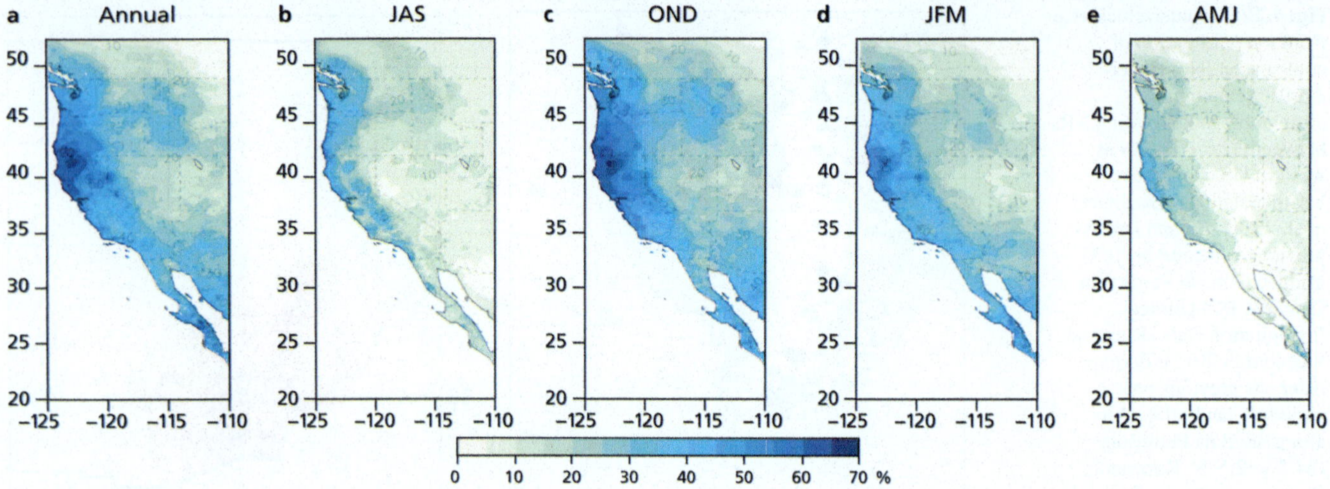

Fig. 4.23 Seasonal contribution of AR-related precipitation to total seasonal precipitation based on Livneh et al. (2013) 6 × 6-km gridded daily precipitation data. (Gershunov et al. 2017)

utable to ARs. The top-decile AR fraction is largest (>65%) north of San Francisco Bay, exceeds 50% along and west of the Cascades and Sierra Nevada, and is smallest east of the Rockies. Over the IMW, the top-decile AR fraction is largest over the northwest and southwest, where significant spatial variability attributable to complex topography is apparent, whereas it is smallest downstream of the High Sierra across the central Great Basin.

The seasonal AR fraction (Fig. 4.26d) is defined as the fraction of total cool-season precipitation that is attributable to ARs. Similarly to the top-decile AR fraction, the seasonal AR fraction is largest (>55%) along the US West Coast north of San Francisco Bay, exceeds 40% along and west of the Cascades and Sierra Nevada, and is smallest along and east of the Rockies. Over the IMW, the seasonal AR fraction is largest over the southwest—followed closely by the northwest—and smallest downstream of the High Sierra.

Recent trajectory-based analyses have shown that trajectories associated with inland-penetrating ARs (Rutz et al. 2015) and trajectories associated with heavy cool-season precipitation events over the Interior West (Alexander et al. 2015)—many of which are likely the same—do indeed cluster along the corridors into the IMW described above. The paragraphs below focus on these independent studies which, despite taking different approaches, resulted in very similar conclusions.

The study by Rutz et al. (2015) was designed to differentiate between the characteristics of ARs that decay near the coast and ARs that penetrate into the interior. To do so, they defined a transect of points along the North American west coast (24–52.5°N) and identified ARs at these points, following Rutz et al. (2014). Whenever an AR was identified at a given point, they computed 72-h forward trajectories starting from 950 and 700 hPa and recorded quantities such as water vapor, wind speed, and water vapor flux at each 1-h interpolated time-step along these trajectories. They defined two additional transects, both oriented quasi-parallel to the first: one in the lee of the Pacific Ranges and another along the next major topographical barrier that most trajectories encounter (roughly, the Rocky Mountains). Furthermore, at each time-step, they determined whether or not each trajectory was still located within a grid cell where an AR was simultaneously identified. Trajectories no longer within an AR upon reaching the first transect were classified as coastal-decaying AR trajectories, and those still within an AR upon reaching the second transect were defined as interior-penetrating AR trajectories.

Alexander et al. (2015) examined the general pathways by which moisture reached the IMW in association with the top 150 daily precipitation events that occurred at stations within selected regions between 1979 and 2011. To do so, they computed backward trajectories (from a pressure level 50–100 hPa above the surface) for each of these events, initiated at the four Climate Forecast System Reanalysis (CFSR; Saha et al. 2010) grid points that surround a given station. An example is shown in Fig. 4.27, in which a station in western Idaho recorded exceptionally heavy precipitation (104 mm) on 16 February 1982. The air parcels that arrived that day skirted the northern edge of the Sierra Nevada, traversing the northwestern Great Basin before reaching Idaho. These trajectories varied in pressure from 700 to 800 hPa east of California, but were previously at higher pressures (lower elevations) over California, and likely obtained moisture from evaporation over the Pacific Ocean.

The 950-hPa coastal-decaying and interior-penetrating forward trajectories calculated by Rutz et al. (2015), as well as the number of trajectories that passed through each 1.0°

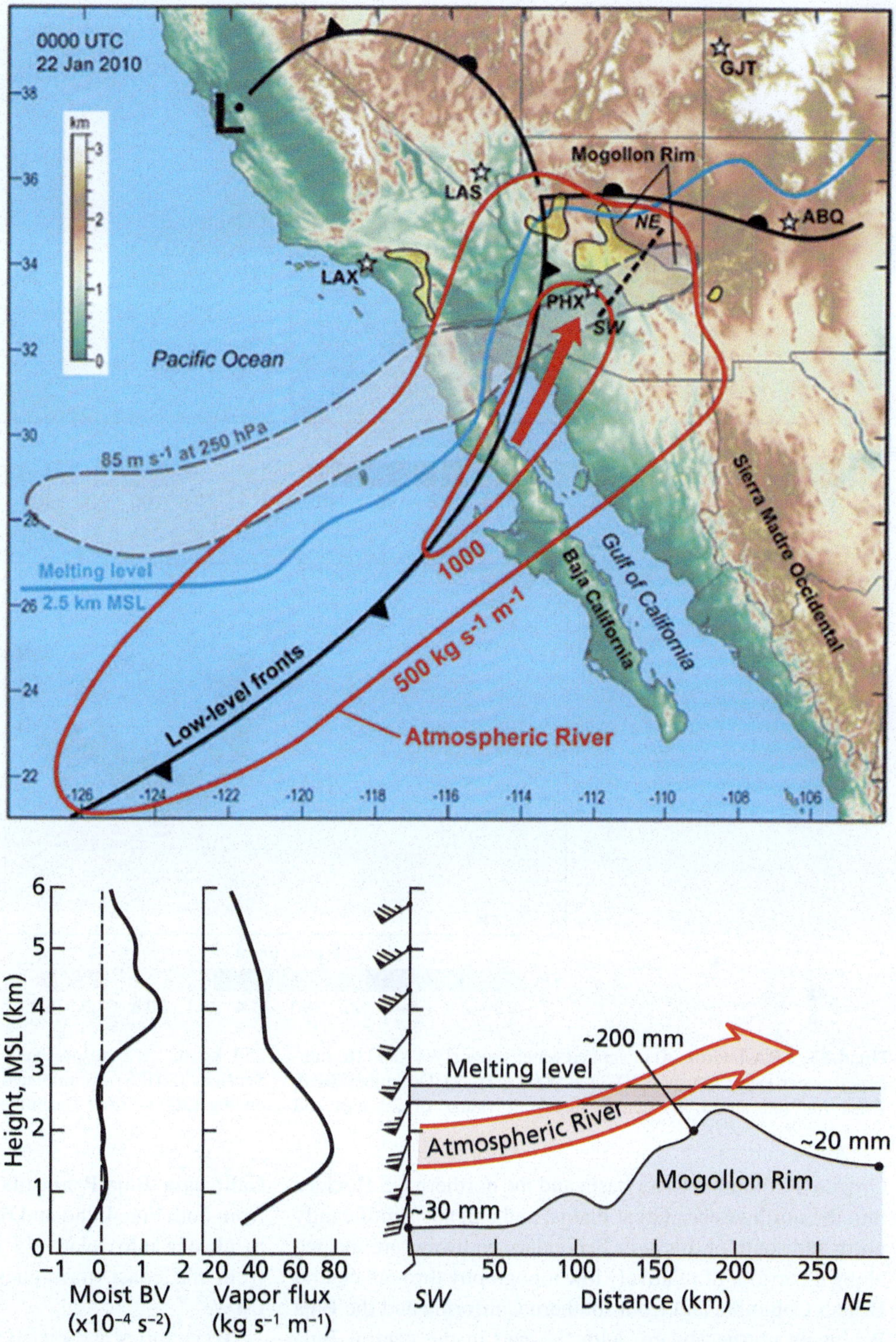

Fig. 4.24 Conceptual representation of the atmosphere at 0000 UTC 22 January, and 24-h precipitation accumulations ending at 1200 UTC 22 January 2010. *Top:* Plan view schematic of integrated water vapor (IVT) magnitude (*red contours*, with units of kg s^{-1} m^{-1}; *bold red arrow* shows the IVT vector direction in the AR core), the 85 m-s^{-1} isotach at 250 hPa (*gray dashed contour*; *interior shading* > 85 m s^{-1}), the melting level at 2.5 km mean sea level (MSL) (*blue contour*; estimated from the Climate Forecast System Reanalysis (CFSR) 0 °C altitude at 2.7 km, with the assumption that the melting level is located ~200 m below the 0 °C isotherm (e.g., Stewart et al. 1984; White et al. 2002), and the 75-mm isohyets (*thin solid contours; interior shading* >75 mm). The *black dashed line* along SW–NE shows the baseline for the cross-section in the bottom panel. Standard notation is used for the near-surface fronts. *Bottom:* Cross-section schematic across the Mogollon Rim (along SW–NE in the top panel) showing the melting level (*gray-shaded bar*), the AR (*red arrow*), and representative 24-h precipitation totals (mm) at three locations (*bold black dots*). The following vertical profiles at the southwest end of the cross-section are also shown: wind velocity and barbs (*flags* = 25 m s^{-1}, *barbs* = 5 m s^{-1}, *half-barbs* = 2.5 m s^{-1}), water vapor flux (kg s^{-1} m^{-1}; directed from 220°), and moist Brunt–Väisälä frequency squared (×104 s^{-2})

grid square, are shown in Fig. 4.28. Similarly, results from Alexander et al. (2015) show the number of backward trajectories that passed through each 0.5° CFSR grid square during the 5 days before their heavy precipitation events over four IMW regions: eastern Washington and northern Idaho; eastern Oregon and southern Idaho; Utah and western Colorado; and Arizona and New Mexico (Fig. 4.29).

Whereas the distribution of coastal-decaying trajectories shows a general decrease from north to south, the distribution of interior-penetrating trajectories clearly shows the influences of topography, highlighting three preferred corridors of AR penetration into the interior (Fig. 4.28c, d). The most prominent pathway of AR penetration into the interior is centered from the Pacific Northwest eastward through

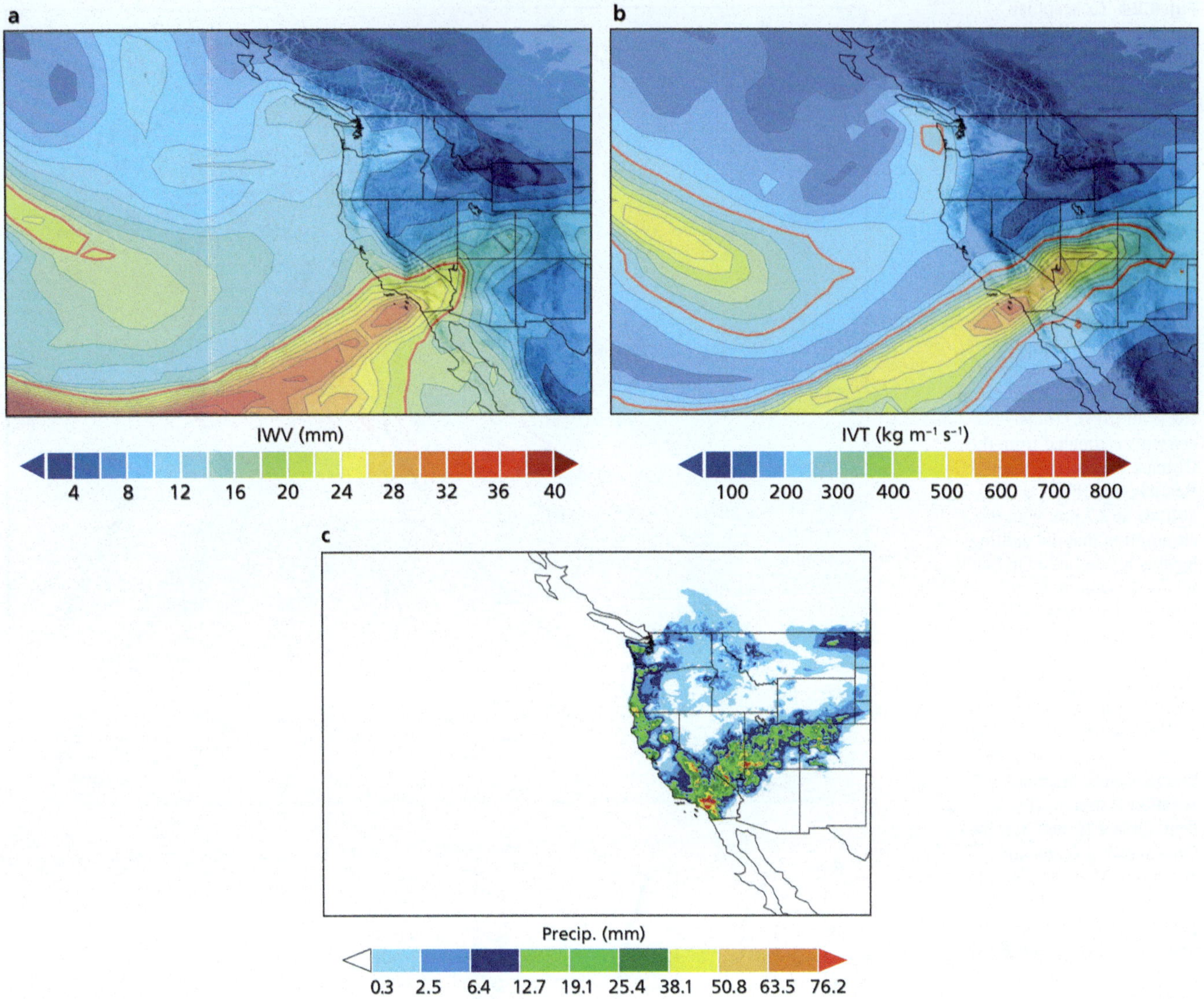

Fig. 4.25 ERA-Interim (**a**) integrated water vapor (IWV) and (**b**) integrated water vapor transport (IVT) at 0000 UTC 21 December 2010. *Thick red line* in (**a**, **b**) denote threshold values of 20 mm and 250 kg m^{-1} s^{-1}, respectively. (**c**) Advanced Hydrologic Prediction Services (AHPS) accumulated precipitation analysis for 24-h period ending 1200 UTC 21 December 2010

Oregon's Columbia River basin and the northern US Rockies into the northwestern Great Plains, and extends significantly north and south of this axis. Secondary pathways are located along a corridor of relatively low topography through central British Columbia and from southern California and the Baja Peninsula northeastward, with the gap in the central Baja Peninsula being highlighted by Hughes et al. (2014). Likewise, trajectories associated with major precipitation events across the IMW also tend to follow these same pathways (Fig. 4.29), with the exact pathway followed depending on the region in which heavy precipitation is occurring. For IMW precipitation events north of ~42°N, most trajectories passed through some portion of the Pacific Northwest pathway. In contrast, for precipitation events south of ~42°N, trajectories tended to follow a variation of the southern California–Baja Peninsula pathway. Furthermore, it is clear from both Fig. 4.28d and Fig. 4.30 that these trajectories tend to cluster into relatively low-elevation corridors that lead from the coast, through the Pacific Ranges, and into the IMW.

Trajectories associated with interior-penetrating ARs and heavy IMW precipitation events rarely pass over the High Sierra (approximately 35° to 38°N), because of factors that include flow blocking and deflection upstream of the barrier, as well as water vapor depletion as moist parcels ascend and precipitate. The dearth of such trajectories passing over this range indicates frequent AR decay in this region and explains the low AR frequency, duration, and fraction of seasonal precipitation associated with ARs across the central Great Basin. However, some trajectories pass north or south of the High

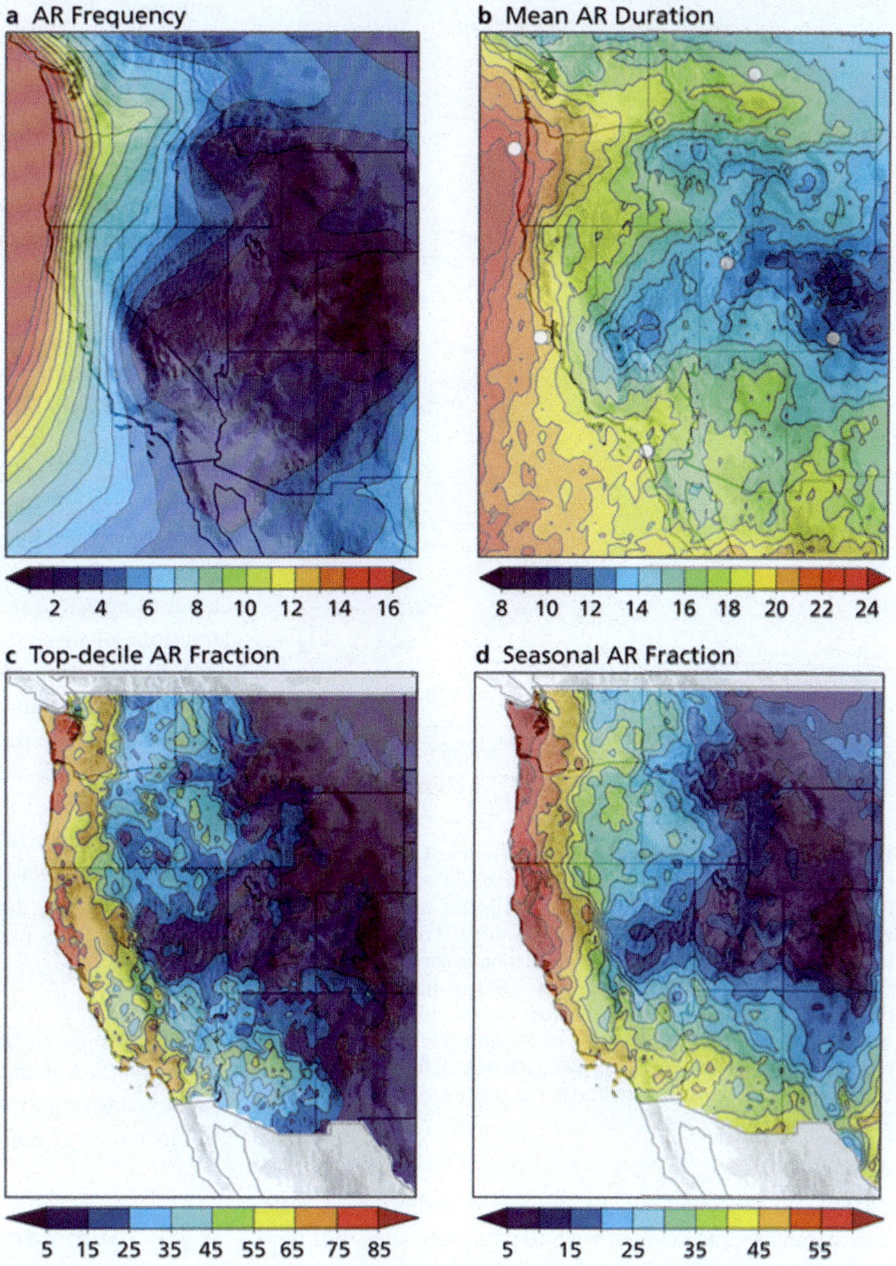

Fig. 4.26 The (**a**) AR frequency, (**b**) mean AR duration, (**c**) top-decile AR fraction, and (**d**) seasonal AR fraction. (Rutz et al. 2014)

Sierra, with those passing north sometimes taking a unique route—often reaching land in central California, going north through the central valley, entering the IMW via the gap in elevation between the southern Cascades and northern Sierra, and continuing through northwestern Nevada and into southern Idaho (Figs. 4.27 and 4.30c, d). The trajectories passing south of the Sierra make landfall north of the US–Mexico border and can bring moisture to the Wasatch Mountains in central Utah and the San Juan Mountains in southwest Colorado (Fig. 4.30g, h).

Landfalling ARs exhibit different characteristics depending on the location of landfall, and whether or not they penetrate into the interior, as shown using synoptic composites by Rutz et al. (2015). In general, interior-penetrating ARs are associated with a much more amplified atmospheric pattern, increased southwesterly flow, and greater IVT than coastal-decaying ARs (Fig. 4.31). Note how the northeastward extension of the subtropical ridge is more noticeable for landfalls further north (Fig. 4.31b), and the trough along the coast is more noticeable for landfalls further south (Fig. 4.31f), but both act to create a more amplified pattern. Also note that interior-penetrating ARs making landfall upstream of the High Sierra are typically associated with very large values of IVT, because this makes the AR much more likely to penetrate over or around this barrier and into the IMW (Fig. 4.31d). Rutz et al.

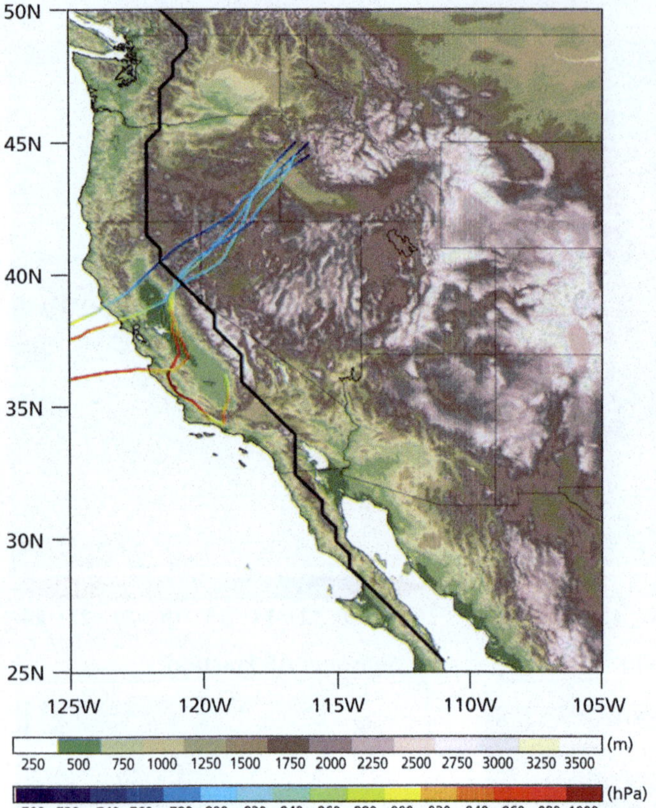

Fig. 4.27 Backward trajectories that were initiated near the top of the boundary layer (50–100 hPa above the surface) at the four Climate Forecast System Reanalysis (CFSR) grid points around a station in western Idaho at 00Z 16 Feb 1982, the day that 104 mm of precipitation fell. This was the largest 1-day precipitation event that occurred at a station in a region in eastern Oregon–southern Idaho. The pressure (hPa) along a trajectory segment is shown by the color (*blue-red*) scale and the terrain height (m) by the (*green-white*) scale, both shown at *bottom*. The *black curve*—extending along the crest of the Cascade, Sierra, and Peninsular mountains—indicates the position of the cross-section shown in Fig. 4.29.

(2015) did not examine the vertical distribution of water vapor transport associated with these composites, but work by Backes et al. (2015) has highlighted the importance of mid-level transport in projecting moisture over and downstream of topographic barriers.

Methodological differences between Rutz et al. (2015) and Alexander et al. (2015) produce complementary results. The study by Rutz et al. (2015) presents results that pertain to all of the AR-related trajectories that affect a broad swath of the North American west coast. This provides a sense of where AR-related trajectories are most concentrated over the western US, but information about the exact paths of trajectories across this region are not presented and must be inferred. In contrast, the study by Alexander et al. (2015) presents results related to heavy precipitation events at a set of individual points. This provides information about the pathways most likely travelled by trajectories that reach a specific point, but does not highlight that some points will receive more ARs and more heavy precipitation events than others. Taken together, these two studies provide a holistic view of the spatial distribution of moisture penetration into the IMW.

The schematic in Fig. 4.29, as mentioned earlier, summarizes the key findings of Rutz et al. (2015) and Alexander et al. (2015). As shown in Figs. 4.28 and 4.31, the pathways most often taken (and not taken) by trajectories associated with interior-penetrating ARs and heavy IMW precipitation events are remarkably similar, suggesting a strong relationship between these phenomena. These pathways are located along relatively low-elevation corridors that extend from the coast, through the Pacific Ranges, and into the IMW. In contrast, they rarely pass over major topographical barriers, such as the High Sierra. These characteristics can be readily observed in the schematic plan view (Fig. 4.29) and cross-section through the crest of the Pacific Ranges (Fig. 4.29; crest shown as a black line in Fig. 4.27). These pathways can be broken into three regimes based on the total number of landfalling ARs (and hence, calculated trajectories) and the fractions of these that go on to become interior-penetrating AR trajectories. Interestingly, these fractions are largest where AR landfalls are most rare (along the Baja Peninsula) as a result of the relatively lower and narrower topographic barriers over this region. Finally, many interior-penetrating AR trajectories from both Regimes 1 and 2 tend to track toward the same relatively low-elevation corridor centered over the Columbia River basin, contributing to the most preferred pathway for inland penetration, highlighted in Fig. 4.28d. Note that the spatial resolutions of the analyses used to identify these pathways tend to blur the complex topography, and, consequently, the water vapor transport, across portions of the western US. Hence, only large, broad pathways are clear, but not some of the smaller gaps that favor water vapor transport, such as those found near the San Francisco Bay (White et al. 2015).

4.5.1 Summary of Sect. 4.5

In summary, ARs can and do penetrate into the interior western US, often bringing high-impact weather, as in the case of the January 2010 event discussed above. Considering the whole western US, the climatological characteristics of ARs—their frequency, duration, and contribution to top-decile 24-h precipitation events and total cool-season precipitation—are generally largest along the coasts of Oregon and Washington, decrease southward along the coast and eastward toward the interior, and are smallest over the Great Basin. This climatology is naturally influenced by the frequency, intensity, duration, and orientation of ARs making landfall along the US West Coast, some of which are presented earlier in this section. As the large-scale synoptic patterns that support these ARs evolve and progress eastward, they interact with the complex topography of the western

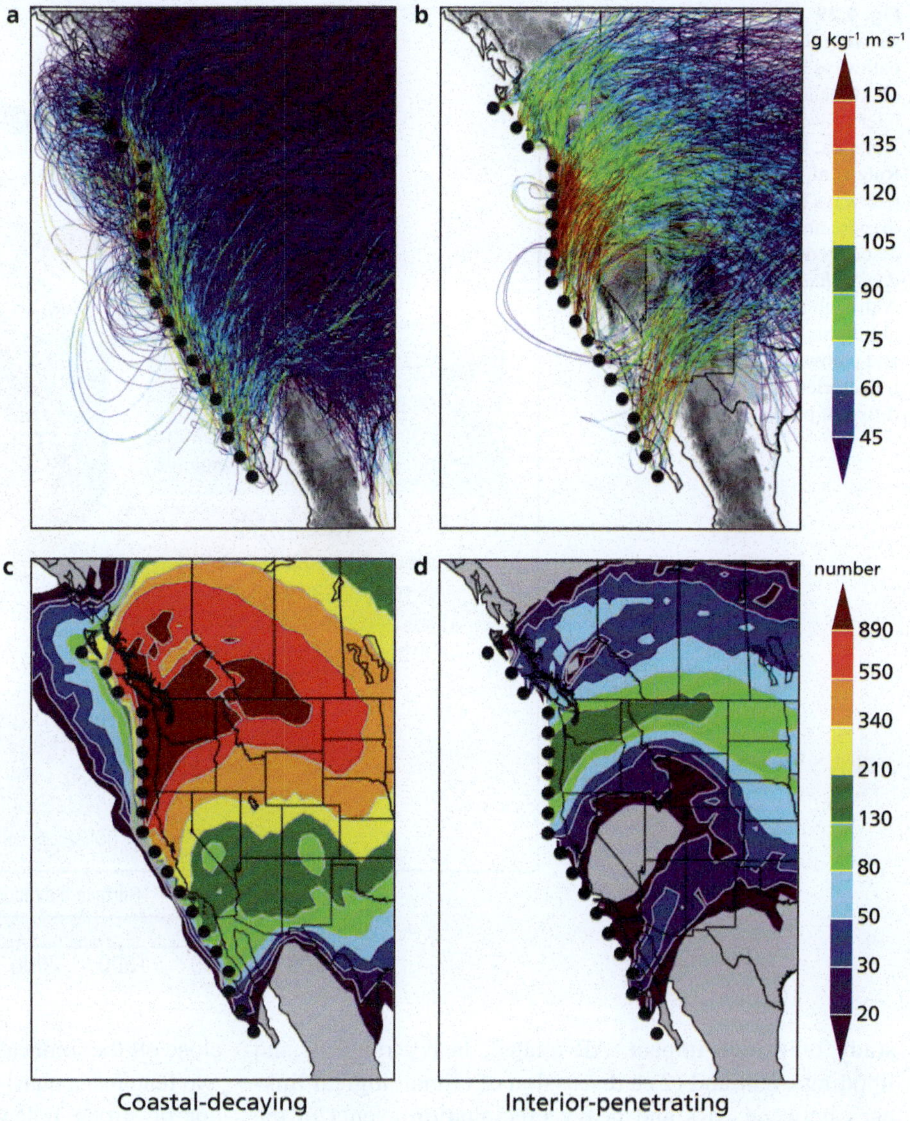

Fig. 4.28 (**a**) Coastal-decaying and (**b**) interior-penetrating 950-hPa AR trajectories with color indicating water vapor flux (*scale at right*). (**c**, **d**) as in (**a**, **b**), but for trajectory count. *Black circles* indicate points from which trajectories are initiated

US, and these interactions modulate the ARs in any number of ways. The influence of topography is clearly evident, because relatively low-elevation corridors serve as conduits for AR penetration into the interior, whereas topographical barriers often cause AR decay because they block and deflect flow, and deplete water vapor following ascent and precipitation.

4.6 ARs in the Southeastern US

Kelly M. Mahoney and Benjamin J. Moore

The southeastern US can experience heavy precipitation events throughout the year within a wide range of scenarios and in conjunction with various atmospheric phenomena, such as mid-latitude (baroclinic) cyclones, tropical cyclones (TCs), and mesoscale convective systems (MCSs). The synoptic-scale conditions over the southeastern US vary considerably from event to event and with time of year, but all support the basic ingredients for heavy precipitation over this region: lift, moisture supply, and reduced static stability (Moore et al. 2015 and references therein). Recent research has demonstrated that in some southeastern US heavy precipitation scenarios—most often those in the winter and the transition seasons—moisture is supplied by ARs (Moore et al. 2012; Mahoney et al. 2016; Debbage et al. 2017) or AR-like or AR-related features (e.g., Pfahl et al. 2014; Moore et al. 2015) linked to strong synoptic-scale weather systems. In some cases, ARs can support extraordinarily heavy rainfall, resulting in significant flooding (e.g., Moore et al. 2012). Given these potential effects, understanding the climatology, dynamics, and hydrometeorological effects of ARs in the southeastern US is critical. This section highlights our current understanding of ARs in the area through (1) a brief case

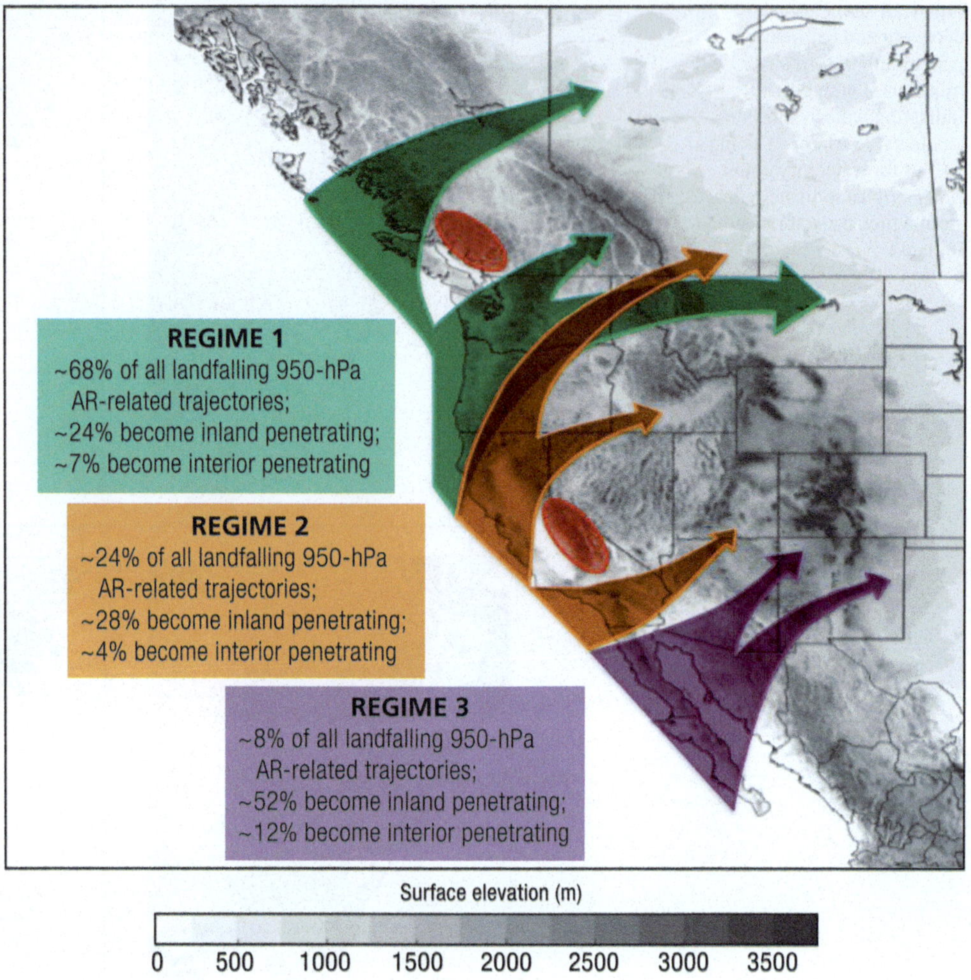

Fig. 4.29 Schematic showing the primary pathways for the penetration of AR-related trajectories into interior western North America. Plan view based on Rutz et al. (2015) with pathways shown as *black arrows*, and regions associated with frequent AR decay *shaded in red*. *Note:* Although this schematic highlights common regimes and pathways, individual trajectories follow many different paths

study of a high-impact, AR-related, heavy rainfall, and flooding event and (2) a discussion of climatological linkages between ARs and heavy precipitation events in the southeastern US.

An exemplary southeastern US AR-related, high-impact, flood-producing, heavy precipitation event occurred during 1–3 May 2010 over Tennessee and Kentucky, including the Nashville, Tennessee, metropolitan area. During this event, which Moore et al. (2012) investigated in detail, widespread extreme rainfall amounts (>400 mm) resulted from near-continuous, heavy convective rainfall linked to a strong AR that was rooted in the tropics. During the ~8 days leading up to the event, a pronounced upper-level Rossby wave train extended across the North Pacific and North America (Fig. 4.32, left), culminating in the formation of a high-amplitude and slow-moving upper-level trough–ridge pattern over North America between 1 and 3 May (denoted in Fig. 4.33 by the 2-PVU contour on the 320-K isentropic surface). This upper-level wave pattern was linked to the formation and maintenance of key quasi-stationary, lower-tropospheric circulation features over North America, namely a lee trough along the eastern coast of Mexico and a pronounced anticyclone off the southeastern US coast (Fig. 4.33). Between these two features, a quasi-stationary, meridionally elongated corridor of strong poleward flow extended across the Gulf of Mexico and the southern US. This corridor was linked to an intense AR, characterized by strong (>1000 kg m^{-1} s^{-1}) IVT and high (>50 mm) IWV (Fig. 4.33a), along which moist air was transported from over the Caribbean Sea, the eastern tropical Pacific, and the Gulf of Mexico into the flood region over Tennessee and Kentucky. Trajectory calculations demonstrate that the AR was associated with extensive poleward transports of moist air parcels from the tropics into the flood region (Fig. 4.33b), indicative of a tropical moisture export (TME) (Knippertz and Wernli 2010). The AR remained nearly stationary during 1–3 May, maintaining strong IVT and large IWV, and producing heavy rainfall over the flood region (Fig. 4.32, right).

Mahoney et al. (2016) examined the relevance of ARs to heavy precipitation events over the southeastern US. In particular, using the NCEP CFSR (Saha et al. 2010), they utilized an objective ARDM (Wick 2014) to identify ARs in this area during 2002–2011. They then compared the ARs to heavy (>100 mm (24 h)$^{-1}$) precipitation events identified in a

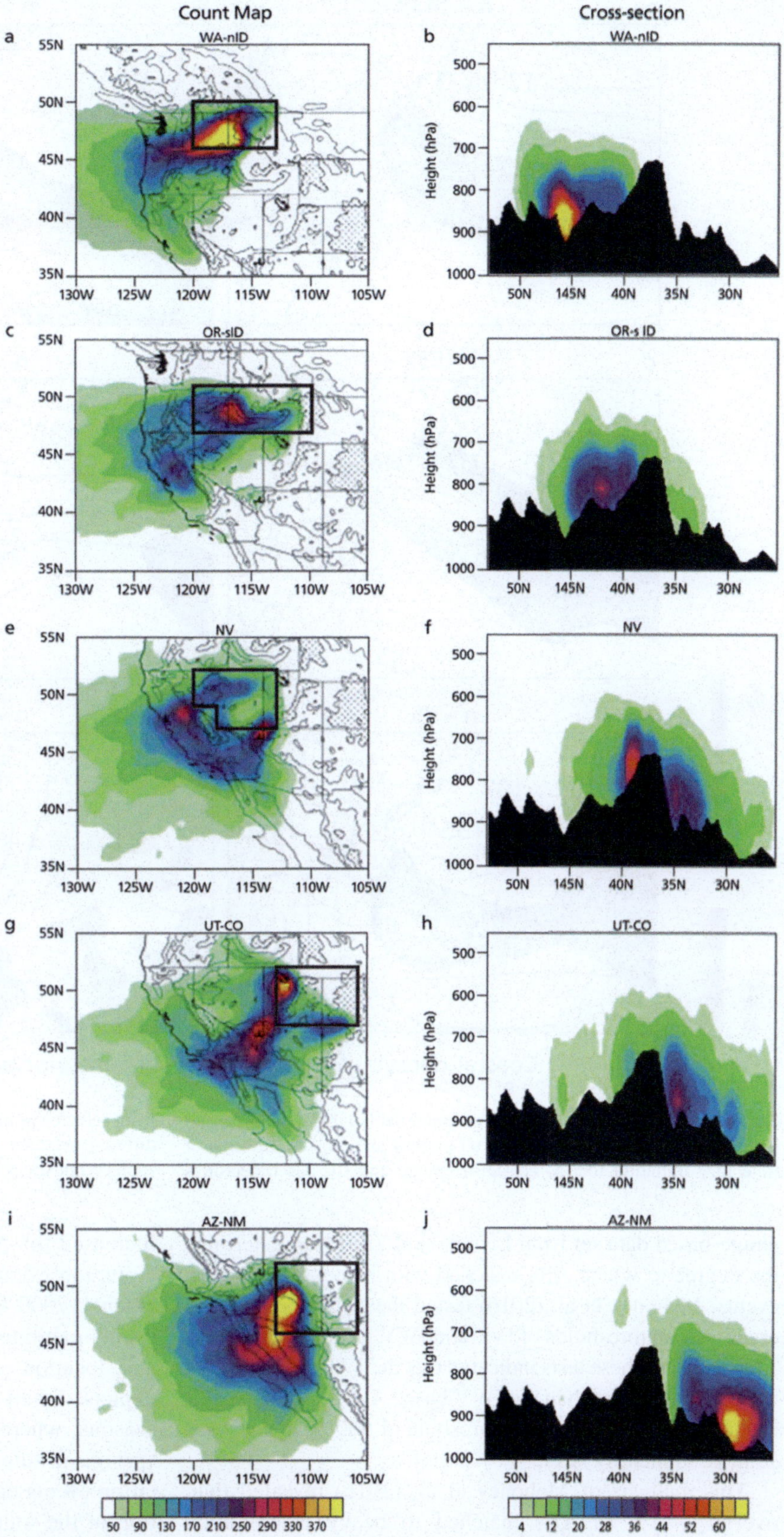

Fig. 4.30 Count maps *left* and vertical cross-sections *right* indicating the number of back trajectories that pass through (**a**) Climate Forecast System Reanalysis (CFSR) grid column that originates in the following regions: (**a**, **b**) Washington–northern Idaho, (**c**, **d**) Oregon–southern Idaho, (**e**, **f**) Nevada, (**g**, **h**) Utah–Colorado, and (**i**, **j**) Arizona–New Mexico. A total of 2400 trajectories were initiated in each region. The position of a trajectory is estimated at 1-h intervals over the five previous days using the 6-hourly 3-D CFSR wind fields. Topography is indicated by *contours* at 1000 m (3281 ft), 1500 m (4921 ft), and 2300 m (7546 ft) and *stippling* above 2300 m. The cross-sections are aligned along the crest of the Cascade, Sierra, and Peninsular mountains (*black curve* in Fig. 4.31), with the terrain shown in *black*

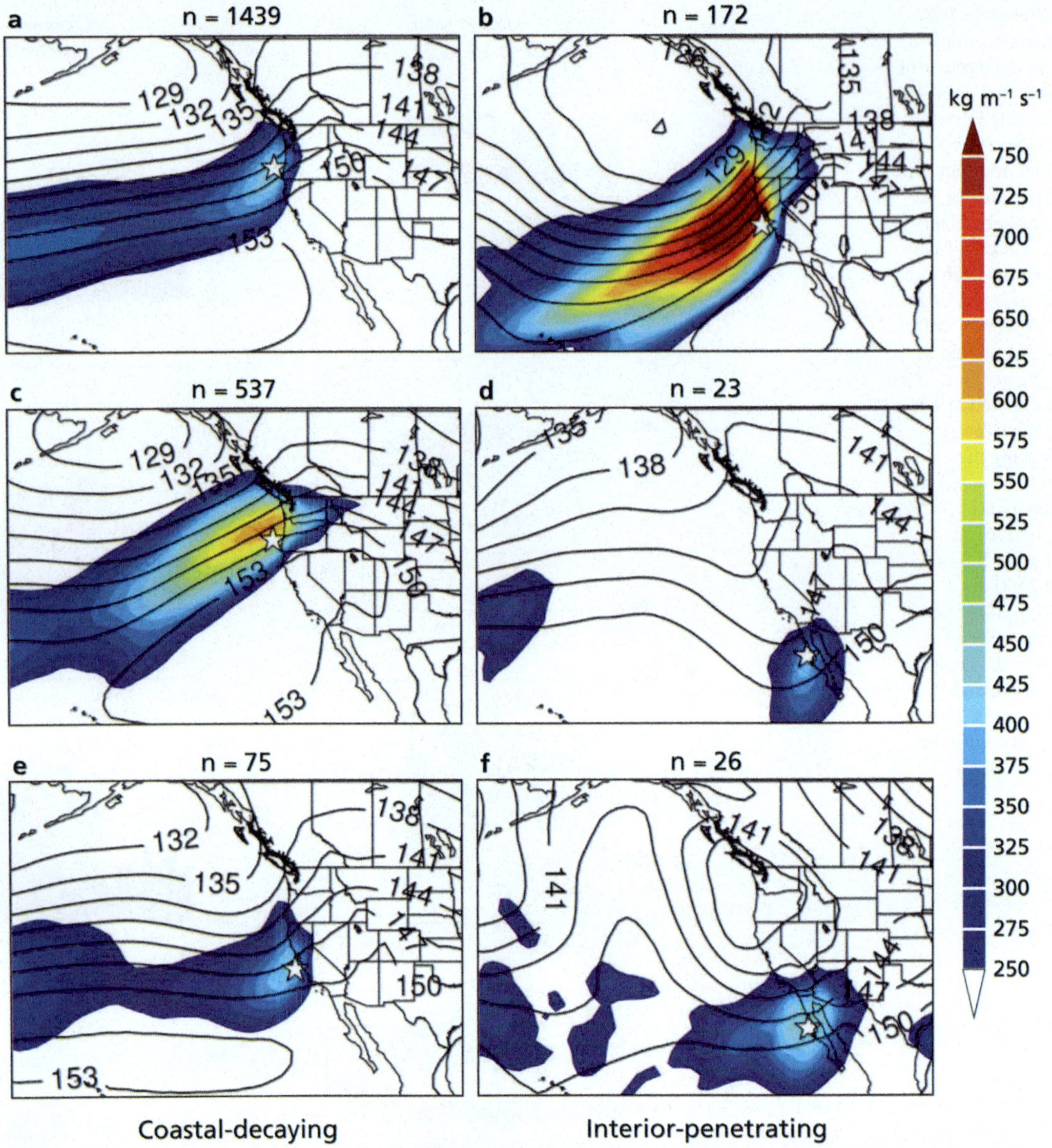

Fig. 4.31 Composites of 850-hPa geopotential heights (*contours*) and integrated water vapor transport (IVT) (*color shading*) at the time of trajectory initiation for (**a**) coastal-decaying and (**b**) AR trajectories from selected points (*starred locations*). (**c**, **d**), (**e**, **f**) as in (**a**, **b**), but for different selected locations. Number of observations (*n*) contributing to each composite shown in lower left

gauge-based data set from Livneh et al. (2013) to determine the degree to which ARs matched with heavy precipitation events. Mahoney et al. (2016) tested the ARDM using several different thresholds of IVT and IWV for to identify ARs. The results of these tests indicated that the IVT-based ARDM was more effective than the IWV-based ARDM in the southeastern US, and that an IVT threshold of 500 kg m^{-1} s^{-1} was suitable to identify ARs.

This analysis of Mahoney et al. (2016) revealed that, overall, ~41% of ARs matched to heavy precipitation events. They found the match rate to be higher in the winter and transition seasons, when strong water vapor transport forced by strong synoptic-scale weather systems is common, relative to the summer. The overall match rate increased to ~52% when they considered only heavy precipitation events that exhibited a large spatial scale (areal extent >7000 km^2). The AR-matched heavy precipitation events exhibited large variability in spatial scale, season, and location within the southeastern US, as Fig. 4.34 depicts. Small- and medium-scale events occurred in all seasons, whereas large-scale events occurred only during spring and autumn. In addition, AR-matched heavy precipitation events occurred more frequently in the western portion of the southeastern US (i.e., west of the Appalachian Mountains) than in the eastern portion. In general, the relatively strong connection between ARs and heavy precipitation events found by Mahoney et al. (2016) is consistent with the connections to heavy precipitation found for

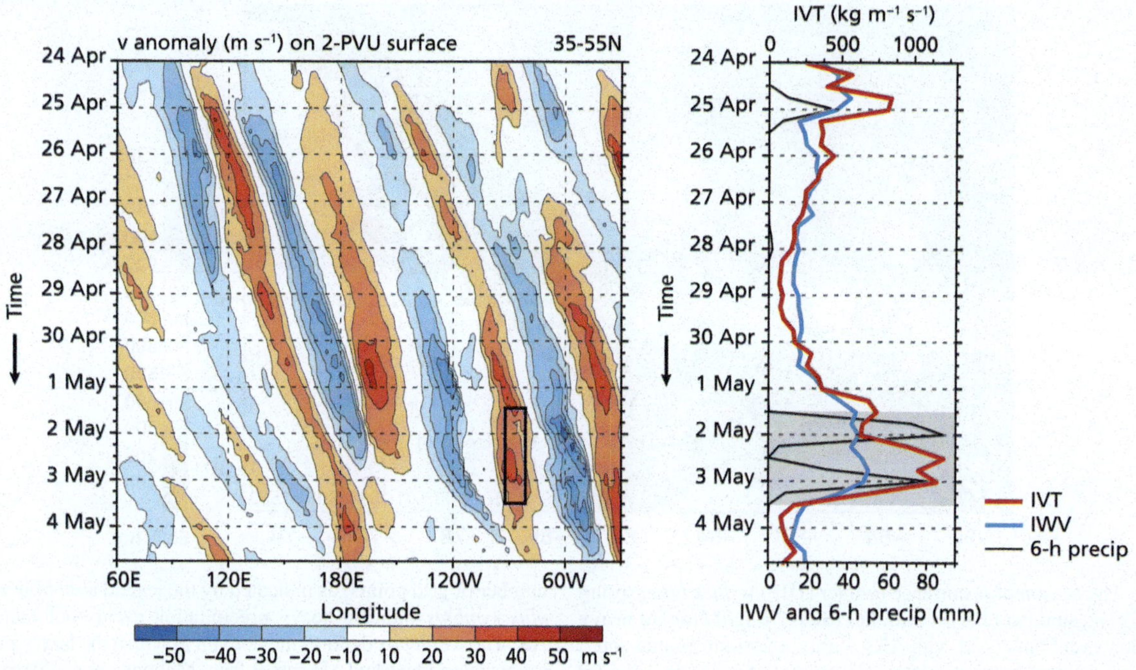

Fig. 4.32 (**left**) Hovmöller diagram of 35°–55°N averaged meridional wind anomalies (m s^{-1}, *shaded* according to the color bar) on the dynamic tropopause (two-potential vorticity unit [PVU] surface) from the National Centers for Environmental Prediction Climate Forecast System Reanalysis (NCEP CFSR). Anomalies are relative to a long-term (1979–2009) daily climatology. The *green box* denotes the approximate time-period and location of the heavy precipitation event over Tennessee and Kentucky. (**right**) Time-series of 1000–300-hPa integrated water vapor transport (IVT) (*red*; *top abscissa*) and total column integrated water vapor (IWV) (*blue*; *bottom abscissa*) from the NCEP CFSR and of 6-h precipitation (*black, bottom abscissa*) from the NCEP Stage-IV data set (Lin and Mitchell 2005) at the grid point closest to Nashville, Tennessee. The *gray box* denotes the time-period of the heavy precipitation event

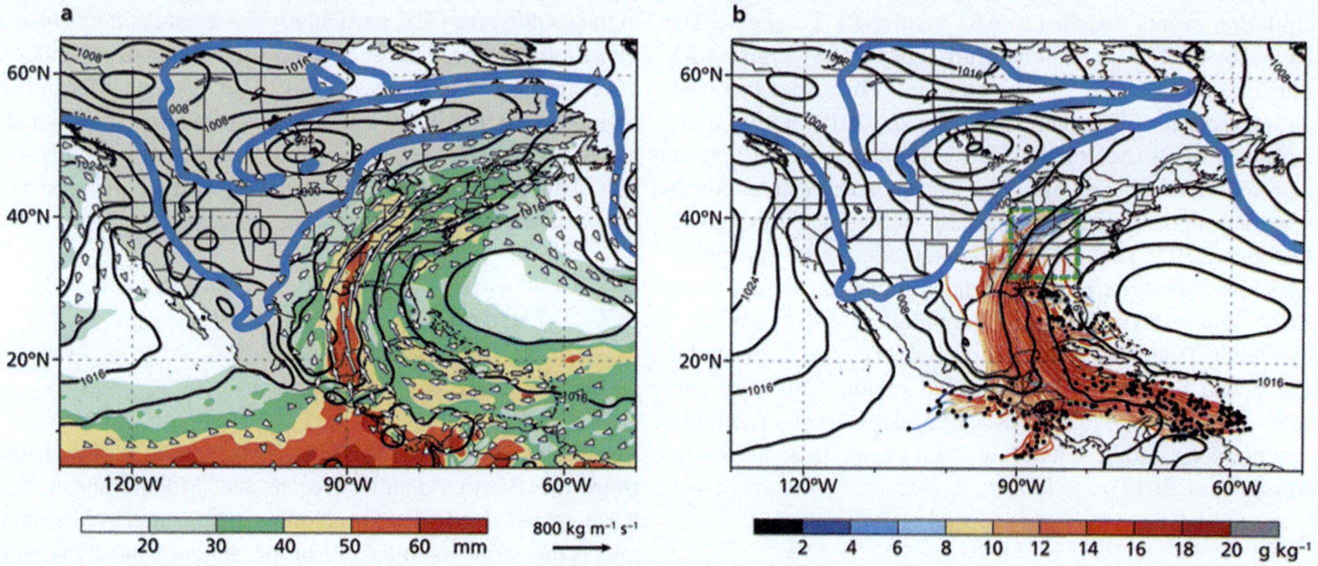

Fig. 4.33 (**a**) Total column integrated water vapor (IWV) (mm, *shaded according to the color bar*), 1000–300-hPa integrated water vapor transport (IVT) vectors (kg m^{-1} s^{-1}, reference vector in *lower right*), sea level pressure (*black contours* every 4 hPa), and the two-potential vorticity unit (PVU) contour on the 320-K isentropic surface at 1200 UTC 2 May 2010 from the National Centers for Environmental Prediction Climate Forecast System Reanalysis (NCEP CFSR). (**b**) 96-h backward trajectories released at 1200 UTC 2 May 2010 from grid points between 1000 and 200 hPa within the *green box* with >90% relative humidity. Only those trajectories exhibiting a specific humidity decrease of at least 5 g kg^{-1} in the final 24 h are plotted. Trajectories are shaded according to the parcel-specific humidity value (g kg^{-1}; *see color bar*), and starting locations for the trajectories are marked by *black dots*. Trajectories were calculated using the NOAA Hybrid Single Particle Lagrangian Integrated Trajectory Model (HYSPLIT) model (Stein et al. 2015) with the NCEP CFSR. For reference, time-mean sea level pressure (*black contours* every 4 hPa) and 2-PVU contour on the 320-K isentropic surface for 0000 UTC 1 May–0000 UTC 3 May 2010 from the NCEP CFSR are overlaid

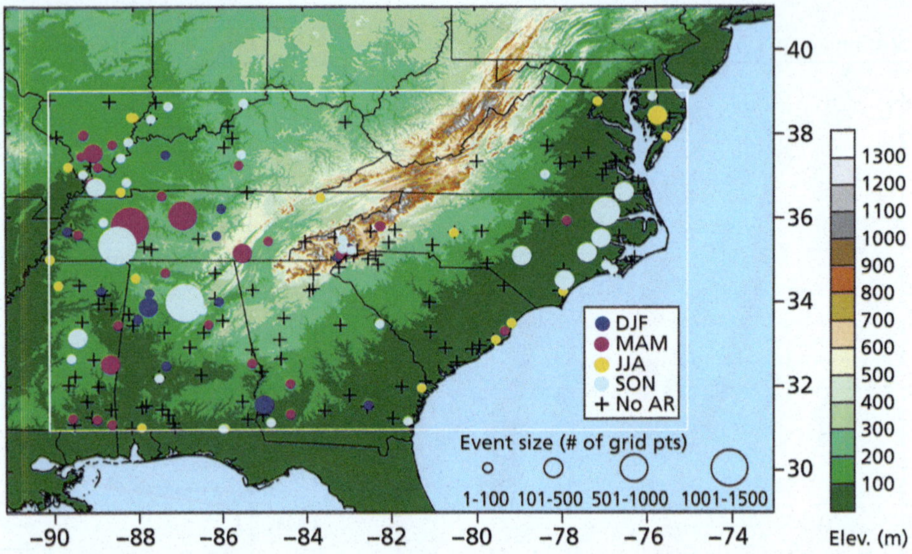

Fig. 4.34 The season of occurrence (winter (DJF) = *dark blue*, spring (MAM) = *pink*, summer (JJA) = *gold*, fall (SON) = *light blue*) of heavy precipitation events matched with ARs within a 250-km radius and a 24-h period, plotted over a terrain elevation basemap (m, *shaded according to the color bar*). Each marker denotes the center location of a heavy precipitation event. *Circle size* corresponds to area extent (in number of grid points) as indicated by the legend at *bottom right*. *Black plus symbols* indicate heavy precipitation events not matched to an AR. The *white box* denotes the domain in which the heavy precipitation events were identified. (Figure 9 from Mahoney et al. 2016)

AR-related phenomena in the southeastern US, such as mid-latitude cyclones (Pfahl and Wernli 2012) and warm conveyor belts (WCBs) (Pfahl et al. 2014).

Notably, a majority of the large-spatial-scale heavy precipitation events matched to ARs involved a TC (i.e., a TC named by the NOAA National Hurricane Center). Figure 4.35 illustrates an example of such an event. This event involved an interaction of Tropical Storm Nicole (2010) with a mid-latitude trough (not shown) that culminated in widespread extreme rainfall amounts (>500 mm) over eastern North Carolina and Virginia during 28 September through 2 October 2010. The TC–trough interaction was associated with the formation of a contiguous quasi-linear region of IWV values >60 mm and IVT magnitudes >1500 kg m^{-1} s^{-1} extending from the eastern portion of TC Nicole over the Caribbean Sea into a heavy rainfall region. This case and other similar TC-related AR cases also closely resemble "predecessor rain events" (e.g., Galarneau Jr et al. 2010; Moore et al. 2013).

4.6.1 Summary of Sect. 4.6

In summary, ARs play a key role in the production of heavy precipitation and, in some cases, flooding in the southeastern US. An examination of an AR case in May 2010 highlighted a linkage between the persistence of strong AR conditions—facilitated in conjunction with quasi-stationary synoptic-scale circulation features—and extreme rainfall and flooding. Moreover, the case study illustrates the interrelationship that can exist among ARs, TMEs, and extreme rainfall. The results of Mahoney et al. (2016) revealed a relatively strong climatological linkage between ARs and heavy precipitation events in the southeastern US, particularly for events in the cool and transition seasons, and for events with a large spatial scale. A novel finding of Mahoney et al. (2016) was that ARs in this region—particularly those associated with large-spatial-scale heavy precipitation events—can sometimes directly involve TCs. This connection to TCs appears to be a unique aspect of the phenomenology of ARs in the southeastern US.

4.7 Europe

David Lavers and Alexandre Ramos

European precipitation is affected by the transport of moisture from the North Atlantic Ocean and Mediterranean Sea (Gimeno et al. 2013). The North Atlantic moisture source affects the entire continent of Europe, although mostly in winter, while the Mediterranean source largely affects southern Europe, but all throughout the year. One of the major mechanisms associated with moisture transport from the North Atlantic toward Europe is an AR (e.g., Ramos et al. 2016), embedded in the trailing fronts of extratropical cyclones.

The connection between ARs and precipitation in Europe has been investigated less than in North America, despite the importance of ARs in the European hydrological cycle. In

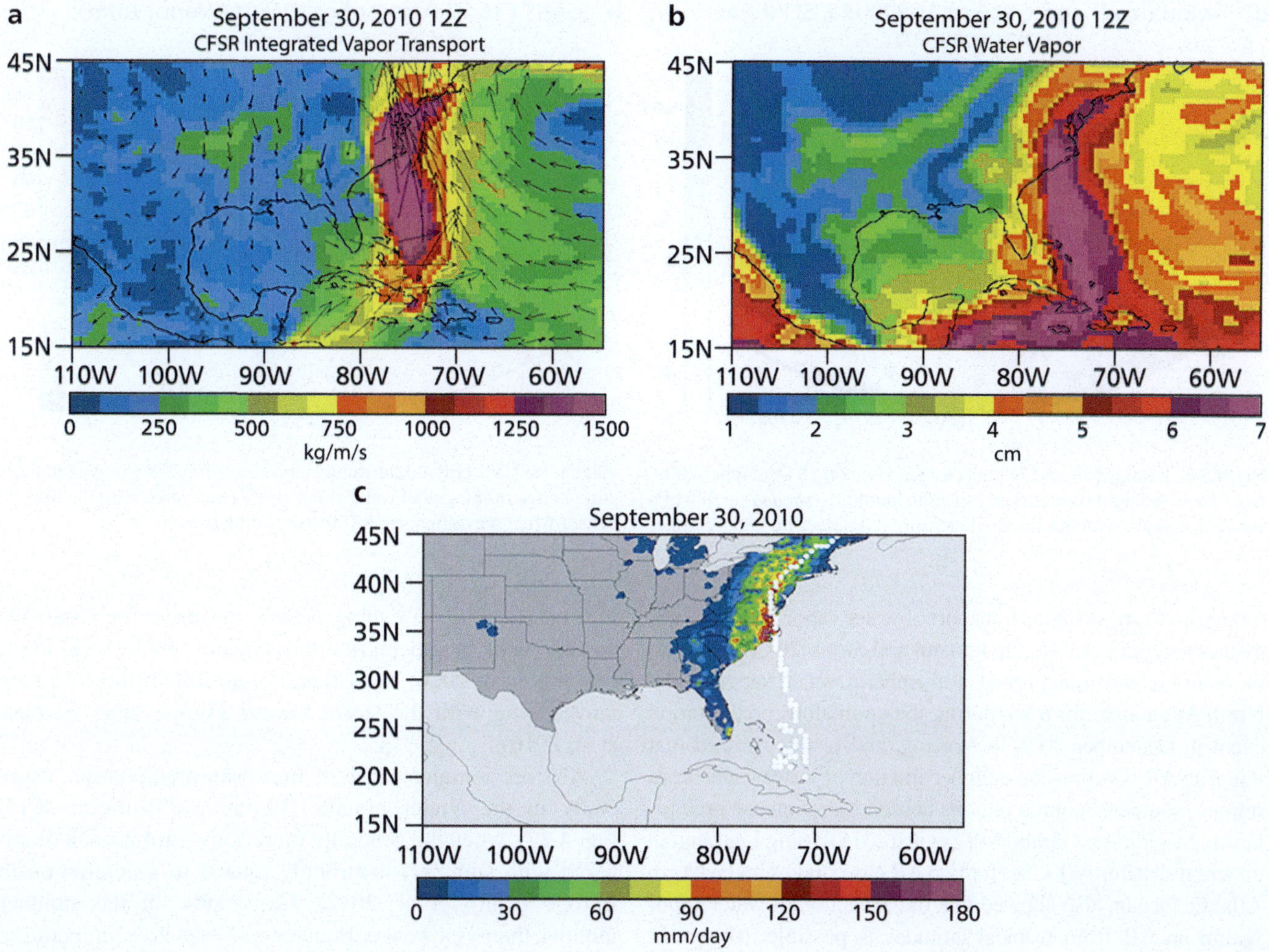

Fig. 4.35 (**a**) integrated water vapor transport (IVT) (kg m^{-1} s^{-1}, magnitude shaded according to the *color bar with vectors overlaid*; reference vector in *lower left* represents IVT magnitude of 250 kg m^{-1} s^{-1}) at 1200 UTC 30 September 2010 during the extratropical transition of Tropical Storm (TS) Nicole (2010) from the National Centers for Environmental Prediction Climate Forecast System Reanalysis (NCEP CFSR). (**b**) as in (**a**), but for integrated water vapor (IWV) (mm). (**c**) The 24-h precipitation ending at 1200 UTC 30 September 2010 from the Livneh et al. (2013) data set (mm, *shaded according to the color bar*) with *white points* denoting the location of the AR as identified by the Automated Atmospheric River Detection (ARDT–IVT) at 1200 UTC 30 September 2010. (Figure 10 from Mahoney et al. 2016)

fact, one of the first studies relating ARs and extreme precipitation in Europe (Norway) was only published by Stohl et al. in 2008. This study found that extreme precipitation in September 2005 on the southwest Norwegian coast was associated with the occurrence of an AR. Since 2010, there has been an increasing interest in understanding the nature of ARs in the North Atlantic basin and their possible effects in Western Europe. Several papers have been presented, focusing not only on a few specific extreme events (e.g., Liberato et al. 2013; Pereira et al. 2016), but also using several long-term climatological studies in different European domains such as the Iberian Peninsula (e.g., Ramos et al. 2015), the UK (Lavers et al. 2011, 2012), and also across Europe (Lavers and Villarini 2013b; Brands et al. 2017).

This section offers a regional perspective of the present knowledge on ARs in the North Atlantic Ocean that affect Western Europe. Figure 4.36 shows an example of an AR that affected the Iberian Peninsula on 28 December 2009. In the left panel, the wind field (vectors) and specific humidity (shaded) at 900 hPa are shown, along with the sea level pressure (contours) on 28 December 2009 at 00 UTC; the right panel shows the North Atlantic satellite image of IWV measured with the SSM/I (morning passes).

Over the North Atlantic Ocean, Dacre et al. (2015) analyzed selected cases of the transport of water vapor within a climatology of wintertime North Atlantic extratropical cyclones. In this particular study, AR formation is considered to be a result of the cold front that sweeps up water vapor in the warm sector as it catches up with the warm front. This, in turn, causes a narrow band of high water vapor content to form ahead of the cold front at the base of the WCB airflow. Thus, according to Dacre et al. (2015), water vapor in the warm sector of the cyclone—

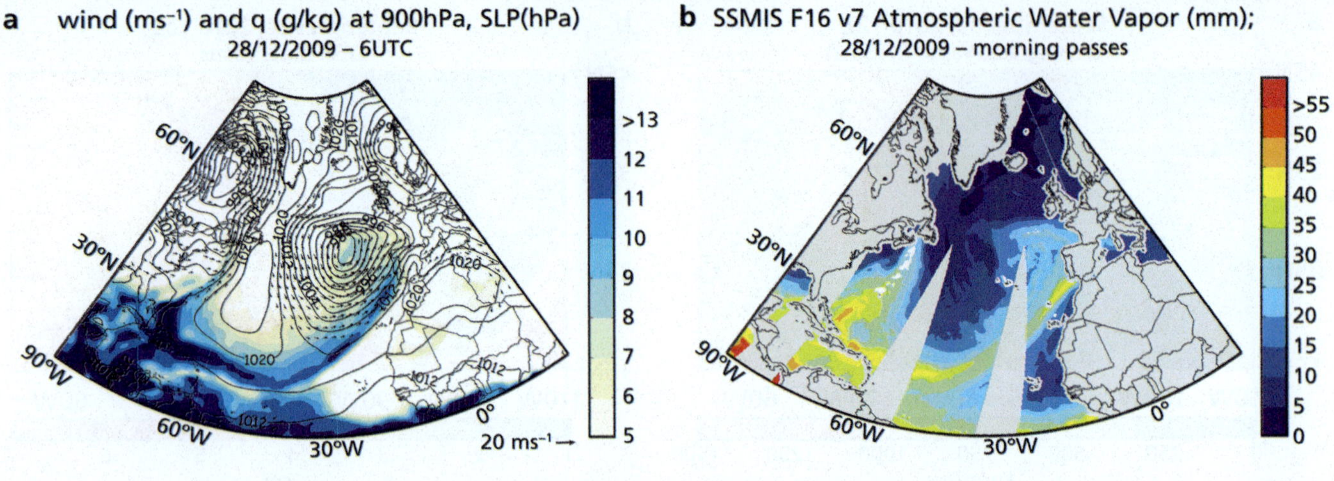

Fig. 4.36 Example of an AR impacting Europe on 28 December 2009. *Left:* The wind field (*vectors*) and specific humidity (*shaded*) at 900 hPa are shown along with the sea level pressure (*contours*) on 28 December 2009 at 00 UTC (ERA-Interim reanalysis) *right* the North Atlantic satellite image of integrated water vapor (IWV) measured with the Special Sensor Microwave/Imager (SSM/I) (morning passes)

rather than long-distance transport of water vapor from the subtropics—generates ARs. Sodemann and Stohl (2013) analyzed the sources and transport of atmospheric water vapor in the North Atlantic storm track during the anomalous precipitation event in December 2006 in Norway, and results showed that when an AR was present, a higher fraction of water vapor from remote, southerly source regions caused more intense precipitation. In addition, Garaboa-Paz et al. (2015), using Lagrangian coherent structures (LCSs) for two AR case studies in the North Atlantic Ocean, also showed that the advection of water vapor within an AR from tropical latitudes is possible. Moreover, Ramos et al. (2016) analyzed the moisture sources (from a Lagrangian perspective) of the ARs that arrived in Western Europe between 1979 and 2012 for the extended winter months (October to March). The authors showed that for the ARs that make landfall on the Western European coast, the anomalous moisture linked with the ARs comes mainly from subtropical areas and, to a lesser extent, from the tropical and mid-latitude regions. Therefore, it seems that the poleward transport of tropical moisture combined with mid-latitude moisture sources along the AR paths is responsible for the generation and evolution of ARs in the North Atlantic Ocean.

Given the role of latent heat release in the development of explosive cyclogenesis (e.g., Pinto et al. 2009; Fink et al. 2012; Pirret et al. 2016) in the North Atlantic basin, the presence of a strong AR is hypothesized to influence the deepening of the extratropical cyclone if it passes nearby. This explosive cyclogenesis can later cause a high-impact weather event in Europe, such as the Gong storm in January 2013 (Liberato 2014). For the North Atlantic basin, different studies show that the presence of an AR—that is, a subtropical moisture export over the western and central subtropical Atlantic, which converges into the cyclogenesis region, leading to an explosive cyclogenesis—can drive the explosive development in the mid-latitudes (Zhu and Newell 1994; Ferreira et al. 2016; Eiras-Barca et al. 2017); the ARs then move along with the storm toward Europe (e.g., Ferreira et al. 2016).

ARs are a major cause of European precipitation, especially in the winter months (Lavers and Villarini 2015; Fig. 4.37), when the generally more active storm track (compared with summer) is strongly related to precipitation in Europe (Catto et al. 2012). Conversely, in the summer months, there is a weaker Equator-to-North Pole temperature gradient and a weaker North Atlantic storm track, resulting in fewer extratropical cyclones occurring, and thus the AR effect on precipitation is weaker. In summer, precipitation tends to be more related to smaller-scale convective weather systems (e.g., Berg et al. 2009). Furthermore, Champion et al. (2015) showed that ARs do not have large explanatory power of UK summer extreme precipitation.

In Europe, the main regions affected by the ARs are located in the Iberian Peninsula (Ramos et al. 2015; Eiras-Barca et al. 2016; Couto et al. 2015; Ramos et al. 2016), France (Lavers and Villarini 2013b), the UK (Lavers et al. 2011, 2012; Lavers and Villarini 2013b), and western Scandinavia (Stohl et al. 2008; Lavers and Villarini 2013b). Figure 4.38 shows for five different domains—namely the Iberian Peninsula (9.75°W; 36°N–43.75°N), France (4.5°W; 43.75°N–50°N), the UK (4.5°W; 50°N–59°N), southern Scandinavia and the Netherlands (5.25°E; 50°N–59°N), and northern Scandinavia (5.25°E; 59°N–70°N)—the median, 90th, and tenth percentile of the maximum IVT positions of the different ARs along their first-guess trajectories.

Note that in Europe there is an indication that ARs can affect regions further inland (e.g., Poland; Lavers and

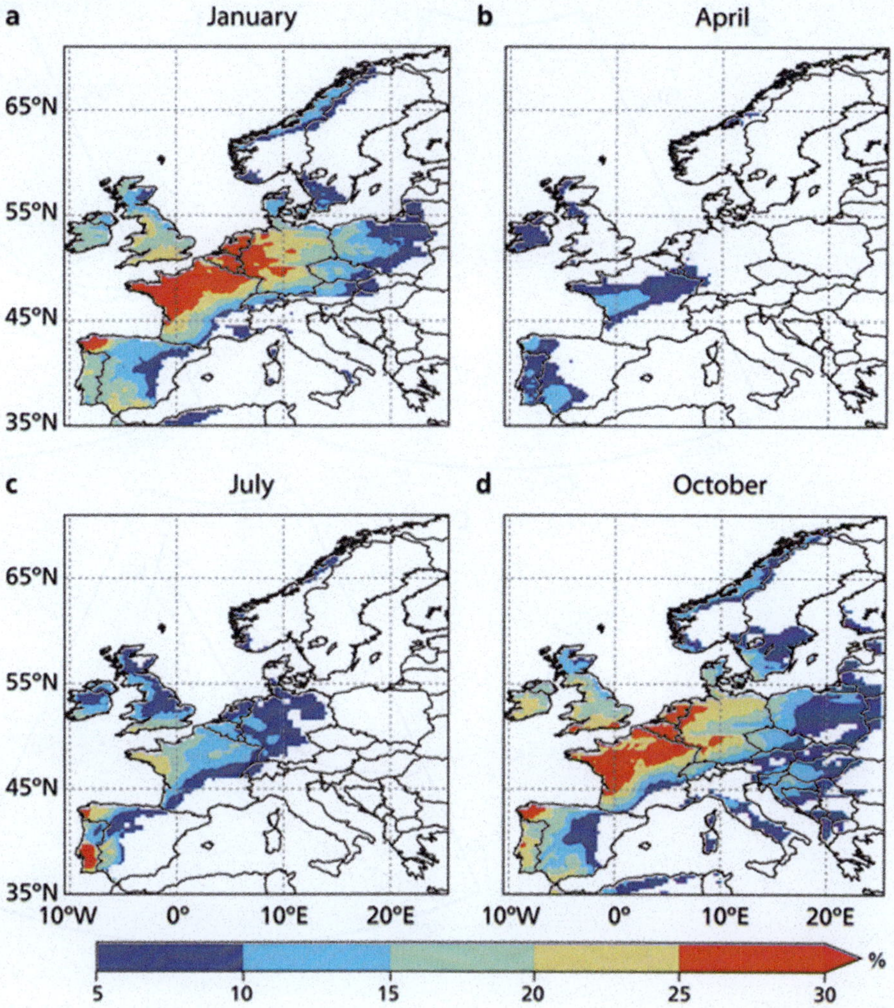

Fig. 4.37 The average AR fraction (in %) across Europe for (**a**) January, (**b**) April, (**c**) July, and (**d**) October over the period 1979–2012. (Results from Lavers and Villarini 2015)

Villarini 2013b) than in western North America. Although Rutz et al. (2014) has shown that ARs can reach into the interior USA, their effect is likely to be less than in Europe because the more mountainous terrain in the USA acts as a barrier to inland penetration.

Research has also shown that a strong relationship exists between ARs and annual maximum precipitation days in Western Europe, which is particularly pronounced along the coast, where some areas had eight of their top ten annual maximum precipitation days related to ARs—in particular in areas of the northwestern Iberian Peninsula and southwestern Norway (Lavers and Villarini 2013b).

The variability of ARs that affect Europe is driven by large-scale atmospheric patterns that occur over the Atlantic and European sectors. Lavers and Villarini (2013b) showed that the NAO plays an important role in the occurrence of ARs throughout Europe. In addition, the frequency of winter ARs for the British Isles was found to also depend (negative correlation) on the Scandinavian pattern (Lavers et al. 2012); for the Iberian Peninsula, a positive correlation was found between ARs frequency and the Eastern Atlantic pattern (Ramos et al. 2015).

4.7.1 Summary of Sect. 4.7

The ARs that make landfall in Western Europe play a major role in explaining winter precipitation and are also connected with extreme precipitation in different domains in Europe (e.g., the UK, the Iberian Peninsula, and Norway). In Europe, over the last few years, there have been several papers that investigated the effects of ARs and their relationship with the major large-scale atmospheric patterns, which has increased

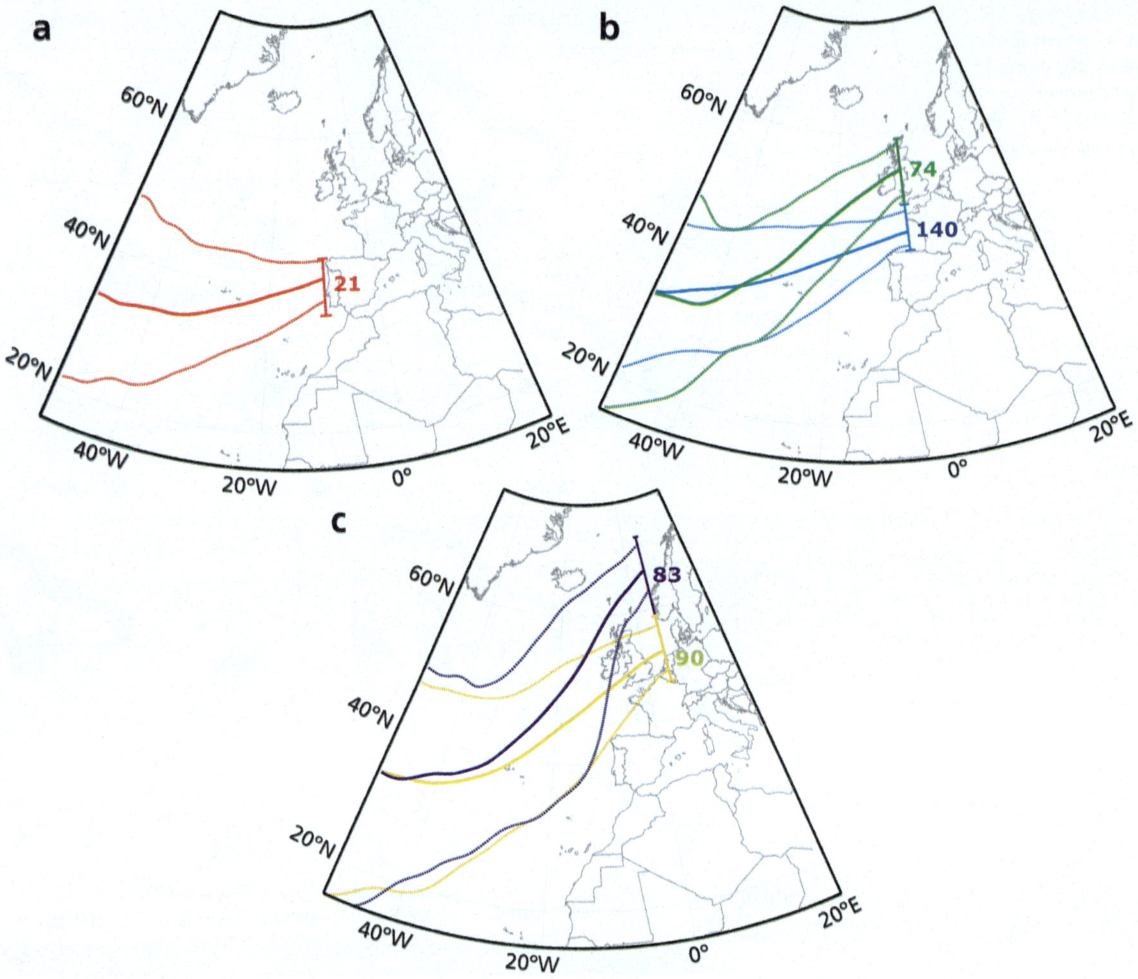

Fig. 4.38 The median position (*colored line*) and the respective 90th and tenth percentile (*dashed line*) of the AR path along the North Atlantic Ocean before arriving in each studied domain: (**a**) Iberian Peninsula (*red*), (**b**) France (*blue*) and the UK (*green*), and (**c**) southern Scandinavia and the Netherlands (*yellow*) and northern Scandinavia (*purple*). In addition, the number of persistent ARs in each domain during the 1979–2012 period is also highlighted. (Adapted from Ramos et al. 2016)

understanding of this phenomenon. There is, however, a lack of studies in the North Atlantic basin and Western Europe that use "real" observations of ARs from aircraft dropsonde data or Doppler wind profilers; these would help the water vapor transport of ARs in this region to be better understood.

4.8 Southern South America

Maximiliano Viale and Raúl Valenzuela

ARs and their effects on precipitation have been less investigated in South America than in North America. Similarities between AR effects on both continents include frequent AR landfall along their mid-latitude west coasts, both of which feature large-scale (> 1000 km) mountain ranges (the Cascades and Sierra Nevada in North America and the Andes in South America) oriented mostly perpendicular to the water vapor transport of ARs. These characteristics normally favor the enhancement of rainfall and snow events through orographic precipitation processes. Despite these similarities, local topographic differences between North America and South America mountain ranges (e.g., altitude, orientation and width), as well as the presence of small-scale (<1000 km) upstream coastal ranges, can lead to differences in the orographic modulation of landfalling ARs and the resulting orographic precipitation on both continental west coasts, although the details of this comparison still remain largely unknown.

Climatological features associated with landfalling ARs in South America are described in Viale and Nuñez (2011). Most of the annual total rain over the subtropical west coast of South America and snow over the Andes (30°S–37°S) accumulate through a few heavy wintertime precipitation events normally associated with landfalling ARs. This portion of the South American west coast, as with similar latitudes along the North American west coast (i.e., southern California), accumulates most of the annual total precipitation almost exclusively during wintertime. On average, four or five heavy wintertime precipitation events are responsible for around 70% of annual total precipitation. From a total of 46 heavy precipitation events occurring during seven consecutive winters (1970–1976), 40 events (~87%) were associated with landfalling ARs. (The period 1970–1976 was chosen because it represents an older, yet denser data network; most recent periods comprise a sparser data network.) The presence of an AR in each precipitation event was established through the IWV criteria (Ralph et al. 2004) by using reanalysis and radiosonde data.

Further inspection of the 46 heaviest precipitation events occurring between 1970 and 1976 reveals that those associated with stronger winds and deeper cyclones featured greater accumulated precipitation, and all were linked to a landfalling AR. The 46 heaviest precipitation events—defined from days with precipitation in the fourth quartile at representative surface stations—were subdivided into two subcategories: extreme and intense events. Extreme and intense events were defined as daily precipitation within the 95th–100th and 75th–95th percentiles, respectively. Extreme events were associated with deeper cyclones and stronger plumes of water vapor transport toward the Andes than intense events. All extreme events were linked to landfalling ARs, and most of them produced floods and landslides that caused serious damages and fatalities in central Chile (see Chap. 5, Effects of ARs). These findings denote the crucial role of ARs in not only providing most of the water supply in southwestern South America, but also in leading to extreme orographic precipitation events that normally caused floods, landslides, and serious damages.

The typical synoptic and weather conditions during heavy precipitation events on the subtropical west coast of South America are summarized in the conceptual representation of Fig. 4.39. A deep extratropical cyclone is located on the Pacific coast of South America, with an AR within the prefrontal environment impinging and raining-out on the corrugated coastal terrain and the Andes. Upstream of the Andes (Chile), low-level flow is typically blocked, forming a barrier jet; downstream of the Andes (Argentina), strong downslope winds usually occur and cause an abrupt drying of the air mass with little or no spill-over precipitation.

Because winter precipitation events that occur under warmer conditions elevate the freezing level and increase the pluvial area over the Andean River basins, setting the stage for hydrometeorological hazards, warm winter storms have also been investigated. For example, Garreaud (2013) documented that around 80% of warm rain events in the subtropical west coast of South America that represent nearly one-third of total winter events are linked to ARs. The synoptic-scale conditions during

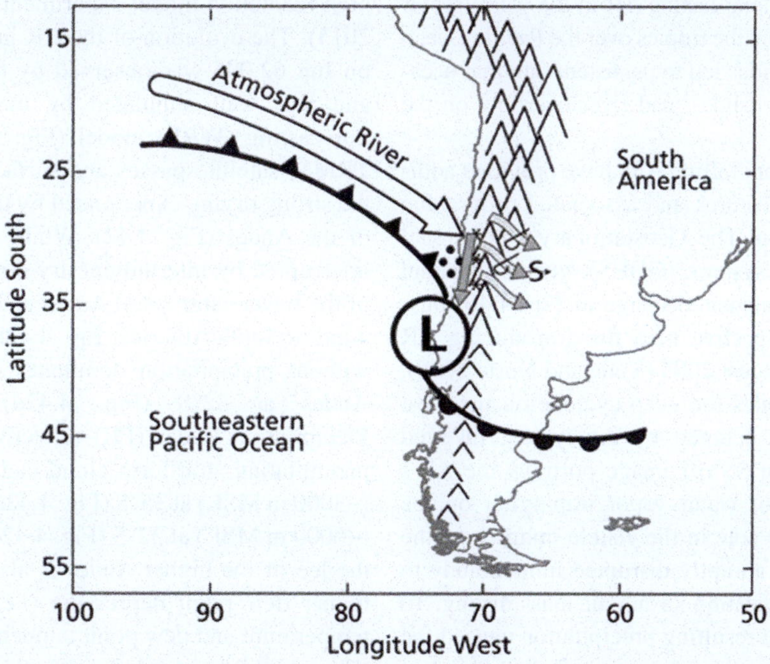

Fig. 4.39 Conceptual representation of the typical meteorological conditions during the heavy precipitation events over the subtropical west coast of South America. The *long and narrow white arrow* along the cold front associated with the extratropical cyclone corresponds to the AR making landfall and impacting the Andes. Typical airflow and weather conditions in the windward and lee sides of the Andes are indicated by *gray filled arrows* and *weather symbols*. In the windward side, an along-barrier jet, rain, and snowstorm are typically observed; in the lee side, a downslope windstorm and orographic clouds denoted the strong air mass drying that typically occurs. (Adapted from Viale and Nuñez 2011)

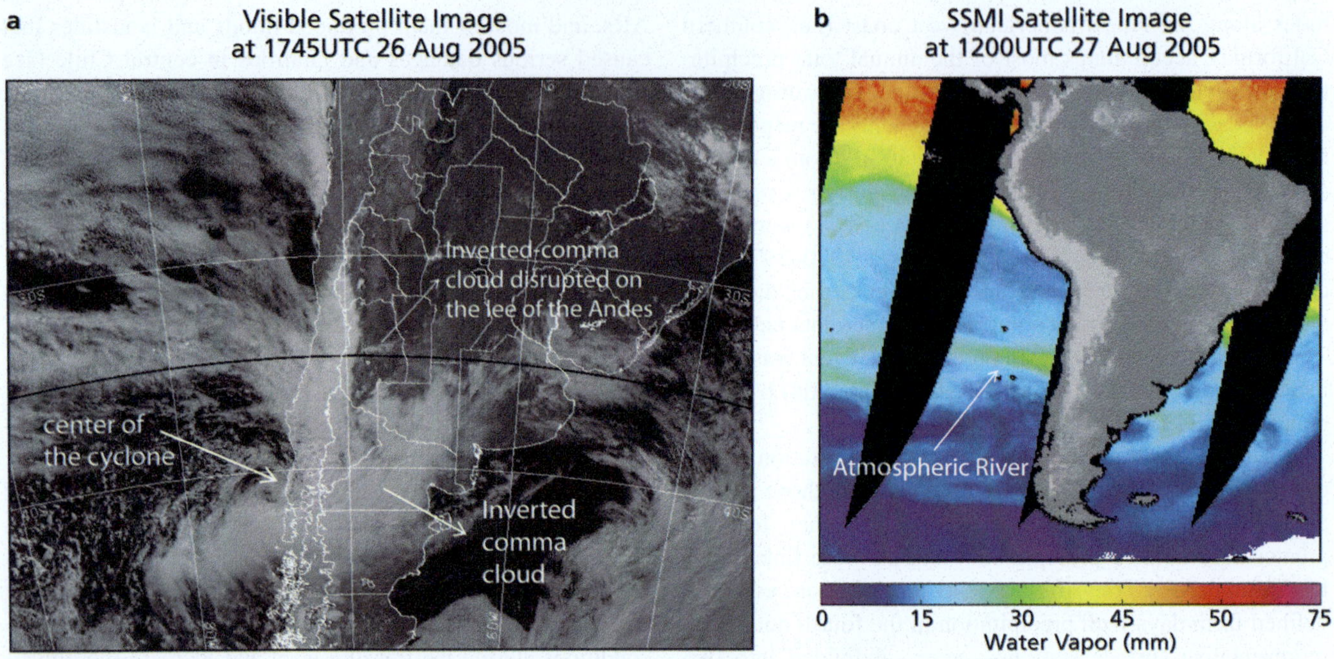

Fig. 4.40 (**a**) Geostationary Operational Environmental Satellites (GOES) image in the visible channel at 1745 UTC 26 August 2005 showing the inverted comma-shaped cloud associated with the extratropical cyclone on the west coast of South America. The inverted comma-shaped cloud is abruptly disrupted immediately lee of the Andes by downslope flow (adapted from Viale and Norte 2009). (**b**) Special Sensor Microwave/Imager (SSM/I) composited image around 1200 UTC 27 August 2005 showing the plume of integrated water vapor (IWV) (mm) that represents a landfalling AR

these events are characterized by a slow-moving and zonally oriented cold front with a long-lasting AR in the warm sector that transports moisture from the tropics over the Pacific Ocean to the Andes. These conditions led to persistent rain that occasionally caused floods, landslides, and river overflow on the slopes of the Andes.

A few case studies of landfalling ARs have provided additional details in airflow features and associated orographic precipitation in the region. The Geostationary Operational Environmental Satellite system (GOES) visible channel image and the SSM/I composited image of Fig. 4.40 illustrate the large-scale perspective of a flood-producing AR storm that occurred on August 2005 (Viale and Norte 2009). The visible image illustrates the deep cyclone on the west coast of South America denoted by the typical inverted comma-shaped cloud; the SSM/I image portrays the AR's zonally oriented plume of water vapor impinging on the Andes. It is interesting to note in the visible image how the inverted-comma cloud is abruptly disrupted immediately to the east of the Andes, indicating strong air mass drying. To the west of the Andes, the resulting precipitation caused the river that runs through the city of Santiago—the capital of Chile—to overflow, leading to serious damages and losses. Viale and Norte (2009) describe this AR case in more detail.

Another case study of an AR storm on the west coast of South America occurred on June 2006 and highlights a few distinctive regional features by using SSM/I and Tropical Rainfall Measuring Mission (TRMM) imagery as well as high-resolution model experiments (Viale 2010; Viale et al. 2013). The evolution of the AR just before making landfall on Jun 62,006 was observed by the SSM/I satellite passes and was well simulated by the Weather Research and Forecasting (WRF) model (Fig. 4.41). At the same time, TRMM satellite passes and surface observations registered the strong drying experienced by the air mass on the lee side of the Andes (Fig. 4.42). While spill-over precipitation is interrupted by intermittent dry downslope winds on the lee of the lower subtropical Andes at 35°S (referred to as Zonda wind in South America; Fig. 4.42b), a unique Zonda period without precipitation dominates on the lee of the higher Andes at 32°S (Fig. 4.42d). Consistently, TRMM Precipitation Radar (PR) reflectivity fields reveal that the precipitating stratiform cloud can surpass the lower Andes (~3000 m MSL) at 35°S (Fig. 4.42a) but not the higher Andes (~6000 m MSL) at 32°S (Fig. 4.42d). The stronger drying on the lee of the higher Andes is also denoted by the notably higher dew point depression (i.e., the difference between temperature and dew point temperature) at Mendoza Station (Fig. 4.42d) compared with that observed at Malargue Station on the lee of the lower Andes (Fig. 4.42c). This finding suggests that AR penetrations to the lee of the subtropical Andes are uncommon, but further investigation is required.

Because the Andes mountain range is higher in the subtropical latitudes, it exerts strong control on the precipitation

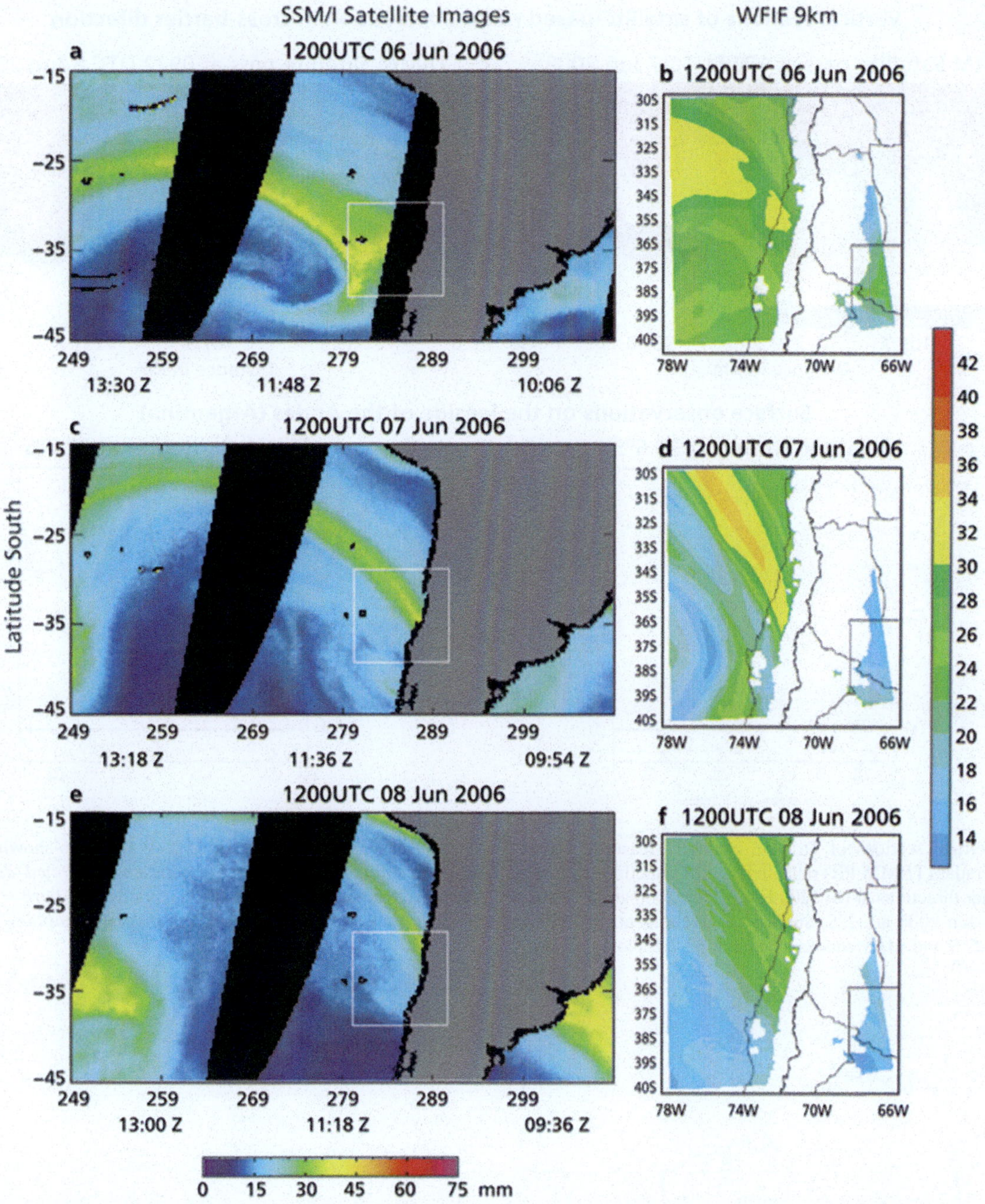

Fig. 4.41 Special Sensor Microwave/Imager (SSM/I) satellite images showing the evolution of an AR after it made landfall on the west coast of South America at around (**a**) 1200 UTC 06 Jun, (**c**) 1200 UTC 07 Jun, and (**d**) 1200 UTC 08 Jun 2006. The panels (**b, d, f**) show the same times of the satellite observations but, for the Weather Research and Forecasting (WRF) model output, configured with a grid spacing of 9 km. (Adapted from Viale 2010)

over the mountains and their adjacent lowlands when ARs make landfall on the west coast of South America. The schematic illustration of Fig. 4.43 depicts the typical kinematic and microphysical behavior of landfalling ARs. Upstream of the Andes, the mountains typically block and divert poleward the AR's low-level flow in the region ahead of the cold front. This upstream blocked flow then leads to higher rain accumulations by low- and mid-level orographic processes over lowlands upstream of the Andes than over the open ocean (Fig. 4.43b). At low levels, the blocked flow enhances the low-level convergence in the cold front, and thus strengthens the updrafts and the water content in clouds upstream of the Andes, creating a feeder cloud (e.g., Bergeron 1965). At mid-level, ascent of the moist AR over the blocked airflow

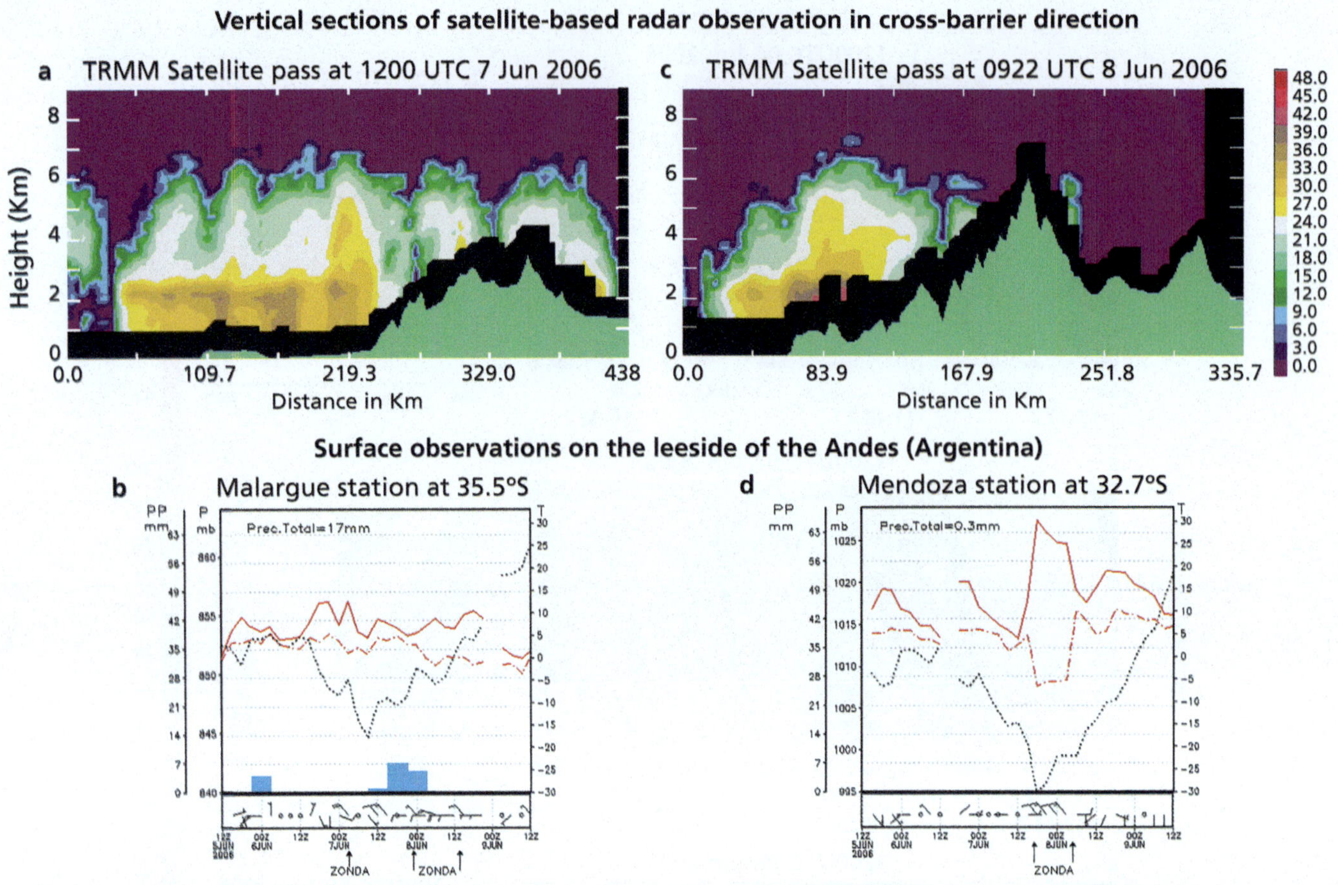

Fig. 4.42 Vertical sections of Tropical Rainfall Measuring Mission precipitation radar (TRMM PR) reflectivity (dBZ) satellite observation in cross-barrier directions at (**a**) 1200 UTC 7 Jun 2006 at 35°S and (**b**) 0922 UTC 8 Jun 2006 at 32°S. Surface observations at (**c**) Malargue Station at 35.5°S and (**d**) Mendoza Station at 32.7°S plotted every 3 h from 1200 UTC 5 Jun to 1200 UTC 9 Jun 2006 showing temperature (°C *red solid line*), dew point temperature (°C, *dotted–dashed red line*), sea level pressure (hPa, *black dots*), winds (*full barb* = 10 m s^{-1}), and 6-h accumulated precipitation (mm, *shaded light blue*). (Adapted from Viale 2010)

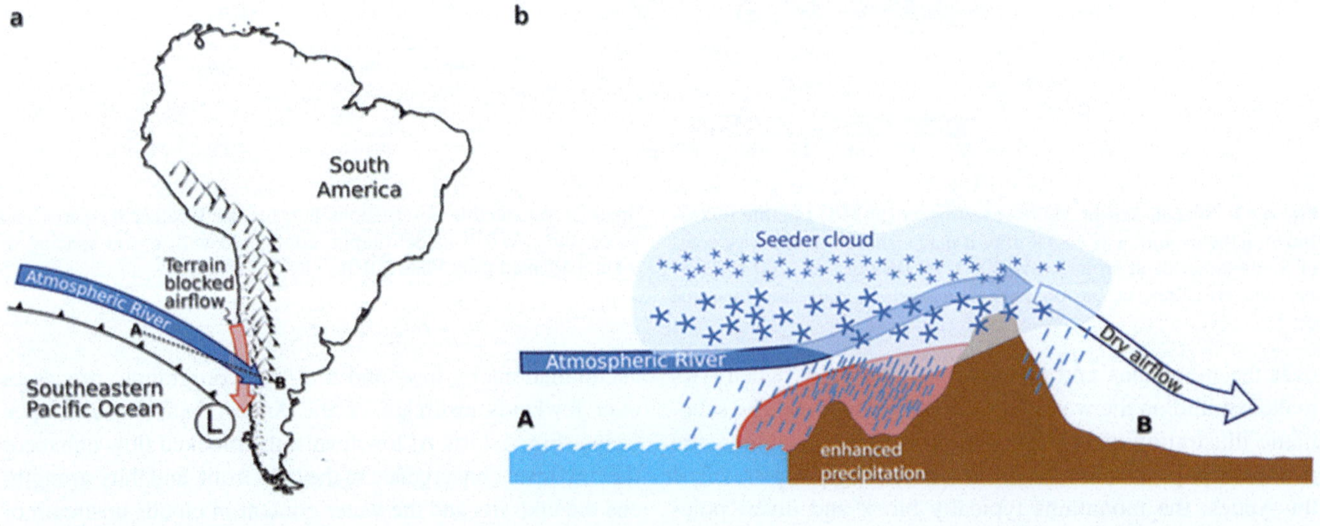

Fig. 4.43 Schematic representation of the kinematic and microphysical behavior of the AR impacting against the mountainous west coast of South America: (**a**) plan view and (**b**) cross-barrier view

layer produces snow aloft and light rain (seeder cloud) in the region ahead of the front. A seeder–feeder mechanism then becomes active when the front moves into the orographically enhanced mid-level cloud layer. This mechanism accelerates the microphysical growth processes of rain droplets and gradually increases the precipitation rates and accumulations as the distance from the Andes is reduced. More details in this upstream orographic precipitation enhancement are described by Viale et al. (2013), who also performed a sensitivity simulation with a 50% reduction in the Andes topography, for further comparison to various west coast mountain ranges of North America—which suggests the extreme height of the actual Andes strengthens the frontogenesis and upstream blocking. Over the windward slopes of the Andes, the precipitation is also enhanced because one portion of the mid-level precipitating particles—which are orographically enhanced upstream of the barrier—is then transported by strong mid-level west winds to the windward slope of the Andes, and because additional ascent of a mid-level moist AR directly over the windward slope of the Andes can increase snowfall accumulation. Over leeward slopes and downstream of the Andes, strong dry downslope winds markedly reduce precipitation results.

4.8.1 Summary of Sect. 4.8

Despite there being few studies of ARs in South America (compared to North America and Europe), these studies reveal that—similarly to these western continental regions—ARs play a crucial role on the west coast of South America by providing most of the water supply through heavy orographic precipitation, which can, in turn, set the stage for hydrometeorological hazards that occasionally produce floods, landslides, and serious damages and losses. Most of the few heavy wintertime precipitation events per year that accumulate above the 75th percentile of the annual total precipitation on the subtropical west coast of South America are linked to landfalling ARs. In particular, most warm winter storms, which are potentially hazardous because of the increased pluvial area over Andean basins, are linked to landfalling ARs. Additionally, distinctive regional features of landfalling ARs over the mountainous terrain of the Andes have been described. The high Andes in the subtropical latitudes induce a strong upstream blocking effect on landfalling ARs, which enhances precipitation over coastal lowlands upstream of the Andes, as well as causes an abrupt drying of the air mass downstream of the barrier, removing most of the AR's moisture.

Since all AR studies of South America have been conducted in the subtropical latitudes where most of the population of Chile and western Argentina lives, how ARs affect the extratropical west coast has yet to be understood. Exploring how ARs affect this remote region—from 37°S to the southern tip of the continent—points to the direction for future research.

4.9 ARs in the Polar Regions

Irina Gorodetskaya, William Neff, Deniz Bozkurt, Hans Christian Steen-Larsen, and Maria Tsukernik

Atmospheric rivers represent an efficient mechanism for the poleward transport of warm moist air (Newell et al. 1992) to the Arctic (Bonne et al. 2015; Neff et al. 2014) and the Antarctic (Gorodetskaya et al. 2014). ARs that reach the Greenland or Antarctic ice sheets are subject to orographic uplift because of the relatively steep slopes of the ice sheets' escarpment regions. In the case of Greenland, the ice sheet elevation increases ~3200 m over ~500 km from the west coast to Summit Station; in East Antarctica, the topography increases to 1500 m over ~200 km in the escarpment zone, with the mountain ranges protruding through the ice at some locations. These landfalling ARs are expected to strongly influence polar precipitation, similarly to the ARs that reach the coasts of California or South America that have mountainous terrain (see Chap. 5 "Effects of ARs"). In addition to anomalous precipitation, ARs also bring strong warm-air advection. In July 2012, an AR with associated moisture and heat advection, together with thin liquid-containing clouds, were responsible for the melt event in central Greenland (Neff et al. 2014; Bennartz et al. 2013). Further, Bozkurt et al. (2018) demonstrated that an AR intensified the foehn winds, which produced record high temperatures over the Antarctic Peninsula in 2015. Recently, enhanced winter precipitation over Spitzbergen was also shown to be associated with an AR (Serreze et al. 2015). The importance of ARs in polar regions is demonstrated below, based on events such as the Greenland melt event of July 2012 (Neff et al. 2014), the anomalously high snowfall in East Antarctica during 2009 and 2011 (Gorodetskaya et al. 2014), and the record high temperatures in the Antarctic Peninsula (Bozkurt et al. 2018). Such events in polar regions are also worthy of further investigation, given the accelerating rate of melting of the Greenland ice sheet since the early 1990s (Graeter et al. 2018) as well as accelerating ice mass loss from the West Antarctic ice sheet and the Antarctic Peninsula (Rignot et al. 2004; Cook and Vaughan 2010; Mouginot et al. 2014). These increases in the melting rate also coincide with increasing moisture transport to the Greenland ice sheet (Mattingly et al. 2016), warming in recent decades as seen in $\delta^{18}O$ signatures in ice cores for northwestern Greenland (Masson-Delmotte et al. 2015), and coincident sea level increases along the US East Coast associated with increasing Greenland melting (Davis and Vinogradova 2017).

4.9.1 Arctic

In July 2012, virtually the entire surface of the Greenland ice sheet melted (Nghiem et al. 2012). Neff et al. (2014) examined this melt episode in the context of ARs and compared it to the last similar melt episode, which happened in the summer of 1889 (as seen in ice core records; Meese et al. 1994), and found similar ARs at work using the twentieth Century Reanalysis (20CR; Compo et al. 2011). In July 2012, an AR transported warm air from a mid-continental heat wave in North America over the Atlantic Ocean and then northward to the west coast of Greenland, spreading over the ice sheet (Fig. 4.44). In July 1889, similar transport occurred together with significant black carbon amounts associated with massive forest fires in both the USA and Canada (Neff et al. 2014). It was suggested that the black carbon contributed to the melting over Greenland in 1889. Back trajectories using the 20CR for 1889 showed origins of air masses arriving on the west coast of Greenland from both the mid-continent fire areas (below 800 hPa) as well as from the Caribbean at 700 and 600 hPa.

During both episodes, the expression of poleward moisture transport as an AR was associated with continental heat anomalies in the trajectory source regions as well as a trough–ridge pattern that focused transport along the west coast of Greenland. The latter consisted of a trough of low pressure situated to the west—generally over Baffin Island and Hudson Bay—and a high-pressure ridge that impinged on the southeast coast of Greenland (Fig. 4.45). This type of trough–ridge pattern was present in a major rain event in late summer of 2011 along the western margin of the Greenland ice sheet and accelerated the flow of ice into the ocean. This event, which began on 24 August 2011 and featured maximum precipitation on 27 August, is described in detail by Doyle et al. (2015). Figure 4.46a shows the initial phase of this event as an AR approaches, as seen in satellite microwave imagery for 12–24Z on 24 August. Figure 4.46b shows the expansion of the surface melt area over 3 days. During this event, IWV increased on 27 August to 37 kg m^{-2} (ERA-Interim at 60°N, −50°W) as seen initially in Fig. 4.46c. Figure 4.46d shows associated wind vectors at 850 hPa: here, the wind has no upslope component, which may account for the lack of inland penetration of the melt region. In Fig. 4.46e, f both the 20CR and the NCEP–NCAR reanalyses show similar distributions of IWV.

Extreme moisture transport into the Arctic in winter along the west coast of Greenland (with an AR-like structure) has been identified as an artifact of Rossby wave breaking (Liu and Barnes 2015). In particular, as concluded by Liu and Barnes (2015), cyclonic wave breaking during the negative phase of the NAO favors moisture transport along the west coast of Greenland; conversely, anticyclonic wave breaking favors moisture transport into the North Atlantic. ARs have also been identified in the transport of warm moist air across the entire Arctic during the Surface Heat Budget of the Arctic Ocean (SHEBA) experiment (Persson et al. 2016). The NCEP–NCAR reanalysis has been used to identify the significant contribution of low-frequency variability to moisture transport to the Arctic (Newman et al. 2012), identifying a major pathway from the extratropics to Greenland through the western Atlantic during summer. This pathway is consistent with the results of Bonne et al. (2015) and Neff et al. (2014) for the 2012 Greenland melt episode and the AR structure associated with the moisture transport.

Thanks to the recent evolution in analytical capability, the isotopic composition of the AR moisture during the 12 July 2012 melt event in Greenland could be matched with the source region conditions of the marine boundary layer (Bonne et al. 2015). The evolution in the water vapor isotopic composition of the AR could be tracked as it moved northward because of a network of in situ atmospheric water vapor monitoring stations. These stations carried out observations on top of the Greenland ice sheet, in south Greenland, and at Bermuda. From water isotope theory, Jouzel et al. (2013) hypothesized that the second-order isotopic parameters (d-excess = δD − 8 × δ^{18}O) of the water vapor in an air mass would be conserved during transport and, hence, provide information about the conditions of the source regions. For the first time ever, this was confirmed by measuring within uncertainty the same level of the atmospheric d-excess along all three stations. The atmospheric water vapor isotopes measured at the three stations before, during, and after the AR's occurrence were also compared to simulations using isotope-enabled General Circulation Models (Laboratoire de Météorologie Dynamique Zoom–iso (LMDZ–isotope) in Risi et al. 2010; European Centre [from ECMWF] Hamburg ECHAM5-water isotopes in Werner et al. 2011) nudged to reanalysis products. This comparison showed that during the AR event the isotope-enabled GCMs were able to capture the correct d-excess level measured on top of the Greenland ice sheet. Compared to the rest of the season, this was only possible during the AR event. This was interpreted as an indication that the GCMs captured the correct atmospheric moisture transport during the AR event because of relatively simplified transport physics. The conservation of the d-excess signal in the atmospheric water vapor of the AR was also interpreted to indicate a lack of snow crystal formation in the cloud, which is in accordance with the observations of rain on top of the Greenland ice sheet (Bonne et al. 2015).

4.9.2 Antarctic

As Zhu and Newell (1998) noted, three to four ARs can be present in mid-latitudes at any moment. Figure 4.47 shows

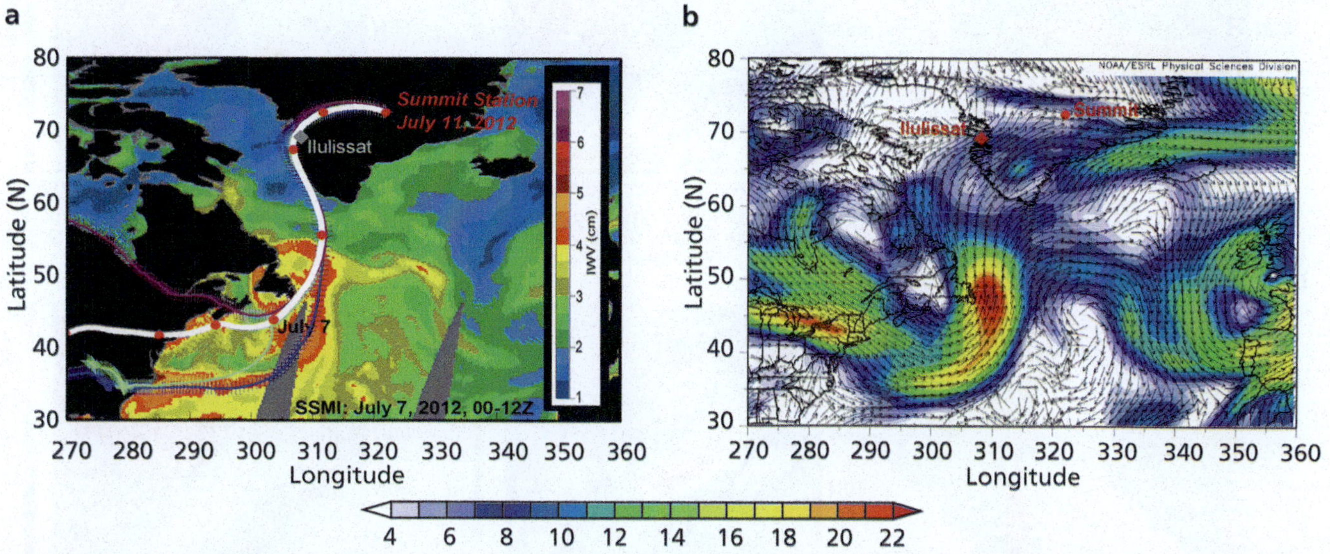

Fig. 4.44 (**a**) Special Sensor Microwave/Imager (SSM/I) image showing the developing AR and associated water vapor filaments that extend from the USA toward the southwest coast of Greenland on 7 July 2012 together with back trajectories. (**b**) ERA-Interim wind vectors (700 hPa) and speeds (ms^{-1}, scale below) for 7 July 2012

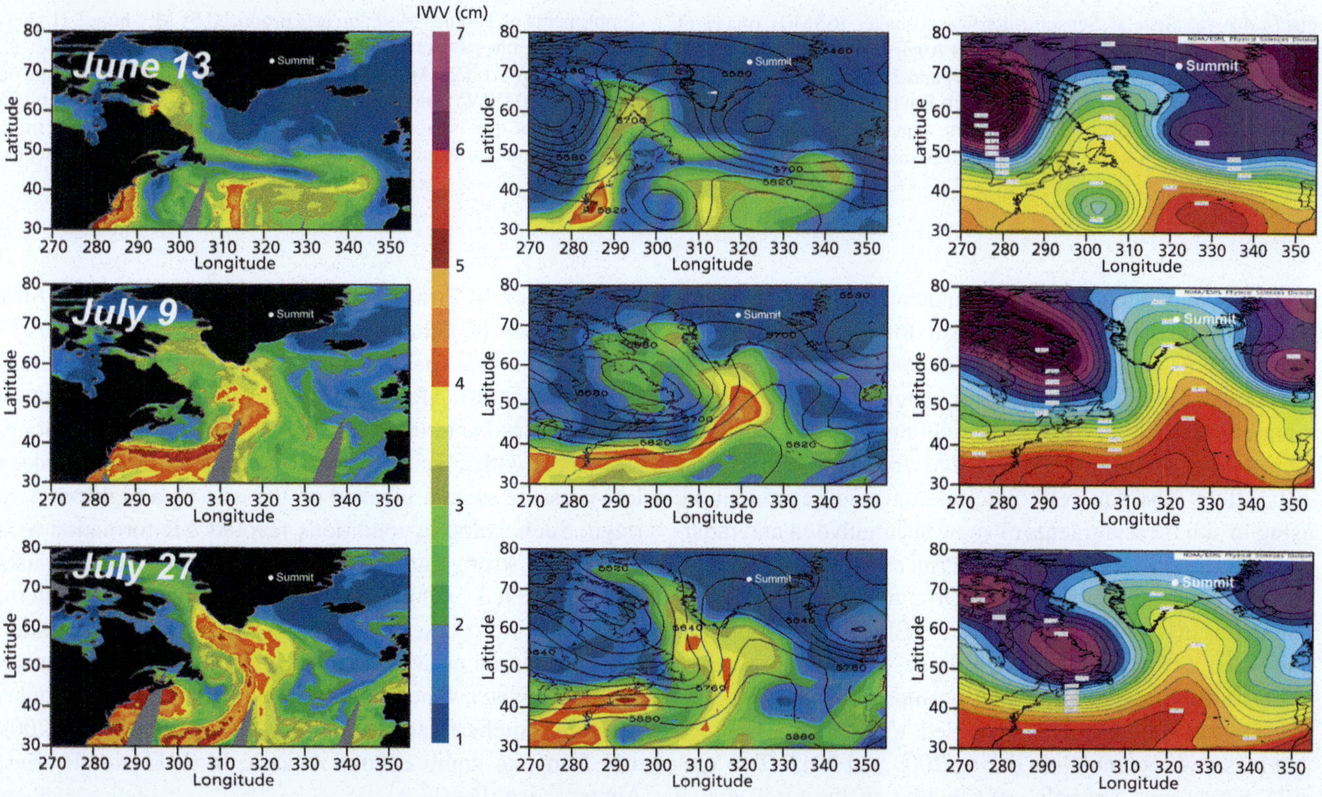

Fig. 4.45 Comparison of *left* Special Sensor Microwave/Imager (SSM/I)-derived integrated water vapor (IWV) with that from the *middle* Global Forecast System (GFS) analysis and *right* 500-hPa-height fields for 2012

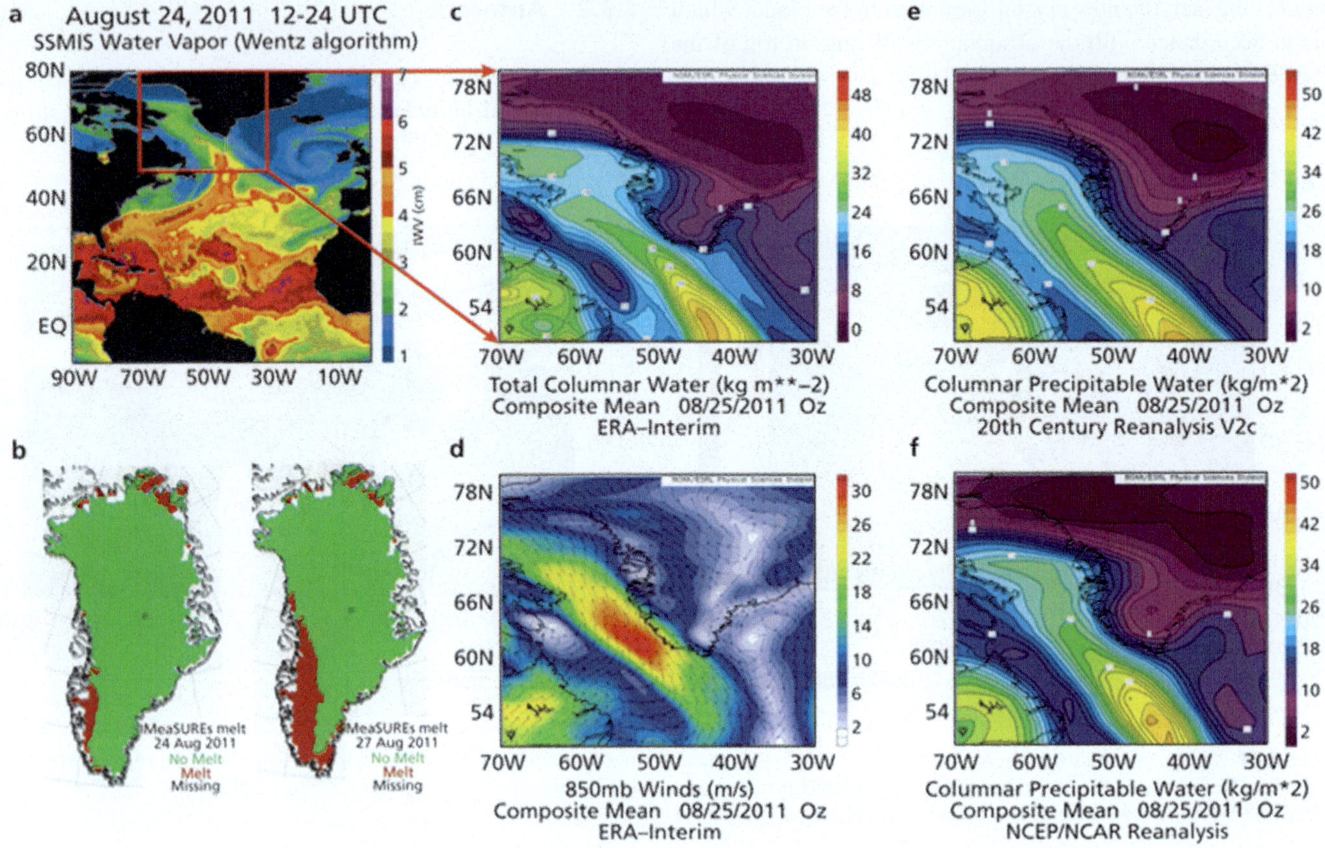

Fig. 4.46 (**a**) Special Sensor Microwave/Imager (SSM/I) image of total integrated water vapor (IWV) on 24 August 2011 (compliments of G. Wick), (**b**) surface melt patterns for 24 and 27 August. Making Earth System data records for Use in Research Environments (MEaSUREs) Greenland Surface Melt Daily 25 km Equal-Area Scalable Earth (EASE)-Grid 2.0, V1 derived from satellite microwave measurements, compliments of Thomas Mote; images produced by M. Shupe). Highest precipitation rates occurred on 27 August as noted by Doyle et al. (2015). (**c**) IWV on 25 August from ERA-Interim. (**d**) Wind vectors on 25 August. (**e**) IWV from 20CR reanalysis on 25 August. (**f**) IWV on 25 August from the National Centers for Environmental Prediction–National Center for Atmospheric Research (NCEP–NCAR) reanalysis

vertically integrated meridional moisture transport over the Southern Hemisphere at an instant for 19 May 2009, 00 UTC, where several corridors of intense moisture transport can be identified. One of them extends all the way to Antarctica, reaching the ice sheet along the 60°E meridian. Gorodetskaya et al. (2014) analyzed, for the first time, how ARs affect East Antarctic precipitation and accumulation, using in situ measurements of snow accumulation and radar-derived snowfall estimates from Princess Elisabeth Station, located in the escarpment area of Dronning Maud Land, East Antarctica. Measurements at Princess Elisabeth Station were combined with the ERA-Interim large-scale meteorological fields, to investigate the relationship between ARs and intense snowfall events, which led to anomalously high annual snow accumulation during 2009 and 2011. Two specific cases, each strongly contributing to the total annual accumulation at Princess Elisabeth Station, on 18–19 May 2009 and 14–16 February 2011, were analyzed in terms of moisture transport patterns and associated large-scale circulation (Fig. 4.48). ARs that brought anomalous precipitation events in East Antarctica in 2009 and 2011 were commonly associated with specific synoptic conditions, namely a deep low-pressure system blocked on the east by a high-pressure ridge. Such synoptic conditions favored AR formation as a narrow and strong moisture transport along the eastern flanks of the blocked cyclone. A large-scale circulation pattern, similar for both cases, steered an intense narrow moisture flux into Dronning Maud Land, with the band of enhanced integrated water vapor stretching from subtropical latitudes (from the southern Indian Ocean east of Madagascar in 2009 and from the south central Atlantic in 2011) to the East Antarctic ice sheet.

The importance of such blocking events in the heat and moisture transport toward the Antarctic interior was emphasized in earlier studies (e.g., Hirasawa et al. 2000, 2013; Frezzotti et al. 2007). Gorodetskaya et al. (2014) showed that such synoptic and associated moisture flux patterns of meridional nature are characteristics of ARs, and introduced a modified definition for ARs in Antarctica, which considers a lower saturation capacity of the polar troposphere. Applying this definition to the entire 2009–2012 period showed that the anomalously high accumulation observed in Dronning Maud Land in 2009 and 2011 can be attributed to a few extreme accumulation events (four to five per year), all of which were associated with ARs that reached the Antarctic coast within the 7°E–60°E longitudinal sector.

It is important to note that the AR events of 2009 and 2011 stand out in the long-term record as meridional moisture transport anomalies. As shown by ERA-Interim data (1979–2014; Tsukernik and Lynch 2013; Gorodetskaya et al. 2014), meridional moisture transport time-series have no significant trend throughout the last 35 years—but years 2009 and 2011 stand out as anomalous in Dronning Maud Land. These results correspond to ground observations at Princess Elisabeth Station, as well as at Syowa Station and with traverse snow stakes data. Regional processes most likely govern the abundance of ARs in Dronning Maud Land in 2009 and 2011. The Southern Annular Mode (SAM)—the dominant large-scale mode of variability of the Southern Hemisphere—experienced near-neutral conditions during both 2009 and 2011. In fact, no significant correlation was revealed between SAM and meridional moisture transport in either the East or West Antarctic region. Although positive SAM corresponds to more cyclones formed in the Southern

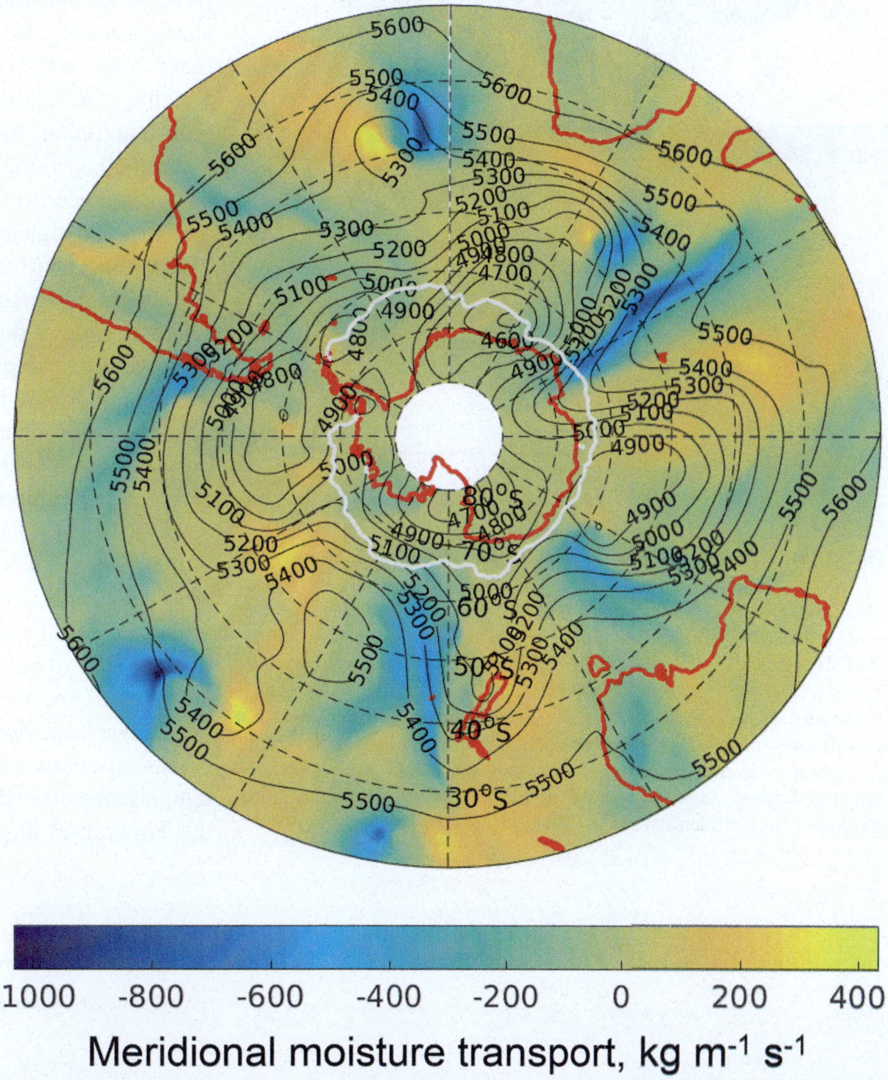

Fig. 4.47 Vertically integrated meridional moisture transport (*shading*, kg s^{-1} m^{-1}), 500-hPa geopotential height (*black contours*) and sea ice edge (*white contour*) for 20ºS-80ºS on 19 May 2009, 0000 UTC. Figure adapted from Gorodetskaya et al (2014)

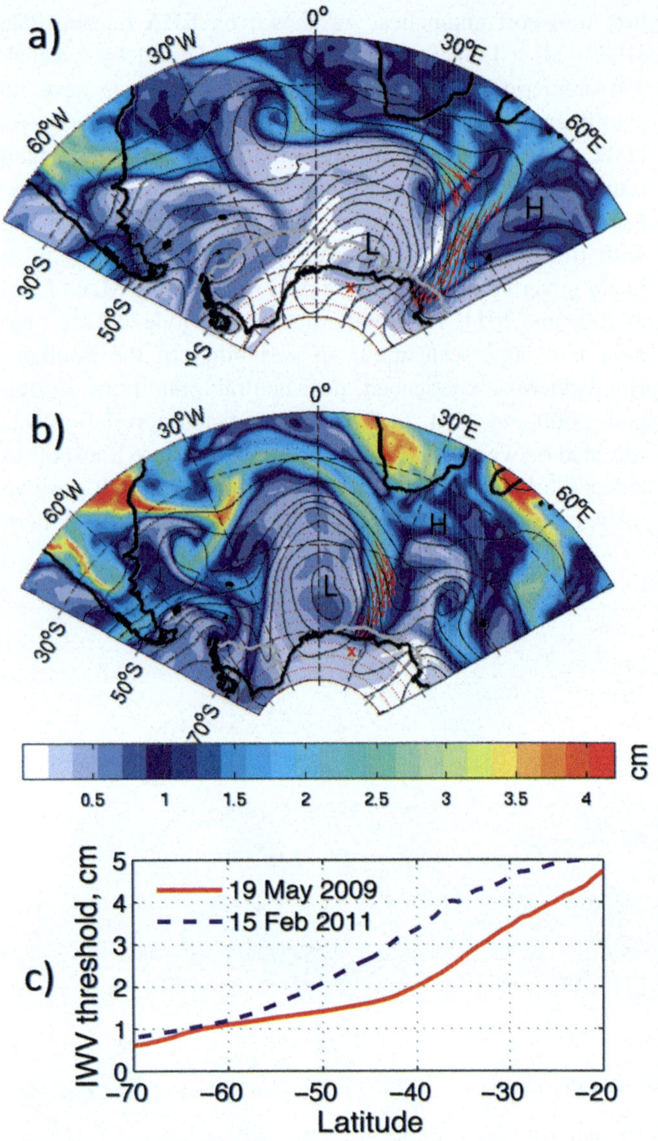

Fig. 4.48 Vertically integrated water vapor (*shading*, cm) and total horizontal moisture transport (*red arrows:* kg m^{-1} s^{-1}) within each AR as identified using the definition adapted for Antarctica by Gorodetskaya et al (2014) during (**a**) 19 May 2009 00 UTC and (**b**) 15 February 2011 00 UTC. Black contours are 500-hPa geopotential heights, where L shows a closed trough at 500 hPa influencing Dronning Maud Land and H shows the blocking high-pressure ridge downstream of the low. Red cross shows the location of the Princess Elisabeth station, where high precipitation events associated with the ARs were measured. (**c**) Integrated water vapor threshold as a function of latitude. Figure adapted from Gorodetskaya et al (2014)

Ocean, the strong westerly winds associated with positive SAM tend to govern storms along the shores of the Antarctic continent, rather than promote meridional propagation (Uotila et al. 2013).

The continental slope of Antarctica, combined with strong katabatic winds, act as a barrier, similar to other mountain ranges, such as the Sierra Nevada in California. Moisture transport during the AR events overcomes this barrier transporting the largest amounts of moisture at higher levels above the katabatic boundary layer. Analysis using radiosonde profiling at the coastal Antarctic stations showed that the moisture flux during AR events frequently attains the highest values at the levels between 950 and 800 hPa, and corresponds to the peak in the wind speed and elevated specific humidity within this layer (Fig. 4.49; Silva et al. 2017). This LLJ that assists the anomalous moisture transport onto the ice sheet is of synoptic origin, and can be strictly meridional (poleward) or contain a zonal component. This flow differs from the katabatic flow from the high Antarctic plateau that is characterized by local wind maxima, especially in the escarpment zone, and typically occupies the near-surface layer. Specific case studies showed that while the LLJ coincides with maximum moisture transport core away from the Antarctic coast, the moist air advection reaching Antarctica gets decoupled from the LLJ when approaching the ice sheet and occurs at higher levels (Terpstra and Gorodetskaya 2017).

An important feature of the ARs that affect Antarctica is the isentropic ascent of the large amounts of moisture as it typically follows the WCB in the vicinity of an extratropical cyclone. For example, during the 18–19 May 2009 event, the 275 K and 285 K isentropic surfaces were extending from near the subtropical ocean surface up to 800- to 600-hPa pressure levels over the East Antarctic escarpment zone and carrying large amounts of moisture (Fig. 4.50; Gorodetskaya et al. 2011). As discussed in Sect. 2.3, the maximum integrated ascent of WCB trajectories (along-isentropic and cross-isentropic enhanced by diabatic heating) can be as strong as 600-hPa in <2 days (Wernli and Davies 1997).

The ARs that reach Antarctica not only bring large amounts of moisture and precipitation, as well as strong heat advection, but can also trigger local mesoscale processes. Bozkurt et al. (2018) examined a record-setting temperature of 17.5 °C that occurred on 24 March 2015 at the Esperanza Base located near the northern tip of the

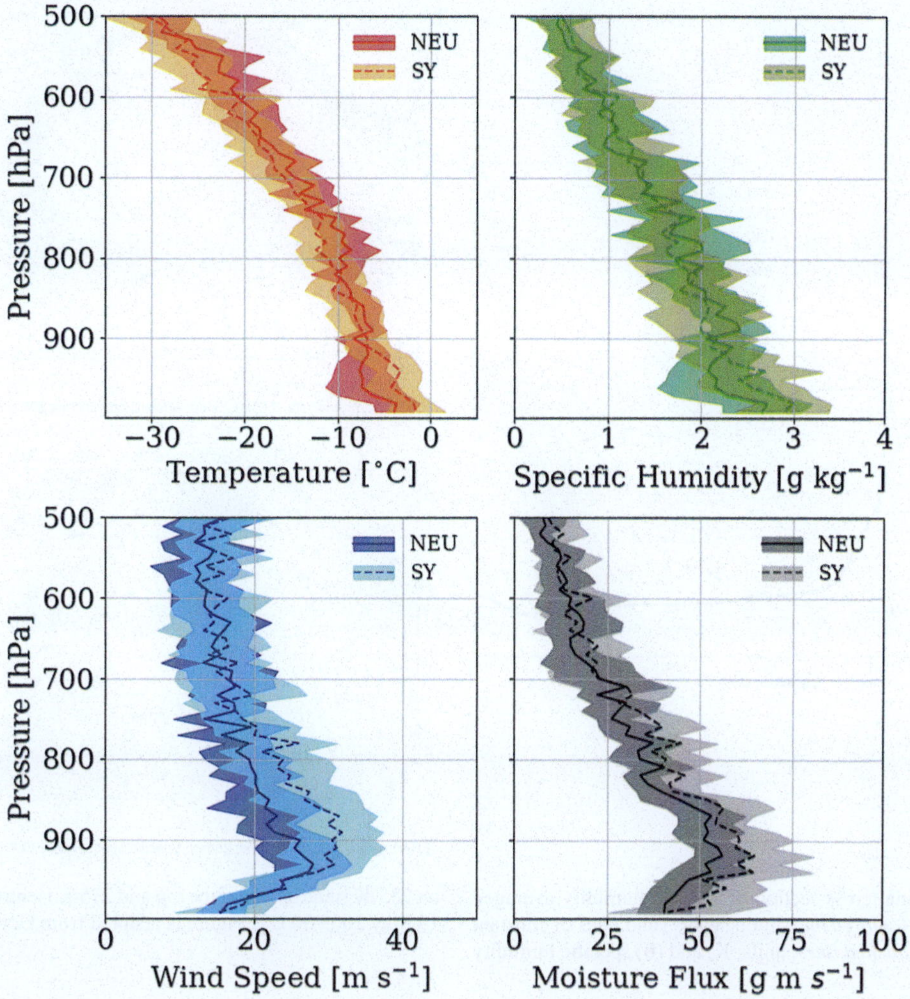

Fig. 4.49 Composite profiles (from near the surface to 500 hPa) for temperature, specific humidity, wind speed and moisture flux during the enhanced moisture transport events (with integrated water vapor transport greater than 100 kg m^{-1} s^{-1}, and a peak in the moisture flux along the profile exceeding 50 g kg^{-1} m s^{-1}). Based on radiosonde measurements at two coastal stations in Dronning Maud Land, East Antarctica: Syowa (SY, *dashed lines*) and Neumayer (NEU, *solid lines*) during 2009–2012. Mean values are shown by lines and spread (±one standard deviation) is shown by color shading. The data are interpolated to 10-hPa height steps. Figure is adapted from Silva et al. (2017)

Antarctic Peninsula. An AR event resulted in warm and moist air advection toward the peninsula before the record-setting temperature event, eventually triggering a foehn episode on the leeward side of the central and northern Antarctic Peninsula on 24 March 2015. The Moderate Resolution Imaging Spectroradiometer (MODIS) Antarctic Ice Shelf Image Archive of Larsen B and Larsen A embayments (Fig. 4.51) provides clear evidence for disintegration and advection of sea ice, as well as the formation of melt ponds on the fast ice and glacier surface at the base of the Antarctic Peninsula mountain range. Bozkurt et al. (2018) suggested a link between local-scale forcing (i.e., foehn effect warming) and large-scale forcing (i.e., an AR) in explaining the record-setting temperature that occurred on 24 March 2015 at the Esperanza Base. The synoptic-scale analysis based on the ERA-Interim reanalysis data shows a deep low-pressure center over the Amundsen–Bellingshausen Sea, and a blocking ridge over the southeast Pacific, which provided favorable conditions for the development of an AR with a northwest–southeast orientation, directing warm and moist air toward the Antarctic Peninsula and triggering a widespread foehn episode. The

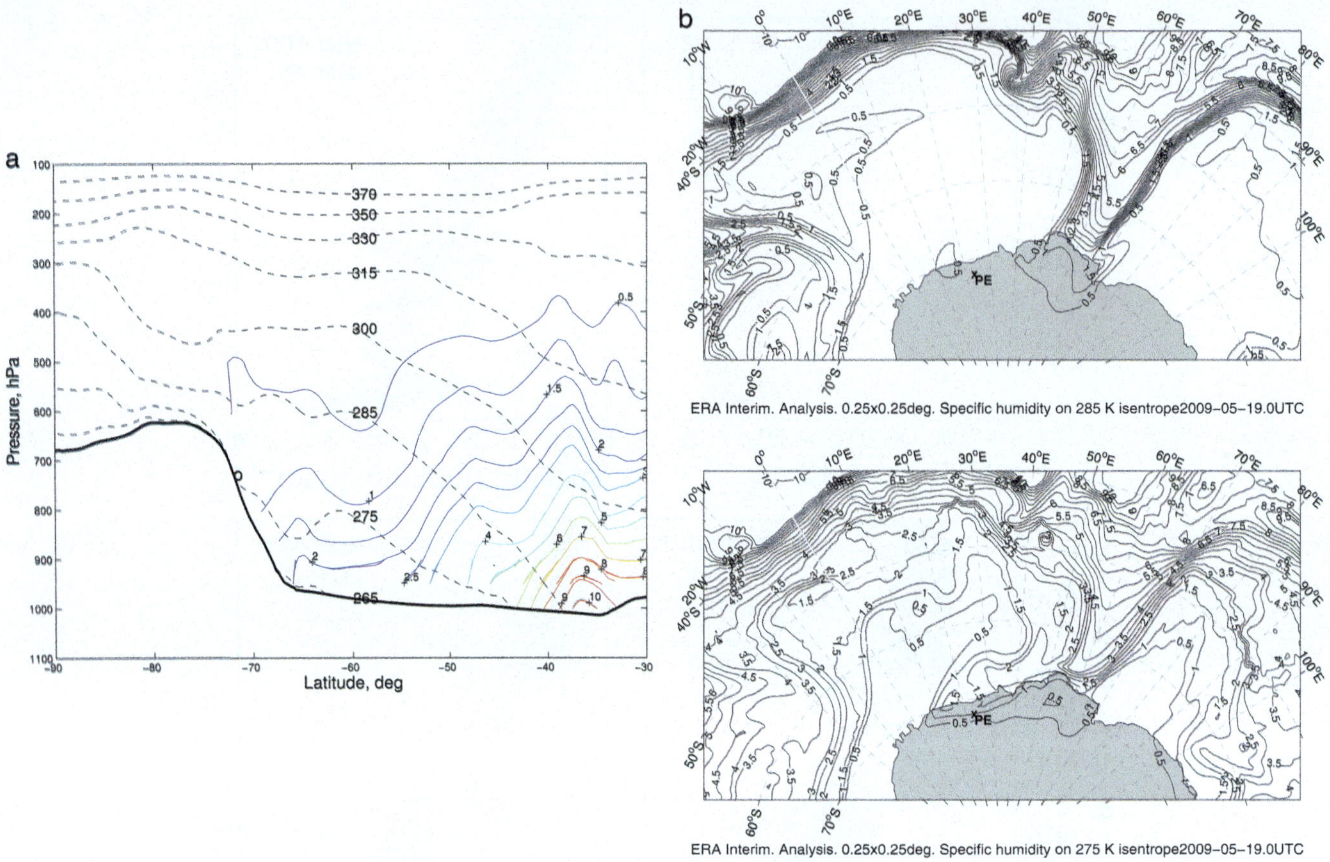

Fig. 4.50 (**a**) Latitudinal cross-section of specific humidity averaged over a sector 20°-60°E (*colored lines*, units: g kg^{-1}) and lines of constant potential temperature (*dashed lines*, units: K) and (**b**) specific humidity on 285-K isentropic surface *top* and 275 K isentropic surface *bottom* for 19 May 2009 00 UT. Figure is adapted from Gorodetskaya et al. (2011)

large amounts of water vapor that reached the windward side of the peninsula from the AR was an important precursor to the orographic precipitation enhancement and latent heat release on the windward side, suggesting that the main foehn mechanism was at work during this episode. An anomalous source of water vapor that enhances the moisture transport toward the Antarctic Peninsula through ARs may, arguably, play a crucial role in the occurrence of foehn events, as well as in determining the extremity of these events and their role in the ice shelf melt (Cape et al. 2015).

4.9.3 Summary of Sect. 4.9

The above examples demonstrate how significantly ARs affect climate and also the surface mass and energy balance of polar regions. These effects can be different, depending on an intricate balance between various processes: on one hand, large amounts of moisture transported with ARs can bring a lot of snowfall, and thus an increase in the regional ice/snow mass of the ice sheets, as was observed in East Antarctica in 2009 and 2011. However, heat advection that is also associated with the ARs can result in more frequent rain rather than snowfall, with the opposite effect on the regional ice/snow mass balance, decreasing the surface albedo and accelerating melt. Similarly, the July 2012 melt event in Greenland was triggered by warm advection and specific cloud properties that promoted surface heating. Although light rain was observed only on the coast during the 2012 event, an earlier AR event in August 2011 produced a major rain event along the western margin of the Greenland ice sheet, accelerating the loss of ice into the ocean. Also, the March 2015 record high temperatures and melt event east of the Antarctic Peninsula was linked to an AR that impinged on the mountain range and caused an orographically enhanced precipitation formation with associated strong latent heating, which triggered the local foehn effect and the associated temperature rise and melt on the eastward side of the peninsula. Thus, ARs are critical in understanding cryosphere changes, given the associated large warming amplification and anomalous precipitation events.

Acknowledgements Alexandre M. Ramos was supported by the Scientific Employment Stimulus 2017 from the Portuguese Science Foundation (Fundação para a Ciência e a Tecnologia, FCT), Portugal (CEECIND/00027/2017). Irina V. Gorodetskaya thanks FCT/ MCTES for the financial support to CESAM (UIDP/50017/2020+UIDB/50017/2020) through national funds.

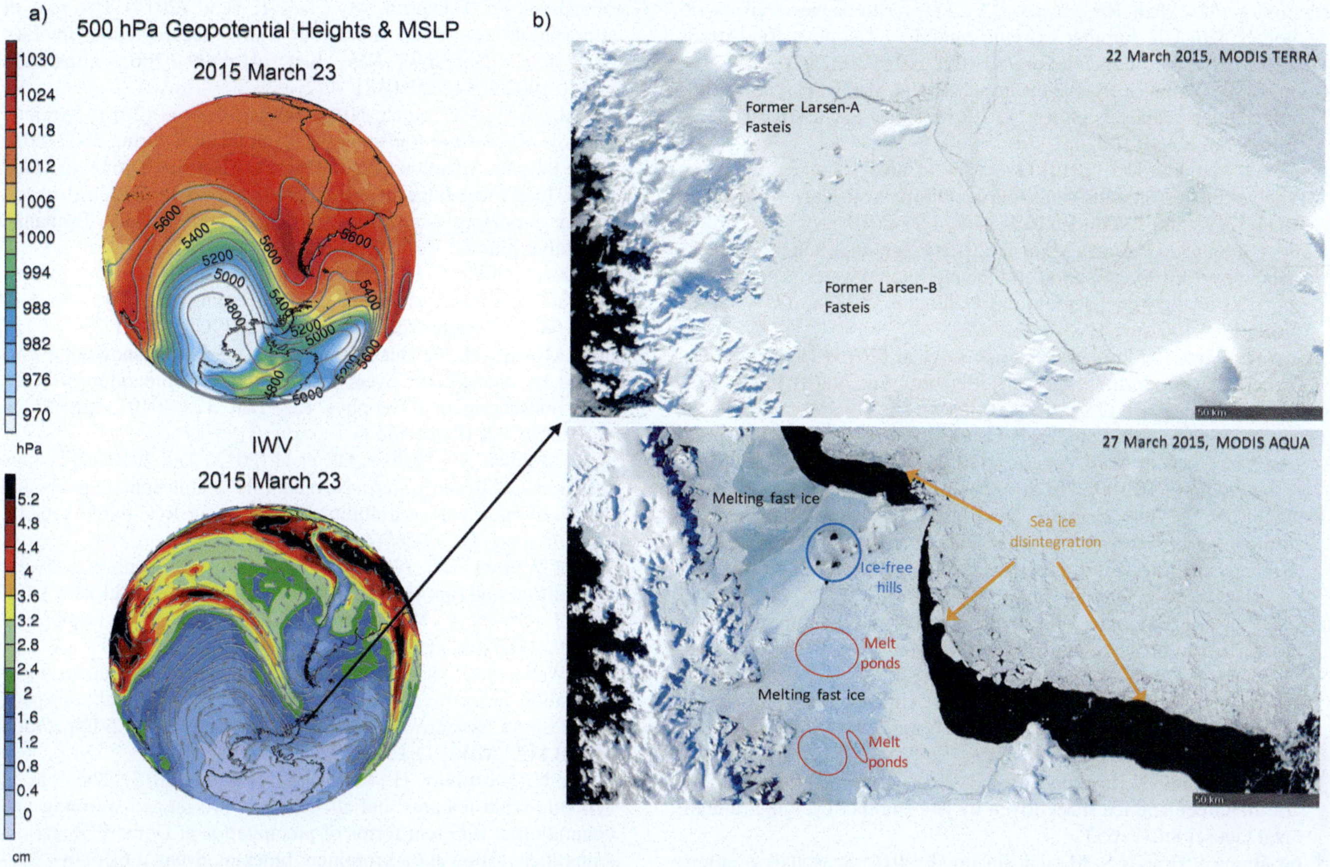

Fig. 4.51 (**a**) The 23 March 2015 mean sea level pressure (hPa; *shaded*) and 500-hPa geopotential heights (m; contour lines at 100-m intervals) from ERA-Interim *top*. Also shown is 23 March 2015 integrated water vapor (IWV) (cm; *shaded*) and 850-hPa wind vectors from ERA-Interim *bottom*. (**b**) Moderate Resolution Imaging Spectroradiometer (MODIS) images of Larsen B and Larsen A embayments before (22 March 2015) (*top*) and after (27 March 2015) (*bottom*) the record high-temperature event. Orange arrows indicate areas of sea ice disintegration and offshore advection. Red and blue circles contain melt ponds and ice-free hills, respectively. Figure adapted from Bozkurt et al. (2018)

References

Abatzoglou JT (2016) Contribution of cutoff lows to precipitation across the United States. J Appl Meteorol Climatol 55:893–899

Alexander MA, Scott JD, Swales D et al (2015) Moisture pathways into the U.S. intermountain west associated with heavy winter precipitation events. J Hydrometeorol 16:1184–1206

Alijani B, O'Brien J, Yarnal B (2008) Spatial analysis of precipitation intensity and concentration in Iran. Theor Appl Climatol 94:107–124. https://doi.org/10.1007/s00704-007-0344-y

Ambaum MHP, Hoskins BJ, Stephenson DB (2001) Arctic oscillation or North Atlantic oscillation? J Clim 14:3495–3507. https://doi.org/10.1175/1520-0442(2001)014<3495:AOONAO>2.0.CO;2

Backes TM, Kaplan ML, Schumer R et al (2015) A climatology of the vertical structure of water vapor transport to the Sierra Nevada in cool season atmospheric river precipitation events. J Hydrometeorol 16:1029–1047. https://doi.org/10.1175/JHM-D-14-0077.1

Beebe RG, Bates FC (1955) A mechanism for assisting in the release of convective instability. Mon Weather Rev 83:1–10

Bennartz R, Shupe MD, Turner DD et al (2013) July 2012 Greenland melt extent enhanced by low-level liquid clouds. Nature 496:83–86

Berg P, Haerter JO, Thejll P et al (2009) Seasonal characteristics of the relationship between daily precipitation intensity and surface temperature. J Geophys Res 114:D18102. https://doi.org/10.1029/2009JD012008

Bergeron T (1965) On the low-level redistribution of atmospheric water caused by orography. In: Proc Int Conf Cloud Physics (Suppl), Tokyo, Japan, International Association of Meteorology and Atmospheric Physics. pp 96–100

Bernhardt D (2006) Glacier National Park flooding November 2006. NWS Western region tech. Attachment 08-23. NWS, Great Falls, 15 pp

Bonne J et al (2015) The summer 2012 Greenland heat wave: in situ and remote sensing observations of water vapor isotopic composition during an atmospheric river event. J Geophys Res-Atmos 120:2970–2989. https://doi.org/10.1002/2014JD022602

Bozkurt D, Rondanelli R, Marin JC et al (2018) Foehn event triggered by an atmospheric river underlies record-setting temperature along continental Antarctica. J Geophys Res 123:3871–3892. https://doi.org/10.1002/2017JD027796

Brands S, Gutiérrez JM, San-Martín D (2017) Twentieth-century atmospheric river activity along the west coasts of Europe and North America: algorithm formulation, reanalysis uncertainty and links to atmospheric circulation patterns. Clim Dyn 48(9–10):2771–2795. https://doi.org/10.1007/s00382-016-3095-6

Cape MR, Vernet M, Skvarca P et al (2015) Foehn winds link climate-driven warming to ice shelf evolution in Antarctica. J Geophys Res-Atmos 120:11,037–11,057. https://doi.org/10.1002/2015JD023465

Catto JL, Jakob C, Berry G et al (2012) Relating global precipitation to atmospheric fronts. Geophys Res Lett 39:L10805. https://doi.org/10.1029/2012GL051736

Champion AJ, Allan RP, Lavers DA (2015) Atmospheric rivers do not explain UK summer extreme rainfall. J Geophys Res-Atmos 120:6731–6741. https://doi.org/10.1002/2014JD022863

Compo GP, Whitaker JS, Sardeshmukh PD et al (2011) The twentieth century reanalysis project. Q J R Meteorol Soc 137(654):1–28. https://doi.org/10.1002/qj.776

Cook AJ, Vaughan DG (2010) Overview of areal changes of the ice shelves on the Antarctic peninsula over the past 50 years. Cryosphere 4(1):77–98. https://doi.org/10.5194/tc-4-77-2010

Couto FT, Salgado R, Costa MJ et al (2015) Precipitation in the Madeira Island over a 10-year period and the meridional water vapour transport during the winter seasons. Int J Climatol 35:3748–3759. https://doi.org/10.1002/joc.4243

Dacre H, Clark P, Martinez-Alvarado O et al (2015) How do atmospheric rivers form? Bull Am Meteorol Soc 96(8):1243–1255. https://doi.org/10.1175/BAMSD1400031.1

Davis JL, Vinogradova NT (2017) Causes of accelerating sea level on the East Coast of North America. Geophys Res Lett 44:5133–5141. https://doi.org/10.1002/2017gl072845

Debbage N, Miller P, Poore S et al (2017) A climatology of atmospheric river interactions with the southeastern United States coastline. Int J Climatol 37(11):4077–4091. https://doi.org/10.1002/joc.5000

Dee DP et al (2011) The ERA-Interim reanalysis: configuration and performance of the data assimilation system. Q J R Meteorol Soc 137:553–597. https://doi.org/10.1002/qj.828

Dettinger MD, Ralph FM, Das T et al (2011) Atmospheric rivers, floods, and the water resources of California. Water 3:445–478. https://doi.org/10.3390/w3020445

Doyle SH, Lesins G, Thackray CP et al (2015) Amplified melt and flow of the Greenland ice sheet driven by late-summer cyclonic rainfall. Nat Geosci 8:647–653

Eiras-Barca J, Brands S, Miguel-Macho G (2016) Seasonal variations in North Atlantic atmospheric river activity and associations with anomalous precipitation over the Iberian Atlantic Margin. J Geophys Res-Atmos 121:931–948. https://doi.org/10.1002/2015JD023379

Eiras-Barca J, Ramos AM, Liberato MLR et al (2017) The importance of atmospheric rivers in the development of explosive cyclogenesis in the North Atlantic and North Pacific Basins. Earth Syst. Dynam., 9, 91–102, https://doi.org/10.5194/esd-9-91-2018, 2018.

Ferreira JA, Liberato MLR, Ramos AM (2016) On the relationship between atmospheric water vapour transport and extra-tropical cyclones development. Phys Chem Earth 94:56–65. https://doi.org/10.1016/j.pce.2016.01.001

Fink AH, Pohle S, Pinto JG et al (2012) Diagnosing the influence of diabatic processes on the explosive deepening of extratropical cyclones. Geophys Res Lett 39:L07803. https://doi.org/10.1029/2012GL051025

Frezzotti M, Urbini S, Proposito M et al (2007) Spatial and temporal variability of surface mass balance near Talos Dome, East Antarctica. J Geophys Res 112(F2):F02032. https://doi.org/10.1029/2006JF000638

Galarneau T Jr, Bosart LF, Schumacher RS (2010) Predecessor rain events ahead of tropical cyclones. Mon Weather Rev 138:3272–3297. https://doi.org/10.1175/2010MWR3243.1

Garaboa-Paz D, Eiras-Barca J, Huhn F et al (2015) Lagrangian coherent structures along atmospheric rivers. Chaos 25:063105. https://doi.org/10.1063/1.4919768

Garreaud RD (2013) Warm winter storms over central Chile. J Hydrometeorol 14:1515–1534

Gershunov A, Shulgina TM, Ralph FM et al (2017) Assessing climate-scale behavior of atmospheric rivers affecting the west coast of North America. Geophys Res Lett 44(15):7900–7908

Gimeno L, Nieto R, Drumond A et al (2013) Influence of the intensification of the major oceanic moisture sources on continental precipitation. Geophys Res Lett 40:1443–1450. https://doi.org/10.1002/grl.50338

Gorodetskaya IV, Tsukernik M, Claes K et al (2014) The role of atmospheric rivers in anomalous snow accumulation in East Antarctica. Geophys Res Lett 41:6199–6206. https://doi.org/10.1002/2014GL060881

Gorodetskaya IV, Van Lipzig NPM, Scheere L et al (2011) Synoptic controls of moisture transport and accumulation during 2009–2010 at the Belgian Antarctic Station Princess Elisabeth, Dronning Maud Land. In: Proceedings of the 6th Antarctic meteorological observation, modeling and forecasting workshop, Hobart, Tasmania, Australia, June 22–24, 2011

Graeter KA, Osterberg EC, Ferris DG et al (2018) Ice core records of West Greenland melt and climate forcing. Geophys Res Lett 45:3164–3172. https://doi.org/10.1002/2017GL076641

Guan B, Molotch N, Waliser DE et al (2010) Extreme snowfall events linked to atmospheric rivers and surface air temperature via satellite measurements. Geophys Res Lett 37:L20401. https://doi.org/10.1029/2010GL044696

Guan B, Molotch NP, Waliser DE et al (2013) The 2010/2011 snow season in California's Sierra Nevada: role of atmospheric rivers and modes of large-scale variability. Water Resour Res 49:6731–6743. https://doi.org/10.1002/wrcr.20537

Guan B, Waliser DE (2015) Detection of atmospheric rivers: evaluation and application of an algorithm for global studies. J Geophys Res-Atmos 120:12514–12535. https://doi.org/10.1002/2015JD024257

Guan B, Waliser DE, Molotch NP et al (2012) Does the Madden–Julian oscillation influence wintertime atmospheric rivers and snowpack in the Sierra Nevada? Mon Weather Rev 140:325–342. https://doi.org/10.1175/MWR-D-11-00087.1

Hirasawa N, Nakamura H, Motoyama H et al (2013) The role of synoptic-scale features and advection in pro-longed warming and generation of different forms of precipitation at Dome Fuji station, Antarctica, following a prominent blocking event. J Geophys Res-Atmos 118(13):6916–6928. https://doi.org/10.1002/jgrd.50532

Hirasawa N, Nakamura H, Yamanouchi T (2000) Abrupt changes in meteorological conditions observed at an inland Antarctic station in association with wintertime blocking. Geophys Res Lett 27(13):1911–1914. https://doi.org/10.1029/1999GL011039

Hirota N, Takayabu YN, Kato M et al (2016) Roles of an atmospheric river and a cutoff low in the extreme precipitation event in Hiroshima on 19 August 2014. Mon Weather Rev 144:1145–1160. https://doi.org/10.1175/MWR-D-15-0299.1

Hollinger JP, Peirce JL, Poe GA (1990) SSM/I instrument evaluation. IEEE Trans Geosci Remote Sens 28:781–790

Huffman GJ, Adler RF, Morrissey MM et al (2001) Global precipitation at one-degree daily resolution from multisatellite observations. J Hydrometeorol 2:36–50. https://doi.org/10.1175/1525-7541(2001)002<0036:GPAODD>2.0.CO;2

Hughes M, Mahoney KM, Neiman PJ et al (2014) The landfall and inland penetration of a flood-producing atmospheric river in Arizona. Part II: sensitivity of modeled precipitation to terrain height and atmospheric river orientation. J Hydrometeorol 15:1954–1974

Jeong J-H, Ho C-H (2005) Changes in occurrence of cold surges over East Asia in association with Arctic oscillation. Geophys Res Lett 32:L14704. https://doi.org/10.1029/2005GL023024

Jouzel J, Delaygue G, Landais A et al (2013) Water isotopes as tools to document oceanic sources of precipitation. Water Resour Res 49(11):7469–7486. https://doi.org/10.1002/2013WR013508

Kingston DG, Lavers DA, Hannah DM (2016) Floods in the Southern Alps of New Zealand: the importance of atmospheric rivers. Hydrol Process 30:5063–5070. https://doi.org/10.1002/hyp.10982

Knippertz P, Wernli H (2010) A Lagrangian climatology of tropical moisture exports to the Northern Hemispheric extratropics. J Clim 23:987–1003. https://doi.org/10.1175/2009JCLI3333.1

Lavers DA, Allan RP, Wood EF et al (2011) Winter floods in Britain are connected to atmospheric rivers. Geophys Res Lett 38:L23803. https://doi.org/10.1029/2011GL049783

Lavers DA, Villarini G (2013a) Atmospheric rivers and flooding over the Central United States. J Clim 26:7829–7836. https://doi.org/10.1175/JCLI-D-13-00212.1

Lavers DA, Villarini G (2013b) The nexus between atmospheric rivers and extreme precipitation across Europe. Geophys Res Lett 40:3259–3264

Lavers DA, Villarini G (2015) The contribution of atmospheric rivers to precipitation in Europe and the United States. J Hydrol 522:382–390. https://doi.org/10.1016/j.jhydrol.2014.12.010

Lavers DA, Villarini G, Allan RP et al (2012) The detection of atmospheric rivers in atmospheric reanalyses and their links to British winter floods and the large-scale climatic circulation. J Geophys Res 117:D20106. https://doi.org/10.1029/2012JD018027

Liberato MLR (2014) The 19 January 2013 windstorm over the North Atlantic: large-scale dynamics and impacts on Iberia. Weather Clim Extreme 5–6:16–28. https://doi.org/10.1016/j.wace.2014.06.002

Liberato MLR, Ramos AM, Trigo RM et al (2013) Moisture sources and large-scale dynamics associated with a flash flood event. In: Lin J, Brunner D, Gerbig C et al (eds) Lagrangian modeling of the atmosphere. American Geophysical Union, Washington, DC, pp 111–126

Lin Y, Mitchell KE (2005) The NCEP Stage II/IV hourly precipitation analyses: development and applications. In: Preprints, 19th Conf. on Hydrology, San Diego, CA, Amer. Meteor. Soc. 1.2. http://ams.confex.com/ams/pdfpapers/83847.pdf

Liu CJ, Barnes EA (2015) Extreme moisture transport into the Arctic linked to Rossby wave breaking. J Geophys Res-Atmos 120:3774–3788

Livneh B, Rosenberg EA, Lin C et al (2013) A long-term hydrologically based dataset of land surface fluxes and states for the conterminous United States: update and extensions. J Clim 26:9384–9392. https://doi.org/10.1175/JCLI-D-12-00508.1

Madden RA, Julian PR (1971) Detection of a 40–50 day oscillation in the zonal wind in the tropical Pacific. J Atmos Sci 28:702–708. https://doi.org/10.1175/1520-0469(1971)028<0702:DOADOI>2.0.CO;2

Madden RA, Julian PR (1972) Description of global-scale circulation cells in the tropics with a 40–50 day period. J Atmos Sci 29:1109–1123. https://doi.org/10.1175/1520-0469(1972)029<1109:DOGSCC>2.0.CO;2

Mahoney K et al (2016) Understanding the role of atmospheric rivers in heavy precipitation in the Southeast United States. Mon Weather Rev 144:1617–1632. https://doi.org/10.1175/MWR-D-15-0279.1

Masson-Delmotte V et al (2015) Recent changes in north-west Greenland climate documented by NEEM shallow ice core data and simulations, and implications for past-temperature reconstructions. Cryosphere 9:1481–1504. https://doi.org/10.5194/tc-9-1481-2015

Matthews AJ (2000) Propagation mechanisms for the Madden–Julian oscillation. Q J R Meteorol Soc 126:2637–2651. https://doi.org/10.1002/qj.49712656902

Mattingly KS, Ramseyer CA, Rosen JJ et al (2016) Increasing water vapor transport to the Greenland Ice Sheet revealed using self-organizing maps. Geophys Res Lett 43:9250–9258. https://doi.org/10.1002/2016gl070424

Meese DA, Gow AJ, Grootes P et al (1994) The accumulation record from the GISP2 core as an indicator of climate-change throughout the Holocene. Science 266:1680–1682. https://doi.org/10.1126/science.266.5191.1680

Moore BJ, Bosart LF, Keyser D et al (2013) Synoptic-scale environments of predecessor rain events occurring east of the Rocky Mountains in association with Atlantic basin tropical cyclones. Mon Weather Rev 141:1022–1047. https://doi.org/10.1175/MWR-D-12-00178.1

Moore BJ, Mahoney KM, Sukovich EM et al (2015) Climatology and environmental characteristics of extreme precipitation events in the southeastern United States. Mon Weather Rev 143:718–741. https://doi.org/10.1175/MWR-D-14-00065.1

Moore BJ, Neiman PJ, Ralph FM et al (2012) Physical processes associated with heavy flooding rainfall in Nashville, Tennessee, and vicinity during 1–2 May 2010: the role of an atmospheric river and mesoscale convective systems. Mon Weather Rev 140:358–378. https://doi.org/10.1175/MWR-D-11-00126.1

Mouginot J, Rignot E, Scheuchl B (2014) Sustained increase in ice discharge from the Amundsen Sea Embayment, West Antarctica, from 1973 to 2013. Geophys Res Lett 41:1576–1584. https://doi.org/10.1002/2013GL059069

Mueller M, Mahoney KM, Hughes MR (2017) High-resolution model-based investigation of moisture transport into the Pacific Northwest during a strong atmospheric river event. Mon Weather Rev 145(9):3861–3879. https://doi.org/10.1175/MWR-D-16-0466.1

Mundhenk BD, Barnes EA, Maloney ED (2016) All-season climatology and variability of atmospheric river frequencies over the North Pacific. J Clim 29(13):4885–4903. https://doi.org/10.1175/JCLI-D-15-0655.1

NCDC (2010) Storm Data 52(1). 250 pp

Neff W, Compo GP, Ralph FM et al (2014) Continental heat anomalies and the extreme melting of the Greenland ice surface in 2012 and 1889. J Geophys Res Atmos 119:6520–6536. https://doi.org/10.1002/2014JD021470

Neiman PJ, Ralph FM, Moore BJ et al (2013) The landfall and inland penetration of a flood-producing atmospheric river in Arizona. Part 1: observed synoptic-scale, orographic, and hydrometeorological characteristics. J Hydrometeorol 14:460–484

Neiman PJ, Ralph FM, Wick GA et al (2008a) Meteorological characteristics and overland precipitation impacts of atmospheric rivers affecting the West Coast of North America based on eight years of SSM/I satellite observations. J Hydrometeorol 9:22–47. https://doi.org/10.1175/2007JHM855.1

Neiman PJ, Ralph FM, Wick GA et al (2008b) Diagnosis of an intense atmospheric river impacting the Pacific Northwest: storm summary and offshore vertical structure observed with COSMIC satellite retrievals. Mon Weather Rev 136:4398–4420

Newell RE, Newell NE, Zhu Y et al (1992) Tropospheric Rivers? – a pilot study. Geophys Res Lett 19:2401–2404

Newman M, Kiladis GN, Weickmann KM et al (2012) Relative contributions of synoptic and low-frequency eddies to time-mean atmospheric moisture transport, including the role of atmospheric rivers. J Clim 25:7341–7361

Nghiem SV, Hall DK, Mote TL et al (2012) The extreme melt across the Greenland ice sheet in 2012. Geophys Res Lett 39:L20502. https://doi.org/10.1029/2012GL053611

Oakley NS, Redmond KT (2014) A climatology of 500-hPa closed lows in the northeastern Pacific Ocean, 1948–2011. J Appl Meteorol Climatol 53:1578–1592

Payne AE, Magnusdottir G (2014) Dynamics of landfalling atmospheric rivers over the North Pacific in 30 years of MERRA reanalysis. J Clim 27:7133–7150. https://doi.org/10.1175/JCLI-D-14-000h34.1

Pereira S, Ramos AM, Zêzere JL et al (2016) Spatial impact and triggering conditions of the exceptional hydro-geomorphological event of December 1909 in Iberia. Nat Hazards Earth Syst Sci 16:371–390. https://doi.org/10.5194/nhess-16-371-2016

Persson POG, Shupe MD, Perovich D et al (2016) Linking atmospheric synoptic transport, cloud phase, surface energy fluxes, and sea-ice growth: observations of midwinter SHEBA conditions. Clim Dyn 49(4):1341–1364

Pfahl S, Madonna E, Boettcher M et al (2014) Warm conveyor belts in the ERA-Interim dataset (1979–2010). Part II: moisture origin and relevance for precipitation. J Clim 27:27–40. https://doi.org/10.1175/JCLI-D-13-00223.1

Pfahl S, Wernli H (2012) Quantifying the relevance of cyclones for precipitation extremes. J Clim 25:6770–6780. https://doi.org/10.1175/JCLI-D-11-00705.1

Pinto JG, Zacharias S, Fink AH et al (2009) Factors contributing to the development of extreme North Atlantic cyclones and their relationship with the NAO. Clim Dyn 32:711–737. https://doi.org/10.1007/s00382-008-0396-4

Pirret JSR, Knippertz P, Trzeciak TM (2016) Drivers for the deepening of severe European windstorms and their impacts on forecast quality. Q J R Meteorol Soc 143:309–320. https://doi.org/10.1002/qj.2923

Ralph FM, Coleman T, Neiman PJ et al (2013) Observed impacts of duration and seasonality of atmospheric-river landfalls on soil moisture and runoff in coastal northern California. J Hydrometeorol 14:443–459. https://doi.org/10.1175/JHM-D-12-076.1

Ralph FM, Neiman PJ, Rotunno R (2005) Dropsonde observations in low-level jets over the Northeastern Pacific Ocean from CALJET-1998 and PACJET-2001: mean vertical-profile and atmospheric-river characteristics. Mon Weather Rev 133:889–910

Ralph FM, Neiman PJ, Wick GA (2004) Satellite and CALJET aircraft observations of atmospheric rivers over the eastern North Pacific Ocean during the winter of 1997/98. Mon Weather Rev 132:1721–1745

Ramos AM, Nieto R, Tomé R et al (2016) Atmospheric rivers moisture sources from a Lagrangian perspective. Earth Syst Dynam 7:371–384. https://doi.org/10.5194/esd-7-371-2016

Ramos AM, Trigo RM, Liberato MLR, Tomé R (2015) Daily precipitation extreme events in the Iberian Peninsula and its association with atmospheric rivers. J Hydrometeorol 16:579–597. https://doi.org/10.1175/JHM-D-14-0103.1

Raziei T, Mofidi A, Santos JA et al (2012) Spatial patterns and regimes of daily precipitation in Iran in relation to large-scale atmospheric circulation. Int J Climatol 32:1226–1237. https://doi.org/10.1002/joc.2347

Rignot E, Casassa G, Gogineni P et al (2004) Accelerated ice discharge from the Antarctic Peninsula following the collapse of Larsen B Ice Shelf. Geophys Res Lett 31:L18401. https://doi.org/10.1029/2004GL020697

Risi C, Bony S, Vimeux F et al (2010) Water-stable isotopes in the LMDZ4 general circulation model: model evaluation for present-day and past climates and applications to climatic interpretations of tropical isotopic records. J Geophys Res 115:D12118. https://doi.org/10.1029/2009JD013255

Rivera ER, Dominguez F, Castro CL (2014) Atmospheric rivers and cool season extreme precipitation events in the Verde River Basin of Arizona. J Hydrometeorol 15:813–829

Rutz JJ, Steenburgh WJ (2012) Quantifying the role of atmospheric rivers in the interior western United States. Atmos Sci Lett 13:257–261. https://doi.org/10.1002/asl.392

Rutz JJ, Steenburgh WJ, Ralph FM (2014) Climatological characteristics of atmospheric rivers and their inland penetration over the western United States. Mon Weather Rev 142:905–921. https://doi.org/10.1175/MWR-D-13-00168.1

Rutz JJ, Steenburgh WJ, Ralph FM (2015) The inland penetration of atmospheric rivers over western North America: a Lagrangian analysis. Mon Weather Rev 143:1924–1944

Saha S, Moorthi S, Pan H-L et al (2010) The NCEP climate forecast system reanalysis. Bull Am Meteorol Soc 91(8):1015–1057. https://doi.org/10.1175/2010BAMS3001.1

Serreze MC, Crawford AD, Barrett AP (2015) Extreme daily precipitation events at Spitsbergen, an Arctic Island. Int J Climatol 35:4574–4588. https://doi.org/10.1002/joc.4308

Silva T, Gorodetskaya IV, Terpstra A et al (2017) Role of low-level jets in the enhanced moisture transport towards Antarctica from radiosonde measurements at the coastal stations and Southern Ocean. In: 12th workshop on Antarctic meteorology and climate, Boulder, Colorado, USA, June 26–28, 2017

Sodemann H, Stohl A (2013) Moisture origin and meridional transport in atmospheric rivers and their association with multiple cyclones. Mon Weather Rev 141:2850–2868

Stein AF, Draxler RR, Rolph GD et al (2015) NOAA's HYSPLIT atmospheric transport and dispersion modeling system. Bull Am Meteorol Soc 96:2059–2077. https://doi.org/10.1175/BAMS-D-14-00110.1

Stewart RE, Marwitz JD, Pace JC et al (1984) Characteristics through the melting layer of stratiform clouds. J Atmos Sci 41:3227–3237

Stohl A, Forster C, Sodemann H (2008) Remote sources of water vapor forming precipitation on the Norwegian west coast at 60°N: a tale of hurricanes and an atmospheric river. J Geophys Res 113:D05102. https://doi.org/10.1029/2007JD009006

Terpstra A, Gorodetskaya IV (2017) A case study of intense moisture transport and precipitation over the Southern Ocean and East Antarctic ice sheet. In: 12th workshop on Antarctic meteorology and climate. Boulder, Colorado, USA, 26–28.

Thompson DWJ, Wallace JM (1998) The Arctic oscillation signature in the wintertime geopotential height and temperature fields. Geophys Res Lett 25:1297–1300. https://doi.org/10.1029/98GL00950

Tsukernik M, Lynch AH (2013) Atmospheric meridional moisture flux over the Southern Ocean: a story of the Amundsen Sea. J Clim 26(20):8055–8064. https://doi.org/10.1175/JCLI-D-12-00381.1

Uotila P, Vihma T, Tsukernik M (2013) Close interactions between the Antarctic cyclone budget and large-scale atmospheric circulation. Geophys Res Lett 40:3237–3241. https://doi.org/10.1002/grl.50560

Viale M (2010) Characteristics of winter orographic precipitation over the subtropical Andes. PhD Thesis (in Spanish), University of Buenos Aires, Argentina, 163 p. http://digital.bl.fcen.uba.ar/Download/Tesis/Tesis_4707_Viale.pdf

Viale M, Houze RA Jr, Rasmussen KL (2013) Upstream orographic enhancement of a narrow cold frontal rainband approaching the Andes. Mon Weather Rev 141:1708–1730

Viale M, Norte FA (2009) Strong cross-barrier flow under stable conditions producing intense winter orographic precipitation: a case study over Subtropical Central Andes. Weather Forecast 24:1009–1031

Viale M, Nuñez MN (2011) Climatology of winter orographic precipitation over the Subtropical Central Andes and associated synoptic and regional characteristics. J Hydrometeorol 12:481–507

Wallace JM, Gutzler DS (1981) Teleconnections in the geopotential height field during the Northern Hemisphere winter. Mon Weather Rev 109:784–812. https://doi.org/10.1175/1520-0493(1981)109<0784:TITGHF>2.0.CO;2

Werner M, Langebroek PM, Carlsen T et al (2011) Stable water isotopes in the ECHAM5 general circulation model: toward high-resolution isotope modeling on a global scale. J Geophys Res 116:D15109. https://doi.org/10.1029/2011JD015681

Wernli H, Davies HC (1997) A Lagrangian-based analysis of extratropical cyclones. I: the method and some applications. Q J R Meteorol Soc 123:467–489

Wheeler M, Hendon HH (2004) An all-season real-time multivariate MJO index: development of an index for monitoring and prediction. Mon Weather Rev 132:1917–1932. https://doi.org/10.1175/1520-0493(2004)132<1917:AARMMI>2.0.CO;2

White AB, Gottas DJ, Strem E et al (2002) An automated bright-band height detection algorithm for use with Doppler radar vertical spectral moments. J Ocean Atmos Technol 19:687–697

White AB, Neiman PJ, Creamean JM et al (2015) The impacts of California's San Francisco Bay Area gap on precipitation observed in the Sierra Nevada during HMT and CalWater. J Hydrometeorol 16:1048–1069. https://doi.org/10.1175/JHM-D-14-0160.1

Wick GA (2014) Implementation and initial application of an atmospheric river detection tool based on integrated vapor transport. In: 2014 Fall meeting, San Francisco, CA, Amer Geophys Union Abstract A34E-06

Zhu Y, Newell RE (1994) Atmospheric rivers and bombs. Geophys Res Lett 21(18):1999–2002. https://doi.org/10.1029/94GL01710

Zhu Y, Newell RE (1998) A proposed algorithm for moisture fluxes from atmospheric rivers. Mon Weather Rev 126:725–735. https://doi.org/10.1175/1520-0493(1998)126<0725:APAFMF>2.0.CO;2

Effects of Atmospheric Rivers

5

Michael D. Dettinger, David A. Lavers, Gilbert P. Compo, Irina V. Gorodetskaya, William Neff, Paul J. Neiman, Alexandre M. Ramos, Jonathan J. Rutz, Maximiliano Viale, Andrew J. Wade, and Allen B. White

5.1 Introduction

The extremes of precipitation that large ARs bring as they interact with mountain topography and the warm-conveyor belts (WCBs) in mid-latitude cyclones are intense enough, and often sustained enough, that they play important roles in flooding, water resources, and a variety of related processes and events in the areas they pass over. The effects of ARs range from modest to large and from beneficial to hazardous. They are a natural part of the climates, landscapes, societies, and ecosystems that ARs affect and interact with. This chapter surveys the wide range of AR effects identified to date, and Table 5.1 offers a sense of the range and diversity of AR effects that have been identified in the scientific literature along the US West Coast.

In large part, research on AR effects to date has focused on the floods ARs can cause, with AR contributions to water resources receiving somewhat less attention. Thus, after a detailed discussion below of how ARs interact with mountains and other topography to produce precipitation and precipitation enhancements, this chapter covers in detail how intense that precipitation can be, as well as AR contributions to flooding and water resources. Other effects will then be covered in less detail, including AR contributions to snowpacks; snow avalanches; landslides and debris flows; groundwater recharge; aquatic, floodplain, and estuarine water quality and ecosystems; and terrestrial ecosystems with emphasis on vegetation productivity and wildfire risks. Most of these effects arise from the large volumes of precipitation ARs can deliver. More recently, studies that consider the links between ARs and underlying patterns of damaging winds have begun to emerge, suggesting that ARs may have effects beyond their role in precipitation extremes.

M. D. Dettinger (✉)
Retired, U.S. Geological Survey, Carson City, NV, USA
e-mail: mdettinger@ucsd.edu

D. A. Lavers
European Centre for Medium-Range Weather Forecasts (ECMWF), Reading, UK
e-mail: david.lavers@ecmwf.int

G. P. Compo · W. Neff
Cooperative Institute for Research in Environmental Sciences, University of Colorado–Boulder, Boulder, CO, USA

Physical Sciences Laboratory, National Oceanic and Atmospheric Administration (NOAA), Earth System Research Laboratory, Boulder, CO, USA
e-mail: gilbert.p.compo@noaa.gov; william.neff@colorado.edu

I. V. Gorodetskaya
Department of Physics, Centre for Environmental and Marine Studies, University of Aveiro, Aveiro, Portugal
e-mail: irina.gorodetskaya@ua.pt

P. J. Neiman · A. B. White
National Oceanic and Atmospheric Administration (NOAA), Earth System Research Laboratory, Boulder, CO, USA
e-mail: Paul.J.Neiman@noaa.gov; allen.b.white@noaa.gov

A. M. Ramos
Instituto Dom Luiz, Faculdade de Ciências, Universidade de Lisboa, Lisboa, Portugal
e-mail: amramos@fc.ul.pt

J. J. Rutz
Science and Technology Infusion Division, National Weather Service, Salt Lake City, UT, USA
e-mail: jonathan.rutz@noaa.gov

M. Viale
Argentinean Institute for Snow, Glaciology and Environmental Science, (IANIGLA - CONICET), Mendoza, Argentina
e-mail: maxiviale@mendoza-conicet.gob.ar

A. J. Wade
University of Reading, Reading, UK
e-mail: a.j.wade@reading.ac.uk

F. M. Ralph et al. (eds.), *Atmospheric Rivers*, https://doi.org/10.1007/978-3-030-28906-5_5

Table 5.1 Examples of effects of ARs on the US West Coast; within *red border* for mostly hazardous and *green border* for mostly beneficial

	Atmospheric rivers (ARs) on the US West Coast	Quantitative finding	References
Mostly hazardous	Cause the heaviest rains	92% of West Coast's heaviest 3-day rain events fed by ARs	Ralph and Dettinger (2012)
	Bring warmest storms (less snow, more rain)	Average >50% more precipitation and 2.5 °C warmer than other storms in the Sierra	Dettinger et al. (2009), Backes et al. (2015), Guan et al. (2016)
	Cause West Coast floods	40–90% of major floods in West Coast rivers have been fed by ARs	Ralph et al. (2006), Neiman et al. (2011), Konrad and Dettinger (2017)
	Cause storm surges in coastal areas	15–50% of annual sea-level maxima are associated with ARs	Khouakhi and Villarini (2016)
	Yield extreme coastal winds	20–50% of extreme coastal-wind episodes associated with ARs	Waliser and Guan (2017)
	Breach levees	81% of Central Valley levee breaks happened during ARs	Florsheim and Dettinger (2015)
	Cause landslides, debris flows, and avalanches	E.g., ARs cause 68% of post-fire debris flows in southern California	Oakley et al. (2017), Young et al. (2017), Hatchett et al. (2017)
	Bring cycles of wet and dry years	Account for 85% of multi-year precipitation variance in California	Dettinger (2016)
Mostly beneficial	Fill reservoirs and provide water supplies	30–50% of California rain, snow streamflow from ARs	Guan et al. (2010), Dettinger et al. (2011)
	End West Coast droughts	40–75% of droughts on West Coast ended by an AR	Dettinger (2013)
	Sustain wetlands, floodplains, and fisheries	77% of ecologically significant inundations of Yolo bypass floodplain, Sacramento River, initiated by ARs	Florsheim and Dettinger (2015)
	Water deserts and forests far inland, modulate wildfire risks	Statistically significant relations found between summer NDVI (greenness) and areas burned in parts of the interior Southwest	Albano et al. (2017)
	Freshen estuaries but sometimes threaten estuarine fauna	March 2011 ARs freshened San Francisco Bay by 60%, resulting in wild oyster kill rate of 97–100%	Cheng et al. (2016)
	Modify banks and bottom sediments, modulating aquatic fauna in mountain streams	More invertabrate densities and diversity after major AR flooding; 10× more in pre-disturbed settings	Herbst and Cooper (2010)

5.2 ARs and Orographic Precipitation

ARs can produce widespread precipitation via synoptic-scale and frontal-scale lift associated with the parent cyclone. However, ARs often focus heavier precipitation on the windward slopes of hills or mountains, where elevated terrain provides the lift required to produce precipitation. This process increases precipitation efficiency, and often results in an orographic precipitation enhancement. Thus, to address the most studied effects of ARs globally, the conditions and processes that most commonly produce orographic precipitation from ARs must first be reviewed.

When wind approaches a topographic barrier such as an isolated mountain or an organized range, three factors determine whether the wind will be carried over or around the barrier:

1. The static stability (defined below) of the air approaching the barrier
2. The wind speed approaching the barrier
3. The topographic characteristics of the underlying terrain (e.g., horizontal and vertical extent of the terrain, terrain orientation for incoming wind direction, and surface roughness)

Static stability is a measure of the atmosphere's resistance to vertical motion. The more stable the air, the more likely it will flow *around*—or in the case of a mountain range—*along* the barrier. Unstable air or neutrally stable air will flow *over* the barrier. Here, the issue is the wind and thermodynamic structure that accompany ARs, which are often characterized by near-neutral stability. How close to neutral stability the air is can be quantified by calculating the Brunt–Väisälä frequency, defined as

$$N = \sqrt{g / \theta d\theta / dz},$$

where g is the local acceleration of gravity, θ is the potential temperature, and z is the geometric height. When $N^2 = 0$, the flow is neutrally stable, whereas $N^2 < 0$ indicates unstable flow, and $N^2 > 0$ indicates stable flow. Potential (or convective) instability is defined as a layer or column of unsaturated air where θ increases with height, and equivalent potential temperature (θ_e) decreases with height. If such a layer is lifted until it reaches saturation, it will become unstable. This often occurs when ARs encounter terrain upon landfall.

ARs approaching the US West Coast, for example, first encounter coastal mountain ranges that are generally below 2 km in elevation. For these terrain features, the speed of the flow and the near-neutral static stability is usually sufficient to cause the flow to pass over the barrier. However, particularly along the US West Coast, the second major terrain feature an AR encounters is much higher and longer: the Cascades of Washington through Northern California, and California's lofty Sierra Nevada. In the Sierra, a phenomenon known as the Sierra Barrier Jet (SBJ) often occurs as winter storms strike the US West Coast. In particular, an SBJ forms in response to decelerating stably stratified flow as it approaches the western Sierra foothills. The Coriolis force causes the flow to turn to the left (northward) along the Sierra. The resulting corridor of blocked flow, which is maintained by a quasi-stationary, statically stable pressure ridge dammed up against the windward slope of the Sierra, parallels the mountain range's long axis on its west side below crest level (e.g., Neiman et al. 2013a). In this way, the SBJ acts as a terrain-induced atmospheric barrier, causing the moist air in an AR to rise upstream from the Sierra (e.g., as depicted in Figs. 5.1, 5.2, 5.3, 5.4, 5.5, and 5.6). The SBJ depth and horizontal extent (normal to the terrain) thereby determine where along the slope of the Sierra the maximum orographic precipitation enhancement occurs. SBJs were first documented during the Sierra Cooperative Pilot Project (SCPP; Reynolds and Dennis 1996; Parish 1982; Marwitz 1983, 1987) and in a multi-winter study that used rawinsondes launched from the western base of the Sierra (Smutz 1986). Figure 5.1 illustrates examples of an SBJ observed with aircraft during the SCPP, along with the evolution of a stronger barrier jet observed with a Doppler wind profiler (Carter et al. 1995). Both the cross-section example in Fig. 5.1a and the time–height example in Fig. 5.1b show the enhanced terrain parallel flow associated with the SBJ. In Fig. 5.1a, the SBJ was sampled in situ using a gust probe on a research aircraft during a 43-min flight track. In Fig. 5.1b, the SBJ is sampled remotely by the ground-based wind profiler, which provides height-resolved winds above the instrument at hourly resolution, as the wind barbs depict.

Similar barrier jet flows have been documented along the windward slopes of other prominent mountain ranges across North America (e.g., Bell and Bosart 1988; Colle and Mass 1995; Loescher et al. 2006; Braun et al. 1997; Yu and Smull 2000) and for other mountain ranges around the world.

5.2.1 Precipitation Formation

Rising air over a mountain barrier causes uncondensed air to cool at a rate of 9.8 °C km^{-1}, i.e., the dry adiabatic lapse rate. When the air cools such that it can no longer maintain water as vapor, the vapor condenses onto cloud condensation nuclei (CCN) to form cloud droplets. CCN particles are the result of a ubiquitous distribution of aerosols in the atmosphere that result from both biogenic and anthropogenic processes. Condensation releases heat, which increases the buoyancy of the air, causing it to rise more rapidly. Now under saturated conditions, the air cools at a slower rate—i.e., the moist adiabatic lapse rate (a typical value of ~6 °C km^{-1}). Eventually, the cloudy air cools to temperatures below freezing, but the droplets often do not freeze until they reach much colder temperatures (as cold as −40 °C, dispelling the commonly held notion that water freezes in the atmosphere at 0 °C). In other words, the population of cloud particles can exist as supercooled cloud droplets to temperatures well below freezing (Hu et al. 2010; Zawadzki et al. 2001; Rosenfeld and Woodley 2000).

For precipitation to occur, the cloud particles must grow large enough that the acceleration from gravity far exceeds the acceleration from drag (which results from the cloud particles moving through the air). Two processes allow precipitation-sized particles to develop within clouds. The first process occurs in clouds where temperatures throughout the cloud are warmer than about −15 °C, and requires a distribution of droplet sizes in the cloud. Larger drops fall faster than smaller drops because drag on the larger drops is reduced. This allows the larger drops to collide and coalesce with the smaller drops, which further increases their size and fall speed to the point where they fall out of the cloud. When the sub-cloud layer is dry, the precipitation can evaporate before it reaches the surface (virga). This precipitation-generation mechanism is referred to as a warm rain process because it does not require the presence of ice to induce the precipitation. The distribution of drop sizes required for this process to work occurs when hygroscopic aerosols of differing sizes are present (Whiteman 2000).

A second process that involves ice crystal growth is often referred to as the Wegener–Bergeron–Findeisen process (Storelvmo and Tan 2015). This process occurs in mixed-phase clouds (where ice and supercooled liquid water are both present) in regions where the ambient vapor pressure falls between the saturation vapor pressure over water and the lesser saturation vapor pressure over ice. This results in vapor deposition, where water vapor is diffused directly onto the ice particles. This allows the ice particles to grow preferentially at the expense of the cloud droplets. As the ice particles descend through the cloud, they can collide with supercooled droplets, which freeze on the surface of the ice

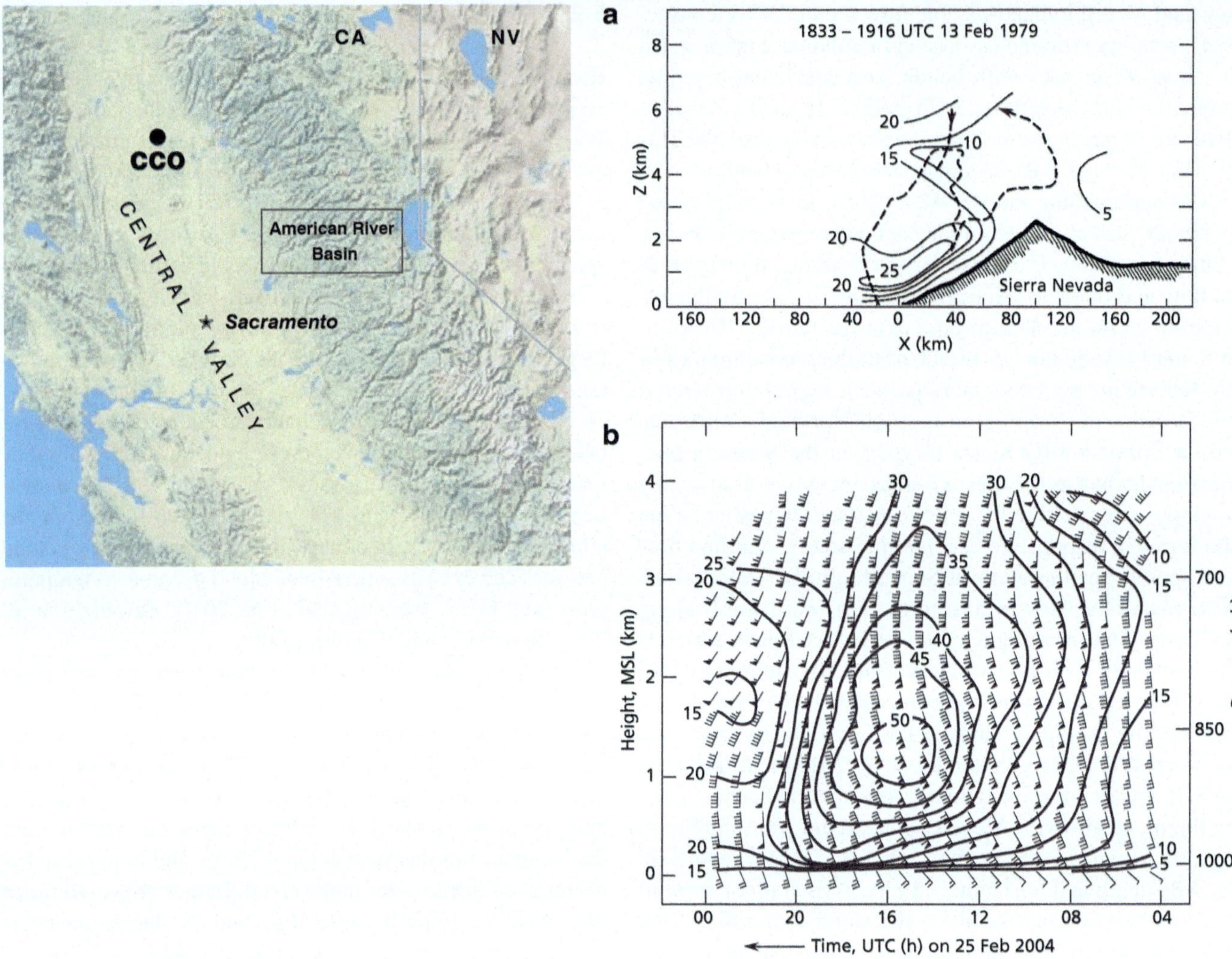

Fig. 5.1 *Left* Base map of north-central California showing the locations of a Doppler wind profiler in the northern Central Valley at Chico (CCO) and the American River basin studied during Sierra Cooperative Pilot Project (SCPP). *Right* (**a**) Cross-section of barrier-parallel isotachs (m s^{-1}; directed toward 340°) observed by the Wyoming King Air (*dashed line*) over the American River basin along the western slope of the Sierra Nevada on 13 Feb 1979 (from Parish 1982), (**b**) Time–height section of hourly averaged wind profiles (*every other range gate shown*) and barrier-parallel isotachs (m s^{-1}; directed toward 340°) at CCO on 25 Feb 2004 (*wind flags* = 25 m s^{-1}, *barbs* = 5 m s^{-1}, *half barbs* = 2.5 m s^{-1}) (Neiman et al. 2010)

particles, further increasing their size and fall speeds. Eventually, precipitation-sized particles are generated. If the ice particles fall through a freezing level, the precipitation arrives at the surface as raindrops. If sufficient melting does not occur to produce rain, the precipitation can fall as snow or graupel. The presence of in-cloud ice required for this process to work can occur (1) when ice crystals fall from higher clouds, (2) when supercooled cloud droplets within the cloud freeze, or (3) when ice particles form on small aerosols called ice nuclei (IN). These IN occur at much smaller concentrations than CCN, which helps explain why cloud droplets do not freeze at temperatures below freezing (Whiteman 2000).

5.2.2 Orographic Precipitation Enhancement

With terrain helping to generate clouds by causing the air flow to rise before the barrier crest, the same clouds might be expected to enhance precipitation over the windward slopes of mountain ranges. In the Western US, this is absolutely the case, as is shown climatologically in Fig. 5.2. Orographically enhanced precipitation is evident for the Olympic Mountains, Coast Ranges, Cascade Range, Sierra Nevada, Mogollon Rim, and the Rockies.

The degree to which orography enhances precipitation varies by geographic location and many parameters of the particular storm, including the storm's duration, the amount of

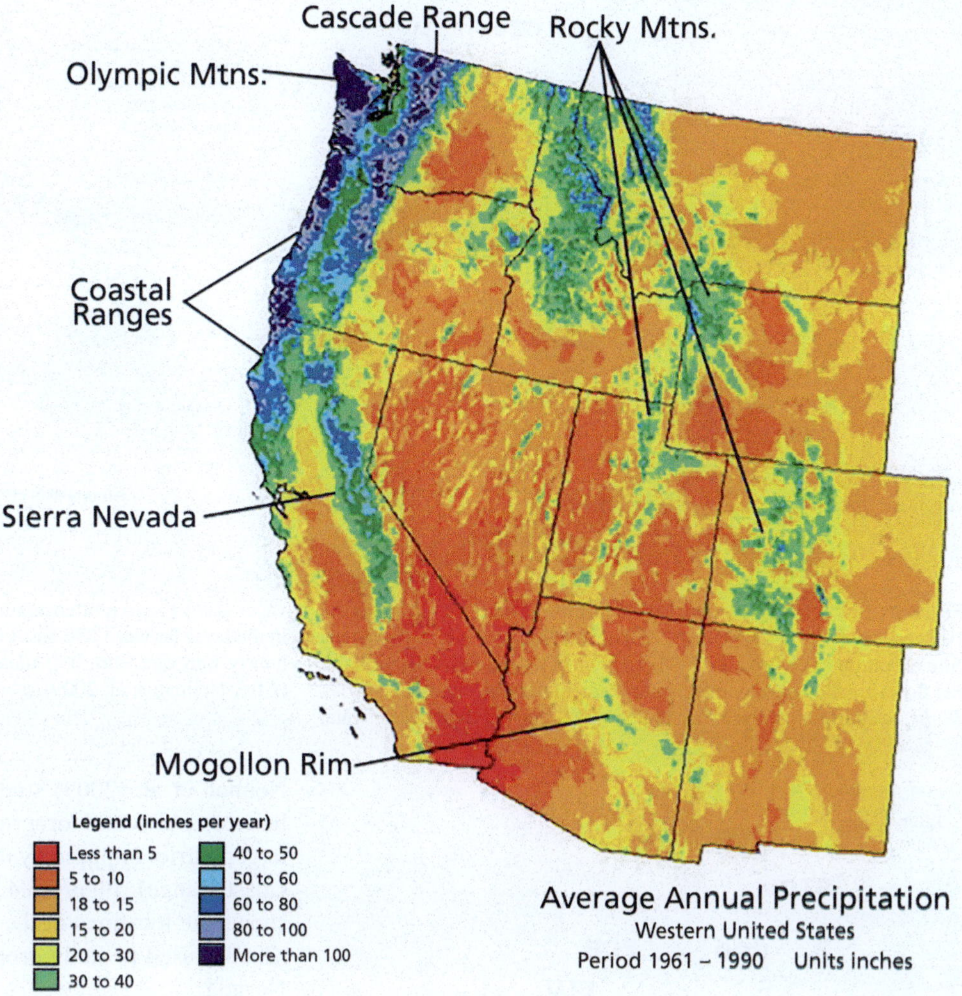

Fig. 5.2 Climatology of precipitation observed in the western US for the 30-year period between 1961 and 1990. Courtesy of the Western Regional Climate Center using the Parameter-elevation Relationships on Independent Slopes Model (PRISM) data set generated by the Oregon Climate Service (Daly et al. 1994)

water vapor available, precipitation efficiency (related to how effectively the available water vapor is converted into precipitation) wind direction, and thermodynamic stability, among other factors. This variation continues to make quantitative precipitation forecasting (QPF) of orographic precipitation using numerical weather prediction a challenge.

Neiman et al. (2002) began studying orographic precipitation and its effects in the coastal ranges of California using observations from strategically placed coastal wind profilers and rain gauges deployed at the coast and in the coastal mountains. They found that the upslope winds (i.e., perpendicular to the axis of the terrain crest) at ~1 km above mean sea level (MSL) were most highly correlated with the precipitation that fell in the coastal mountains, and that the orographic rainfall ratio (i.e., the ratio of rainfall observed at a coastal mountain location versus rainfall observed at the coast) varied, depending on geographic location. A later study by Neiman et al. (2009) expanded upon earlier results to include measurements from a long-term coastal observing couplet north of San Francisco for four winter seasons ending in 2001, 2004, 2005, and 2006. (The full complement of observations was not available in 2002 and 2003.) Figure 5.3 illustrates a key finding from this study: the statistically robust relationship between upslope flow, GPS-derived vertically integrated water vapor (IWV; Bevis et al. 1992; Gutman et al. 2004), and orographically enhanced rain rate. In particular, to produce a rain rate ≥ 10 mm h^{-1} in the coastal mountains requires an upslope flow ≥ 12.5 m s^{-1} and an IWV ≥ 2 cm. Because these heavier rain rates are often associated with ARs in this region, the thresholds in upslope flow and IWV defined here were used to identify ARs in subsequent studies, and also in the water vapor flux tool associated with the AR observatories described in Sect. 3.3.

The Neiman et al. (2002, 2009) studies were geographically limited to a coastal observing couplet in Northern California. More recently, researchers have been investigating how ARs affect orographic precipitation, particularly in the Sierra Nevada (Kingsmill et al. 2013; Neiman et al. 2013a, b, 2014a; White et al. 2015) and elsewhere in the Intermountain West (IMW; Rutz et al. 2014; Alexander et al. 2015), as well as other locations in the USA (Hughes et al. 2014; Mahoney et al. 2016) and around the world (Stohl

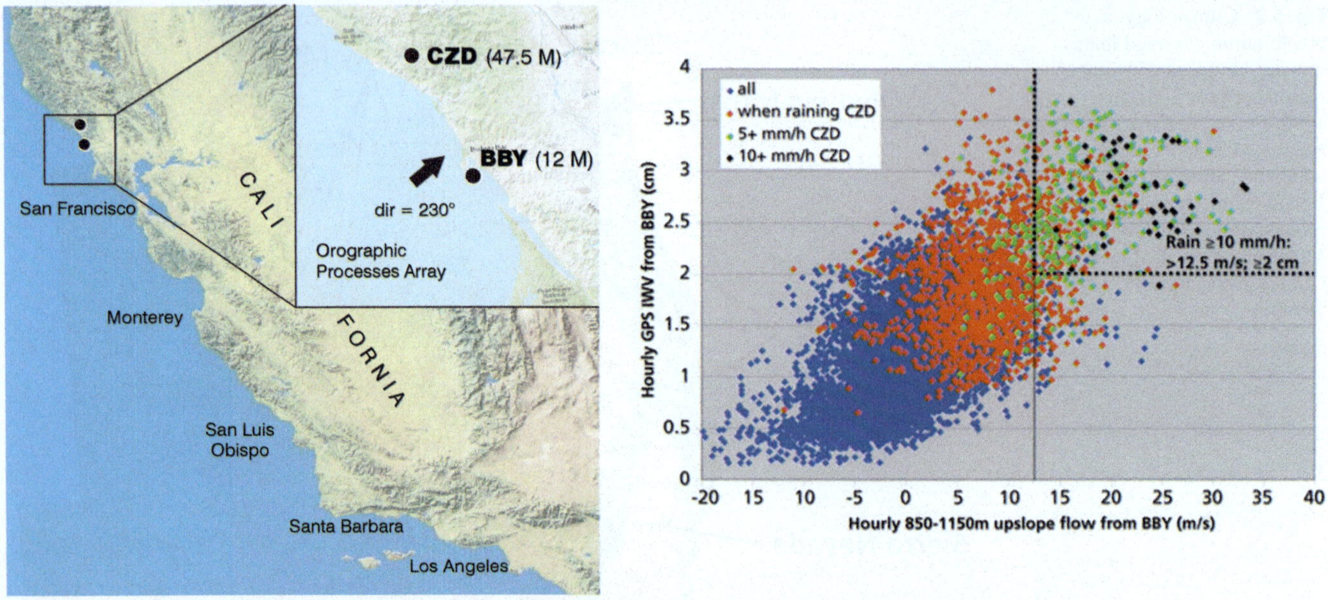

Fig. 5.3 *Left* Terrain base map (m) of California and inset showing the Bodega Bay (BBY)–Cazadero (CZD) orographic processes subdomain. Site elevations mean sea level (MSL) are labeled, and the *arrow* shows the flow direction approximately perpendicular to the mountain barrier. *Right* Scatterplot analyses of hourly GPS-derived integrated water vapor (IWV) (cm) plotted against hourly upslope flow (m s^{-1}) measured in the layer between 850 and 1150 m MSL at BBY and as a function of hourly rain rate (mm h^{-1}) measured at CZD (*scale in the upper left*) (After Neiman et al. 2009)

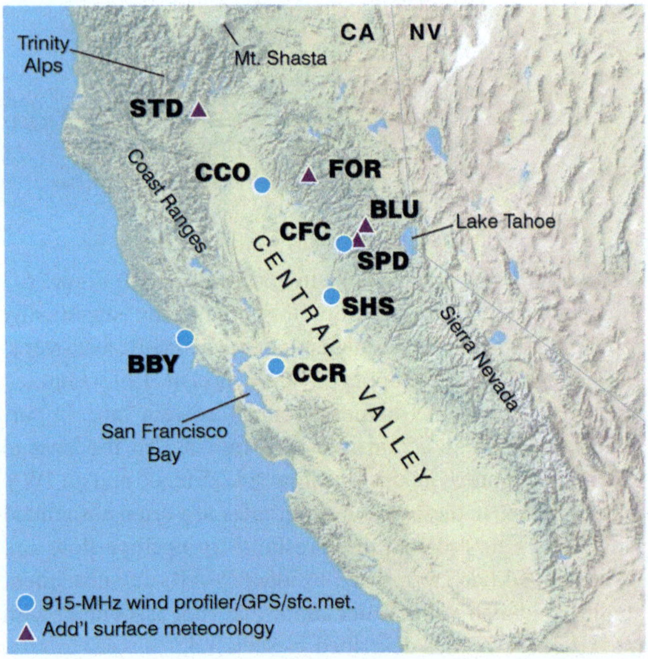

Fig. 5.4 Terrain base map of California that shows the locations of five 915-MHz wind profilers (*blue circles*) and four surface meteorological stations (*purple triangles*). (Neiman et al. 2013a)

et al. 2008; Viale and Nuñez 2011; Lavers et al. 2012; Lavers and Villarini 2013b; Garreaud 2013; Neiman et al. 2014b).

Here, the results from Neiman et al. (2013a) are briefly discussed because they complement the findings of the Neiman et al. (2009) study by contrasting the relationship between ARs and orographic precipitation for the much larger barrier imposed by the Sierra Nevada (compared to the Coast Range). Figure 5.4 is a base map of California that shows the locations of the observing couplets Neiman et al. (2013a) used to study orographic precipitation in the Sierra Nevada.

Figure 5.5 shows the composite structure of the SBJ (left panel) and AR flow aloft (right panel) measured by a Doppler wind profiler located at the east end of the Central Valley at Sloughhouse (SHS). The composite was produced from the 20 strongest SBJs measured over the course of six cool seasons (i.e., November into April) from 2005 to 2011. The SBJ acts to extend the barrier westward and causes the AR to ascend before it reaches the actual topographic barrier. The depth and width of the SBJ thereby modulates the altitude in the Sierra Nevada where the maximum orographic precipitation enhancement occurs.

Figure 5.6 is a conceptual representation of how an AR interacts with the SBJ. Note that at the northern end of the Central Valley, the shallow SBJ ascends the Shasta–Trinity Alps from the southeast and also slopes up the Sierra Nevada foothills. The upper portion of the AR ascends the SBJ and Sierra Nevada from the southwest, while the lower portion is either rained out along the coast ranges or is funneled through the San Francisco Bay (SFB) gap, joins the dry SBJ that originates from the southeast (denoted by blue), and veers northwestward while adding water vapor to the SBJ over the northern Central Valley (denoted by purple).

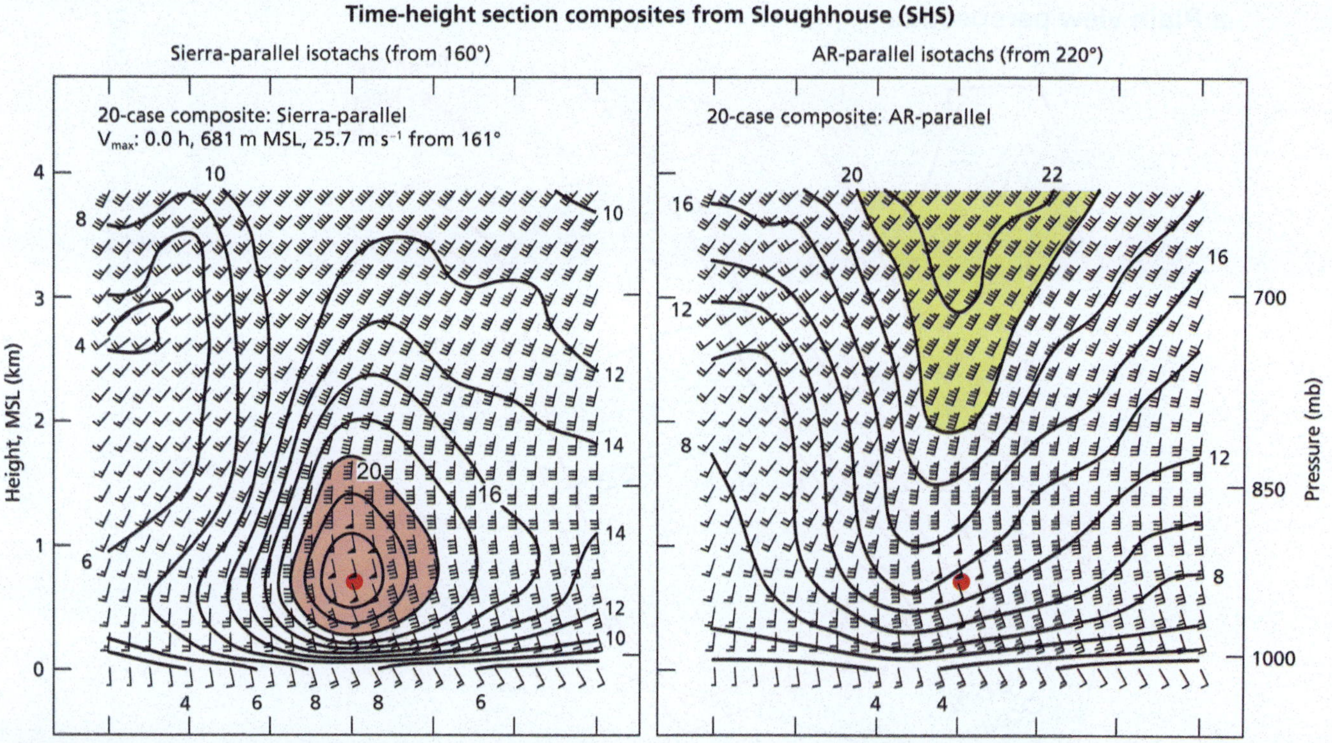

Fig. 5.5 Composite 24-h duration time–height sections of hourly averaged wind profiles (*flags*, 25 m s^{-1}; *barbs*, 5 m s^{-1}; *half barbs*, 2.5 m s^{-1}) and isotach components (m s^{-1}) during SBJs observed at SHS: (**a**) 20-case Sierra parallel (directed from 160°), (**b**) 20-case AR parallel (directed from 220°). *Red* and *yellow shading* correspond to >20 m s^{-1} Sierra- and AR-parallel flow, respectively. Time = 0 h corresponds to the time of each SBJ core (i.e., Vmax) observed at SHS. The *red dot* in each panel marks the time and altitude of V_{max}, the attributes of which are also given. Time increases from right to left to portray the advection of synoptic features from west to east (Neiman et al. 2013a)

Finally, Neiman et al. (2013a) used three Central Valley/Sierra Nevada/Trinity Alps observing couplets shown in Fig. 5.4 to determine if the relationship between upslope flow, IWV, and orographic precipitation was as statistically robust as it was for the Coast Ranges. These sites collected data simultaneously only during the cool seasons of 2009–2010 and 2010–2011. However, 13 of the 20 strongest SBJ cases occurred during these two cool seasons, so the following results are composited over those 13 corresponding events. Figure 5.7 shows the results of this analysis, which confirms a strong correlation between the AR forcing (upslope wind and IWV multiplied together to produce upslope IWV flux) and the resulting orographic precipitation that occurs in the Sierra Nevada and Trinity Alps.

5.3 ARs, Floods and Water Resources

As noted in previous sections, ARs transport vast quantities of water vapor and, when they are uplifted by mountains, can yield very large amounts of precipitation. To put these extremes into perspective, Ralph and Dettinger (2012) searched for the largest 3-day precipitation totals ever recorded at each of over 6000 long-term weather stations across the conterminous US. The result was a map of the largest 3-day totals, as categorized into four rainfall category ("R-Cat") bins, which highlighted only two US regions where 3-day precipitation totals >400 mm (16 inches) have ever been recorded (Fig. 5.8). The larger of the two regions stretched from Texas to Florida and as far north as parts of Oklahoma and South Carolina, but mostly was the area that experiences large-scale hurricane and tropical storm landfalls (although sometimes ARs called Mayan Expresses make landfall there, too; Moore et al. 2012). The other, smaller region with these extreme precipitation totals comprises the Coastal Ranges and Sierra Nevada of California. Of the 48 occasions when 3-day precipitation totals >400 mm were recorded in California from 1948 to 2010, 44 (92%) resulted from landfalling ARs. Indeed, in California, a number of stations have surpassed this precipitation threshold multiple times, whereas in the Southeastern US multiple "sightings" of such precipitation are very rare. Thus, the largest ARs (as measured by precipitation totals) in California yield precipitation comparable to precipitation totals associated with landfalling hurricanes and tropical storms on the Gulf of Mexico coast. AR-storm totals are as large as any recorded anywhere in the USA, historically.

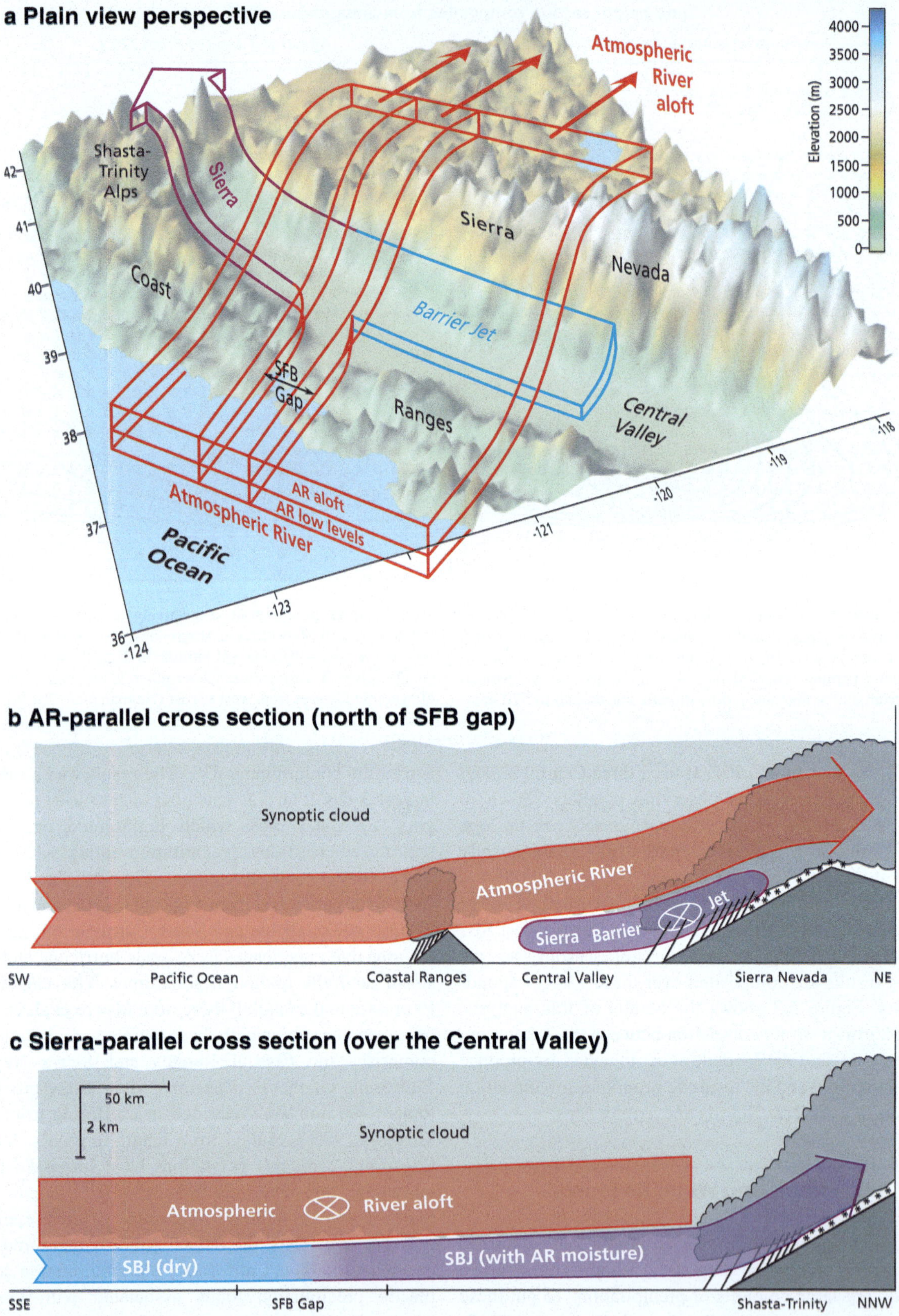

Fig. 5.6 Conceptual representation of key Sierra Barrier Jet (SBJ) and AR characteristics based on the 13-case composite analysis. (**a**) A 3-D plan-view perspective of the SBJ over the Central Valley (*blue/purple airstream*) and the AR making landfall (*red airstream*). (**b**, **c**) AR- and Sierra-parallel cross-sectional perspectives of the SBJ and AR, respectively (*color coding as in* **a**). A schematic representation of the orographically-enhanced clouds (*medium gray shade, dark outline*) and precipitation over the Sierra Nevada, the Shasta–Trinity Alps, and the Coast Ranges; and the synoptic cloud field (*light gray shade*). The SBJ deepens poleward of the SFB gap as the low-level portion of the AR contributes to the SBJ airstream there. (Neiman et al. 2013a)

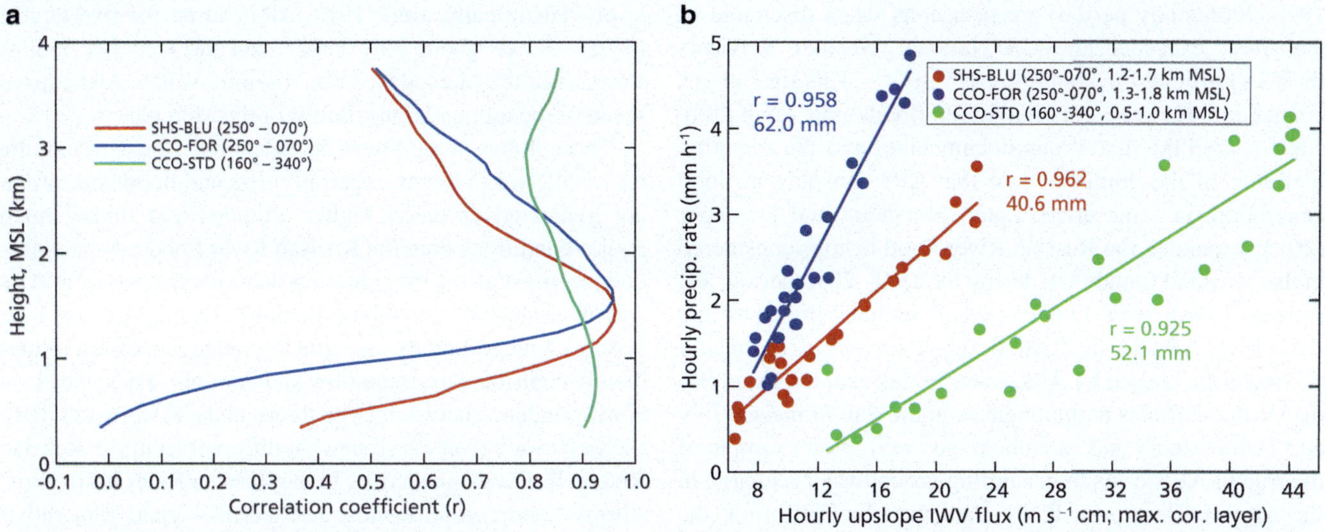

Fig. 5.7 Composite, 13-case, 24-h-duration orographic precipitation analysis from the wind profiler–precipitation gauge couplets at SHS–BLU (*red curves*; upslope direction from 250°), CCO–FOR (*blue curves*; upslope direction from 250°), and CCO–STD (*green curves*; upslope direction from 160°). (**a**) Vertical profiles of linear correlation coefficient, based on hourly averaged profiles of upslope integrated water vapor (IWV) flux vs. hourly precipitation rate. (**b**) Scatterplot analyses and linear regression fits in the layer of maximum correlation coefficient (1.2–1.7 km MSL at SHS–BLU, 1.3–1.8 km MSL at CCO–FOR, 0.5–1.0 km MSL at CCO–STD). Numerical values of correlation coefficient r and composite accumulated precipitation are given (Neiman et al. 2013a)

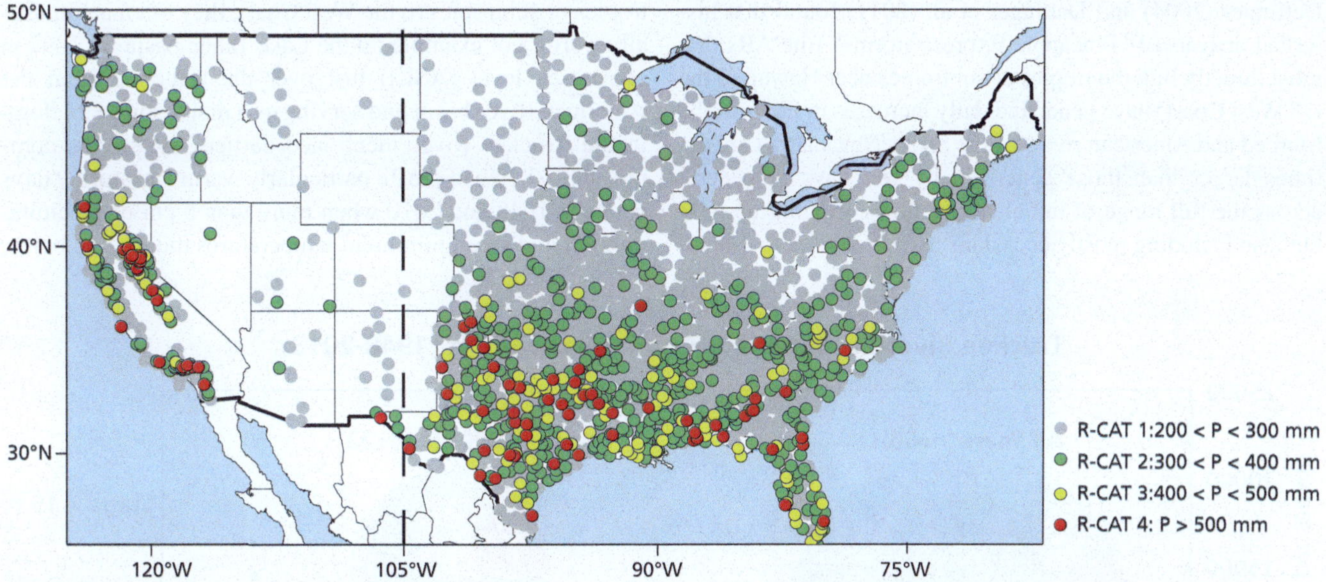

Fig. 5.8 Historical maximum 3-day precipitation extremes at 5877 Cooperative Observer Program (COOP) weather stations in the conterminous US (Ralph and Dettinger 2012)

This is the context in which the role of ARs in flooding and water resources must be understood.

5.3.1 Flooding

Along the Pacific coast of the USA and in westernmost Europe, a number of studies have documented the historical role of the large precipitation totals that landfalling ARs sometimes deposit as they cause major floods. Other, more scattered studies now also are documenting AR-caused floods in other parts of the globe. Notably, Ralph et al. (2006) showed that in the Russian River basin in the Coastal Ranges of Northern California, north of San Francisco Bay, the arrival of significant ARs and their intense rains caused all officially "declared" floods of the Russian River during the Special Sensor Microwave/Imager (SSM/I) period. These declared or "monitor-stage" floods (seven in their 8-year

1998–2006 study period) are occasions when discharge of that river exceeded the US National Weather Service's (NWS's) official threshold of 1075 $m^3 s^{-1}$. This unanimous connection between Russian River floods and landfalling ARs marked the first strong documentation in the scientific literature of the dominant role that ARs can play in flood generation, in some areas. Later, Florsheim and Dettinger (2015) expanded the Russian River flood history considered in this way and found that, during the 1948–2013 period, AR storms caused 34 of 39 (87%) of all declared floods of that river. Most ARs do not cause floods in this basin, but nearly all floods are caused by ARs. Later in this chapter, examples of AR flood effects and influences in the San Francisco Bay and Delta estuary and catchment—as well as an example of the role of AR storms in generating groundwater recharge in the Mojave Desert—will illustrate that, in California, the dominance of AR floods in the Russian River basin is not unique to that basin.

In the Sierra Nevada of California, a different relation exists between ARs and flooding. Since 1948, the top ten flood flows of the Truckee River on the rain-shadowed side of the Sierra Nevada have all been caused by AR precipitation spilling over the >3000-m ridgeline of that range (Fig. 5.9). Dettinger (2004) and Dettinger et al. (2011) found that historical arrivals of "Pineapple Express storms" (the ARs that most directly link the tropics or subtropics near Hawaii to the US West Coast) have generated daily increases in flows of the Merced and American rivers of the Sierra Nevada that are ten times larger than those generated by wet days in general across the full range of initial winter flows, ranging up to—but also including mostly occasions that do not rise to—flood levels. Historically, since 1913, ARs caused the five largest floods—floods about 20% larger than the next five largest storms—of the Merced River in Yosemite Valley; ARs caused seven of the top ten largest floods of the Merced.

These floods in the Sierra Nevada differ from those of the Russian River in that the Sierran rivers and floods occur and are generated at much higher altitudes and under much cooler conditions than the Russian River floods. As a consequence, most of the precipitation that falls there—from ARs and other mechanisms—falls as snow. Precipitation that falls as snow, unless it melts very quickly indeed, does not immediately contribute to streamflow and flooding. Thus, the close correspondence between large floods and AR storms in these Sierra Nevada rivers is somewhat different from the Russian River AR–flood connection. In the Russian River basin, most storms—and certainly most AR storms—yield rain rather than snow nearly everywhere, so if the storm and precipitation (i.e., rainfall) are intense enough, flooding can often result. In the Sierra Nevada, for AR storms to cause the floods they do, much of the precipitation they deposit must also be rainfall, but this requires the AR storms that cause the floods to be unusually warm. Indeed, because many ARs (and especially those "Pineapple Express" ARs) connect the tropics or subtropics to the West Coast, they often are unusually warm. For example, in the Lake Tahoe basin (at 1897 m above sea level [ASL]) just over the ridgeline from the American River basin and serving as a major source of flows in the Truckee River mentioned earlier, AR storms comprised 31 (81%) of the 38 particularly warm and stormy days between 1948 and 2010 when more than 5 cm of precipitation fell and when minimum temperatures did not fall below

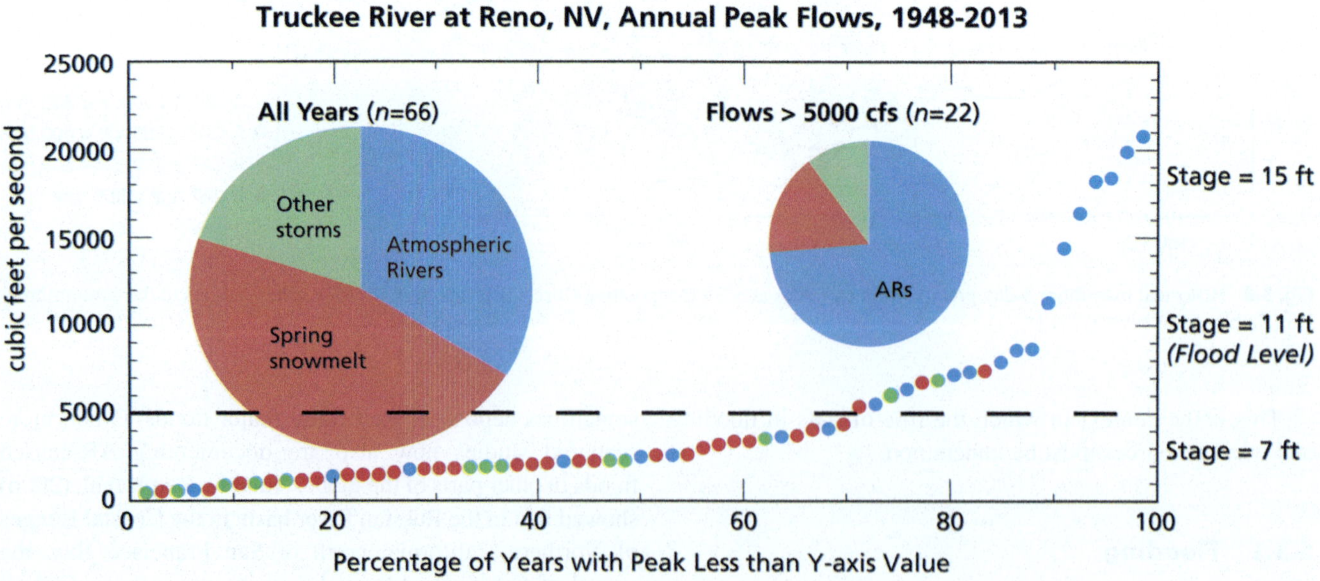

Fig. 5.9 Percentages and causes of annual peak flows in the Truckee River on the eastern, leeward side of the Sierra Nevada of California (Albano et al. 2016)

freezing. On average, between 1998 and 2010, days when ARs arrived at Lake Tahoe were 2 °C warmer and 85% wetter than wet days in general. Thus, AR storms in the Sierra Nevada tend to be warmer and often are considerably wetter than wet days in general. Guan et al. (2010) note, however, in their analysis of Sierra Nevada snowpacks and air temperatures during winters from 2004 to 2010, that the wettest (largest IWV and largest snowfall amounts) ARs tended to be cooler than weaker ARs, overall.

However, the extra warmth of ARs more generally yields, on average, a snowline (the altitude above which precipitation falls as snow and below which it falls as rain) in the Sierra Nevada that is about 400 m higher than the average for all storms. In many basins, the extra catchment area that receives rain rather than snow when the snowline is higher can be substantial; for example, in the Merced River basin, a 400-m rise in snowline elevation means an increase in rain-receiving catchment area of over 25%. This extra runoff-contributing area can mean a 25% increase in the total rainfall that contributes to flood flows. When snowlines are elevated in an AR storm, rain can fall on the lower-elevation parts of an already extant snow cover, thereby yielding extra snowmelt through combinations of advection of heat into the snowpacks by the rain itself and deposition (condensation) of latent heat from the rainy atmosphere directly onto the snow surfaces (Marks 1998). This rain-on-snow phenomenon is relatively common in flood situations in the Sierra Nevada and across many western North American mountain ranges (McCabe et al. 2007) and is often associated with landfalling ARs (Wayand et al. 2015). In particular, Guan et al. (2016) have found that, although ARs accompanied only 17% of all precipitation events in the Sierra Nevada from 1998 to 2014, they were associated with 50% of all rain-on-snow episodes. Between larger rain-collecting areas and more rain-on-snow, ARs are particularly well suited to cause floods even in the high altitudes of the Sierra Nevada, and historically have caused the very largest floods from that range.

Farther north along the US West Coast, in Western Washington, Neiman et al. (2011) analyzed annual peak daily flows in four major river basins from 1998 to 2009, and found that AR landfalls accompanied the peak flows in 44 of 48 (92%) cases. A longer-term analysis, spanning 1980–2009, indicated that the top ten annual peak flows resulted from heavy precipitation in the basins under conditions warm enough to ensure that the snow–rain transition averaged about 1900 m above sea level—high enough so that the entire basins would have been receiving rainfall, rather than snowfall, almost throughout, yielding enhanced runoff. The basins differed in their topographic orientations, with two presenting primarily westward-facing slopes and two presenting more southwestern-facing vantages. The orientations of the ARs that yielded the peak flows differed accordingly, with ARs that approached the slopes from the most perpendicular direction causing the largest flows in each. Farther north, ARs frequently make landfall in south-central Alaska where they cause floods that are among the largest and most damaging in the region—other than ice-jam floods during the spring breakup season (Papineau and Holloway 2011). In Southeast Alaska and British Columbia, ARs are important sources of flood risk (Pacific Climate Impacts Consortium 2013).

Inland from the US West Coast, Rutz et al. (2014; see Fig. 4.26) has shown that substantial fractions of winter precipitation have historically occurred during landfall and inland penetration of ARs beyond the first major mountain ranges on the West Coast—the Sierra Nevada of California and the Cascade Mountains of Washington and Oregon. Neiman et al. (2013b) documented the inland penetration of an AR north-northeastward across the Baja California peninsula and Sea of Cortez into Central Arizona, where heavy precipitation and flooding resulted; in a historical assessment, Kim (2015) reported that over 50% of floods in some Arizona basins resulted from ARs. Other inland-penetrating ARs that resulted in flooding in Northeastern Utah (Ralph and Dettinger 2012) and in Glacier National Park in Montana (Bernhardt 2006) have been well documented. These and other AR-induced floods in the interior Western US indicate that ARs contribute importantly to flood regimes well beyond the Pacific coast.

However, not all ARs enter North America from over the Pacific Ocean. ARs often arise over the Gulf of Mexico to pass into the Southeastern US (Moore et al. 2012; Durkee et al. 2012; Mahoney et al. 2016), Central US (e.g., Lavers and Villarini 2013a; Nayak et al. 2016), and as far north as Alberta (Milrad et al. 2015). Other ARs skirt the east coast of the USA to generate important floods (e.g., Grumm 2011). In these areas, other storm mechanisms, in addition to ARs, also bring extreme precipitation and cause floods, but ARs still pose significant risks.

Early on, Lavers et al. (2011) found that ARs from over the North Atlantic are a critical mechanism for winter floods in the UK, with the top ten most extreme floods since 1970 in a collection of four rivers that span the UK north to south all being preceded by—and then occurring simultaneously with—landfalling ARs. Smaller floods (as in North America) are not always associated with ARs; other mechanisms also contribute to the smaller events. In 2012, Lavers et al. (2012) pushed the analysis farther, and showed that, for streamflow peaks above a threshold exceeded on average once per year in nine rivers in the UK, ARs were at work between 40% and 80% of the time, with floods in basins at higher latitudes generally being more routinely associated with AR landfalls. Archer and Fowler (2015) also attributed several major flash floods in Britain to the arrival of intense ARs.

As far north as 60°N on the west coast of Norway, Stohl et al. (2008) documented major flooding in September 2005, triggered by an AR that carried moisture from a recent hurricane from the tropics and subtropics up nearly to the Arctic Circle, where Norway's coastal mountains lifted the moist air, extracting strong orographic precipitation from it. Far to the south, the Iberian Peninsula (and associated islands) has also experienced major AR-caused flooding (Liberato et al. 2012; Levizzani et al. 2013; Trigo et al. 2014). As in the USA, ARs also penetrate inland in Europe well beyond the first coastal mountain ranges to bring intense precipitation and major floods across Central Europe (Lavers and Villarini 2013b; Lavers et al. 2014), and ARs have been documented instigating rain-on-snow flooding in the Alps (Rössler et al. 2014). Some ARs even wind their way deep into the Mediterranean basin, with reports of AR-caused floods in Italy (Malguzzi et al. 2006; Lang and Pierce 2012) and Greece (Mita et al. 2014).

Farther afield, AR-caused floods have recently been documented in Hawaii (Neiman et al. 2014b), New Zealand (Kingston et al. 2016), Japan (Hirota et al. 2016), Chile (Viale 2013), and the Himalaya of northern India (Rao 2016).

Together, these reports and analyses illustrate the dangerous and important role ARs play in flood generation in regions all around the world. Nearly all of the flood examples listed reflect runoff from the extravagantly orographically enhanced precipitation that ARs produce when they interact, to greater or lesser extent, with continental or island mountain ranges. In some cold and high-altitude settings, the warmth that is typical of ARs plays an important role in flood generation by increasing the amount of rainfall (rather than snow) that occurs; in other lower-latitude or otherwise warmer locales, the intense rains associated with strong ARs provide all the fuel necessary to generate floods.

5.3.2 Water Resources

Although ARs are important flood generators in many parts of the world, and although floods are the AR effect that has been studied most, ARs also play positive and crucial roles in providing water resources to most land areas that they reach with any great frequency. Although ARs can be uncommon in some areas, they also can be large and intense, with the most intense ARs yielding higher instantaneous precipitation rates and storm totals than other storms. As a result, significant fractions of total precipitation, and especially of AR-season precipitation, come from ARs in many settings. This precipitation is an ultimate source of water resources in these regions (listed below), and—to the extent it accumulates and then runs off into the region's rivers or recharges its aquifers—it becomes the area's water resources. Thus, to the extent that water is derived from AR storms, ARs are—however hazardous they may appear from the perspective of flood-risk management—crucial to the survival of societies and ecosystems.

As with so many other aspects of AR research, the first indications of this important role were reported in and around the heavily instrumented Hydrometeorological Testbed (HMT) in coastal Central California, where Neiman (2008) reported that precipitation during observed AR-landfall conditions yielded 44% of all precipitation in the winters of 2001 and 2004–2006. Guan et al. (2010) followed up by analyzing Sierra Nevada snowpack water contents deposited by all winter storms during 2004–2010. They found that AR storms typically deposited four times more snow water than other non-AR storms and, on the whole, contributed an average of 38% of all snowfall water content during the 7 years studied (ranging from 24% to 53% in individual years). Dettinger et al. (2011) then analyzed precipitation records from several hundred long-term cooperative weather stations (generally at lower altitudes than the snow-instrumentation sites analyzed by Guan et al. (2010)), and estimated that ARs provided from an average of 25% to 50% of water-year precipitation in California during just a few events per year in the 1998–2008 and 1948–2008 periods (using two different AR chronologies). The percentages of water-year precipitation that fell in AR storms are smaller in Oregon and Washington, where more precipitation falls during months when ARs are less common (outside of the December–March peak season). In that area, ARs contribute about 20% to 35% of all precipitation.

Focusing on cool-season (November–April 1988–2011) precipitation and with a greater attention to the interior Western US, Rutz et al. (2014; Fig. 5.10) extended such analyses to estimate that ARs contribute between 40% and 50% of cool-season precipitation along the US West Coast and from 25% to 35% in the next tier of states inland (Arizona, Nevada, Idaho). A prominent AR-shadow is apparent in the lee of the Sierra Nevada, in Rutz et al.'s (2014) iconic AR-fraction map, where AR-precipitation fractions drop to about 10% over much of Nevada and all of Utah. In this study that focused on the effects of ARs that arrive from over the North Pacific, even farther inland, such low but non-zero AR-precipitation fractions are typical.

In 2015, Lavers and Villarini (2015) extended this kind of analysis still farther east to consider the Central and Eastern US and Western Europe. They found that AR-precipitation fractions ranged, in 1979–2012, from 15% to 30% of all precipitation in the east-central US (peaking in Western Kentucky and Tennesee), and from about 15% to 25% in westernmost Europe (Iberia, the Western UK, and Western France).

Dettinger et al. (2011) extended these analyses of ARs as resource further by also considering how streamflows in river basins with few human modifications (as close to "natural" streamflows as the Western US offers) respond to the arrival of AR storms. In California, AR contributions to overall streamflow were between 34% and 52% of total streamflow at most of the sites, where "AR contributions" were

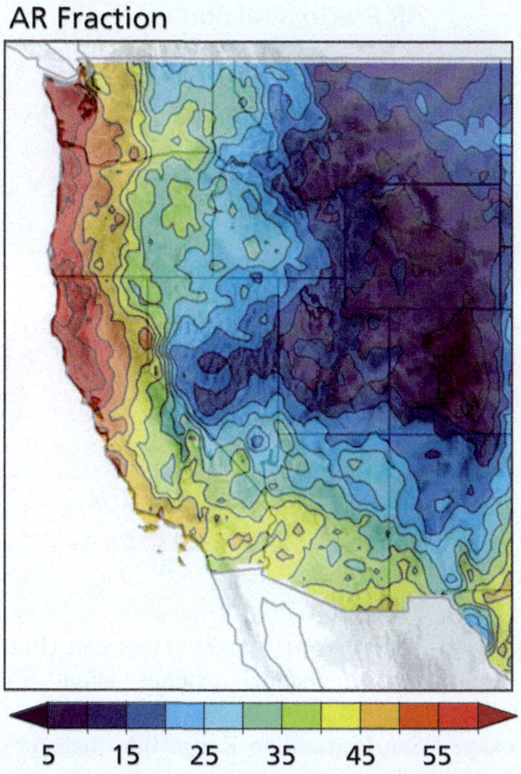

Fig. 5.10 Fractions of water precipitation provided by landfalling and inland-penetrating ARs (Rutz et al. 2014)

defined to be the total flow on the day of the AR's arrival and several subsequent days. (Subsequent days were included to allow for the lags that occur between precipitation and streamflow in many basins.) Notably, AR fractions of streamflow for rivers in the snowbound Sierra Nevada ranged from <10% to over 30%. These percentages are significant because they occur despite the tendency for much AR precipitation to fall as snow (Guan et al. 2010) rather than rain, and thus to be tied up in the seasonal snowpacks there for much longer periods until eventual spring snowmelt. Thus, even though most AR-derived snowpack in Sierra Nevada rivers does not contribute directly to immediate runoff during the storm or the first several days afterward, enough AR-precipitation falls as rain and enough snow melts in association with the arrival of ARs (Guan et al. 2016) that a still significant fraction of Sierra Nevada streamflow is associated immediately (in time) with ARs. Nonetheless, these estimates of contributions of AR precipitation to streamflow in California, and especially in the Sierra Nevada, are certainly under-estimates of the overall contributions of those storms, because most of the AR precipitation that will eventually run off runs off weeks to months later. This "delayed runoff" results from the slow pathways that runoff follows through the subsurface, and to the delays between snowfall and eventual snowmelt and runoff. This delayed runoff is not labeled in terms of how much is from ARs versus from non-AR storms in these purely observations-based analyses; however, recent augmentations of a hydrologic model to "label" precipitation from ARs as it flows through the various pathways and processes in western watersheds offers hope that future studies will be able to more definitively estimate the overall contributions of ARs to total streamflow (Hu et al. 2016).

More recently, Guan et al. (2016; their Fig. 5.11) produced a global assessment of the fractions of precipitation that come from ARs, which places these earlier studies focused on the Western US into perspective. Globally, AR precipitation exceeds 30% of the total precipitation in extratropical regions where ARs are relatively common, including Western North America, as well as in the Central and Eastern US and Western Europe (much as in Lavers and Villarini 2015). Guan et al. (2016) found that ARs may provide similar fractional precipitation contributions on the easternmost ramparts of Asia, in the Middle East as far east as Afghanistan, in northwesternmost Africa (e.g., Sudan), and in southeastern Greenland in the Northern Hemisphere. In the Southern Hemisphere, Southern Australia, New Zealand, Chile and Argentina, Uruguay and Southern Brazil, and South Africa receive a similar "third" of their precipitation from ARs.

At this global scale, strong seasonal patterns in AR contributions to water resources are evident in Guan and Waliser's (2015) results. Along the west coasts of North and South America, in keeping with higher AR frequency and stronger AR precipitation there in winter, AR precipitation fractions are much larger in their respective winter seasons than in summers. Seasonal contrasts also occur in east coast areas like easternmost Asia, where the peak seasons may reflect competition between AR and monsoon contributions where both are present.

Thus, ARs make important (20% to as much as 50%) contributions to overall precipitation in regions around the world where ARs commonly make landfall. Moving inland past the first major mountain ranges, fractions generally fall off, but so does total precipitation in many of those settings, so even the smaller inland fractions are crucial to water resources in arid to semi-arid regions. Even with AR-precipitation fractions of as little as 20%, if those contributions from ARs did not occur, many regions of the globe would be in states of perennial drought.

The significant average contributions of ARs to water resources extend also to very significant contributions to multi-year water-resources variability on the west coast of the USA and, especially, in California. In terms of droughts, on the west coast, months during which extended historical droughts ended (as measured by extended monthly Palmer drought severity index [PDSI] reversals to normal-or-wet conditions from extended PDSI drought conditions) were almost always unusually wet—and abruptly so. These "drought-busting" months historically included one or more unusually large storm, and indeed, on average, the wettest 3-day periods in the

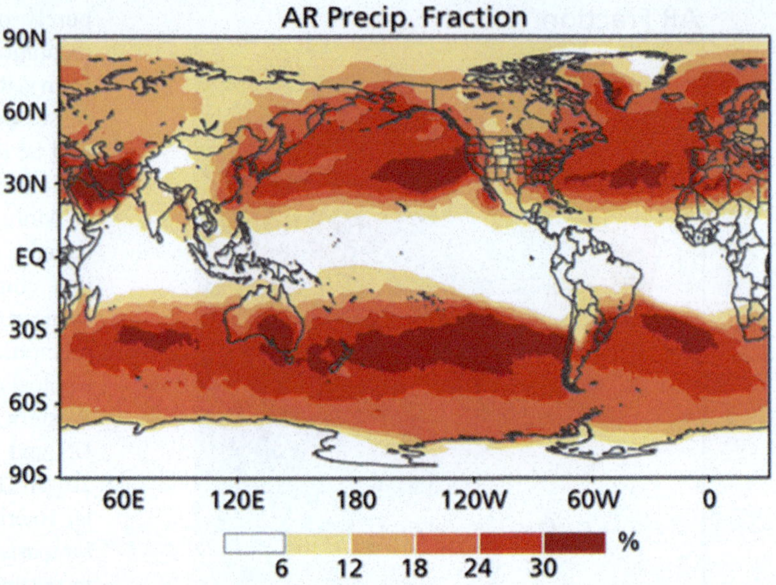

Fig. 5.11 Percentage of total precipitation associated with ARs, 1997–2014 (Guan and Waliser 2015; their Fig. 9i)

months when the droughts ended constituted 48% of those wet-month totals. Those 3-day drought-busting periods proved to coincide with the arrival of AR storms between 75% (in the Pacific Northwest) and 35% (in Southern California) of the time, historically (Dettinger 2013). In a more inclusive analysis, Dettinger (2016) found that extremely large storms (mostly ARs) contribute about one-third of the total precipitation in California on average, but contribute about two-thirds of the year-to-year precipitation-total variance. The year-to-year variance of these large-storm contributions is the source of a large majority of the large fluctuations between very wet (multi-year) periods and very dry periods (droughts) that define California's extremely variable hydroclimate. Those multi-year fluctuations closely follow ($r^2 = 75\%$) the multi-year fluctuations in the number of major "Pineapple Express" ARs on the California coast (Dettinger 2016). This strong dependence of year-to-year variance of total precipitation on just the largest storms and on AR storms only appears to occur, within the continental US, in California, with its Mediterranean climate and limited number of storms overall. Together, these studies indicate that, in at least some regions, the presence and absence of ARs may essentially determine the occurrence—or not—of extended droughts.

No study published to date has addressed the contributions of ARs to groundwater resources. However, in many areas of the world, groundwater is a key water resource (Wada et al. 2010). Groundwater recharge is notoriously difficult to observe and track through time, and its connections to individual storms are usually difficult or impossible to document. However, in the Mojave Desert of Southern California, conditions conspired to allow a brief illustration of an AR-recharge connection: The Mojave Desert occupies much of the southeastern corner of California, in the pronounced rain shadow of the southernmost Coastal and Transverse ranges of that state (Fig. 5.12). The Mojave River channel extends about 170 km (110 miles) eastward from its headwaters in the (often) snow-capped San Bernardino Range into the arid desert. The river is, at best, intermittent, and generally is entirely dry at the surface along nearly all of its length, except during occasional large storms in the headwaters and during some unusually large snowmelt seasons. The channel is situated mostly on alluvial deposits, ranging from silt to gravel, that form a broad alluvial aquifer that nearby communities along its length have much exploited as a groundwater resource. In the late twentieth century, water levels in that aquifer generally declined as groundwater was pumped from it at rates faster than the general recharge to it. However, about 100 km out into the desert near the city of Barstow, the general long-term decline of groundwater levels (by about 0.7 m per year) was interrupted by occasional abrupt rises that were associated with recharge from the river when it flowed. Because they could be quite substantial. These abrupt recharge and groundwater-rise events were nearly the only thing that kept the aquifer from being dewatered at catastrophic levels. For example, in 1993, water levels in a well in Hinkley Valley (roughly 5 km from the river channel) rose almost 24 m in a few days after a high discharge in the river, which resulted from leakage down through the river bed rapidly recharging the alluvial aquifer (Stamos et al. 2001). In this setting, when the river flows more than about a few thousand cubic feet per second, the river channel fills and recharge occurs; otherwise, no recharge happens.

This being a desert, these high-flow events are relatively rare and thus, in this setting, the recharge episodes—marked by the availability of large riverine flows as far out as Barstow—can be cataloged and compared to synoptic

Fig. 5.12 Location of Mojave Desert and Mojave River in western US; riverbed is dry and sandy most of the time but when it flows (4200 cubic feet per second on 16 January 1993 is shown) it recharges the alluvial aquifer for miles around (Stamos et al. 2001)

meteorological conditions on the same day. Such a survey of all eight such occasions (with flows >50 $m^3 s^{-1}$) from 1948 to 2016 indicates that five (63%) of those high-flow—and thus recharge—events occurred as a result of AR landfalls reaching the river's headwaters. (The other three episodes were associated with closed–low storms passing overhead.) Fifty-seven percent of all recharge in the Mojave River aquifer cannot be said to be AR-derived, but most of the recharge in the alluvial aquifer in that desert derives from the few and strongest AR storms in its headwaters.

In most wetter settings, quantifying a connection between ARs and groundwater recharge is difficult or impossible, because recharge is diffusely distributed, and because many different storms and inputs will contribute unevenly to recharge in ways that are difficult to disentangle. However, in settings where floods are particularly important recharge sources, the tendency for ARs to be particularly significant flood generators means they are likely also to be important recharge sources.

5.4 Other Effects of ARs

5.4.1 Aquatic Ecosystems

Floods and droughts play critical roles in many aquatic ecosystems in continental settings influenced by ARs. ARs play important roles as flood generators globally, and are implicated as major mechanisms for initiating, sustaining, and ending droughts on at least the west coast of the USA (e.g., Dettinger 2013; Dettinger 2016). Some examples of the ecosystem risks and benefits associated with ARs from various parts of California and the Western US are discussed here.

Estuarine Effects

The San Francisco Bay and Sacramento–San Joaquin Delta estuary is the largest estuary on the west coast of North America, and once was home to vibrant aquatic, tidal, and wetland ecosystems. Today, its ecosystems are degraded, having lost over 95% of its natural habitats to dredging, development, water-quality changes, and channelization. It is still home to several endangered native fish species, but they are on the edge of extinction (Moyle et al. 2016). Furthermore, large-scale California state and US federal water infrastructures use the inland-Delta part of the estuary as a principal conduit to transport and extract freshwater from the wet, northern one-third of California to its drier, water-thirsty southern two-thirds. Under these circumstances, considerable attention and management effort have gone into maintaining salinities in the estuary (to the extent practicable) in configurations that allow north-to-south freshwater flows through the Delta, and that sustain various beleaguered native plant and animal species in the area.

The traditional metric to categorize and track salinity in the estuary is X2, which is the distance from the ocean entrance of San Francisco Bay, in km, to the point upstream in the estuary where the near-bottom salinity drops to two parts per thousand. Historically, X2 has varied between about 40 km (when the estuary is fresher than normal) and 90 km (when the estuary is salty; Fig. 5.13). X2 reflects the continual tension between (1) salty ocean waters that gravity and tides push inland and that dilute freshwater outflows of rivers from the Central Valley and (2) surrounding mountain ranges, including the Sierra Nevada of California. X2 generally reflects California's Mediterranean climate with its wet winter season, moderate mountain snowpacks, and very dry summers. Thus, seasonally, X2 is smaller (the estuary more dilute) during winter and especially spring, in response to enhanced river outflows during major winter storms and springtime snowmelt

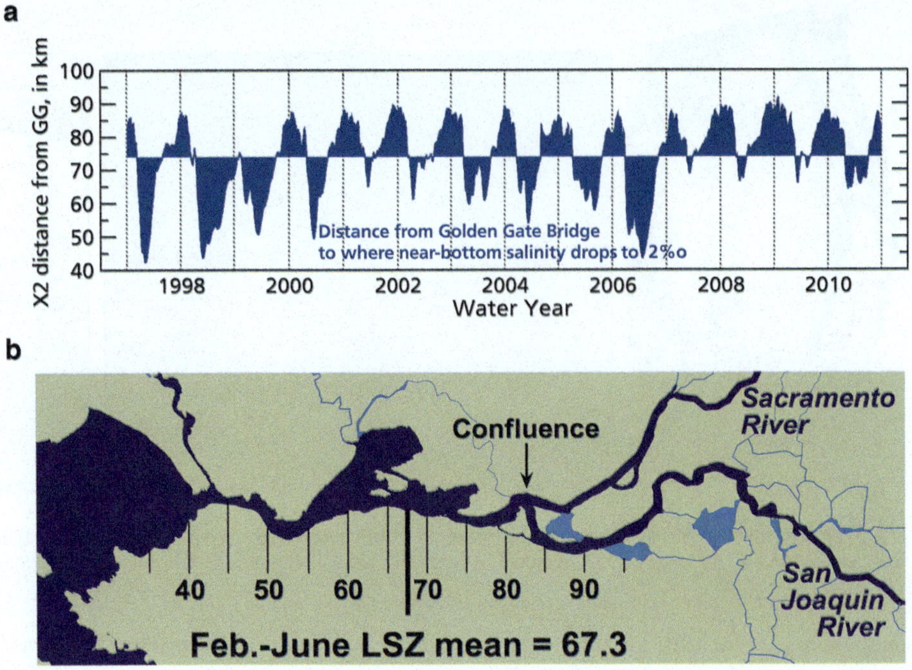

Fig. 5.13 X2 estuarine-salinity metric for location of low-salinity zone (**a**; LSZ) for in the San Francisco Bay–Delta estuary (**b**)

seasons in the contributing mountain catchments; X2 is largest in late summer and fall when river flows decline.

Atop this seasonal cycle are superimposed the influences of droughts (that reduce river outflows and increase X2) and large storms. The large storms and their attendant flood flows send dilute waters rushing into the estuary and force X2 far down the estuary. At the time-scale of individual storms and floods, X2 can move by a km or more downstream in a single day. Not surprisingly, given the dominant role that ARs play in causing and amplifying California's largest floods, AR storms have also dominated these largest X2 excursions. Between 1997 and 2010, landfalling ARs caused 19 (83%) of the 23 occasions when X2 retreated by 0.85 km in a single day. In the same time-period but on longer time-scales, the numbers of AR landfalls per year explain 49% of year-to-year variance of annual minimum X2 distance, 38% of the variance of end-of-summer X2 distance, and 30% of the variance of numbers of calendar-year days with X2 greater than (arbitrarily chosen) 80 km.

The flow and salinity fluctuations that X2 reflects play important roles in fisheries and other parts of the estuarine ecosystem, by (1) determining where chemical conditions are most suitable or most lethal for various species (Brown et al. 2016), and by (2) influencing the density of the estuary's waters and thereby its mixing and turbidity conditions (Schoellhamer et al. 2016), and the pathways for fish migration and residence through and within the estuary (MacWilliams et al. 2016), especially in the labyrinth channel system of the inland Delta. A recent example of the critical role ARs can play in this estuary is the March 2011 mass oyster kill (nearly 100% mortality) in parts of the northern end of San Francisco Bay (Cheng et al. 2016), near where X2 would lie when equal to about 40 km. In that month, ARs provided 69% of unusually large precipitation totals in the Sierra Nevada basins that drain to the estuary, resulting in freshwater inflows to the estuary that rose to >10 times the normal flows (to flows expected roughly once in 3 years). This river discharge caused low salinities that persisted for 8 days and that matched oysters' known critical salinity tolerances. The result was a mass mortality of Olympia oysters (*Ostrea lurida*) in one of its primary (most populated) habitats on the west coast of the USA. This sort of mass mortality is particularly striking in a commercial species that has experienced an 88% decline regionally over the last century.

The flows to, and through, the San Francisco estuary are constrained by 1100 km of levees that have been constructed by natural processes, dredging operations, and some limited modern engineering efforts. These levees prevent the estuarine waters from spreading widely across the estuarine lowlands and riverine floodplains as they did under natural conditions. Over the more than 150 years of their history, these levees have occasionally been breached by structural failures and overtopping by the waters they channel (Florsheim and Dettinger 2007), in response to various combinations of high riverine flow (stands), high tides and storm surges, and various structural defects; the levees are also known to be susceptible to earthquake damage (Deverel et al. 2016). The kinds of floods and, ultimately, AR storms that so alter the salinity of the estuarine waters also challenge the integrity of these levees. Florsheim and Dettinger (2015) considered some 128 levee breaks between 1951 and 2006 in the Delta and in the Central Valley above the Delta, and

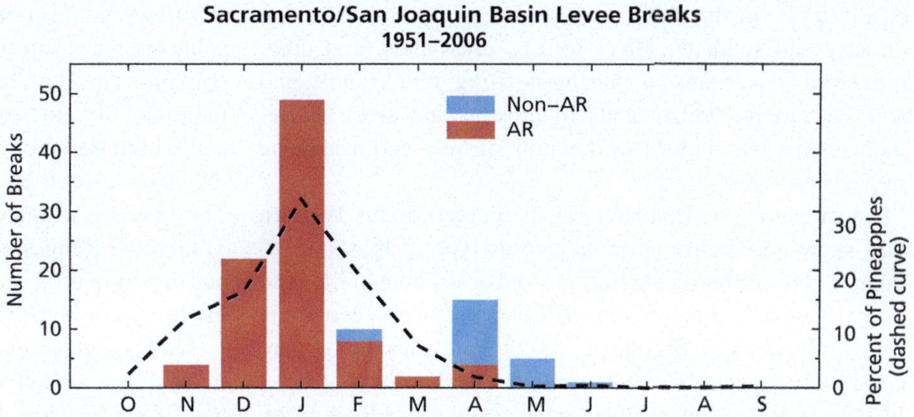

Fig. 5.14 Number of levee breaks in the Sacramento–San Joaquin rivers (Delta) drainages, 1951–2006, as histogram, and percentage of long-term mean "Pineapple Express" storms (intense ARs that extend directly—linearly—from the subtropics to the central California coast) by month of water year (Dettinger 2004); all breaks with exact dates reported were included (Florsheim and Dettinger 2015)

determined the weather conditions associated with those breaks. Of those 128 levee breaks (chosen because a specific date and location for each break was reported in various historical records), 104 (81%) occurred with the landfall of an AR storm, 19 (15%) were associated with high river flows fed by major snowmelt episodes, and five were apparently not connected with specific weather events (Fig. 5.14). Given the observation that ARs cause such a large fraction of levee breaks, it is not surprising that the seasonal timing of the breaks overall has strongly reflected the seasonality of major, tropically sourced ARs in California (Fig. 5.14).

Levee breaks in this estuary system (including in its drainage) cause many problems, including flooding of agricultural lands; disruption of some of California's major water-supply conveyances; and major changes to the geometry of the Delta, Central Valley floodplains, and other sensitive sea-level habitats. However, flooding of the lowlands (floodplains) that border the estuary and riverine channels is a natural and important part of this region's ecosystems. Flooding of floodplains helps to disperse and ameliorate downstream flood effects and damages, and provides important and restorative deliveries of sediments and nutrients to the near-channel environments—and the inundated floodplains regularly serve as fish and faunal "cafeterias" and "nurseries" for native terrestrial and aquatic species in the area. Sommer et al. (2004) have reported that native fish species derive vital resources and life-stage opportunities from inundated Central Valley floodplains if those inundations are sustained long enough (ideally >2–4 weeks) and during late winter and spring months. Florsheim and Dettinger (2015) consequently evaluated the historical record of inundations of the largest of the (extant) floodplains in the Delta setting: the Yolo Bypass just west of the Sacramento River as it passes alongside the city of Sacramento. From 1956 to 2010, the Yolo Bypass received overflows from the Sacramento River and was inundated 108 times. Of these inundations, 60 were fed by flood flows from AR landfalls. Of the 30 ecologically most valuable inundations that lasted more than 3 weeks, ARs fed about 80%. This percentage once again mimics the fraction of major floods and levee breaks associated with ARs, so ARs are likely to play a similar role in ecologically beneficial floodplain inundations across much of the Central Valley drainage.

In 14 mountain streams just beyond the Central Valley drainage, on the eastern side of the Sierra Nevada but probably at least representative of conditions in the western Sierra Nevada rivers and streams as well, Herbst and Cooper (2010) evaluated how a major AR storm affects aquatic invertebrate communities. With the major AR storm of New Year's 1997 (which resulted in "floods of record" in many rivers in Central California), they found that densities of invertebrates and many taxa increased by over ten times in flooded reaches of the streams compared to control reaches. Stream banks and riparian vegetation eroded, and many fine sediments were flushed out and replaced by fine particles of organic matter that supported the expansion of filtering and small-particle gatherers among the invertebrate taxa.

In the long run, these various initial findings about the roles of AR storms and floods in California's aquatic ecosystems suggest that, as would be expected, ARs are a natural part of these ecosystems and probably in aquatic ecosystems wherever ARs are a major factor. Many ecosystem inhabitants and communities have evolved to accommodate and even benefit from them. "Protecting" landscapes that have developed naturally to benefit from AR extremes puts these ecosystems—and humans—at peril. On shorter time-scales, management of reservoirs, floods, and water resources in basins with ARs should aim to carefully accommodate these storms and their relations with local fisheries and ecological communities (Jasperse et al. 2017; "new perspective 5" in Healey et al. 2016; and see Sect. 7.3 in this book).

5.4.2 Terrestrial Landscapes

ARs also affect terrestrial (dryland) ecosystems, as individual storms and through the cumulative effects of multiple storms over the course of whole seasons to years (Albano et al. 2017). In the short term, the most extreme ARs can directly and suddenly affect forests, grasslands, and other terrestrial ecosystems by causing flooding, rapid runoff, and soil saturation—which leads to erosion and slope movements—and by wind-throw downing of trees and avalanche impacts on forests.

In mountain environments of the conterminous Western US, snow avalanches are a dangerous type of mass movement that pose significant hazards to forests, human life, and property, resulting in millions of dollars per year in economic losses (Mock and Birkeland 2000). Fatal avalanches are commonly triggered by human actions but also occur because of natural release mechanisms that result from loading by newly fallen snow. Hatchett et al. (2017) have recently reported that AR conditions preceded, by no more than 4 days, some 105 avalanche incidents between 1998 and 2014, and resulted in 123 fatalities. These AR-induced events comprised 31% of avalanche fatalities in the Western US. ARs apparently cause the worst of these avalanche episodes when they rapidly deposit heavy precipitation and snow onto existing weak snowpacks.

ARs, with their intense precipitation, can also play major roles in erosion, debris flows, landslides, and other "mass movements." These mass movements often threaten human lives and property, and are a fundamental part of landscape evolution in many settings worldwide. Ongoing studies in California are beginning to document these AR roles and their underlying mechanisms. In the Transverse Ranges of Southern California, a fairly regular cycle of fire and flood occurs, with wildfires weakening slopes and creating large masses of debris that subsequent rains mobilize in dangerous post-fire debris flows. The combination of steep topography, rock and soil composition, and wildfire—coupled with intense rainfall—initiate post-fire debris flows, even in cases of low storm rainfall totals. Instead of storm totals being the driving mechanisms for the debris flows, short-term maximum storm intensities drive the events, with precipitation intensities >7 mm per hour, for example, being a typical deluge threshold associated with debris-flow formation. Previous research has developed precipitation intensity–duration thresholds for post-fire debris flows in the Transverse Ranges. The required highly intense precipitation rates—even over just minutes—are enough to saturate, weaken, and pressurize soils and debris, and to generate the rapid erosive runoff needed to initiate debris flows. Convective cells within storms are well established as the mechanism to produce rainfall rates that exceed these thresholds.

A compilation of 100 post-fire debris flow occurrence records—triggered by 19 storm events within Ventura, Los Angeles, and San Bernardino counties between 1980 and 2014—was developed by Oakley et al. (2017) through assessment of publications, news articles, and public records. They reviewed precipitation, radar, and satellite imagery records to confirm or revise debris flow trigger times. They analyzed each episode to determine the meteorological conditions most often shared by these debris flows, as well as the range of values of meteorological variables that accompanied the initiation of the post-fire debris flows. ARs were the large-scale storm conditions in 14 of the 19 storm events. The episodes also shared low-level winds orthogonal to the Transverse Ranges, strong upper-level jets over the region, and moist-neutral stability. Several events included closed low-pressure systems or narrow cold frontal rain bands.

On less catastrophic levels, ARs can play important roles in erosion, avulsion, and sediment transport. For example, in the Upper Truckee River of the Lake Tahoe basin in the Sierra Nevada, erosion and sediment transports into the lake are a highly episodic process, with fully 12% of all sediments delivered to the lake in just 36 days between 1971 and 1992. Of those 36 days, with their much higher than normal sediment transports, two-thirds were on the occasions of large AR storms with their intense rains; most of the others were driven by high snowmelt-fed flows, which in the Sierra Nevada most often contain a great deal of "delayed (until the spring snowmelt)" contributions from snows laid down by winter ARs.

Over the longer term, seasonal accumulations of precipitation from multiple ARs affect water availability—effectively controlling seasonal to multi-year availability of water in some areas (Dettinger 2016). This can lead to enhanced soil-moisture storage (Ralph et al. 2013) that can, in turn, months later enhance vegetation growth and abundance, particularly in dryland, arid-to-semi-arid grassland, and shrubland landscapes. Thus far, the role of ARs in terrestrial ecosystems has been little studied, but an example from the US Southwest provides a sense of the role of these storms in dryland processes.

Albano et al. (2017) analyzed the maximum values of the normalized dimensionless vegetation index (NDVI, a satellite-measured index of vegetation greenness and thus, indirectly, abundance) across the Southwestern US in the summers that follow winters with large and small AR contributions to total precipitation. Correlations between winter AR precipitation and annual-maximum NDVI differed between forested and dryland ecosystems, forming patterns that also reflected rain shadowing of AR penetrations into the interior parts of the region. Maximum NDVI responded most, and with positive correlations, to the abundance of AR precipitation earlier in the year in the desert and shrubland landscapes of the Mojave Desert, Sonoran Desert, and Basin and Range ecoregions. Forested ecoregions of the US Southwest responded considerably less to fluctuations of AR precipitation, in accordance with the considerations summarized above.

Wildfires pose major risks in many of these same dryland environments, especially in the Western US, Iberia and the Mediterranean basin, Chile, and other areas where ARs can be particularly important. However, because wildfires are a natural process in these areas, they are typically important parts of these ecosystems. Especially in these drier landscapes, the vegetation growth associated with the greater water availability brought by multiple ARs earlier in the year can be quite rapid and significant. Once the growing season arrives, that greater water availability leads to increased abundance of ready fuels for wildfires later, when the new vegetation has inevitably dried out. In forests, vegetation growth in response to short-term fluctuations of water availability tends to be less pronounced, but more soil moisture and water availability (in part from ARs months earlier) results in a moister landscape and fuels for wildfire that are less susceptible to fire ignition and spread. Indeed, in the American Southwest, Albano et al. (2017) found that highest correlations between AR precipitation and annual areas burned by wildfire occurred in arid areas, where AR–NDVI correlations also were highest. In forested regions, correlations between AR precipitation and areas burned were generally low.

5.4.3 Surface Winds

Given that ARs are defined by integrated water vapor transport (IVT), and wind is a necessary ingredient in IVT calculation, high winds can occur in conjunction with ARs. However, the largest vapor-transport rates associated with ARs are typically a kilometer or more above sea level, and transport rates near the surface have been shown to be uncorrelated (in general) with those aloft, so it is not obvious what the association is between extreme IVT—and thus ARs—and extreme near-surface winds. However, Waliser and Guan (2017) have recently evaluated the occurrences of coastal (surface) wind extremes associated with AR landfalls (Fig. 5.15).

Across broad areas of the mid-latitude oceans and on the western edges of mid-latitude continents, where ARs are relatively common, ARs are highly associated with surface wind extremes. In these regions, about 30% to 60% of all wind extremes (see caption of Fig. 5.15 for definition of "extreme") coincide with AR passage. Moreover, the association of ARs with wind extremes extended well inland (albeit in slightly muted fashion) across all mid-latitude continents and even onto the northern ramparts of Antarctica.

Waliser and Guan (2017) note how these extreme winds affect humans: ARs were associated with 14 of 19 (74%) large European wind-caused insurance losses between 1997 and 2013. Some of these incidents involved losses greater than US $2 billion.

5.4.4 Coastal Sea Level

Those coastal high winds can also result in high coastal sea-level stands. Storm surges and extremely high sea-level stands along mid-latitude continental coastlines are often caused by wind fields that push the water toward the coast (geostrophically or ageostrophically) and with inverse-barometric effect of low atmospheric pressures, both often associated with extratropical cyclones and somewhat remotely by the ARs associated with these extratropical storms. High sea-level stands increase when they coincide with high astronomical tides. The high sea-level stands can, in turn, result in coastal erosion and flooding; increased salinity encroachments into estuaries, coastal wetlands, and lowlands; and increased risks of flooding in near-coastal river reaches.

Khouakhi and Villarini (2016) surveyed connections between ARs and extreme high sea-level stands at 15 tide gauges on the US West Coast to find that some 15–50% of all

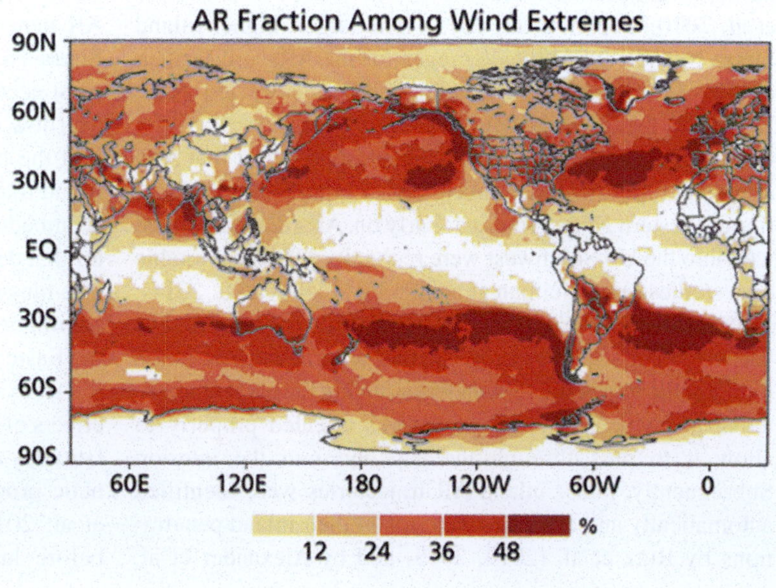

Fig. 5.15 Frequency of surface wind extremes that have been coincident with ARs; values are shaded only if statistically significant at the 99% level. Extreme winds are defined as occasions when surface (10 m) winds exceeded the 98th percentile of wind speed during the period of 1997–2014 (Waliser and Guan 2017)

annual sea-level maxima have been associated with AR landfalls. Those associations have been strongest along the coastline between Southern Washington and Central California, and have been weaker north and south of that range. When astronomical tides are corrected for (removed from) the analyses, the associations of ARs with high sea-level stands increased to 22–65%.

5.5 Regional Perspectives on AR Effects

5.5.1 North America

Many examples presented earlier in this chapter have been drawn from scientific literature that addresses the effects of ARs in various parts of the USA (and a few from Canada). Rather than reiterate those examples, this section briefly overviews the distinct patterns of AR impact across North America.

Until recently, most research and observations on ARs has focused on the US West Coast and in the Pacific coast states of California, Oregon, and Washington. In these states, ARs have caused about 80% or more of historical floods, have contributed 30% to 50% of all precipitation (especially in California), and have played important roles in ending many historical droughts. In these states, ARs making landfall from over the North Pacific Ocean first encounter low coastal ranges, then High Sierra Nevada and Cascade mountain ranges, with both the coastal and the larger inland ranges receiving large amounts of AR-fed precipitation. The ARs tend to be quite heavily moisture-laden when they first make landfall, and the coastal ranges provide sufficient orographic uplift to strip the ARs of significant fractions of their moisture (as illustrated, for example, by the stable-isotopic signatures of coastal AR precipitation that show substantial rainout; Coplen et al. 2008; Yoshimura et al. 2010; Coplen et al. 2015). The larger, higher-inland Sierra Nevada and Cascade ranges lift the ARs still farther, and even more AR precipitation can be extracted (e.g., Ralph and Dettinger 2012).

In recent years, it has become increasingly clear that ARs also sometimes penetrate much farther inland, into the interior of Western North America. Early on, ARs that penetrated well into the US Southwest were recognized in some notable case studies (e.g., Neiman et al. 2013b; Hughes et al. 2014), and a primary determinant of whether an AR penetrated or not was recognized as being whether a particular AR met a topographic gap in the first major mountain ranges it encountered, and whether its long axis was oriented properly to allow it to pass through the gap and into the interior. Subsequently, many inland AR trajectories were identified automatically and classified in terms of their inland penetrations by Rutz et al. (2014, 2015) and by Alexander et al. (2015). Alexander et al. (2015) found that winter extremes of precipitation in the US IMW were generally fed by vapor that enters the region through gaps in the mountain ranges bounding that region's western edge, with those vapor penetrations being "consistent" with AR processes. Rutz et al.'s (2014, 2015) exhaustive study of AR penetrations have shown that, within the IMW, the fractions of cool-season precipitation and intense precipitation events attributable to ARs are largest over the northwest and southwest interior, with a pronounced "AR-rain shadow" east of the highest parts of the Sierra Nevada. Vapor depletion or rainout over major topographic barriers is a key contributor to the decay of AR influences in the interior region. Throughout much of the interior region, inland penetrations are more common and influential during autumn, except in the extreme US Southwest where winter AR arrivals edge them out. The fraction of winter precipitation fed by penetrating ARs ranges from about 30% immediately inland of the Sierra Nevada, Cascades, and the Peninsular ranges of Southern California and Baja California, to 10% or less (after falling off progressively farther inland), in Northern New Mexico, Utah, Wyoming, and Montana. The fractions of intense-precipitation episodes (top-decile of daily precipitation) attributable to penetrating ARs follow much the same geographic pattern, with maximum values that range from 25% to 45% just east of the Sierra Nevada, Cascades, and Peninsular ranges, to minimal contributions in the rain shadow of the High Sierra, and in Northeast Arizona, Utah, Eastern Idaho, and Montana. Some local maxima (approaching 50%) in the fraction of intense-precipitation episodes appear in mountain ranges farther east of the Cascades and Sierra Nevada, including near the Mogollon Rim in Central Arizona, and the Sawtooth and Bitterroot ranges in Central and Northern Idaho, where the ranges are oriented so as to intercept AR penetrations through major topographic gaps farther west. Backes et al. (2015) have examined specifics of AR penetration over (not necessarily through gaps in) the Sierra Nevada. They found that mid-level vapor transport must accompany intense low-level transports for maximum overflow of AR precipitation into the Lake Tahoe basin (just past the primary ridgeline of the Sierra Nevada). They also found that static stability of the atmosphere in which an AR is embedded on its approach to the range is an important factor that determines the amount and extent of penetration over that range. These AR overflows of the Sierra Nevada crest are important contributors to the overall hydroclimatology of the Eastern Sierra and areas just east of the range.

An example from 19 January 2010 shows the hydrologic effects of such penetrations: an AR formed in the lower midlatitudes over the northeastern Pacific Ocean via frontogenetic processes and sea-surface latent-heat fluxes (Neiman et al. 2013b). By 21 January, this AR intensified greatly before landfall across Southern California and Northern

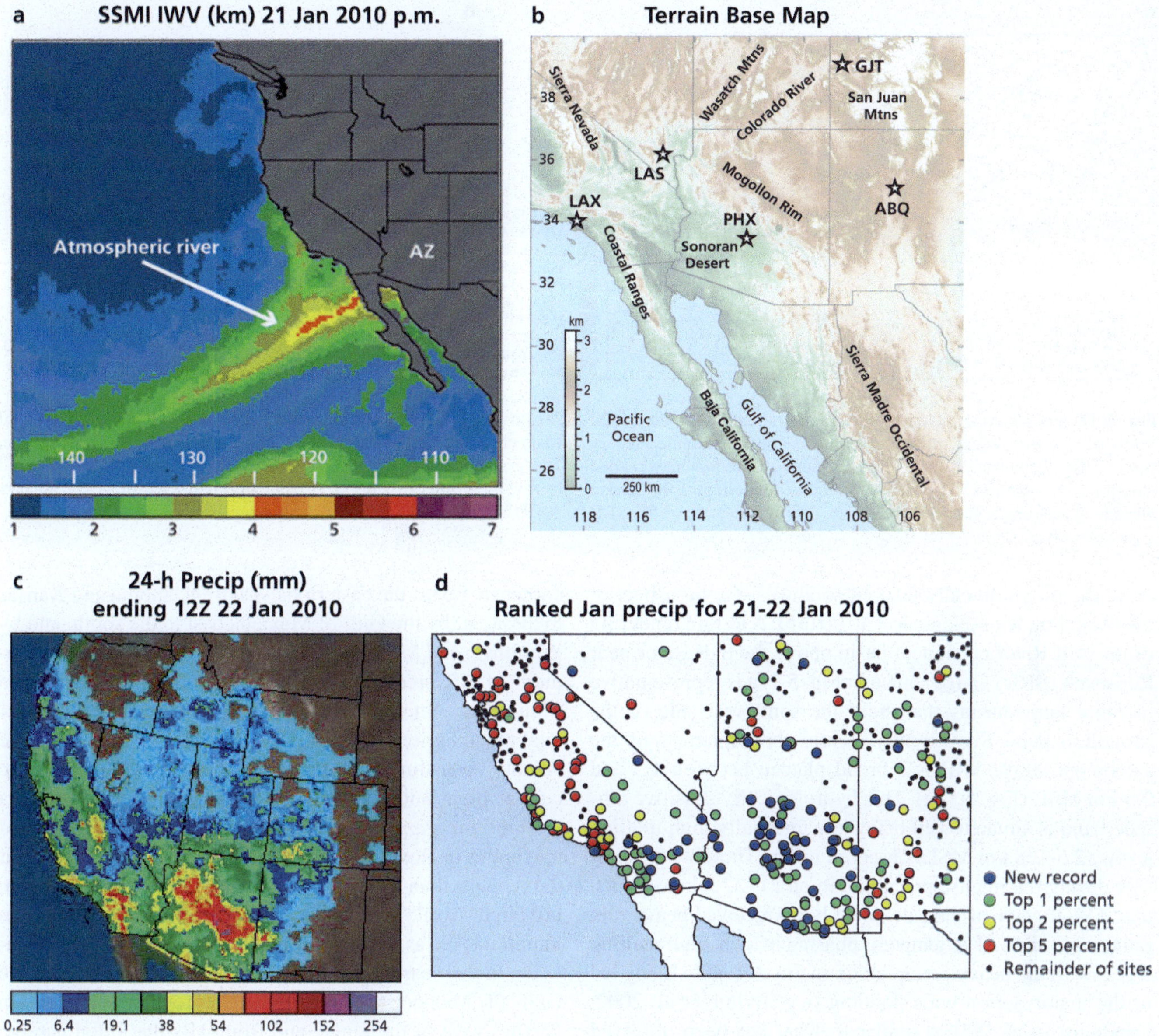

Fig. 5.16 (**a**) Composite Special Sensor Microwave Imager/Sounder (SSM/IS) satellite imagery of integrated water vapor (IWV) (cm) constructed from polar-orbiting observation swaths between ~1200 and 2359 UTC 21 January 2010. The AR of interest is labeled. (**b**) Regional terrain base map (km) of the southwestern US. Selected geographic features and cities (LAX: Los Angeles, CA; LAS: Las Vegas, NV; PHX: Phoenix, AZ; ABQ: Albuquerque, NM; GJT: Grand Junction, CO) are labeled. (**c**) Southwestern US 24-h quantitative precipitation estimation (mm) from the Advanced Hydrologic Prediction Services (AHPS) ending 1200 UTC 22 January 2010. (**d**) Ranking of accumulated Cooperative Observer Program (COOP) precipitation observations on 21–22 January 2010 (percentile, *color coded*) relative to all available January pairs of days between 1950 and 2009 for those COOP gauges having recorded data for ≥25 Januaries. (Neiman et al. 2013b)

Mexico, with a core value of IWV that exceeded 4 cm (Fig. 5.16a) and corresponding IVT that exceeded 1200 kg s^{-1} m^{-1}—which represented an unprecedented ten standard deviations above normal IVT values for that region. The AR and its embedded low-level southwesterly flow were oriented orthogonal to Arizona's Mogollon Rim (Fig. 5.16b), thus optimizing orographic precipitation forcing along this northwest–southeast-oriented 2-km-high escarpment. Consequently, while most of Arizona received at least 13 mm of precipitation during the 24-h period ending 1200 UTC 22 January 2010 (Fig. 5.16c), the Mogollon Rim region was inundated with 50–250 mm. A ranking of precipitation accumulation on 21–22 January 2010 was assessed using gauge sites that collected data for at least 25 of the last 50 Januaries (Fig. 5.16d), and it showed a large number of record and near-record 2-day accumulations over Arizona, with the greatest concentration in the Mogollon Rim region.

A companion ranking of 2-day runoff observed from gauges on unregulated streams (Fig. 5.17a) also shows record to near-record values, especially in those basins that

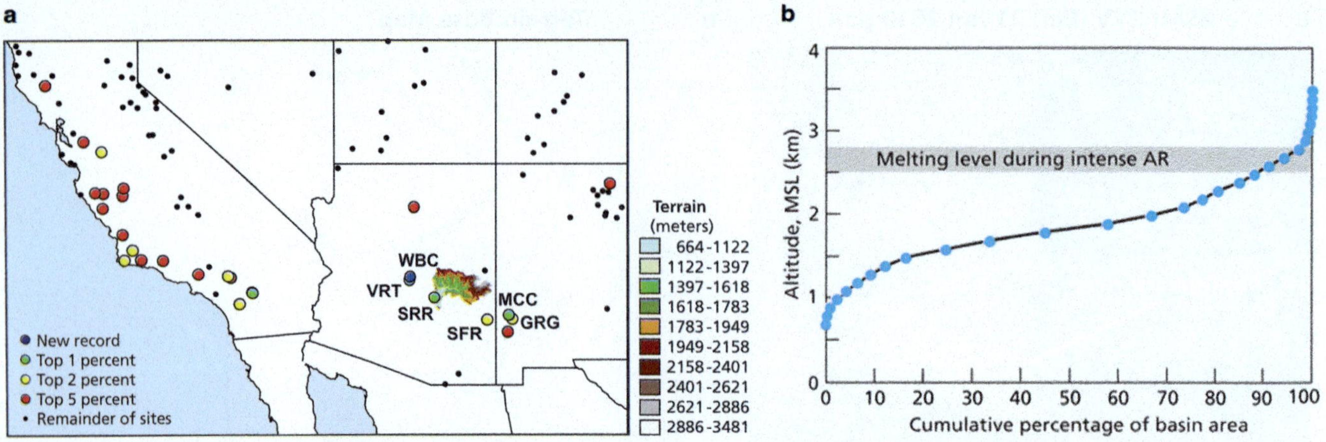

Fig. 5.17 (**a**) Ranking of daily unregulated streamflows on 21–22 January 2010 (percentile, *color coded*) relative to all available January pairs of days between 1901 and 2009 for those gauges having recorded data for ≥25 Januaries. Also shown is a color rendering of the terrain altitude in Arizona's Salt River basin (*m*, *color coded*) above the SRR stream gauge. (**b**) Cumulative percentage of basin area as a function of basin elevation upstream of the Salt River near Roosevelt (SRR) gauge. The *gray-shaded bar* marks the melting-level altitude range recorded by a wind profiler in Tucson, Arizona during the AR of 21–22 January 2010 (Neiman et al. 2013b)

drain the orographically favored southwest-facing aspect of the Mogollon Rim (Neiman et al. 2013b). A terrain rendering of the Salt River catchment basin above the Salt River near Roosevelt (SRR) gauge shown in Fig. 5.17a is representative of other large basins that drain the southwest side of the Mogollon Rim. Significantly, nearly three-quarters of the catchment above SRR is a broad plateau between 1.5 and 2.5 km MSL (Fig. 5.17b). This hydrologically sensitive altitude band is situated well below the unusually high melting levels (2.5–2.8 km MSL) observed by the Tucson wind profiler during the intense precipitation on 21–22 January, such that 90% to 98% of the catchment basin received heavy rain rather than snow. The temporal phasing of very high melting levels with the heavy precipitation in this AR quite likely led to the major region-wide flooding (e.g., White et al. 2002; Lundquist et al. 2008), similar to what has been observed with AR landfalls along the US West Coast (e.g., Neiman et al. 2008, 2011; Dettinger et al. 2009).

East of the Rocky Mountains, research and understanding have progressed similarly. Initially, early studies of ARs in the Eastern US were mostly case studies (e.g., Moore et al. 2012; Nayak et al. 2016) that showed the importance of ARs that made landfall and penetrated far into the less mountainous Eastern US in some major storms and floods. More recently, though, more comprehensive evaluations of many ARs over the Eastern US have been conducted by Villarini and colleagues (Lavers and Villarini 2015; Nayak and Villarini 2017), Mahoney et al. (2016), and others. These broad surveys of the impacts of many ARs (e.g., Lavers and Villarini, 2015) have shown that AR fractions of precipitation east of the Rocky Mountains reach their maximum of about 25–30% of total precipitation over Western Kentucky and Tennessee, declining to 10% or less by the northeastern corner of Texas, the eastern halves of Oklahoma and Kansas to the west, by the Gulf of Mexico coast to the south, and by Wisconsin and Michigan to the north. East of a notable minimum of AR-precipitation fractions along the Appalachian Mountains centered over West Virginia, about 15–20% of total precipitation derives from ARs that sometimes skirt the Atlantic coastline and travel over the northern Atlantic coastal plain and New England. The AR-fraction maxima west of the Appalachian Mountains reach their seasonal maximum in November–December; the AR fractions east of the Appalachian Mountains reach their seasonal maximum in March–April. Thus, these Appalachian Mountains play an important role as a barrier between AR influences on precipitation to their east and west, with ARs arriving from over the Gulf of Mexico (and even the subtropical northeastern Pacific) west of the range, and from ARs that often parallel the Atlantic coast and Gulf Stream to the east. Nayak and Villarini (2017) revisited these analyses, and found that 70% of the annual peak streamflows over the Central US have historically been associated with winter and spring AR storms. In the Southeastern US, Mahoney et al. (2016) found that heavy-precipitation days (>100 mm/day) between 2002 and 2014 were associated with AR landfalls about 40% of the time, mostly in two seasons: October–December and March–May, separated by a pronounced lull in Februaries.

Taken together, these various studies provide a continent-wide picture of AR impacts, with:

1. ARs arriving at the US West Coast from over the Pacific and dominating the precipitation extremes, floods, and, ultimately, water resources of the Pacific coast states;
2. A smaller number, albeit still significant for the region, of these Pacific ARs penetrating into the mountainous inte-

rior Western US where they contribute significantly to winter and total precipitation, except in the rain shadows of the highest parts of the Sierra Nevada as far east as the western slopes of the Rocky Mountains;
3. ARs arriving across the Gulf Coast of the Southeastern US from over the Gulf of Mexico and sometimes the subtropical eastern Pacific, and often penetrating very far northward into the mid-continent between the eastern slopes of the Rocky Mountains and the western slopes of the Appalachian Mountains, to feed many extreme-weather and extreme precipitation events, with some resulting floods; and.
4. ARs that arrive along the Atlantic Coast often transiting parts of the Eastern US along trajectories that are broadly parallel to the US East Coast and Gulf Stream, carrying moisture that upon occasion yields extreme precipitation and flooding there.

Thus, the sources and reach of ARs in North America are broad and often dictated by the continent's largest-scale topography. Where ARs make landfall and/or penetrate to, they can be the moisture sources for major storms, extreme precipitation rates, and sometimes flooding. Where ARs penetrate frequently, they become important contributors to annual precipitation and, ultimately, water supplies for humans and landscapes.

5.5.2 Europe

AR effects have been analyzed extensively along the western coast of North America (e.g., Ralph et al. 2006; Neiman et al. 2008; Ralph and Dettinger 2011; and references therein). Conversely, in Western Europe, AR effects only started to be studied in the late 2000s, with the work of Stohl et al. (2008), where the extreme precipitation in September 2005 on the Norwegian southwest coast was found to be associated with an AR.

As a general overview, in Western Europe, ARs are responsible for approximately 20–30% of all precipitation with a strong seasonality, where the maximum is recorded in autumn and winter (mainly in France and the Iberian Peninsula). In addition, AR effects in Western Europe are generally felt further inland (e.g., Poland) than in Western North America (Lavers and Villarini 2015; their Fig. 2), which is likely to be associated with the higher mountainous terrain there, which acts as a barrier to inland AR penetration. Moreover, there is a strong relationship between ARs and the occurrence of annual maximum precipitation days in Western Europe, which is particularly pronounced along the western seaboard (Northwestern Iberian Peninsula, western British Isles, and southwestern Norway); some areas had eight of their ten highest annual maximum precipitation days related to ARs (Lavers and Villarini 2013b). Figure 5.18 shows the main areas of AR influence on precipitation in Western Europe, and intense precipitation from ARs associated with floods and landslides. Three main areas emerge from the different studies undertaken hitherto for AR effects: the Iberian Peninsula (e.g., Ramos et al. 2015; Eiras-Barca et al. 2016), the British Isles (e.g., Lavers et al. 2011, 2012), and Norway (e.g., Stohl et al. 2008; Sodemann and Stohl 2013).

From a climatological perspective, Lavers et al. (2012) showed that, for nine river basins in the British Isles, the number of winter floods associated with persistent ARs (ARs lasting more than 18 h) ranged from approximately 40% to 80%. Ramos et al. (2015) showed that, since 1950, the association between ARs and extreme precipitation days in the western regions of the Iberian Peninsula (Portugal and three interna-

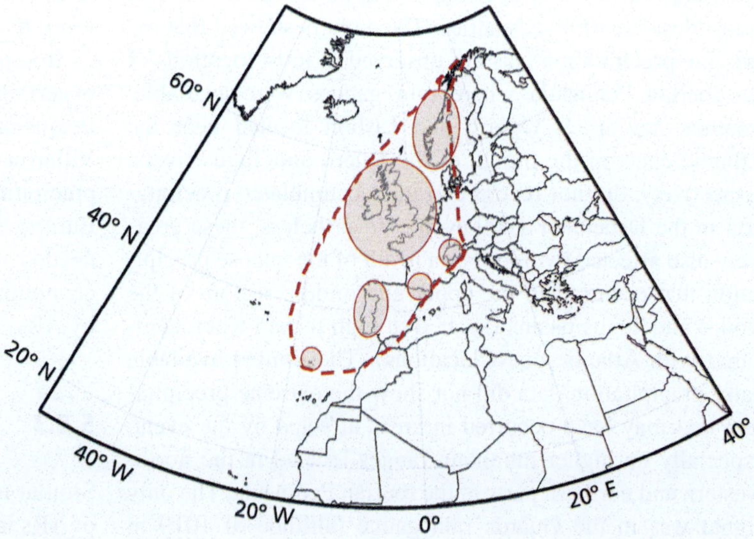

Fig. 5.18 Main areas of AR influence on precipitation in Western Europe (*red dashed outline*), and intense precipitation from ARs associated with floods and landslides (*solid color*). (Adapted and updated from Gimeno et al. 2016). The two largest storms in Central Chile over a 7-year period were associated with ARs and caused floods and deaths

tional river basins: Minho, Tagus, and Duero) is noteworthy; ARs affect the eastern and southern basins (Ebro, Guadiana, and Guadalquivir river basins) less. In addition, ARs do not contribute homogeneously toward the extreme precipitation ranking list for the Iberian Peninsula: they have a striking role for the most extreme precipitation days, but a significantly decreasing role for the least extreme precipitation days.

Despite Western Europe's lower terrain than Western North America's, different studies show that the interaction between ARs and topography can explain extreme precipitation, as the following examples show.

The top of the eastern mountain range on the Island of Madeira—where the orography is dominated by a volcanic landscape, with steep slopes and scarps—reaches a maximum altitude of 1861 m. On 20 February 2010, torrential rainfall hit the island, which triggered catastrophic flash floods that caused 45 fatalities: roughly half of the victims (22) were in the capital city of Funchal; six others were declared missing. It was the deadliest hydrometeorological catastrophe in the Portuguese territory in the last four decades, and the economic damages were estimated to be US $1.9 billion (Fragoso et al. 2012). The rainstorm's exceptionality was further confirmed by the return period associated with the daily precipitation registered at the two long-term record stations: 146.9 mm observed in the city of Funchal, and 333.8 mm on the mountain top—corresponding to an estimated return period of approximately 290 years and 90 years, respectively. Couto et al. (2015) showed that an AR hit the Island of Madeira on 20 February 2010, producing enhanced precipitation when it hit the steep topography.

The second example is a historical case on the Iberian Peninsula from December 1909. This hydro-geomorphologic event had the highest number of floods and landslides that had occurred in Portugal since 1865 (Pereira et al. 2016)—and it resulted from an AR. The Iberian Peninsula was spatially affected during this event along the SW–NE direction, spanning from Lisbon to León, where numerous floods and landslides caused 89 casualties. The authors showed that the extreme precipitation was not observed in most locations of the Iberian Peninsula where rain gauges were available: weather stations at Oporto and Lisbon located near the Atlantic coast in the mouth of the Duero and Tagus rivers, respectively, did not record extreme accumulated precipitation in the December 1909 event. Nevertheless, these areas were also affected by floods as a result of the intense precipitation that occurred in the upper and middle sectors of the Douro and Tagus basins (located in high terrain when compared with Atlantic coast locations). The limited available daily precipitation data did not show the extreme precipitation that may have occurred in areas affected by the event, especially on higher mountain ranges located in the northwestern and northern parts of the Iberian Peninsula. The only signal was in the Guarda rain gauge (altitude of 1019 m above sea level [ASL]), which registered a daily precipitation record of nearly 200 mm.

Other historical cases of intense precipitation from ARs that resulted in flooding-caused major socio-economic impacts in Western Europe are noteworthy. Good examples are the December 1876 event in the western Iberian Peninsula (Trigo et al. 2014) or the November 1982 catastrophic flash flood (44 fatalities and widespread damage) in the Eastern Pyrenees that affected Andorra, France, and Spain—with rainfall accumulations of more than 400 mm in 24 h (Trapero et al. 2013).

Because of the nature of European topography, rain-on-snow events from ARs are not so likely to occur as in Western North America (e.g., Guan et al. 2016). However, a rain-on-snow event on 10 October 2011 from an AR in the Bernese Alps, Switzerland, caused significant damage. The atmospheric drivers of this rain-on-snow flood were (1) sustained snowfall, and (2) following it, the passage of an AR that brought warm and moist air toward the Alps. As a result, intense rainfall (average of 100 mm/day) was accompanied by a temperature increase that shifted the snow line from 1500 to 3200 m ASL in 24 h, with a maximum increase of 9 K in 9 h (Rössler et al. 2014).

These are cases where topography played a major role in enhancing precipitation (e.g., Couto et al. 2015; Trapero et al. 2013; Trigo et al. 2014). However, other cases of ARs featured extreme precipitation that occurred in relatively flat areas, such as the November 1983 flash flood in Lisbon (total of 95.6 mm in 24 h); it was the wettest day during the twentieth century—and one of the wettest registered since 1864 (Liberato et al. 2012). According to the authors, this episode resulted from a combination of favorable thermodynamical conditions, including (1) a lower-than-usual latitudinal location of the jet stream related to the presence of a strong meridional temperature gradient slightly south of its usual latitude, (2) a prolonged formation of cumulonimbus structures fed by intense vertical instability in an area of upper-level divergence, and (3) the presence of an AR.

In summary, the socio-economic effects of ARs are known in certain areas in Western Europe, specifically in the Iberian Peninsula and in the British Isles. According to Lavers and Villarini (2013b), the effects of ARs in the annual maxima precipitation are also important in other areas of Western Europe such as France and Norway. Further studies are needed to confirm if this extreme precipitation has socio-economic effects or not, and which areas are most affected by ARs.

5.5.3 South America

Similar to Western North America, the most important effect of ARs in southwestern South America (SA) is that they lead

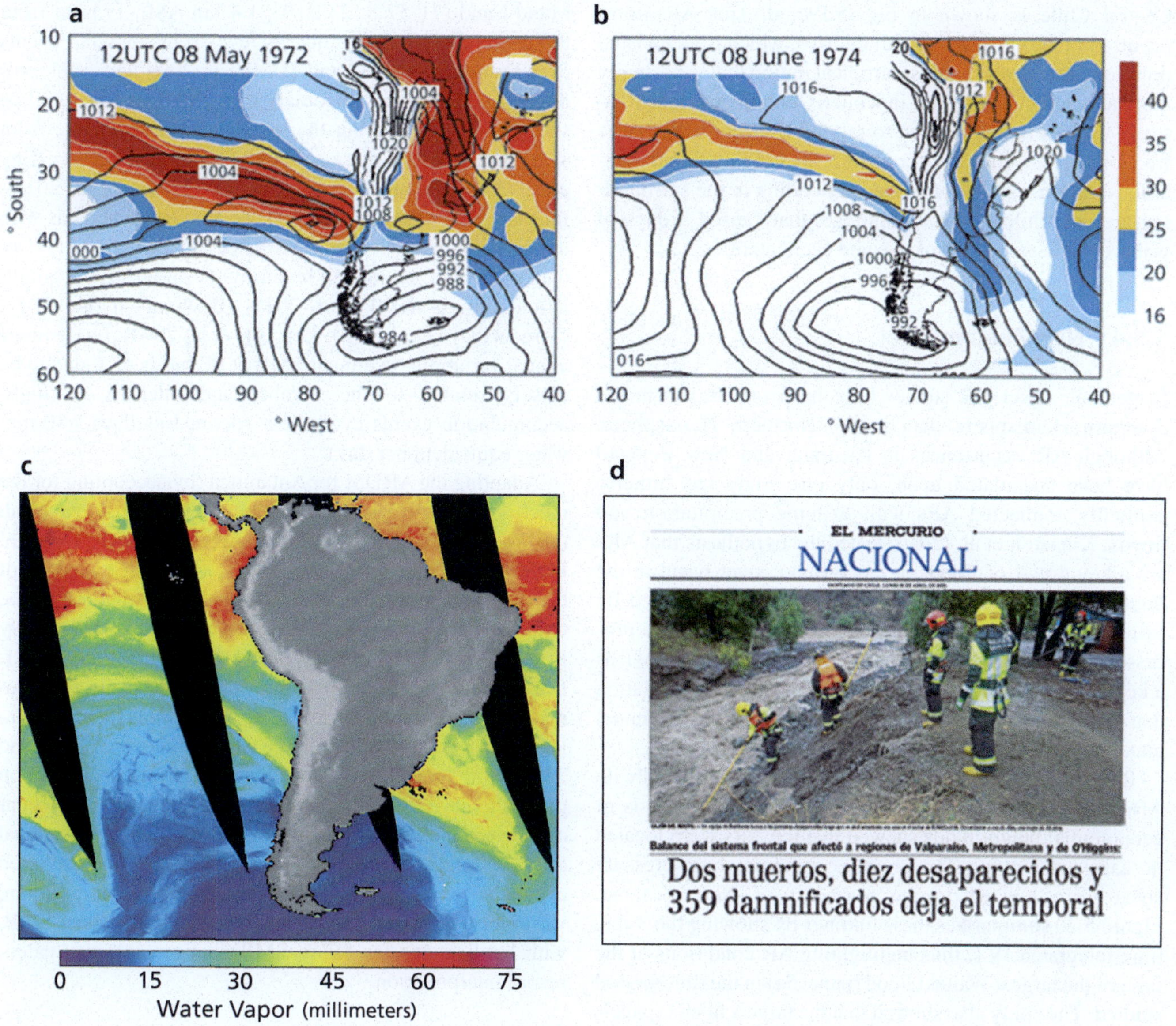

Fig. 5.19 (**a**, **b**) Weather maps showing the sea-level pressure (*lines* in hPa) and the intense plumes of water vapor (*shaded* in mm) associated with the ARs that produced the two largest storms in central Chile over a 7-year period (1970–1976) and caused floods and fatalities (Adapted from Viale and Nuñez 2011). (**c**) Satellite Advanced Microwave Scanning Radiometer (AMSR-2) imagery showing the plume of water vapor content linked to the AR that made landfall in early austral fall in April 2006 over central Chile. (**d**) Front-page Santiago de Chile newspaper story highlighting the damages, debris flow, and fatalities caused by the heavy AR storm of April 2016

to heavy orographic precipitation events that provide most of the region's water resources, but also occasionally produce floods, landslides, and river overflow, which can cause irrecoverable losses and fatalities. That ARs produce floods and damages in SA has been mostly documented in Central Chile and the subtropical Andes where most people live (30°–37°S); but these effects have also been observed in the remote and less populated Western Patagonia, even in the leeside (Argentina) of the low extratropical Andes through the inland penetration of ARs. As stated in Sect. 4.8, 40 of the 46 heaviest precipitation events that occurred in Central Chile over a 7-year period (1970–1976) were linked to ARs making landfall on the west coast of SA (Viale and Nuñez 2011). Figure 5.19a, b highlights the intense plumes of water vapor that reached the west coast of SA during the two heaviest precipitation events of that period, in which major disaster declarations were declared in the area, and 20 and seven fatalities, respectively, were reported.

In the winter of 2005 and 2006, two cases of heavy precipitation events linked to landfalling ARs were documented by Viale and Norte (2009) and Viale et al. (2013), respectively. These events caused floods and damages in the two most populated cities in Chile: Santiago and Concepcion. The last example of a powerful AR storm making landfall over

Central Chile is shown in Fig. 5.19c, d. This AR storm occurred in the early austral fall of 2016, which is an unusual time for AR arrivals in this subtropical region that is normally more affected by AR storms in winter. Consequently, the relatively warm environment led to rainfall over unusually large fractions of the Andean river basins, favoring rapid and hazardous runoff. This event produced flooding in the surrounding area of Chile's capital, Santiago, that caused 3 days of water outage, serious damages, and even fatalities.

5.5.4 New Zealand

In general, fewer AR studies have been undertaken in the Southern Hemisphere than in the Northern Hemisphere. Although AR occurrences in Australia and New Zealand have been speculated upon, only one study has hitherto explicitly connected ARs with extreme precipitation and floods: Kingston et al. (2016) tested the hypothesis that ARs are a key driver of floods in the Waitaki River basin in the Southern Alps of New Zealand, a region characterized by extremely high precipitation totals from synoptic-scale processes and orographic enhancement. The study used IVT as a key AR diagnostic to evaluate the atmospheric circulation that was in place concurrently with the eight largest annual maxima floods (1979–2012) in the Waitaki River.

Results revealed that the eight floods were associated with ARs positioned over the Waitaki basin; further analysis of geopotential height fields showed that the ARs were located in eastward-moving low-pressure systems; high-pressure areas were located to the northeast of New Zealand. Figure 5.20 summarizes these findings by showing the 3-day time-integrated IVT, thus highlighting AR conditions at the time of the largest Waitaki flood (panel 5a) in the time-period studied. The study also showed that the largest floods (panels a-h) all occurred in the Southern Hemisphere summer (a somewhat surprising result, given the Northern Hemisphere studies), which is a result partly of snowmelt and partly of the large moisture flux in the ARs. This study confirmed the importance of ARs for hydrometeorological extremes on the South Island of New Zealand, and opens up the possibility of future research into, and application of, AR science across New Zealand and, more broadly, in Australia.

5.5.5 Polar Regions

Antarctica

Gorodetskaya et al. (2014) analyzed for the first time how ARs affected the Antarctic ice sheet's precipitation and accumulation, using in situ measurements of snow accumulation and radar-derived snowfall estimates from Princess Elisabeth station (PE) located in the escarpment area of Dronning Maud Land (71°57′S, 23°21′E; 1.4 km ASL, 173 km from the coast) in East Antarctica. A few strong snowfall events over Dronning Maud Land (DML) in 2009 and 2011 produced a positive mass anomaly over the East Antarctic ice sheet, counter-balancing the increasing ice discharge from West Antarctica in these years (Boening et al. 2012; King et al. 2012; Shepherd et al. 2012; McMillan et al. 2014). Inter-annual variability of the PE accumulation accords with the inter-annual surface mass balance (SMB) variability over the larger DML region, all featuring the anomalous accumulation during 2009 and 2011 (Boening et al. 2012; Gorodetskaya et al. 2014; Lenaerts et al. 2013). Large inter-annual variability in annual SMB at PE has been found to be closely related to the number and intensity of high-accumulation events every year (defined as days ≥10-mm water equivalent per day).

Adapting the ARDM for Antarctica (by accounting for the much colder and drier environment), Gorodetskaya et al. (2014) found that four and five ARs that reached coastal DML in 2009 and 2011, respectively, contributed 74–80% of the outstanding SMB at PE. Significant contribution to the total local PE SMB during these 2 years comes particularly from two events: 18–19 May 2009 and 14–16 February 2011. During both events, snow height measured by the PE automatic weather station increased by 48–51-mm water equivalent, thereby contributing a large percentage—up to 21% and 22%—of the total SMB for each year. The large contribution of ARs to DML SMB implies that how frequently ARs occur can determine the difference in the regional total annual SMB. The relationship between ARs and high-accumulation events is very important for understanding inter-annual variability and trends of the total Antarctic ice sheet SMB, with implications for future SMB changes and also paleo-record interpretation.

Arctic

Recently, Neff et al. (2014) examined the 2012 summer Greenland melt episode and compared it to the last episode in 1889 using the Twentieth Century Reanalysis (20CR; Compo et al. 2011), finding similar factors at work. A key factor in 2012 was the presence of an AR that transported warm air from a mid-continent heat wave over the Atlantic Ocean and thence to the west coast of Greenland, and then over the ice sheet with a confirming water vapor isotopic signature (Bonne et al. 2015). ARs represent an efficient poleward transport mechanism for warm moist air (Newell et al. 1992) to the Arctic (Bonne et al. 2015; Neff et al. 2014) and the Antarctic (Gorodetskaya et al. 2014).

Some common characteristics of the events in 1889 and 2012, in addition to the expression of poleward moisture transport as an AR, included continental heat anomalies in the trajectory source regions as well as a trough–ridge pattern that focused transport along the west coast of Greenland.

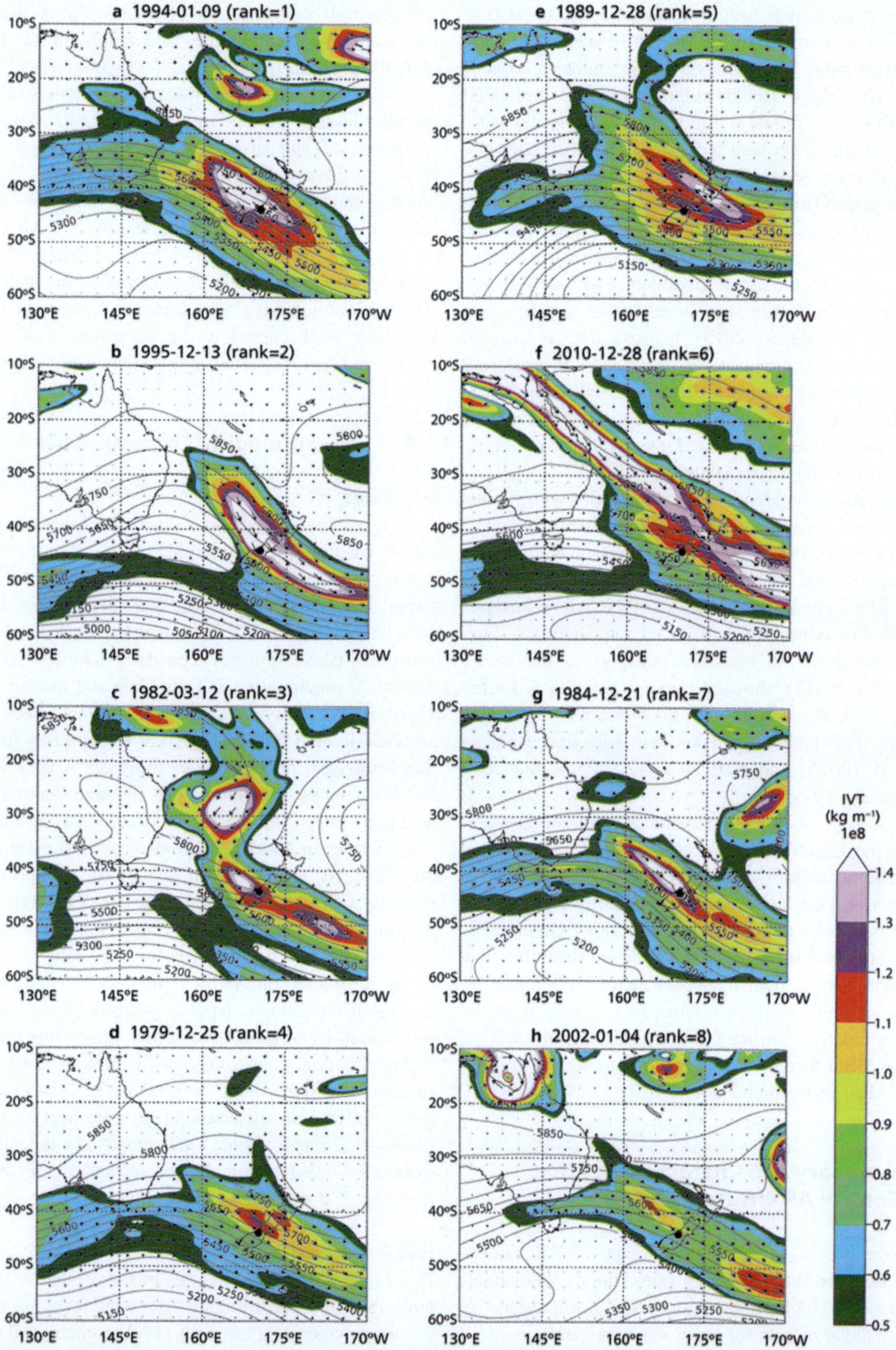

Fig. 5.20 The 3-day time-integrated integrated water vapor (IVT; *shaded*) and 3-day average 500-hPa geopotential heights (*contours* on a 50-m interval) before the largest eight floods in the Waitaki basin of New Zealand (**a-h**) (Kingston et al. 2016)

The latter consisted of a trough of low pressure situated to the west, generally over Baffin Island and Hudson Bay, and a high-pressure ridge that impinged on the southeast coast of Greenland. This type of trough–ridge pattern was implicated in a major rain event in late summer in 2011 along the western margin of the Greenland ice sheet that accelerated the flow of ice into the ocean (Doyle et al. 2015). Similarly, an AR and associated ridge–trough pattern was associated with enhanced winter precipitation over Spitzbergen (Serreze et al. 2015). Extreme moisture transport into the Arctic in winter along the west coast of Greenland (with an AR-like structure) has been identified as an artifact of Rossby wave breaking (Liu and Barnes 2015). In particular, as Liu and Barnes concluded, cyclonic wave breaking (CWB) during the negative phase of the North Atlantic Oscillation (NAO) favors moisture transport along the west coast of Greenland; conversely, anticyclonic wave breaking (AWB) favors moisture transport into the North Atlantic.

ARs have also been identified in transport of warm moist air across the entire Arctic during the Surface Heat Budget of the Arctic Ocean (SHEBA) experiment (Persson et al. 2016). The National Centers for Environmental Prediction/National Center for Atmospheric Research (NCEP/NCAR) reanalysis has been used to identify the significant contribution of low-frequency variability to moisture transport to the Arctic (Newman et al. 2012), showing a major pathway from the extratropics to Greenland through the western Atlantic during summer. This pathway is consistent with the results of Bonne et al. (2015) and Neff et al. (2014) for the 2012 Greenland melt episode.

Although these isolated melt episodes on the highest elevations of the Greenland ice sheet occurred in 2012 and 1889, questions remain about near-misses and trends in events that transport heat and moisture northward into the Arctic. Of particular interest is transport along Greenland's west coast and the lower elevations of the ice sheet that may be complicit in their own increased melting (Doyle et al. 2015). For example, analysis of shallow ice cores from the North Greenland Eemian Ice Drilling (NEEM) site show an increase in δ18O in recent decades, and extremes in 1928 and 2010 (Masson-Delmotte et al. 2015).

5.6 Summary and Characteristics that Control AR Effects

This chapter has discussed a wide range of AR effects, including hazards such as heavy rains, floods, landslides, snow avalanches, high winds, and coastal storm surges, as well as beneficial effects for water resources and terrestrial and aquatic ecosystems. Table 5.1 listed 14 such effects. Different effects—hazardous and beneficial—have been recognized and studied in different regions of the world. Taken together, this compendium of effects shows that ARs are important to the landscapes and communities they reach, almost regardless of setting.

This chapter, as well as most other chapters of this book, has also described important features of ARs as meteorological processes and structures, where "important" often has meant "important in terms of how they will affect surface weather and weather effects." Summarized below are the key meteorological and land characteristics of ARs that most strongly determine the consequences of their landfalls and inland penetrations. These characteristics and their effects on AR precipitation have been detailed in previous chapters of this book or discussed in the descriptions of various AR effects and regions earlier in this chapter.

5.6.1 Meteorological Characteristics

IVT Rates

A key parameter that determines precipitation rates from ARs—whether in an orographic setting or in the WCB of an extratropical cyclone—is the rate at which the AR transports vapor. The IVT is the variable that encapsulates that parameter (Neiman et al. 2002, 2011; Ralph et al. 2013). Figure 5.21 illustrates correlation between daily weather-station observations of precipitation in California and measured upslope vapor rates (a proxy for IVT that is available from AR observatories on the US West Coast) at Bodega Bay just north of San Francisco; upslope vapor transport is very well correlated ($r > +0.8$) along the narrow linear projection of the AR as it penetrates from the coast to the Sierra Nevada.

IVT rates in ARs range from meager to extremely large, and most AR effects and impacts on the ground scale fairly directly with those rates. In most AR-detection methods, a minimum amount of vapor (or vapor transport) must be concentrated in an IVT pathway before the pathway is "eligible" to be considered an AR; a typical IVT threshold is about 200 kg/m/s, although, depending on location, that threshold might vary by at least +50 kg/m/s, depending on the study and application. The strongest IVT rates also vary with location, but rise to as much as 1500 kg/m/s. As indicated, AR precipitation and impacts tend to scale fairly directly with these IVT rates—from light mostly beneficial rains to extremely heavy, often dangerous and damaging torrents (see Sect. 5.2 and also Chap. 8).

IWV Amounts

IVT rates are vertical integrals of the product of vapor contents and winds. Historically, ARs were recognized in SSM/I satellite imagery of (vertically) IWV content, and this metric is still an important AR descriptor. Large IVT rates can occur in ARs with high winds and very moist air, or with very high winds and moist air. Both can generate substantial precipita-

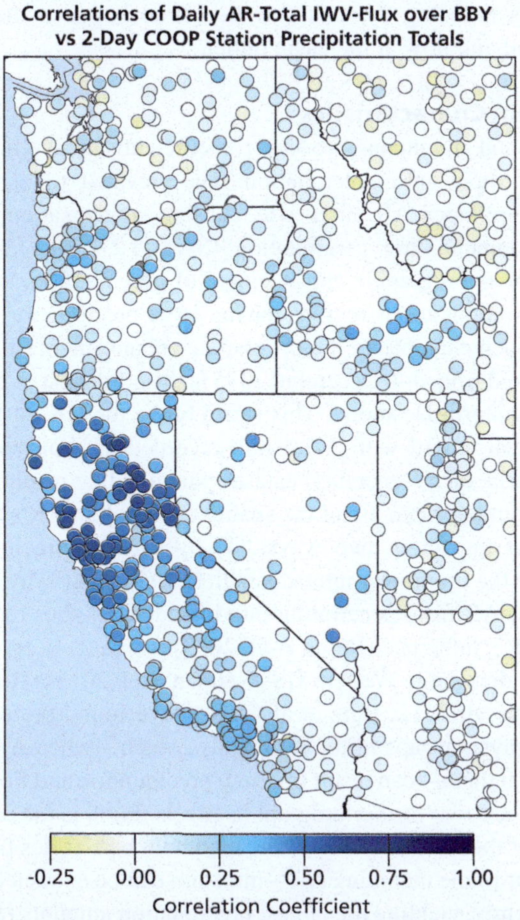

Fig. 5.21 Correlations between vapor fluxes over the Bodega Bay (BBY) AR observatory (ARO) and Cooperative Observer Program (COOP) station precipitation totals in the western US on (and immediately following) days when an AR made landfall at BBY

tion and runoff under the right conditions, but high-IWV ARs are often the most obvious in satellite imagery and weather forecasts.

Rates of AR Translation Across the Landscape

The IVT rate of upslope vapor delivery is the first-order control on precipitation rates in AR storms, but the AR-storm total precipitation amount also depends on the length of time that AR conditions (and precipitation generation) persist overhead of a given location. This duration depends on how quickly the AR passes overhead, which can be a function of how quickly the AR as a whole is moving, but often is a function—for some periods of landfall—of the AR's shape. Meso-scale-wave structures (see Chap. 2) are curves and waves along ARs (and the cold fronts that ARs parallel) that, as the large-scale AR structure translates across a point, can result in a prolongation (stalling) overhead of AR segments that are oriented parallel or near-parallel to the direction of overall AR translation. Ralph et al. (2013) found that, at the Bodega Bay (BBY) AR observatory (ARO), AR conditions, on average, persist overhead for about 20 h, but that their passage can take anywhere from 8 h to over 48 h. If precipitation fell at the same average rate during AR conditions for 8 or 48 h, the latter would yield six times as much precipitation.

Air Temperature

As noted below, altitude plays an important role in determining whether snow or rain falls from an AR at a given location. ARs also differ in their air temperatures, so that snowlines (the altitude above which snow rather than rain falls) vary from storm to storm. On the whole, because ARs often tap into subtropical moisture and air masses, ARs on the west coast of North America (at least) are frequently relatively warm. For example, between 1998 and 2010, AR storms at Tahoe City in the Sierra Nevada at 6200 feet ASL were warmer by 2 °C, and wetter by 85%, than wet days in general. Viewed from another perspective, AR storms arrived on 81% of the 38 days between 1948 and 2010 that were warmer than freezing and on which more than 5 cm of precipitation fell at Tahoe City. These warm temperatures ensure that snowlines during ARs tend to be higher (by about 300 m)—and the area of the range upon which rain falls much broader—than during most other winter storms.

Atmospheric Stability

The stability of the atmospheric profiles in which most ARs are embedded is close to moist–static neutral stratification (Ralph et al. 2004). Neutral stratification imposes only weak stability—relative to vertical motions—on air parcels in the atmospheres, so that the frequent weak stability of ARs allows them to rise up and over mountain barriers with relative ease, and under the right conditions also facilitates the onset of cumulus convection, which can greatly amplify the amounts of precipitation that fall.

Elevation of the AR Jet

Typically, ARs are centered between about 1.5 km and 2.5 km ASL over the oceans, with orographic barriers imparting vertical uplift once the ARs are over land, which raise them to greater altitudes. These mountain barriers will uplift an approaching AR, even if the AR is located at or above the ridgeline altitude, but the AR core's altitude—as well as the distribution of vapor throughout the atmosphere column within which the AR is embedded—plays an important role in determining whether the AR will push past that ridgeline to deposit precipitation on the lee side of the mountains and beyond (Backes et al. 2015). This aspect of AR interactions with terrestrial topography has not been as well explored as it deserves.

Barrier Jets

In their interactions with mountain ranges, ARs often spawn barrier jets, a low-level (even lower than ARs) jet that flows along or ahead of the range an AR is approaching and pushing over. Along the North American margins, these barrier jets form just west of the medium-altitude coastal ranges, often just offshore (Loescher et al. 2006; Neiman et al. 2004) and in front of the High Sierra Nevada of California over the eastern half of the Central Valley. In these settings, given the orientation of the mountain ranges and of most ARs making landfall, the barrier jets transport low-level air, vapor, and pollutants generally northward along and above the western ramparts of the mountain ranges. Neiman et al. (2013a) and Hughes et al. (2012) have shown that, because of the vapor they transport, Sierra Barrier Jets contribute to orographic precipitation at the northern end of the Central Valley, where the barrier jets encounter orographic uplift by east–west-oriented mountain and highlands. Around the AR core, the AR can raise up over the barrier jet, yielding an odd version of orographic precipitation where the AR lifts up over the barrier jet, cooling, condensing, and precipitating ahead of the actual topographic rise into the mountain range.

5.6.2 Land Characteristics

The AR's interaction with the land surface—particularly on the western edges of mid-latitude continents, such as the USA and Europe—largely determines the level of the its hydrological effect. This section briefly lists different aspects of land surface physiography that are vital in translating AR water vapor transport into precipitation, and ultimately into river discharge. However, given the large heterogeneities of the land surface, only a physiographical overview is given because each river basin has unique characteristics that modulate the rainfall–runoff process in different ways, many of which are not fully understood. The characteristics covered here include: antecedent conditions (i.e., soil moisture); terrain height, steepness of slopes, and orientation of orography; underlying bedrock and soil type; land use; and the presence of lakes and reservoirs.

Antecedent Conditions

The initial state of the hydrological system, or antecedent conditions, largely controls the magnitude of the AR's signal on hydrology. If an AR reaches a river basin after a protracted dry period, when low soil moisture levels are likely, these soils will possibly be able to absorb the incoming rainfall and prevent the river basin from flooding. Conversely, if the basin is wet, after previous AR events, for example, less water is likely to infiltrate the soil, thus producing surface runoff, or the 'new' event water will push greater volumes of older 'pre-event' water into the channel network, with both effects reducing the concentration time between the peak rainfall and flow at the basin outlet.

Terrain Characteristics

Hills and mountains are key for efficiently converting AR water vapor into precipitation. The elevated relief forces moisture-laden AR air to rise, causing condensation, cloud development, and precipitation (Sect. 5.2). This process, known as orographic enhancement of precipitation, leads to many mountainous regions having large precipitation totals. One such example of the orographic enhancement of an AR occurred from 4–6 December 2015 in the northwest of Europe during storm Desmond. This storm broke the UK 24-h rainfall total record with 341.4 mm recorded at Honister Pass, Cumbria, on 5 December; catastrophic flooding resulted.

Mountains can affect the spatial extent of an AR's hydrological impact in two ways: (1) by causing precipitation where the orography interacts with the AR, and (2) by blocking the AR from penetrating into inland areas, thus reducing the AR's influence (Ralph et al. 2003).

Studies in the Western US (Neiman et al. 2013b; Rutz and Steenburgh 2012; Rutz et al. 2014) have found preferential pathways, or gaps in the orography, which—when ARs line up with them—can result in heavy precipitation and flooding. There are two further orographic effects. First, is the orientation of the relief, where—if the mountain barrier lies perpendicular to the incoming AR—moisture can be extracted most efficiently, yielding the largest precipitation amounts (e.g., as clearly demonstrated by Neiman et al. (2011) through contrasting flood responses to ARs with different orientations in four western Washington river basins). Second, steeper slopes in a river basin will lead rainwater to be transported more rapidly to the river channel, potentially exacerbating flood risk.

Furthermore, the elevation affects the surface air temperature, which is crucial for rain–snow partitioning, and thus potential attenuation of the precipitation–runoff signal. Without hills and mountains, the water vapor in many ARs would possibly continue to be transported horizontally without affecting the basin hydrology.

Drainage Density

This is defined as the total length of river channel relative to basin area, and influences how rapidly a basin can convey water from contributing surfaces (and subsurface) to the stream channels. Areas with a high drainage density efficiently transport rainwater to the channel. This efficiency can contribute to flood generation if storm totals are sufficient to overwhelm channel flow-conveyance capacities, and runoff (surface and subsurface) is rapid enough to move all that water into the channels quickly. If precipitation from a storm is *not* sufficient to overwhelm channel conveyance capacities, a high-density drainage network can allow runoff to exit a basin quickly and with less flooding risk. Areas with high-density drainage networks also tend to have low-permeability bedrock. Areas with lower-density drainage networks transport rainwater to channels less efficiently, increasing the risk of between-channel flooding during intense or long storms.

Bedrock and Soil Type

Different bedrocks and soils attenuate the precipitation signal to varying degrees. River basins situated on bedrocks with limited fracturing, such as granite, have low permeability, so that rainwater cannot infiltrate deep into the ground, and with the higher drainage density (total length of river channel relative to basin area) being a general property of impermeable basins, a quick rainfall-runoff response occurs. In these basins, a persistent AR can be sufficient to result in flooding conditions. In contrast, in regions of permeable rock (e.g., chalk and sandstone) or areas with thick permeable soils, precipitation can infiltrate deep into the ground, reaching the groundwater reservoir, therefore strongly attenuating the rainfall signal and yielding a lagged rainfall-runoff response. The rainwater here can take months or even years to transit through the basin to reach rivers and outlets. ARs must be much wetter or more intense to generate floods in such settings; before significant surface runoff and flooding can begin, ARs need to deposit precipitation at rates or in volumes much larger than the infiltration capacities of the rocks or soils.

Non-bedrock surfaces also respond to ARs depending on their permeability and porosity characteristics, with gravels and clays exhibiting quick runoff responses: gravels because they allow quick infiltration but also quick transmission of precipitation down basin slopes to rivers, and clays because they resist infiltration, so that precipitation runs off the soil surfaces directly. Intermediate-permeability and intermediate-porosity soils accept infiltration but slow subsurface transmission of underflows (relative to gravel) to yield slower runoff rates and reduced AR-induced flood risks.

Land Use

Land use type and its heterogeneity have a major role in how an AR will affect a river basin. In rural basins that may contain forested areas, tree cover may reduce an AR's impact, through canopy interception and increased infiltration. Conversely, in urban areas where a large proportion of the basin could be covered in impermeable concrete, the rainfall-runoff response can be rapid, as storm drains transport rainwater to the river channel. An example of ARs affecting urban areas was the case of November 1983 (Liberato et al. 2012), when an extreme precipitation event occurred in Western Iberia in Portugal's Lisbon region. Flash flooding, urban inundations, and landslides resulted, causing considerable infrastructure damage (electric power blackouts as well as road and rail links blocked) and ten human fatalities. With a total of ~95 mm in 24 h observed at the longest-serving station in Portugal (Lisbon's Dom Luiz Observatory), this was the rainiest day during the twentieth century—and one of the rainiest registered since 1864. In this case, an AR and favorable thermodynamic conditions in combination with an impermeable urban surface triggered this flash flood event. Conversely, natural and regulated bodies of water (lakes and reservoirs) act to strongly attenuate the effects of AR precipitation.

5.6.3 Some Examples

Lavers et al. (2012) investigated the relationship between ARs and winter floods in nine western British river basins, and this research exemplified how strongly the land surface affects an incoming AR. In particular, two of the study basins (Fig. 5.22), although only on the order of 50 miles apart, had very different AR–flood relationships. Over the 1979–2010 study period, the Dyfi basin at Dyfi Bridge (basin area = 471.3 km^2) had more than 70% of the largest 31 floods caused by ARs, whereas at the Teifi basin at Glan Teifi (basin area = 893.6 km^2) only about 40% of the largest 31 floods were related to ARs (Fig. 5.23). This underscores the importance of the land surface, because given the basins' close proximity they are both likely to be affected by an AR when one occurs. The reasoning for the different relationships is currently an open question and research area, but the presence of a water reservoir in the Teifi together with less steep and lower relief (than the Dyfi) are hypothesized to be possible causes. In addition, the orientation of the river basin could be an important factor in the generation of a flood.

Different historical cases of flood events in the Iberian Peninsula from ARs have been investigated in recent years. In these different studies, the relevance of above-average precipitation before a major event depends on the temporal and spatial scales of the phenomenon. The December 1876 event (Trigo et al. 2014) along with the December 1909 event (Pereira et al. 2016) highlight the key role played by previously accumulated precipitation in floods lasting several days in the major river basins of the western Iberian Peninsula. As shown in Fig. 5.24, the December 1876 event had major socio-economic effects in the southwestern part of the Iberian Peninsula, while the December 1909 event occurred in the Northwestern Iberian Peninsula. The accumulated precipitation before the December 1876 was above the 95th historical percentile for the Lisbon weather station; for the December 1909 event, it was greater than the mean accumulated precipitation (Fig. 5.24b). In addition, these events occurred before the construction of major dams in the 1950s and 1960s. In the Duero and Tagus international basins, this altered river flow regimes by smoothing out larger peak values (Trigo et al. 2014). In contrast, floods occurring after intense bursts of pre-

Fig. 5.22 A map showing the location of the Dyfi and Teifi river basins in western Britain

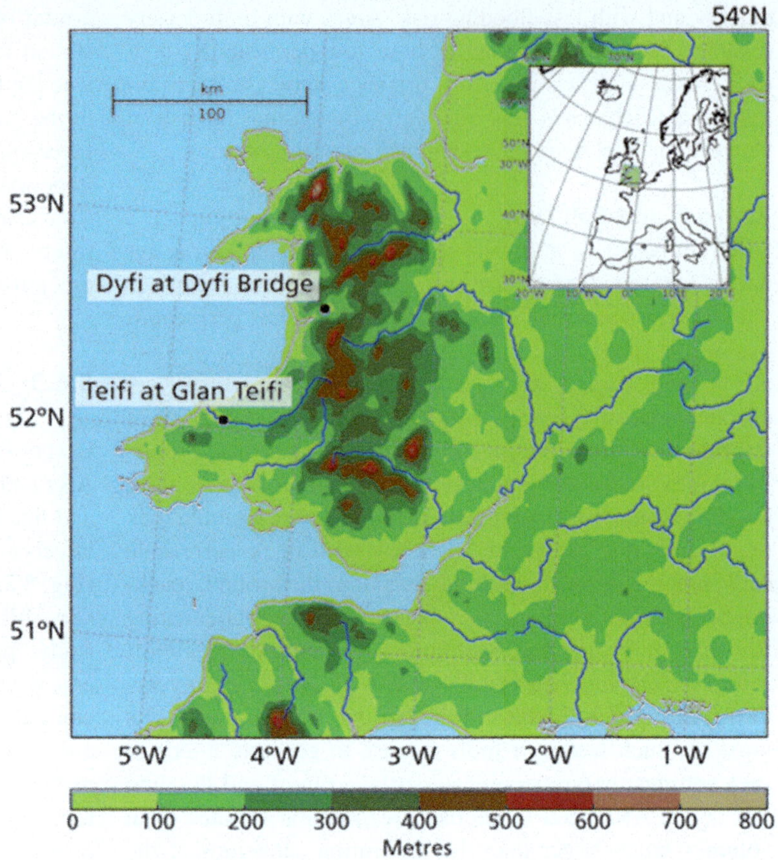

Fig. 5.23 The number of peaks-over-threshold (POT) floods over 1979–2010 related to ARs in five reanalyses in the **a** Dyfi at Dyfi bridge basin and **b** Teifi at Glan Teifi basin

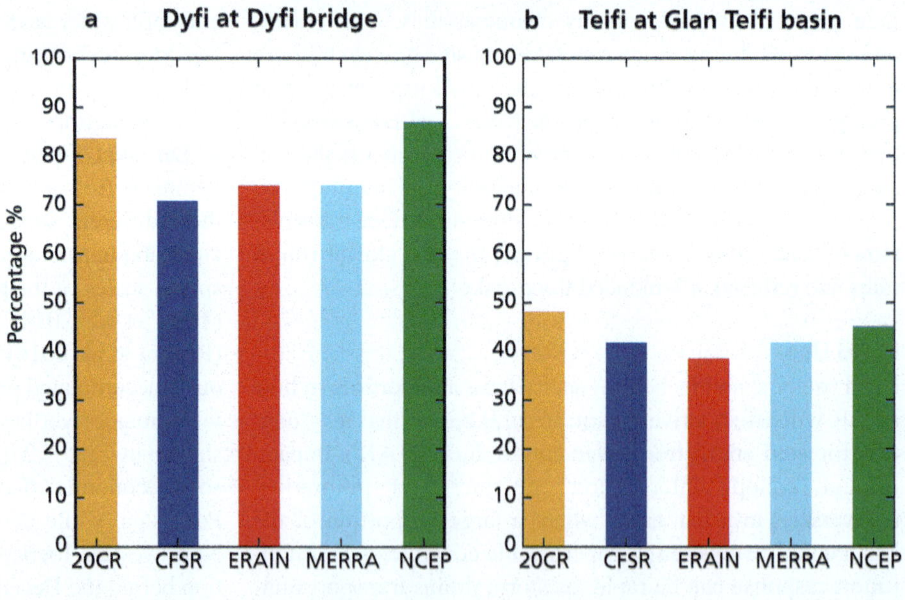

cipitation over relatively confined areas of Western Iberia can induce flash floods, independently of the precipitation that occurred in previous weeks or months, as in the flood 19 November 1983 (Liberato et al. 2012) that occurred during the night between 18 and 19 November 1967. Flooding was confined to the Lisbon metropolitan area (i.e., it did not affect large river basins). The accumulated precipitation before the event was not relevant (being below average), and the flash floods over urban areas occurred mainly as a result of very intense hourly precipitation on impermeable areas of ~32 mm in 8 h between 19 UTC and 3 UTC, and finally ~63 mm in 4 h between 3 UTC and 8 UTC.

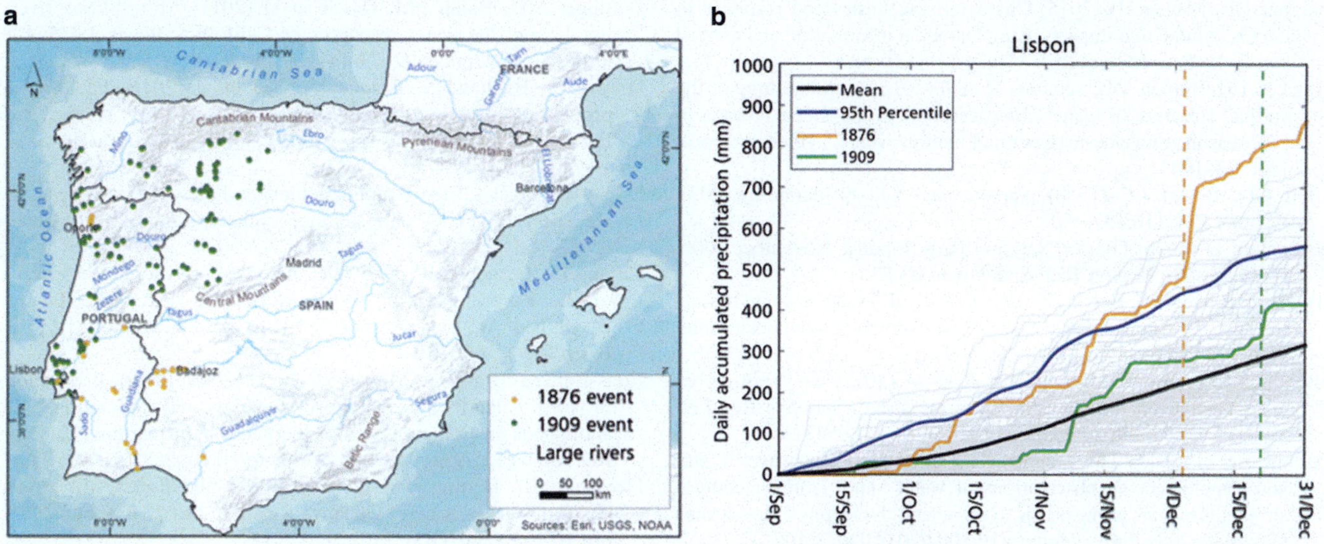

Fig. 5.24 (**a**) Locations affected by the Lisbon floods of December 1876 (*orange dots*) and December 1909 (*green dots*). (**b**) Cumulative precipitation from 1 September to 31 December using historical daily precipitation records from Lisbon between 1864 and 2009. Each year of cumulative precipitation is represented in *grey*; the long-term mean and the 95th percentile of precipitation are represented in *black* and *blue*, respectively. The Lisbon historical floods of December 1876 are in (*orange*) and 1909 (*green*)

5.7 Looking Forward

Despite the long list of effects discussed in this chapter, the numbers of locations where—and the different kinds of—AR effects and impacts have been studied and well documented remain limited. The list of effects—beyond AR floods and some AR water resources—discussed in this chapter (e.g., Table 5.1) will certainly look meager in years to come, because ARs are important parts of every natural and human system they impinge upon, almost certainly in many more ways than recognized and documented thus far. The intensity of the precipitation they often bring—and their unique combinations of durations, geometries, winds, and temperatures—combine to make ARs important and, in their ways, unique pieces of the climatology and landscapes that both natural systems and human societies have evolved to accommodate (to greater or lesser extent). Extreme AR episodes or changes in AR climatologies, however, still push systems past the levels accommodated to yield major impacts, threats, and changes to those same natural systems and human societies.

Looking forward, improvements in the ability to forecast AR storms and conditions offer some of the greatest opportunities for improved ecosystem, resource, and disaster management across all sorts of landscapes and societal needs, as "forewarned is forearmed." However, AR-informed management also requires understanding the roles ARs have played in the histories and evolution of organisms, landscapes, resources, hazards, and the social systems being managed. Without this sense of what these systems are evolved to accommodate and expect, management actions can cause as many problems as they solve. ARs have their own particular properties—physical and statistical—relative to other storms and weather phenomena, and recognition of their special roles in these systems allows more targeted responses to the risks they pose and the benefits they offer. To support the greatest use of those opportunities, to avoid risks where possible, and to more fully understand the natural and social systems involved, the explorations of AR effects begun in this chapter will need to be expanded in the coming years, to better characterize storms and better understand the full range of their effects in human and natural systems.

Acknowledgements Alexandre M. Ramos was supported by the Scientific Employment Stimulus 2017 from the Portuguese Science Foundation (Fundação para a Ciência e a Tecnologia, FCT), Portugal (CEECIND/00027/2017). Irina V. Gorodetskaya thanks FCT/ MCTES for the financial support to CESAM (UIDP/50017/2020+UIDB/50017/2020) through national funds.

References

Albano CM, Dettinger MD, McCarthy MI et al (2016) Application of an extreme winter storm scenario to identify vulnerabilities, mitigation options, and science needs in the Sierra Nevada mountains, USA. Nat Hazards 80:879–900. https://doi.org/10.1007/s11069-015-2003-4

Albano C, Dettinger M, Soulard C (2017) Influence of atmospheric rivers on vegetation productivity and fire patterns in the Southwestern US. J Geophys Res Biogeosci 13:11

Alexander MA, Scott JD, Swales D et al (2015) Moisture pathways into the intermountain west associated with heavy winter precipitation events. J Hydrometeorol 16:1184–1206

Archer DR, Fowler HJ (2015) Characterising flash flood response to intense rainfall and impacts using historical information and gauged data in Britain. J Flood Risk Manag 11:S121–S133

Backes TM, Kaplan ML, Schumer R et al (2015) A climatology of the vertical structure of water vapor transport to the Sierra Nevada in cool season atmospheric river precipitation events. J Hydrometeorol 16:1029–1047

Bell GD, Bosart LF (1988) Appalachian cold-air damming. Mon Weather Rev 116:137–161

Bernhardt D (2006) Glacier National Park flooding November 2006. NWS Western Region Tech Attachment 08-23

Bevis M, Businger S, Herring TA et al (1992) GPS meteorology: remote sensing of the atmospheric water vapor using the global positioning system. J Geophys Res 97(D14):75–94

Boening C, Lebsock M, Landerer F et al (2012) Snowfall-driven mass change on the East Antarctic ice sheet. Geophys Res Lett 39(21):L21501. https://doi.org/10.1029/2012GL053316

Bonne J et al (2015) The summer 2012 Greenland heat wave: in situ and remote sensing observations of water vapor isotopic composition during an atmospheric river event. J Geophys Res-Atmos 120:2970–2989. https://doi.org/10.1002/2014JD022602

Braun SA, Houze RA Jr, Smull BF (1997) Airborne dual-Doppler observations of an intense frontal system approaching the Pacific northwest coast. Mon Weather Rev 125:3131–3156

Brown LR, Komoroske LM, Wagner RW et al (2016) Coupled downscaled climate models and ecophysiological metrics forecast habitat compression for an endangered estuarine fish: PloS ONE 11(1): e0146724. http://doi.org/10.1371/journal.pone.0146724

Carter DA, Gage KS, Ecklund WL et al (1995) Developments in UHF lower tropospheric wind profiling at NOAA's Aeronomy Laboratory. Radio Sci 30:977–1001

Cheng BS, Chang AL, Deck A et al (2016) Atmospheric rivers and the mass mortality of wild oysters: insight into an extreme future? Proc R Soc B 283:20161462. https://doi.org/10.1098/rspb.2016.1462

Colle BA, Mass CF (1995) The structure and evolution of cold surges east of the Rocky Mountains. Mon Weather Rev 123:2577–2610

Compo GP et al (2011) The twentieth century reanalysis project. Q J R Meteorol Soc 137:1–28

Coplen T, Neiman P, White A et al (2008) Extreme changes in stable hydrogen isotopes and precipitation characteristics in a landfalling Pacific storm. Geophys Res Lett 35(21):L21808. https://doi.org/10.1029/2008GL035481

Coplen TB, Neiman PJ, White AB et al (2015) Categorization of northern California rainfall for periods with and without a radar brightband using stable isotopes and a novel automated precipitation collector. Tellus B 67:48. https://doi.org/10.3402/tellusb.v67.28574

Couto FT, Salgado R, Costa MJ et al (2015) Precipitation in the Madeira Island over a 10-year period and the meridional water vapour transport during the winter seasons. Int J Climatol 35:3748–3759. https://doi.org/10.1002/joc.4243

Daly C, Neilson RP, Phillips DL (1994) A statistical-topographic model for mapping climatological precipitation over mountainous terrain. J Appl Meteorol 33:140–158

Dettinger MD (2004) Fifty-two years of pineapple-express storms across the West Coast of North America. California Energy Commission PIER energy-related environmental research report CEC-500-2005-004. 15 p

Dettinger MD (2013) Atmospheric rivers as drought busters on the US west coast. J Hydrometeorol 14:1721–1732. https://doi.org/10.1175/JHM-D-13-02.1

Dettinger MD (2016) Historical and future relations between large storms and droughts in California. San Franc Estuary Watershed Sci 14(2):21

Dettinger MD, Hidalgo H, Das T et al (2009) Projection of potential flood regime changes in California. California Energy Commission Rep. CEC-500-2009-050-D, 68 p

Dettinger MD, Ralph FM, Das T et al (2011) Atmospheric rivers, floods, and the water resources of California. Water 3:455–478. http://www.mdpi.com/2073-4441/3/2/445/

Deverel S, Bachand S, Brandenberg SJ et al (2016) Factors and processes affecting levee system vulnerability. San Franc Estuary Watershed Sci 14(4). https://doi.org/10.15447/sfews.2016v14iss4/art3

Doyle SH et al (2015) Amplified melt and flow of the Greenland ice sheet driven by late-summer cyclonic rainfall. Nat Geosci 8:647–653

Durkee JD, Campbell L, Berry K et al (2012) A Synoptic Perspective of the Record 1-2 May 2010 Mid-South Heavy Precipitation Event. Bull Am Meteorol Soc 93:611–620. https://doi.org/10.1175/BAMS-D-11-00076.1

Eiras-Barca J, Brands S, Miguel-Macho G (2016) Seasonal variations in North Atlantic atmospheric river activity and associations with anomalous precipitation over the Iberian Atlantic margin. J Geophys Res Atmos 121:931–948. https://doi.org/10.1002/2015JD023379

Florsheim JL, Dettinger M (2007) Climate and floods still govern California levee breaks. Geophys Res Lett 34:L22403. https://doi.org/10.1029/2007GL031702

Florsheim J, Dettinger M (2015) Promoting atmospheric-river and snowmelt fueled biogeomorphic processes by restoring river-floodplain connectivity in California's Central Valley. In: Hudson P, Middelkoop H (eds) Geomorphic approaches to integrated floodplain management of lowland fluvial systems in North America and Europe. Springer, New York, pp 119–141. https://doi.org/10.1007/978-1-4939-2380-9_6

Fragoso M, Trigo RM, Pinto JG et al (2012) The 20 February 2010 Madeira flash-floods: synoptic analysis and extreme rainfall assessment. Nat Hazards Earth Syst Sci 12:715–730. https://doi.org/10.5194/nhess-12-715-2012

Garreaud R (2013) Warm winter storms in Central Chile. J Hydrometeorol 14:1515–1534

Gimeno L et al (2016) Major mechanisms of atmospheric moisture transport and their role in extreme precipitation events. Annu Rev Environ Resour 41:117–141. https://doi.org/10.1146/annurev-environ-110615-085558

Gorodetskaya IV, Tsukernik M, Claes K et al (2014) The role of atmospheric rivers in anomalous snow accumulation in East Antarctica. Geophys Res Lett 41:6199–6206. https://doi.org/10.1002/2014GL060881

Grumm RH (2011) New England record maker rain event of 29–30 March 2010. Electron J Oper Meteorol 12(4):1–31

Guan B, Waliser DE (2015) Detection of atmospheric rivers—evaluation and application of an algorithm for global studies. J Geophys Res-Atmos 120:12514–12535. https://doi.org/10.1002/2015JD0245257

Guan B, Molotch NP, Waliser DE et al (2010) Extreme snowfall events linked to atmospheric rivers and surface air temperature via satellite measurements. Geophys Res Lett 37:L20401. https://doi.org/10.1029/2010GL044696

Guan B, Waliser DE, Ralph FM et al (2016) Hydrometeorological characteristics of rain-on-snow events associated with atmospheric rivers. Geophys Res Lett 43:2964–2973. https://doi.org/10.1002/2016GL067978

Gutman SI, Sahm SR, Benjamin SG et al (2004) Rapid retrieval and assimilation of ground based GPS precipitable water observations at the NOAA forecast systems laboratory: impact on weather forecasts. J Meteorol Soc Jpn 82(1B):351–360

Hatchett BJ, Burak S, Rutz JJ et al (2017) Avalanche fatalities during atmospheric river events in the Western United States. J Hydrometeorol 18(5):13591–13374. https://doi.org/10.1175/JHM-D-16-0219.1

Healey M, Dettinger M, Norgaard R (2016) Perspectives on Bay-Delta science and policy. San Franc Estuary Watershed Sci 14(4):25. https://doi.org/10.15447/sfews.2016v14iss4art6

Herbst DB, Cooper SD (2010) Before and after the deluge—rain-on-snow flooding effects on aquatic invertebrate communities of small streams in the Sierra Nevada, California. J N Am Benthol Soc 29:1354–1366

Hirota N, Takayabu YN, Kato M et al (2016) Roles of an atmospheric river and a cutoff low in the extreme precipitation event in Hiroshima on 19 August 2014. Mon Weather Rev 144:1145–1160. https://doi.org/10.1175/MWR-D-15-0299.1

Hu H, Dominguez F, Gochis D (2016) Introducing water tracers in the Noah-MP land surface model (abstract). In: American Geophysical Union Fall Meeting

Hughes M, Neiman PJ, Sukovich D et al (2012) Representation of the sierra barrier jet in 11 years of a high-resolution dynamical reanalysis downscaling compared with long-term wind profiler observations. J Geophys Res 117:D18116. https://doi.org/10.1029/2012JD017869

Hughes M, Mahoney KM, Neiman PJ et al (2014) The landfall and inland penetration of a flood-producing atmospheric river in Arizona. Part II: sensitivity of modeled precipitation to terrain height and atmospheric river orientation. J Hydrometeorol 15:1954–1974

Jasperse J, Ralph M, Anderson M et al (2017) Preliminary viability assessment of Lake Mendocino forecast informed reservoir operations. Center For Western Weather and Water Extremes. 75 pp. Available from: https://pubs.er.usgs.gov/publication/70192184

Khouakhi A, Villarini G (2016) On the relationship between atmospheric rivers and high sea water levels along the US West Coast. Geophys Res Lett 43:8815–8822. https://doi.org/10.1002/2016GL070086

Kim S (2015) An assessment of atmospheric rivers as flood producers in Arizona. MS thesis (Hydrology), University of Arizona, 134 p

King MA, Bingham RJ, Moore P et al (2012) Lower satellite-gravimetry estimates of Antarctic Sea-level contribution. Nature 491(7425):586–589. https://doi.org/10.1038/nature11621

Kingsmill DE, Neiman PJ, Moore BJ et al (2013) Kinematic and thermodynamic structures of Sierra barrier jets and overrunning atmospheric rivers during a land-falling winter storm in northern California. Mon Wea Rev 141:2015–2036. https://doi.org/10.1175/1288MWR-D-12-00277.1

Kingston DG, Lavers DA, Hannah DM (2016) Floods in the southern Alps of New Zealand: the importance of atmospheric rivers. Hydrol Process 30(26):5063–5070. https://doi.org/10.1002/hyp.10982

Konrad CP, Dettinger MD (2017) Flood runoff in relation to water vapor transport by over the Western US, 1949–2015. Geophys Res Lett 44(22):11–456. https://doi.org/10.1002/2017GL075399

Lang S, Pierce H (2012) Heavy rains bring flooding to central and northern Italy. Tropical Rainfall Measuring Mission. Accessed November 19, 2012. https://earthobservatory.nasa.gov/images/79716/heavy-rains-in-northern-italy

Lavers DA, Villarini G (2013a) Atmospheric rivers and flooding over the Central United States. J Clim 26:7829–7836. https://doi.org/10.1175/JCLI-D-13-00212.1

Lavers DA, Villarini G (2013b) The nexus between atmospheric rivers and extreme precipitation across Europe. Geophys Res Lett 40:3259–3264

Lavers DA, Villarini G (2015) The contribution of atmospheric rivers to precipitation in Europe and the United States. J Hydrol 522:382–390

Lavers DA, Allan RP, Wood EF et al (2011) Winter floods in Britain are connected to atmospheric rivers. Geophys Res Lett 38:L23803. https://doi.org/10.1029/2011GL049783

Lavers DA, Villarini G, Allan RP et al (2012) The detection of atmospheric rivers in atmospheric reanalyses and their links to British winter floods and the large-scale climatic circulation. J Geophys Res 117:D20106. https://doi.org/10.1029/2012JD018027

Lavers DA, Pappenberger F, Zsoter E (2014) Extending medium-range predictability of extreme hydrological events in Europe. Nat Commun 5:5382

Lenaerts JTM, van Meijgaard E, van den Broeke MR et al (2013) Recent snowfall anomalies in Dronning Maud Land, East Antarctica, in a historical and future climate perspective. Geophys Res Lett 40(11):2684–2688. https://doi.org/10.1002/grl.50559

Levizzani V, Laviola S, Cattani E et al (2013) Extreme precipitation on the island of Madeira on 20 February 2010 as seen by satellite passive microwave sounders. Eur J Remote Sens 46:475–489

Liberato MLR, Ramos AM, Trigo RM et al (2012) Moisture sources and large-scale dynamics associated with a flash flood event. In: Lin J, Brunner D, Gerbig C et al (eds) Lagrangian modeling of the atmosphere. American Geophysical Union, Washington, DC. https://doi.org/10.1029/2012GM001244

Liu CJ, Barnes EA (2015) Extreme moisture transport into the Arctic linked to Rossby wave breaking. J Geophys Res Atmos 120:3774–3788

Loescher KA, Young GS, Colle BA et al (2006) Climatology of barrier jets along the Alaskan coast. Part I: spatial and temporal distributions. Mon Weather Rev 134:437–453

Lundquist JD, Neiman PJ, Martner BE et al (2008) Rain versus snow in the Sierra Nevada, California: comparing Doppler profiling radar and surface observations of melting level. J Hydrometeorol 9:194–211

MacWilliams ML, Ateljevich E, Monismith SG et al (2016) An overview of multi–dimensional models of the Sacramento–San Joaquin Delta. San Franc Estuary Watershed Sci 14(4):Article 2. https://doi.org/10.15447/sfews.2016v14iss4/art2

Mahoney K, Jackson DL, Neiman P et al (2016) Understanding the role of atmospheric rivers in heavy precipitation in the southeast United States. Mon Weather Rev 144:1617–1632. https://doi.org/10.1175/MWR-D-15-0279.1

Malguzzi P, Grossi G, Buzzi A et al (2006) The 1966 "century" flood in Italy: a meteorological and hydrological revisitation. J Geophys Res 111:D24106. https://doi.org/10.1029/2006JD007111

Marks D, Kimball J, Tingey D, Link T (1998) The sensitivity of snowmelt processes to climate conditions and forest cover during rain-on-snow—a case study of the 1996 Pacific Northwest flood. Hydrol Process 12:1569–1587

Marwitz J (1983) The kinematics of orographic airflow during Sierra storms. J Atmos Sci 40:1218–1227

Marwitz J (1987) Deep orographic storms over the Sierra Nevada. Part I: Thermodynamic and kinematic structure. J Atmos Sci 44:195–173

Masson-Delmotte V et al (2015) Recent changes in north-west Greenland climate documented by NEEM shallow ice core data and simulations, and implications for past-temperature reconstructions. Cryosphere 9:1481–1504

McCabe GJ, Clark MP, Hay LE (2007) Rain-on-snow events in the western United States. Bull Am Meteorol Soc 88(3):319–328

McMillan M, Shepherd A, Sundal A et al (2014) Increased ice losses from Antarctica detected by CryoSat-2. Geophys Res Lett 41(11):3899–3905. https://doi.org/10.1002/2014GL060111

Milrad SM, Gyakum JR, Atallah EH (2015) A meteorological analysis of the 2013 Alberta flood: antecedent large-scale flow pattern and synoptic–dynamic characteristics. Mon Weather Rev 143:2817–2841. https://doi.org/10.1175/MWR-D-14-00236.1

Mita C, Sfetsos A, Gounaris N et al (2014) Investigation of extreme precipitation in Greece: Can this phenomenon be attributed to atmospheric rivers. Conference: 12th International Conference on Meteorology, Climatology and Atmospheric Physics COMECAP 2014At: Heraklion, Greece. https://doi.org/10.13140/2.1.2086.5925

Mock, CJ, Birkeland KW (2000) Snow avalanche climatology of the western United States mountain ranges. Bull. Amer. Meteor. Soc., 81:2367–2392. https://doi.org/10.1175/1520-0477

Moore BJ, Neiman PJ, Ralph FM et al (2012) Physical processes associated with heavy flooding rainfall in Nashville, Tennessee, and vicinity during 1–2 May 2010: The role of an atmospheric river and mesoscale convective systems. Mon Weather Rev 140:358–378. https://doi.org/10.1175/MWR-D-11-00126.1

Nayak MA, Villarini G (2017) A long-term perspective of the hydroclimatological impacts of atmospheric rivers over the central US. Water Resour Res 53(2):1144–1166. https://doi.org/10.1002/2016WR019033

Nayak MA, Villarini G, Bradley AA (2016) Atmospheric rivers and rainfall during NASA's Iowa flood studies (IFloodS) campaign. J Hydrometeorol 17:257–271

Neff W, Compo GP, Ralph FM et al (2014) Continental heat anomalies and the extreme melting of the Greenland ice surface in 2012 and 1889. J Geophys Res-Atmos 119:6520–6536

Neiman PJ (2008) Atmospheric rivers and their role in generating heavy orographic precipitation and flooding along the US West Coast (abs). In: California extreme precipitation symposium. https://cepsym.org/proceedings_2008.php

Neiman PJ, Ralph FM, White AB et al (2002) The statistical relationship between upslope flow and rainfall in California's coastal mountains: observations during CALJET. Mon Weather Rev 130:1468–1492

Neiman PJ, Persson POG, Ralph FM et al (2004) Modification of fronts and precipitation by coastal blocking during an intense landfalling winter storm in Southern California: observations during CALJET. Mon Weather Rev 132:242–273

Neiman PJ, Ralph FM, Wick GA et al (2008) Meteorological characteristics and overland precipitation impacts of atmospheric rivers affecting the West Coast of North America based on eight years of SSM/I satellite observations. J Hydrometeorol 9(1):22–47

Neiman PJ, White AB, Ralph FM et al (2009) A water vapour flux tool for precipitation forecasting. Proc Inst Civ Eng Water Manag 162:83–94. https://doi.org/10.1680/wama.2009.162.2.83

Neiman PJ, Sukovich EM, Ralph FM et al (2010) A seven-year wind profiler–based climatology of the windward barrier jet along California's Northern Sierra Nevada. Mon Weather Rev 138:1206–1233

Neiman PJ, Schick LJ, Ralph FM et al (2011) Flooding in western Washington: the connection to atmospheric rivers. J Hydrometeorol 12:1337–1358. https://doi.org/10.1175/2011JHM1358.1

Neiman PJ, Hughes M, Moore BJ et al (2013a) Sierra barrier jets, atmospheric rivers, and precipitation characteristics in Northern California: a composite perspective based on a network of wind profilers. Mon Weather Rev 141:4211–4233

Neiman PJ, Ralph FM, Moore BJ et al (2013b) The landfall and inland penetration of a flood-producing atmospheric river in Arizona. Part 1: observed synoptic-scale, orographic, and hydrometeorological characteristics. J Hydrometeorol 14:460–484

Neiman PJ, Ralph FM, Moore BJ (2014a) The regional influence of an intense barrier jet and atmospheric river on orographic precipitation in Northern California: a case study. J Hydrometeorol 15:1419–1439

Neiman PJ, Wick GA, Moore BJ et al (2014b) An airborne study of an atmospheric river over the Subtropical Pacific during WISPAR: Dropsonde budget-box diagnostics, and precipitation impacts in Hawaii. Mon Weather Rev 142:3199–3223

Newell RE, Newell NE, Yong Z et al (1992) Tropospheric rivers? – a pilot study. Geophys Res Lett 19(24):2401–2404. https://doi.org/10.1029/92GL02916

Newman M, Kiladis GN, Weickmann KM et al (2012) Relative contributions of synoptic and low-frequency eddies to time-mean atmospheric moisture transport, including the role of atmospheric rivers. J Clim 25:7341–7361

Oakley NS, Lancaster JT, Kaplan ML et al (2017) Synoptic conditions associated with cool season post-fire debris flows in the transverse ranges of southern California. Nat Hazards 88:327–354

Pacific Climate Impacts Consortium (2013) Atmospheric rivers state of knowledge report. In: Proceedings summary of the BC atmospheric river events: state of the knowledge workshop, March 7, 2013, Victoria, BC, Canada. https://www.pacificclimate.org/sites/default/files/publications/Atmospheric%20Report%20Final%20Revised.pdf

Papineau J, Holloway E (2011) The nature of heavy rain and flood events in Alaska, Anchorage Forecast Office research papers, NOAA/NWS/ARH. http://citeseerx.ist.psu.edu/viewdoc/summary?doi=10.1.1.729.6444

Parish TR (1982) Barrier winds along the Sierra Nevada Mountains. J Appl Meteorol 21:925–930

Pereira S, Ramos AM, Zêzere JL et al (2016) Spatial impact and triggering conditions of the exceptional hydro-geomorphological event of December 1909 in Iberia. Nat Hazards Earth Syst Sci 16:371–390. https://doi.org/10.5194/nhess-16-371-2016

Persson POG, Shupe MD, Perovich D et al (2016) Linking atmospheric synoptic transport, cloud phase, surface energy fluxes, and sea-ice growth: observations of midwinter SHEBA conditions. Clim Dyn 49:1341–1364

Ralph FM, Dettinger MD (2011) Storms, floods, and the science of atmospheric rivers. Eos Trans AGU 92(32):265

Ralph FM, Dettinger MD (2012) Historical and national perspectives on extreme west coast precipitation associated with atmospheric Rivers during December 2010. Bull Am Meteorol Soc 93:783–790. https://doi.org/10.1175/BAMS-D-11-00188.1

Ralph FM, Neiman PJ, Kingsmill DE et al (2003) The impact of a prominent rain shadow on flooding in California's Santa Cruz Mountains: a CALJET case study and sensitivity to the ENSO cycle. J Hydrometeorol 4(6):1243–1264. https://doi.org/10.1175/1525-7541(2003)004<1243:TIOAPR>2.0.CO;2

Ralph FM, Neiman PJ, Wick GA (2004) Satellite and CALJET aircraft observations of atmospheric rivers over the eastern North-Pacific Ocean during the winter of 1997/98. Mon Weather Rev 132:1721–1745

Ralph FM, Neiman PJ, Wick GA et al (2006) Flooding on California's Russian River: role of atmospheric rivers. Geophys Res Lett 33:L13801. https://doi.org/10.1029/2006GL026689

Ralph FM, Coleman T, Neiman PJ et al (2013) Observed impacts of duration and seasonality of atmospheric river landfalls on soil moisture and runoff. J Hydrometeorol 14(2):443–459. https://doi.org/10.1175/JHM-D-12-076.1

Ralph FM, Rutz J, Cordeira J et al (2019) A scale to characterize the strength and impacts of atmospheric rivers. Bull Am Meteorol Soc 100:269–289. https://doi.org/10.1175/BAMS-D-18-0023.1

Ramos AM, Trigo RM, Liberato MLR et al (2015) Daily precipitation extreme events in the Iberian Peninsula and its association with atmospheric rivers. J Hydrometeorol 16:579–597. https://doi.org/10.1175/JHM-D-14-0103.1

Rao PC, Archana S, Jaswal AK (2016) Intense precipitation causing floods over Himalayan region of northern India–a case study on role of atmospheric rivers. J Ind Geophys Union 20(2):191–200

Reynolds DW, Dennis AS (1996) A review of the sierra cooperative pilot project. Bull Am Meteorol Soc 67:513–523

Rosenfeld D, Woodley WL (2000) Deep convective clouds with sustained supercooled liquid water down to −37.5 °C. Nature 405:440–442

Rössler O, Froidevaux P, Börst U et al (2014) Retrospective analysis of a nonforecasted rain-on-snow flood in the Alps – a matter of model limitations or unpredictable nature? Hydrol Earth Syst Sci 18:2265–2285. https://doi.org/10.5194/hess-18-2265-2014

Rutz JJ, Steenburgh WJ (2012) Quantifying the role of atmospheric rivers in the interior western United States. Atmos Sci Lett 13:257–261

Rutz JJ, Steenburgh WJ, Ralph FM (2014) Climatological characteristics of atmospheric rivers and their inland penetration over the western United States. Mon Weather Rev 142:905–921

Rutz JJ, Steenburgh WJ, Ralph FM (2015) The inland penetration of atmospheric rivers over western North America: a Lagrangian analysis. Mon Weather Rev 143:1924–1944

Schoellhamer D, Bergamaschi B, Monismith S et al (2016) Recent advances in understanding flow dynamics and transport of water-quality constituents in the Sacramento–San Joaquin River Delta. San Franc Estuary Watershed Sci 14(4):Article 1. https://doi.org/10.15447/sfews.2016v14iss4/art1

Serreze MC, Crawford AD, Barrett AP (2015) Extreme daily precipitation events at Spitsbergen, an Arctic Island. Int. J. Climatol. 35: 4574-4588. https://doi.org/10.1002/joc.4308

Shepherd A et al (2012) A reconciled estimate of ice-sheet mass balance. Science 338(6111):1183–1189. https://doi.org/10.1126/science.1228102

Smutz SW (1986) A climatology of the Sierra Nevada barrier jet. MS thesis, Paper AS-153, Dept. of Atmospheric Science, University of Wyoming, 108 pp

Sodemann H, Stohl A (2013) Moisture origin and meridional transport in atmospheric rivers and their association with multiple cyclones. Mon Weather Rev 141:2850–2868

Sommer T, Jarre WC, Muller Solger A et al (2004) Effects of flow variation on channel and floodplain biota and habitats of the Sacramento River, California, USA. Aquat Conserv Mar Freshwat Ecosyst 14:247–261

Stamos CL, Martin P, Nishikawa T et al (2001) Simulation of groundwater flow in the Mojave River basin, California. In: US Geological Survey Water-Resources Investigations Report 01-4002. 137 p

Stohl A, Forster C, Sodemann H (2008) Remote sources of water vapor forming precipitation on the Norwegian west coast at 60°N–a tale of hurricanes and an atmospheric river. J Geophys Res 113:D05102. https://doi.org/10.1029/2007JD009006

Storelvmo T, Tan I (2015) The Wegener-Bergeron-Findeisen process–its discovery and vital importance for weather and climate. Meteorol Z 24:455–461

Trapero L, Bech J, Duffourg F et al (2013) Mesoscale numerical analysis of the historical November 1982 heavy precipitation event over Andorra (Eastern Pyrenees). Nat Hazards Earth Syst Sci 13:2969–2990. https://doi.org/10.5194/nhess-13-2969-2013

Trigo RM, Varino F, Ramos AM et al (2014) The record precipitation and flood event in Iberia in December 1876: description and synoptic analysis. Front Earth Sci 2:3. https://doi.org/10.3389/feart.2014.00003

Viale M, Norte FA (2009) Strong cross-barrier flow under stable conditions producing intense winter orographic precipitation: a case study over Subtropical Central Andes. Weather Forecast 24:1009–1031

Viale M, Nuñez MN (2011) Climatology of winter orographic precipitation over the Subtropical Central Andes and associated synoptic and regional characteristics. J Hydrometeorol 12:481–507

Viale M, Houze RA Jr, Rasmussen KL (2013) Upstream orographic enhancement of a narrow cold frontal rainband approaching the Andes. Mon Weather Rev 141:1708–1730

Wada Y, van Beek LPH, van Kempen CM et al (2010) Global depletion of groundwater resources. Geophys Res Lett 37:L20402. https://doi.org/10.1029/2010GL044571

Waliser D, Guan B (2017) Extreme winds and precipitation during landfall of atmospheric rivers. Nat Geosci 10:179–183

Wayand NE, Lundquist JD, Clark MP (2015) Modeling the influence of hypsometry, vegetation, and storm energy on snowmelt contributions to basins during rain-on-snow floods. Water Resour Res 51:8551–8569. https://doi.org/10.1002/2014WR016576

White AB, Gottas DJ, Strem E et al (2002) An automated bright-band height detection algorithm for use with Doppler radar vertical spectral moments. J Atmos Ocean Technol 19:687–697

White AB, Neiman PJ, Creamean JM et al (2015) The impacts of California's San Francisco Bay Area gap on precipitation observed in the Sierra Nevada during HMT and CalWater. J Hydrometeorol 16:1048–1069

Whiteman CD (2000) Mountain meteorology: fundamentals and applications. Oxford University Press, Hong Kong, 355 pp

Yoshimura K, Kanamitsu M, Dettinger M (2010) Regional downscaling for stable water isotopes—a case study of an atmospheric river event. J Geophys Res 115:D18114. https://doi.org/10.1029/2010D014032

Young AM, Skelly KT, Cordeira JM (2017) High-impact hydrologic events and atmospheric rivers in California: an investigation using the NCEI storm events database. Geophys Res Lett 44:3393–3401. https://doi.org/10.1002/2017GL073077

Yu C-K, Smull BF (2000) Airborne Doppler observations of a landfalling cold front upstream of steep coastal orography. Mon Weather Rev 128:1577–1603

Zawadzki I, Fabry F, Szyrmer W (2001) Observations of supercooled water and secondary ice generation by a vertically pointing X-band Doppler radar. Atmos Res 59–60:343–359

Atmospheric River Modeling: Forecasts, Climate Simulations, and Climate Projections

6

Duane E. Waliser and Jason M. Cordeira

6.1 Introduction

The previous chapters have discussed and illustrated the importance of ARs to the development and occurrence of weather and hydrology extremes, as well as the critical role they play in helping establish the Earth's climate—namely its water and energy budgets—through their decisive effect on the poleward transport of water vapor. With this in mind, it is imperative that ARs, and their associated dynamics, thermodynamics, and hydrodynamics (herein referred to as "ingredients" in the Regional Models discussion in Sect. 6.3.1), be accurately represented in both space and time in weather and climate models. Although modeling studies in the past have targeted precipitation extremes (e.g., "Pineapple Express" events) or mid-latitude cyclone and extratropical dynamics that are associated with what the meteorological community would now readily refer to as ARs, this chapter targets modeling literature on phenomena specifically identified as "atmospheric rivers."

The demand for accurate weather forecasts, and the associated warnings of hydrologic and high-impact hazardous events (e.g., flooding, landslides, drought relief, high winds; see Chap. 7), comprises the most notable and recurring need for realistic model representations of ARs. Although the synoptic-scale dynamics of mid-latitude baroclinic winter storms are relatively well understood and predicted, their expressions in the form of ARs—including meso-scale dynamics and precipitation processes—are not as well understood or predicted. In comparison with more traditional considerations of mid-latitude weather systems, an information gap remains for quantitative measures of model performance and the associated process-level understanding for ARs: the hydrodynamics of the storm and its means of producing coastal—and, in some cases, inland—precipitation, often in association with complex topography. Although some work has focused on AR forecast skill in limited regional settings, less research has been devoted to understanding and estimating the underlying and intrinsic predictability associated with ARs. The latter is necessary to gauge how well operational systems function, relative to an estimated upper theoretical bound. To further advance AR forecast capabilities, current global forecast skill capabilities must be understood and quantified—and put in the context of analogous estimates of predictability, including their geographic, seasonal, inter-annual, and other multi-scale sensitivities. Section 6.2 reviews research performed to date on topics related to forecasting and predicting ARs for short- to extended-range (i.e., 1- to 15-day) weather forecasts.

Beyond attempting to predict the evolution and details of a specific AR event via the models and associated machinery used in operational weather forecast systems, the character, variability, and effects of ARs in climate simulations and predictions need to be represented. Developing accurate subseasonal-to-seasonal (S2S) predictions that accurately represent AR activity entails first examining and quantifying how well such variability is represented in climate model simulations, both in limited area or regional models and in global climate models (GCMs). In this context, AR evaluation can take advantage of many sorts of observation data products, but typically with the constraint that specific events are not scrutinized against observations but rather that the observed and modeled ARs are compared in terms of their statistical character across many ARs. Demonstrating that climate models exhibit realistic AR behavior enables confidence that the versions of these models used for S2S prediction and long-term climate projections represent key features of the climate, its variability, and its extremes. Section 6.3 reviews the limited amount of research and model-evaluation studies that have examined and quantified how well climate models represent ARs.

D. E. Waliser
Jet Propulsion Laboratory, California Institute of Technology, Pasadena, CA, USA
e-mail: duane.waliser@jpl.nasa.gov

J. M. Cordeira (✉)
Plymouth State University, Plymouth, NH, USA
e-mail: j_cordeira@plymouth.edu

F. M. Ralph et al. (eds.), *Atmospheric Rivers*, https://doi.org/10.1007/978-3-030-28906-5_6

By extension, the more that realism and systematic biases in climate models are understood, the better that uncertainty in climate model projection of ARs and their effects can be characterized and understood. For example, knowing how the spatial and temporal distribution and amplitude of extreme precipitation events will change over the next century is critical. Moreover, how will the poleward transports of moisture—which helps determine the planet's energy and water budgets—change in the next century? The answer to each of these questions strongly depends on how climate models project ARs will change. Section 6.4 reviews research that has examined contemporary sets of climate projections to quantify how ARs and related fields are expected to change. Section 6.5 briefly summarizes and articulates the gaps in AR simulation, multi-time-scale forecasting, and climate projections.

6.2 Forecasting ARs

Annual precipitation over California varies far more than in most US states. Eighty-five percent of the variance in annual precipitation in Northern California results from annual variations in the top 5% wettest days per year (Dettinger and Cayan 2014), which are mostly attributed to orographic-enhanced precipitation that accompanies winter storms in association with ARs. These extreme precipitation events constitute a large portion of annual precipitation (~30–50%), and play a primary role in water resources management and water supply across the Western US (e.g., Dettinger et al. 2011). Similarly, ARs are found to contribute equally large fractions of annual precipitation in many other areas of the globe, particularly mid-latitude portions of Africa, Australia, East Asia, Europe, New Zealand, South America, and southern Greenland (Guan and Waliser 2015). The ability to forecast the occurrence of extreme precipitation associated with ARs over these areas is, therefore, critical to mitigate and plan for both detrimental (flooding) and beneficial (water supply) effects of this precipitation (see also Chap. 5, Sect. 6.1 of this chapter, and Chap. 7; Guan et al. 2010; Dettinger et al. 2011). Using mainly a US West Coast perspective, an "ingredient-based" approach to forecasting ARs is described below, including emerging forecasting tools and the methods and results of evaluating present-day AR forecasts (see also Sect. 6.5).

6.2.1 An Ingredient-Based Approach to Forecasting ARs

In general, various aspects of mid-latitude weather phenomena—including ARs, precipitation, and their effects—are often forecasted through ingredient-based methodologies compiled from observations or from numerical weather prediction (NWP) model guidance. The concept of an ingredient-based methodology was developed for severe convection by Doswell (1987), Johns and Doswell (1992), Doswell et al. (1996), and Schultz and Schumacher (1999), and follows from a breadth of research on conceptualization of observations of multi-scale weather phenomena pioneered by Miller (1946), among others. The ingredient-based methodology derived by Doswell et al. (1996) specifically applies to the amount of precipitation and the potential for flash flooding in any one event that is related to (1) a high precipitation rate and (2) a long duration of precipitation. These necessary—but not always sufficient—ingredients for flash-flood-producing rainfalls require some combination of an abundance and long duration of atmospheric moisture, instability, and upward vertical motion such as that found within a quasi-stationary meso-scale convective system in an air mass characterized by high integrated water vapor (IWV) and large convective available potential energy (CAPE). This methodology of Doswell et al. (1996) lends itself well to forecasting ARs and AR-related precipitation; the abundance and long duration of these three ingredients associated with ARs are discussed below.

The abundance of atmospheric moisture along ARs in the warm sector of winter storms is often characterized by filamentary regions of enhanced IWV values (>20 mm) and enhanced vertically IWV transport (IVT > 250 kg m^{-1} s^{-1}) discussed by Rutz et al. (2014) for locations over the North Pacific and US West Coast. The majority (75%) of IVT within ARs is also confined to the lower 2.25 km of the troposphere, where heavy orographic precipitation may result in regions that intersect mountainous terrain, such as along the US West Coast (Ralph et al. 2005). These characteristic values of IWV and IVT magnitude represent diagnostics that depict regions of concentrated moisture and moisture flux that may influence precipitation production in regions of complex terrain. These diagnostics are, therefore, considered necessary for—but not always sufficient for—precipitation production. In other words, absent a lifting mechanism, enhanced IWV and IVT may not produce precipitation.

The atmospheric stability associated with landfalling ARs is typically characterized, at least over the North Pacific Ocean and across the US West Coast, as moist-neutral and stably stratified (Neiman et al. 2013a). Unlike flash-flood-producing meso-scale convective systems over the central US, buoyant convective ascent is not common in landfalling ARs; there is low CAPE, and there may not be a level of free convection. In these environments, precipitation results from dynamically or topographically forced saturated ascent. The latter is often influenced by strong atmospheric stability and sometimes the production of southerly mountain-parallel barrier jets along the windward slopes of the California Coastal Ranges and Sierra Nevada (Neiman et al. 2010, 2013a, 2014; Kingsmill et al. 2013) that can affect the spatial distribution of precipitation (e.g., Ralph et al. 2016a).

The lifting mechanism associated with US West Coast landfalling ARs is attributed primarily to the rising motion of saturated air induced by orographic ascent (Neiman et al. 2008). The so-called bulk-upslope water vapor flux explains up to 75% of the variance in total precipitation that results from forced saturated ascent along ARs (Ralph et al. 2006). The upslope IVT pertains to the component of water vapor transport that is directed along the gradient of the local terrain. Total precipitation associated with a landfalling AR is, therefore, influenced by the slope and aspect of the terrain and terrain-normal wind speeds. For example, Neiman et al. (2011) illustrated that flooding in watersheds in Western Washington state often occurs in association with landfalling ARs that are oriented in the same direction as the watershed-relative terrain gradient (cf. their Fig. 15). Ralph et al. (2013a) further illustrated that stronger terrain-normal bulk-upslope water vapor flux values are associated with higher precipitation rates in landfalling ARs. Other possible lifting mechanisms associated with a landfalling AR may include rising motion related to synoptic-scale quasi-geostrophic (QG) forcing for ascent that may accompany the parent mid-latitude cyclone that could relate to the isentropic ascent along a warm conveyor belt (Sodemann and Stohl 2013) or, similarly, relate to meso-synoptic-scale forcing for ascent related to frontal circulations (Cordeira et al. 2013; Neiman et al. 2013b). The balance of these and other lifting mechanisms for ARs that occur over other parts of the globe have still received limited study.

The duration of enhanced IWV and terrain-normal IVT strongly influences the duration of juxtaposed AR ingredients and total AR-related precipitation at any one location. The duration of juxtaposed "AR ingredients," which are sometimes referred to as "AR conditions," typically persists for ~12 to 24 h along the US West Coast (Rutz et al. 2014), with an average duration of ~20 h at Bodega Bay, California (Ralph et al. 2013a) that can last >72 h (Rutz et al. 2014). The durations of some ARs are also influenced by the so-called meso-scale frontal waves, which may cause regions of enhanced IWV, terrain-normal IVT, and forced saturated ascent to persist (Ralph et al. 2011).

Forecasts of AR-related precipitation along the US West Coast and other locations around the world thus need to account for the location, intensity, orientation, and vertical and spatial distributions of atmospheric water vapor and water vapor flux. Forecasts of the effects of AR-related precipitation such as flooding or increases in water supply need also to take into account the duration of enhanced IWV and IVT (see Chap. 7). As identified by Doswell et al. (1996), these ingredients—which may vary in relative importance across the globe—are considered necessary, but are not always sufficient conditions for precipitation and effects such as flooding. Additional ingredients related to forcing for ascent, freezing level, and precipitation phase or any local conditions (e.g., antecedent soil moisture) may also influence outcomes associated with individual AR landfalls.

6.2.2 Evaluating Forecasts of Landfalling ARs

Quantitative evaluations of forecasts of landfalling ARs and their mid-latitude cyclones over the northeast Pacific are generally lacking (Wick et al. 2013b). A majority of cool-season precipitation forecasts over the Western US inherently accounts for landfalling ARs through their parent mid-latitude cyclones. The synoptic-scale dynamics of these mid-latitude cyclones—and how well NWP models forecast their sea level pressure and cyclone characteristics—are generally well understood (e.g., McMurdie and Mass 2004; McMurdie and Casola 2009; Froude 2010). As a result, several studies indicate that NWP forecasts of synoptic-scale IVT distributions are skillful for lead times of 7 to 9 days over the North Atlantic (Lavers et al. 2014), over the North Pacific (Lavers et al. 2016), and over the central US (Nayak et al. 2014). Though these NWP forecasts predict mid-latitude cyclones and IVT distributions well, they fail to reliably produce accurate forecasts of the meso-scale features of these mid-latitude cyclones that are crucial to skillful quantitative precipitation forecasts (QPF) associated with landfalling ARs.

Inherent bias and error exist in both NWP analyses and forecasts, in part from a lack of detailed observations of ARs over the upstream ocean basins. For example, Doyle et al. (2014) identified that the largest source of errors in initial conditions for a numerical prediction of a high-impact mid-latitude cyclone over Europe is related to uncertainty in the position of enhanced water vapor downstream of a 700-hPa trough, along a warm conveyor belt (WCB), and in the vicinity of an AR offshore 48 h before verification (Fig. 6.1). Several field campaigns (discussed previously in this book) have tried through observations to validate NWP analyses of water vapor and water vapor transport, and satellite-derived estimates (e.g., Special Sensor Microwave/Imager [SSM/I]-derived IWV) along ARs, including the National Oceanic and Atmospheric Administration (NOAA)-led Winter Storms and Pacific Atmospheric Rivers (WISPAR) experiment in February–March 2011 and the twin CalWater field campaigns in 2009–2011 and 2014–2015, respectively (Ralph et al. 2016b; Cordeira et al. 2017). These field campaigns identified errors in the width, intensity, and total water vapor transport along ARs analyzed in NWP analyses over the ocean as compared to dropsonde observations (Ralph et al. 2017); variability in these errors related directly to NWP model grid spacing (Jackson et al. 2016) and to accurately resolving meso-scale features along ARs.

The meso-scale features associated with landfalling ARs (e.g., width, intensity, and landfall location) are forecast poorly compared to the overall occurrence of ARs (Wick

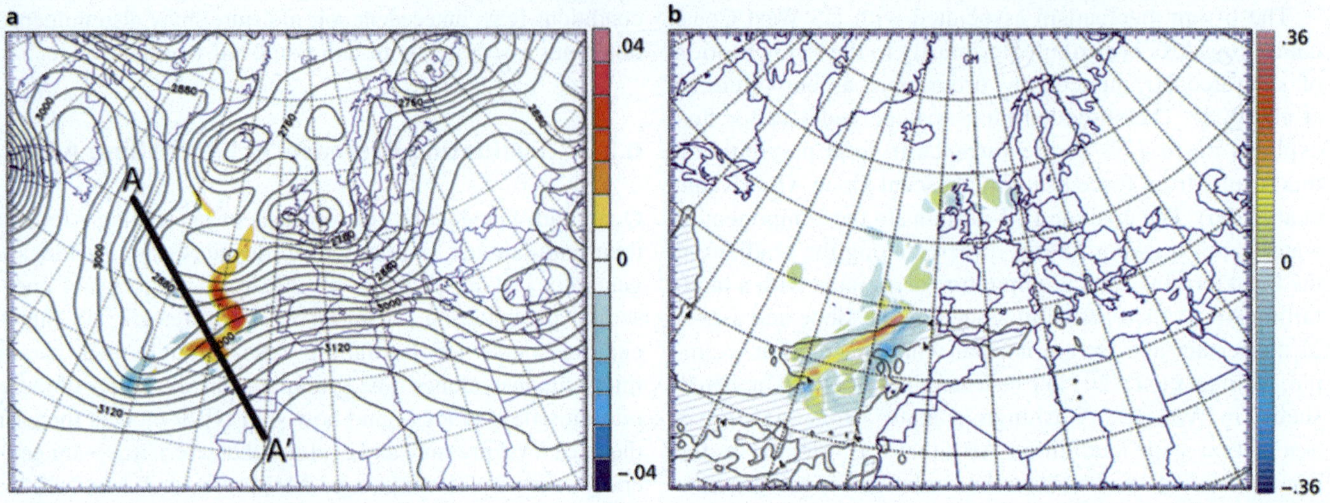

Fig. 6.1 Numerical weather prediction (NWP) model adjoint sensitivity valid at the initial time of 1200 UTC 26 Feb 2010 at 700 hPa for the (**a**) y-wind component (*color scale* with interval every 0.005 m s^{-1}), (**b**) water vapor (*color interval* every 0.01 m^2 s^{-1} [g $kg]^{-1}$). The 700-hPa geopotential height analysis is shown in (**a**) with an interval of 30 m. Water vapor greater than 4 g kg^{-1} is shown by the hatching in (**b**). The sensitivities are scaled by 105 km^{-3}. The location of a vertical cross-section is indicated by the *solid black line* in (**a**), and is not shown in this text (Figure 7a and b in Doyle et al. 2014)

et al. 2013b). The Wick et al. study (2013b) of three cool seasons of ARs over the eastern North Pacific illustrates that the overall occurrence of ARs is well forecast out to a 9- to 10-day lead time (Fig. 6.2a), but landfall occurrence and errors in landfall position are forecast more poorly, with decreasing skill as a function of increasing lead time (Fig. 6.2b). Average errors in the landfall position of ARs along the US West Coast exceed 600–900 km at 7- to 10-day lead times for all NWP models tested (Fig. 6.2c) with a southward bias in landfall latitude of 1°–3° (Fig. 6.2d). These position errors exceed typical length scales of meso-scale phenomena (Fujita 1986). The forecast width of ARs is also generally 40–60 km too wide, resulting primarily from model resolution (Fig. 6.2e); the higher-resolution European Centre for Medium-Range Weather Forecasting (ECMWF) model forecasts contained the least bias in forecast width. The forecast IWV content along the AR axis was also found by Wick et al. (2013b) to possess a slight moist bias (Fig. 6.2f).

These errors in forecast quantities primarily from the location and structure of ARs can have a large influence on model-derived QPF; landfall location errors of 500 km can have vast implications for local orographic forcing and which watersheds receive the highest precipitation. And although it is common to use NWP-model-derived QPF to predict AR-related precipitation extremes, meso-scale features of landfalling ARs and grid- and sub-grid-scale variations in topography (e.g., elevation and aspect) can result in precipitation errors. Analysis of QPF errors by Ralph et al. (2010) demonstrated that some of the largest QPF errors along the US West Coast were associated with landfalling ARs. The QPF forecast statistics of >25.4 mm [24 h] $^{-1}$ produced by human forecasts at the NOAA National Centers for Environmental Prediction (NCEP) Weather Prediction Center (WPC) are well known for their higher skill (higher threat score) compared to NWP-produced QPF forecasts of the same threshold NWP (Sukovich et al. 2014). These successes relate largely to consistent and quality-controlled verification to assess forecast trends and bias, which enables enhanced situational awareness under landfalling AR conditions.

During the "the 2012 Atmospheric River Retrospective Forecasting Experiment" (WPC 2012), enhanced situational awareness related to knowledge of local topography, climate, and seasonal precipitation regimes along the US West Coast was deemed key to successful AR-related QPF. This experiment strove to identify potential methods to improve forecasts of AR-induced extreme precipitation events along the US West Coast. A second key outcome related to situational awareness stated, "Model forecasts of moisture parameters may be helpful in identifying the potential for extreme [precipitation] events, even when the model QPF does not forecast large precipitation amounts."

Correlation analysis of upslope water vapor flux and precipitation by Ralph et al. (2013a) and of IVT magnitude and precipitation by Rutz et al. (2014) has identified that water vapor flux (or IVT) is a key moisture parameter that may help identify the potential for extreme precipitation events associated with landfalling ARs. The IVT vector is also highly useful in identifying the location, orientation, and intensity of ARs before their landfall along the US West Coast. A pair of studies by Lavers et al. (2014, 2016) using the "potential predictability" methodology identified that the NWP forecast of the IVT distribution is potentially more predictive than the corresponding NWP QPF over the

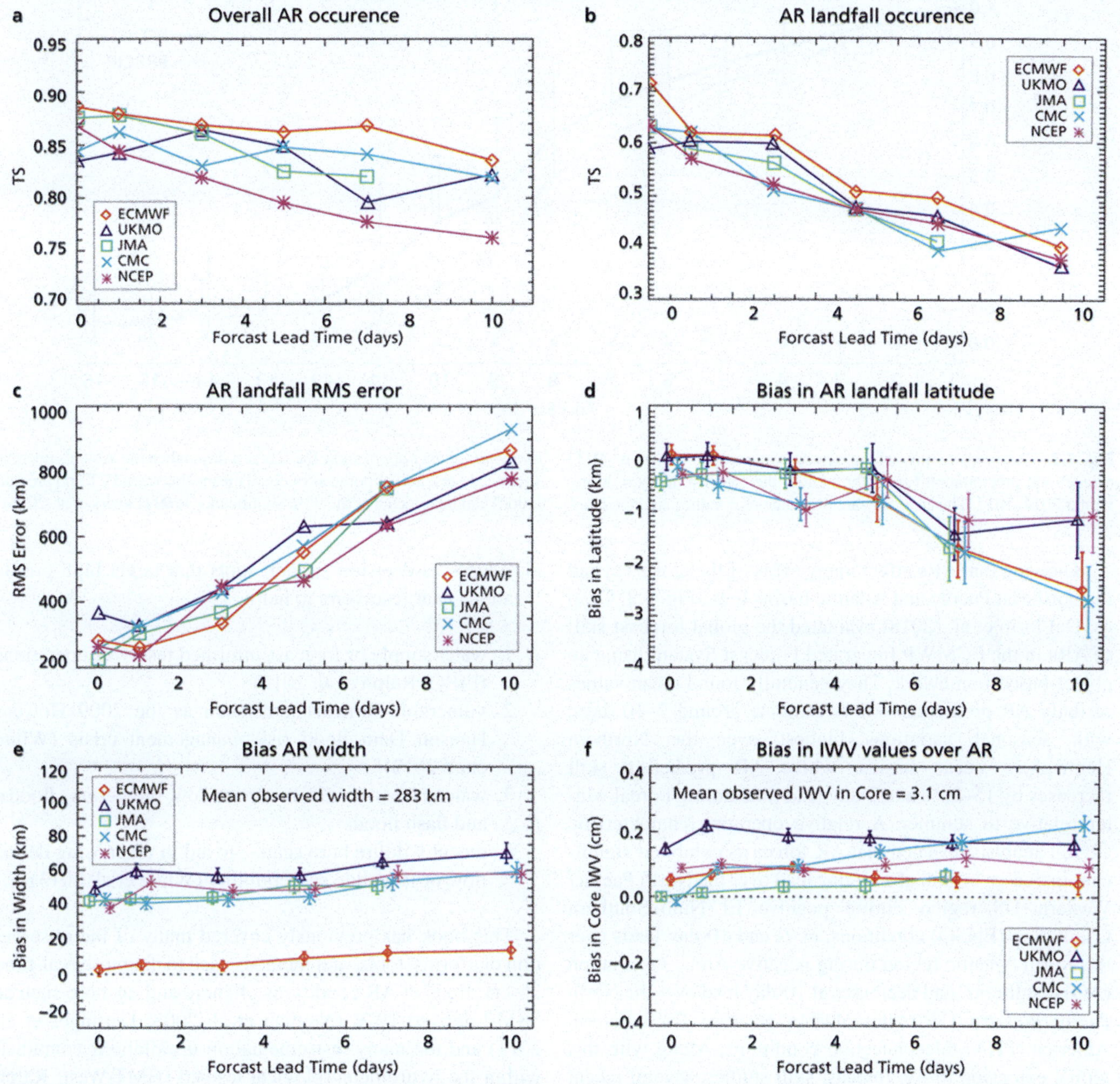

Fig. 6.2 Summary for three cool seasons of forecast verification statistics that reflect the ability of five numerical weather prediction (NWP) models to predict (**a**) the overall occurrence of at least one AR (threat score; TS) somewhere within their North Pacific analysis domain on a given day as a function of forecast lead time, (**b**) the prediction of at least one landfalling AR within the domain on a given day (TS), (**c**) estimates of error in forecast AR landfall location as a function of lead time, (**d**) total root-mean-square (RMS) error (km) in the detected landfall location along the US West Coast, (**e**) bias in the forecast AR width relative to satellite-derived observations computed for an average over the entire length of the AR, and (**f**) bias in the modeled AR strength as represented by the integrated water vapor (IWV) content along the axis of the AR relative to satellite-derived observations computed for an average over the entire length of the AR. Note that (**d**) is annotated with error bars that represent ±1 standard deviation of the mean to reflect uncertainty in the mean value, and with the horizontal position of the points offset slightly to avoid overlap of the error bars (Adapted from Wick et al. 2013b with more information therein)

North Atlantic and North Pacific oceans (Fig. 6.3). The methodology suggests that IVT may, on average, be more predictive than QPF with ~1 to 2 days of lead time (using their measure of skill, r^2, decreasing below 0.5). These results suggest that NWP-forecast IVT might provide enhanced situational awareness for ARs over the North Pacific and North Atlantic before landfall along the US and European west coasts. Section 7.5 in Chap. 7 includes further discussion of the use of these IVT considerations as a forecast index.

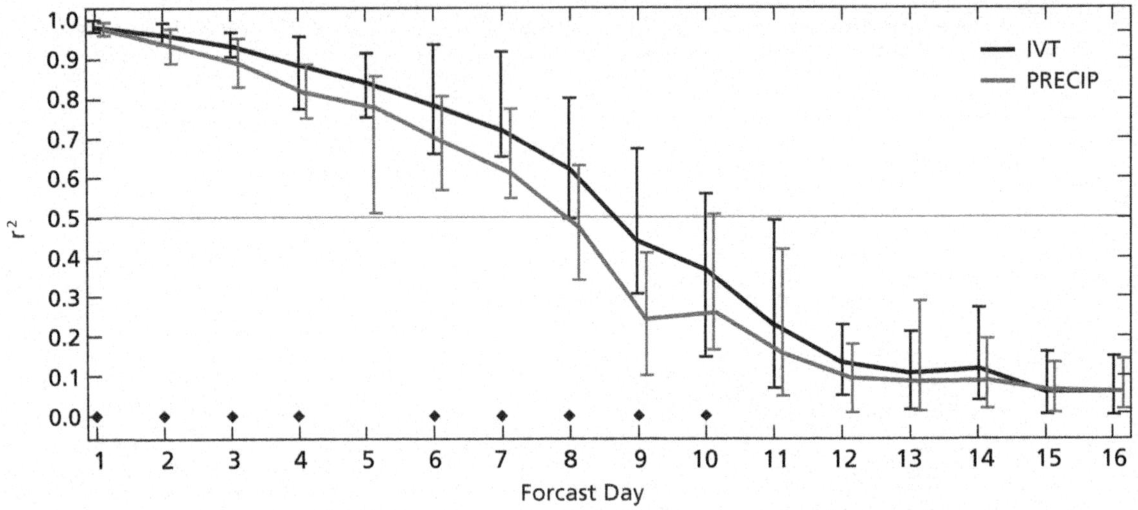

Fig. 6.3 The predictability of integrated water vapor transport (IVT) (*black*) and precipitation (*gray*) throughout the forecast horizon during winter 2014–2015. The error bars are the range of r^2 values, and the *black diamonds at the bottom* show the forecast days when the Welch's unequal variances *t*-test null hypothesis of equal means (between IVT and precipitation) can be rejected at the 95% significance level (Lavers et al. 2016)

Knowing that ARs affect many areas of the globe beyond the northeast Pacific and Atlantic basins (e.g., Fig. 4.2, Chap 4), DeFlorio et al. (2018) evaluated the global forecast skill of ARs in the ECMWF Integrated Forecast System using an object-based framework. They generally found mean values of daily AR prediction skill to saturate around 7–10 days, with seasonal variations highest over the Northern Hemisphere ocean basins, where AR prediction skill increases by 15–20% at a 7-day lead time during boreal winter relative to summer. A relative operating characteristic (ROC) analysis revealed that AR forecasts exhibited significant increases in skill at a 10-day lead over the North Pacific/Western US region during positive El Niño–Southern Oscillation (ENSO) conditions, at 7- and 10-day leads over the North Atlantic region during negative Arctic Oscillation (AO) conditions, and decreases at 10-day lead over the North Pacific/Western US region during negative Pacific/North America (PNA) teleconnection conditions. Along with this global extension to AR forecast skill studies, several recent studies show evidence for some S2S prediction skill of AR activity, based on various patterns of climate variability (Kim et al. 2017; Baggett et al. 2017; DeFlorio et al. 2019).

6.2.3 AR Analysis and Forecasting Tools

AR analysis and forecasting initiatives focus on identifying the location, intensity, orientation, duration, and vertical and spatial distributions of water vapor and water vapor flux along ARs over oceanic regions, susceptible coastlines, and inland locations. This focus is aided by the construction of tools that illustrate both observed and forecast AR conditions (i.e., filamentary regions of enhanced IWV or IVT) that may also serve as decision support tools that target high-profile locations near reservoirs to help:

1. water supply or forecast-informed reservoir operations (FIRO; Ralph et al. 2014)
2. vulnerable infrastructure such as the 2009 Howard Hanson Dam flood risk management crisis (White et al. 2012)
3. watersheds—to aid in streamflow prediction, floods, and flash floods
4. recent wildfire burn scars—to aid in diagnosing debris flow or landslide susceptibility (White et al. 2013)

This book has previously covered many of the observation platforms, tools, and methods used to diagnose real-time observations of AR conditions offshore and onshore such as SSM/I-derived IWV (Neiman et al. 2008; Dettinger et al. 2011) and the many instrumentation installations contained within the hydrometeorological testbed (HMT-West; Ralph et al. 2013b; White et al. 2013), among others. The forecasting tools that follow have been created to better predict ARs, focusing on their application to the US West Coast. Chapter 7 provides additional considerations of AR forecasting for applications.

Thresholding and Skeletonization Image-processing techniques related to "thresholding and skeletonization" can be used to objectively identify global locations that satisfy AR criteria related to IWV spatial characteristics. These techniques led to the development of the "AR Detection Tool" for IWV (ARDT–IWV) by Wick et al. (2013a) and for IVT (ARDT–IVT) by Wick (2014). (The term "AR Detection Tool" is synonymous with the term used throughout this book: "AR detection method" or ARDM.) The ARDT–IWV correctly identifies a large majority (a detection probability of 98.5%) of landfalling ARs along the US West Coast in SSM/I IWV imagery, compared to a manual analysis by Neiman et al. (2008). The ARDT–IWV can be applied to NWP model data (Wick et al. 2013b) to objectively identify locations that satisfy AR criteria in forecast IWV fields. The success of the ARDT–IWV is inherently tied to model resolution (Jackson et al. 2016). The study by Wick et al. (2013) represented the first detailed diagnostic of AR prediction skill on the US West Coast, and provides a key tool in helping to pinpoint the location of ARs offshore, and the possible landfall locations of ARs in gridded model analysis and forecast data. An algorithm with similar objectives but which accommodates global forecast validation and model evaluation applications has recently been developed by Guan and Waliser (2015). These AR identification tools are discussed in more detail in Chaps. 2 and 4.

AR Portal An "AR portal" was developed in late 2013 and early 2014 for various applications along the US West Coast, and was first tested significantly during CalWater 2015 to analyze and forecast the intensity, duration, and landfall of ARs during the experiment (Cordeira et al. 2017). The AR portal represents one location on the Internet where AR-related model analyses and forecast data are available for public access. The AR portal contains archived and real-time observations, gridded analyses, and gridded forecasts of AR-related information over the northeast Pacific and Western US (http://cw3e.ucsd.edu). The gridded analyses and forecasts on the AR portal during CalWater 2015 were created from NCEP Global Forecast System (GFS) and Global Ensemble Forecast System (GEFS) data provided by the NOAA Operational Model Archive and Distribution System (NOMADS). A list of the AR-related GFS and GEFS gridded products that created and supported CalWater 2015 can be found in Table 2 of Cordeira et al. (2017).

The AR-related gridded forecast products on the AR portal focus on identifying and tracking ARs over the northeast Pacific, with attention to their structure, intensity, and orientation at landfall along the US West Coast for lead times of 0–16 days. The gridded forecast products feature plan-view, cross-section, and time-series analyses and forecasts of IWV, horizontal water vapor flux, and IVT, among other synoptic-scale variables. Displays of IVT and other gridded forecast parameters were computed from the deterministic GFS and the 20-member GEFS data. The GEFS IVT forecasts were displayed as thumbnail and probability-over-threshold maps over the northeast Pacific, as multi-member time-series diagrams (e.g., a plume or dispersion diagram) for locations along the US West Coast, and as a probability-over-threshold in a time-latitude framework for locations along the US West Coast. An example of the GEFS IVT multi-member time-series forecast diagrams and time-latitude probability-over-threshold for locations along the US West Coast illustrates the increasing likelihood of AR conditions (i.e., IVT values >250 kg m^{-1} s^{-1}) at coastal locations from two different forecasts initialized during the CalWater 2015 field campaign (Fig. 6.4).

The content on the AR portal has since grown to include analyses and forecasts constructed from the National Center for Atmospheric Research (NCAR) Weather Research and Forecasting (WRF) model that targets the US West Coast, and also problem-specific forecast products related to Western US water resource management. The AR portal now also contains analyses and forecasts from the NCEP North American Model (NAM); there are plans to incorporate additional deterministic and ensemble-based forecast guidance.

6.3 Simulating ARs

Despite the key role that ARs play in the global water and energy cycles via their poleward transport of moisture, and influence on precipitation distribution and its extremes, only a limited number of studies have evaluated AR representation in regional or global climate models. "Evaluated" here means comparison of model fields to observations with a minimum objective of assessing model performance and a maximum objective of identifying and diagnosing model biases. To date, most AR modeling and model-evaluation studies are limited to regional domains (e.g., northeast Pacific/US West Coast), whether global or regional climate models are used, and are focused on AR frequency, distributions, variability, and amplitude of IVT. Few AR modeling studies have been diagnostic and have scrutinized the details of processes such as those associated with water or momentum budgets; cloud, convection, and latent heat processes; meso-scale circulations and landfalling details; etc. Given this lack of diagnostic evaluation of model performance in fundamental AR processes, it is not surprising that not many studies to date have used a model as a simulation "laboratory" to better understand AR process.

Specifically, two main categories of studies have targeted AR simulation and model evaluation: (1) AR simulations with regional models and (2) climate projections of ARs that also include a model-evaluation component. For the latter, these studies are still typically performed with a regional context in mind (e.g., the UK, western North America). Highlighted here are results from a few in Category 1, as well as those from the model-evaluation components of Category

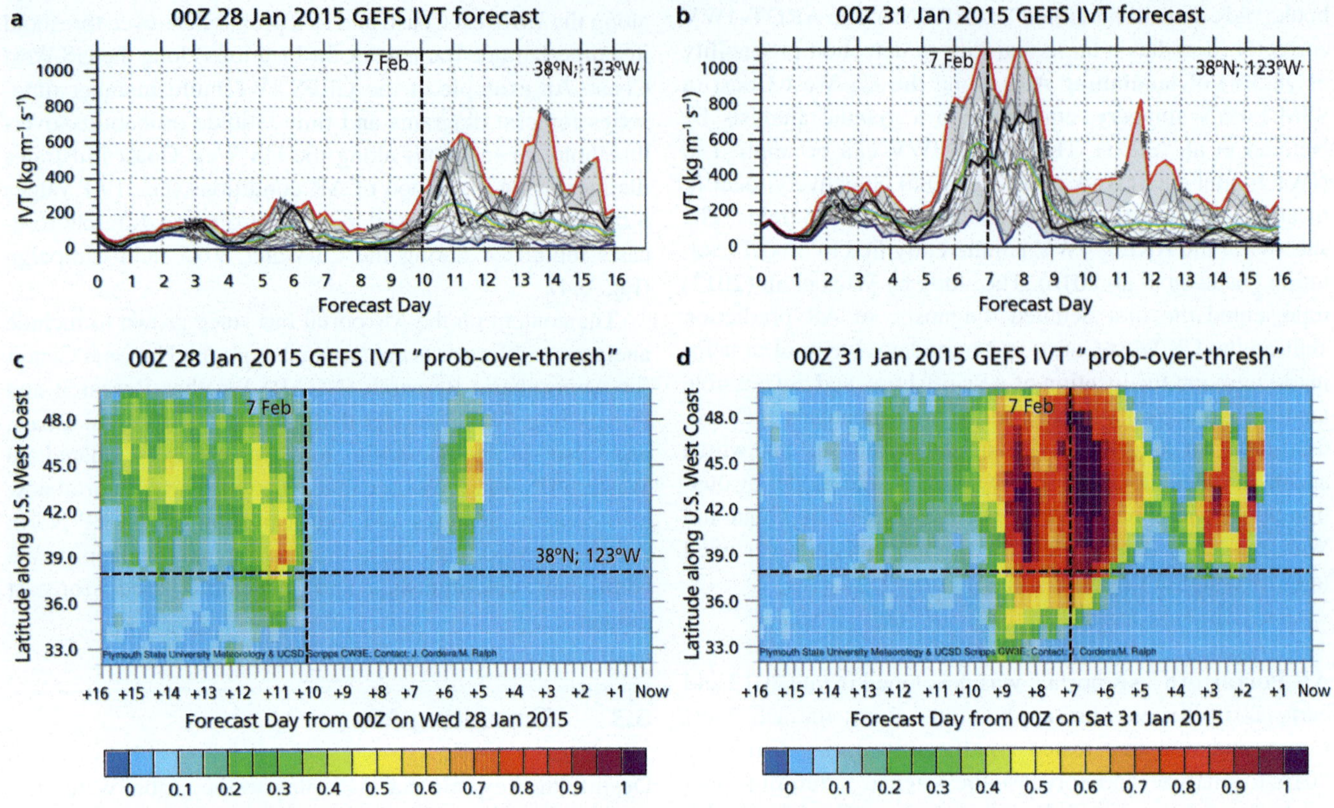

Fig. 6.4 Time-series diagrams of the 16-d forecast of integrated water vapor transport (IVT) magnitude (kg m^{-1} s^{-1}) at 38°N, 123°W initialized at (**a**) 0000 UTC 28 January 2015 and (**b**) 0000 UTC 31 January 2015 for each National Centers for Environmental Prediction Global Ensemble Forecast System (NCEP GEFS) ensemble member (*thin black lines*), the control member (*solid black line*), and the ensemble mean (*green line*). The *red and blue lines* represent the maximum and minimum IVT magnitudes at each forecast hour, whereas the *white shaded regions* represent the spread about the mean (±1 standard deviation) of the ensemble at each forecast hour. A 16-day forecast time "latitude" (where latitude follows the US West Coast) depiction of the fraction of GEFS ensemble members (including the control member) with IVT magnitudes ≥ 250 kg m^{-1} s^{-1} (*shaded* according to scale; *left panels*) initialized at (**c**) 0000 UTC 28 January 2015 and (**d**) 0000 UTC 31 January 2015. The *vertical dashed black lines* denote the time of 0000 UTC 7 February 2015 in panels (**a–d**), whereas the *dashed horizontal line* denotes 38°N in panels (**c, d**) (Cordeira et al. 2017)

2, with the climate projection results of these studies treated in Sect. 6.4. In addition to these two main categories, also briefly mentioned are the very limited amount of process-level modeling of ARs (e.g., aerosol–cloud interaction effects, water vapor flux) and global model evaluation of ARs.

6.3.1 Regional Models

To examine the performance of a regional model in representing ARs and their aggregate effects on California's cold season precipitation, Kim et al. (2013) used WRF v3.1.1 to simulate AR landfalls during ten cold seasons (October–March), from 2001 to 2010. Their outer domain covered the eastern North Pacific Ocean and the Western US, including southern Canada and most of Central America, at a 0.36° horizontal resolution; their inner domain covered a large sector of the far Western US, centered in California, at a 0.09° horizontal resolution. The large-scale forcing data—used only at the boundaries without spectral nudging in the interior—was constructed from the 6-h ERA-Interim reanalysis.

Model performance for the large-scale circulation was measured by computing the pattern correlation between the model fields and the European Centre for Medium-range Weather Forecasts (ECMWF) Re-Analysis-Interim (ERA-Interim) over the outer domain for each day over the 10 cold seasons, and then averaging the 10 time-series of correlation values across the 10 years. Kim et al. (2013) did this for 300- and 700-hPa temperature and geopotential heights and precipitable water. For these measures, the average correlation values remained above 0.9 (0.75) over the entire winter season for geopotential and temperature (precipitable water). This indicated that the model was able to replicate the synoptic patterns within the outer domain. Comparisons between the seasonal mean precipitation and the fraction accounted for by ARs over California were well captured, albeit with the model exhibiting an overall wet bias and a slightly larger AR precipitation fraction in the southern portion of the state. In this study, Kim et al. (2013) defined

AR days using the Neiman et al. (2008) ARDM and database. The model demonstrated an ability to replicate the inter-annual variations of the seasonal total and AR-total precipitation averaged over the southern and northern Sierra Nevada. A key feature the model replicated was the increase in the frequency of all but the lightest precipitation values by ARs relative to the precipitation values over non-AR wet days. Finally, although the model's freezing level in the Sierra Nevada region was generally biased high by ~300–400 m (both for AR and non-AR wet days), it did replicate the increase in freezing-level elevation that typically occurs in AR days relative to non-AR wet days: This difference is on the order of ~400–500 m in observations (based on ERA-Interim), and about 300 m in the model.

Whan and Zwiers (2016) assessed the ability of two Canadian regional climate models (RCMs), CanRCM4 and CRCM5, to simulate North American climate extremes, including ARs, over the period 1989–2009. Both RCMs used lateral boundary conditions derived from the ERA-Interim reanalysis and shared the same dynamical core, but used different nesting strategies, and land-surface and physics Schemes. A few of the experimental elements examined—five model configurations in total—were sensitivities to resolution (0.44° vs 0.22°), boundary-condition forcing (ERA-Interim, NCEP2), and the use of spectral nudging. Whan and Zwiers defined AR days from ERA-Interim daily values of IVT, based on an IVT threshold methodology for the model grid points along the North American coast (from 31.5° to 52.5°N) and threshold values of 250 and 500 kg m^{-1} s^{-1}. They used a relatively high-resolution, gauge-based precipitation record for observation reference, along with precipitation values from ERA-Interim and North American Regional Reanalysis (NARR).

For the AR component of their model evaluation, Whan and Zwiers considered three AR performance measures: They compared the percentage of winter precipitation that comes from ARs, the latitude of the precipitation maximum on AR days, and the intensity of the AR precipitation events. All five model configurations represented the fraction of AR-provided total precipitation relatively well, at least on a regional scale. However, none of the model configurations—nor the ERA-Interim nor NARR—replicated the clear separation between the two maxima in this quantity that occurs over the coastal and Rocky Mountain ranges in the southern tier of Canada. Whan and Zwiers hypothesized that these biases could result from the coarseness of the model topography, or model bias in landfall directions, or their combination. Comparisons between the observed and modeled AR landfall locations indicated an effect from spectral nudging. For example, the nudging case over the non-nudging case for CanCRM4 increased the hit rate from 56% to 76%, using a distance criterion of no more than 200 km between modeled and observed landfall. The probability distribution functions (PDFs) of precipitation for AR days showed varied agreement with reference values, but, with the NARR and RCM, values generally had no difficulty in producing extreme AR precipitation values at frequencies the same or even greater than the observation references used in the study. Moreover, they found that the 1000-km spectral nudging did not unrealistically limit the occurrence of AR-extreme precipitation values.

Fan et al. (2014) used a regional model to study the role of aerosol–cloud interactions in the development and intensity of AR-related precipitation. The investigation focused on two medium-strength ARs that made landfall in California during the 2011 CalWater field experiment with mixed-phase clouds. The investigation emphasized the role of ice-nucleating particles (INP), as influenced by local pollution versus long-range transport of mineral dust. Their model sensitivity experiments indicated that the enhanced INP number concentrations provided by the dust and local (biological) aerosols increased precipitation over the California Central Valley and Sierra Nevada area by 10–20% for both AR events, by increasing (~40%) snow formation. In these cases, the increased snow under the polluted condition resulted from the suppression of warm-rain processes in the Central Valley and foothills by the increased cloud condensation nuclei (CCN) that allowed more droplets to feed the ice-generation regime of the orographic clouds, which increased snowfall through more riming. This study points to the sensitivities of modeled AR precipitation to its microphysics formulation, and of the potential influences from background natural aerosol conditions as well as anthropogenic aerosol inputs.

Additional regional modeling studies have been undertaken to identify and characterize AR interactions with topography, including the development and character of the Sierra Barrier Jet, AR influences on the freezing level, and AR effects on precipitation and flooding (e.g., Kim and Kang 2007; Smith et al. 2010; Minder and Kingsmill 2013).

6.3.2 Global Models

As with regional model studies, few global model studies focus specifically on AR simulation, either to rigorously assess performance and/or to diagnose physical processes that underlie ARs and their variability. Of those global model studies that exist, nearly all target a specific geographical region (e.g., the northeast Pacific), with only one as yet comprehensively evaluating the global features of ARs. In fact, most of the in-depth evaluation studies of AR model fidelity to date have been at the front end of climate change projection studies of ARs, providing some assessment of capabilities and biases in the models' representations before assessing their climate projections of ARs.

In a recent model sensitivity study, Hagos et al. (2015) used two different global models to examine the dependence of AR frequency on grid resolutions and dynamical core. They analyzed "aqua planet" simulations using the Community Atmosphere Model (CAM4) at 240-, 120-, 60-, and 30-km grid spacing, performing each set with the Model for Prediction Across Scales (MPAS) and the High-Order Methods Modeling Environment (HOMME) dynamical core. They used several criteria reminiscent of studies by Ralph et al. (2004) and Ralph and Dettinger (2011, 2012) to define ARs. Criteria included IWV, length, and width thresholds, as well as additional criteria on wind speed and direction, low-level moisture concentration (i.e., fraction of total IWV between surface and 800 hPa), and occurrence poleward of 20° in association with the general transition from easterlies to westerlies at this latitude. From these experimental conditions and AR criteria, AR frequency depended strongly on dynamical core and resolution, with both HOMME and MPAS exhibiting a significant decrease in AR frequency with higher resolution (e.g., factors of ~3–5 reduction in ARs in going from ~100- to 200-km grid spacing to ~25- to 50-km grid spacing). In addition, the HOMME dynamical core produced about twice the number of ARs as the MPAS dynamical core.

The analysis of the model climatologies by Hagos et al. illustrated that three climate features and their related AR criteria depended considerably upon dynamic core and resolution. For example, HOMME developed an atmosphere that contained more moisture than did MPAS, and each exhibited a weak (albeit not so systematic) dependence on resolution. Similarly, there were modest dependencies on low-level moisture concentrations with resolution, and a rather significant systematic difference in the easterly to westerly transition—moving northward with increasing resolution for both dynamical cores. Using a revised definition of ARs based on a simulation-specific percentile threshold of these quantities—e.g., IWV threshold based on a 95% percentile value of the IWV for each model simulation, instead of the common threshold based on absolute IWV—showed that nearly all the resolution and dynamic core dependencies using the original, uniform criteria were largely removed. This left mainly a slight increase in AR frequency by the HOMME dynamical core relative to the MPAS. Hagos et al. also did an additional real-world (i.e., with continents) sensitivity test with the MPAS dynamical core at 120-km and 30-km grid spacing. The results showed evidence that the increase in model resolution decreased AR frequencies in the Pacific sector, namely by producing greater tropical convection. This, in turn, weakened the downstream subtropical westerlies, with the effect more evident in the southeast than in the northeast Pacific. These results indicate the importance and sensitivity of ARDMs when model-evaluation studies are conducted (cf. Shields et al. 2018).

Two additional studies have provided relatively thorough investigations of model fidelity for representing ARs in the northeast Pacific Ocean/western North American coast. Both studies evaluated Coupled Model Intercomparison Project (CMIP) 5 models, and included the response of ARs in this region to projections of global warming (addressed in Sect. 6.4). Payne and Magnusdottir (2015) evaluated historical simulations from 28 CMIP5 models using the period 1980–2004. They used ERA-Interim and Modern-Era Retrospective analysis for Research and Applications–Version 2 (MERRA2) for observational reference. The study targeted ARs that occurred along the North American coast (30°–60°N) and in the extended winter season: October–March. They defined ARs based on the IVT meeting an 85th percentile criterion (defined from each model's own climate) in the targeted landfall region, having IWV greater than 2 cm, the wind speed being greater than 10 m s^{-1} and having non-zero northward and eastward components, and having a zonal extent of more than 2000 km. Their evaluation addressed aspects of frequency and distribution of landfalling ARs in the target region, inter-annual variability, IVT amplitude, and AR-related wind and water vapor patterns.

Evaluating Model Performance for AR Simulations Based on Payne and Magnusdottir (2015)

Figure 6.5 highlights a number of findings from the model-evaluation component of Payne and Magnusdottir's study.

- The *upper plot* illustrates the total number of AR landfalls in the study domain (shown in the *lower left*) identified during the analysis period, 1980–2004. The two reanalyses have ~280 total AR landfalls or ~12 AR landfalls per year. The models exhibit considerable variation around this number, from ~1 to 2 per year to ~20 per year.
- The *lower left plot* shows the multi-model mean AR frequency, which has a unique meaning in this study: It is defined as the number of grid points that meet their AR criteria (*see above*), divided by the number of AR landfalls. Thus, the main objective of the diagram is to describe the typical landfalling AR pattern, as opposed to quantifying how frequently ARs occur at each grid point independent of landfall condition (e.g., Fig. 4.3 from Chap. 4).
- A key feature of this diagram is illustrated by the *red contours*, which show where the models exhibit the greatest inter-model disagreement, which in this case illustrates differences in the main AR landfall location in this region and the typical angle of the AR at landfall.
- The *middle plot* illustrates the bias in AR frequency, as defined in this study, for each model, over the study domain, relative to the two reanalyses products. Most

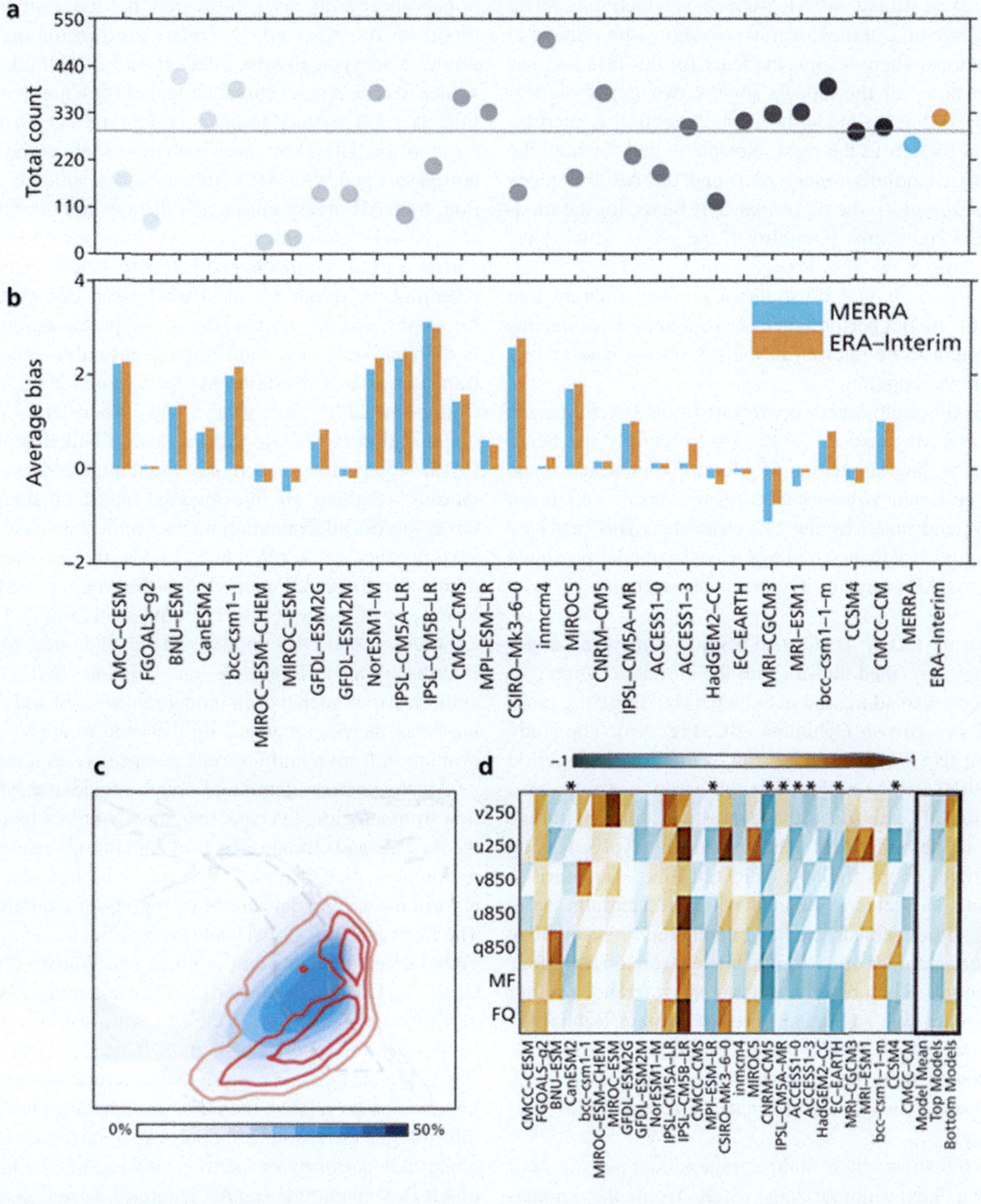

Fig. 6.5 *Top*: The total number of landfalling dates for each model and for the Modern-Era Retrospective analysis for Research and Applications (MERRA) and European Centre for Medium-range Weather Forecasts (ECMWF) Re-Analysis-Interim (ERA-Interim) reanalyses, for the study period 1980–2005. The order of the models from left to right is sorted according to horizontal resolution, with resolution increasing from left to right. *Middle*: The average bias in AR frequency for each model over the study domain (i.e., map on lower left) compared to (*blue*) MERRA and (*orange*) ERA-Interim reanalysis. *Lower left*: *Shading* shows the multi-model mean AR frequency (% of total days) over 28 Coupled Model Intercomparison Project (CMIP5) models, and the *contours* indicate the standard deviation about the mean, i.e., where the models vary the most about the multi-model mean (*contoured* at the interval of 2σ starting from 6σ, where σ is the standard deviation of the mean frequency distribution). *Lower right*: Portrait diagram of the relative error (*brown*: high error; *green*: low error) for each model against each observational data set (MERRA: *top triangle* and ERA-Interim: *bottom triangle*) over (*Y-axis, from top*) 250-hPa meridional wind (v250), 250-hPa zonal wind (u250), 850-hPa meridional wind (v850), 850-hPa zonal wind (u850), 850-hPa specific humidity (q850), medium frequency (MF), and AR frequency (FQ) (Adapted from Payne and Magnusdottir 2015)

models overestimate the AR frequency, which may relate to a positive bias in their overall size and width relative to observations. There is some evidence for this bias because the sort order of the models for the two upper plots is listed according to model horizontal resolution, increasing from the left to the right. Notable is that for both the number of landfalls (*upper plot*) and the AR frequency bias (*middle plot*), the performance is better for the models in the right-most one-third of the plots, which have grid spacing of ~1.3° or less.

- The *lower right plot* illustrates a portrait diagram that highlights model performance across a number of quantities thought to be relevant to the AR phenomena and its model representation.
- By using this multi-variate approach to model performance, Payne and Magnusdottir were able to identify the better models for their simulation of ARs in the present climate (those indicated with negative relative error; i.e., *green shading*, and noted by the "*" character). This provided more confidence in their findings related to model-projected changes in ARs associated with climate warming.

Radić et al. (2015) performed a study somewhat analogous to that described above (with the climate change part of their study also addressed in Sect. 6.4) but targeting landfalling ARs in British Columbia (BC; Fig. 6.6). The study focused on five CMIP5 models, an evaluation time-period of 1979–2010 (1974–2005) for the observations (models), a larger suite of quantities for evaluation, and a unique approach to defining ARs. This approach involved using self-organizing maps (SOMs) to identify the characteristic patterns in IVT fields associated with AR landfalls, from which the authors constructed and compared representative patterns for a number of AR-related fields from observations and the models. They based observational reference on four reanalysis data sets—Climate Forecast System Reanalysis (CFSR), MERRA, NCEP–NCAR, and ERA-Interim—which allowed them to quantify observational uncertainty and measure model performance relative to this uncertainty.

Figure 6.6 shows their multi-variate model performance diagram, a key summarizing result from the model-evaluation component of their study. For each model performance metric, listed along the *Y*-axis, the authors show relative error for each model and reanalysis data set. To compute the relative error, Radić et al. define a "typical reanalysis error" (TRE) as the median of all errors, for a given metric, over the set of four reanalysis products. Then, they define the relative error for a given metric and a given model or reanalysis product as the difference between the model error and the TRE, normalized by the TRE. This manner of normalization and comparison provides a measure of how a given model error compares to the observational uncertainty as represented by the four reanalysis products. As expected, the relative errors of the analysis products are typically the smallest, and exhibit relative error values near 0 in most cases. Note that for a number of quantities it is not atypical to have model errors at least twice as large as the TRE. For some metrics—such as the IVT pattern associated with AR extremes—total autumn precipitation, total AR precipitation, and its map and distribution are all relatively well represented, with most model errors being within ć ±1. The models exhibit more sizable errors in representing the frequency of characteristic IVT patterns, AR frequency, and the ratio of AR to total precipitation. Notable is that the skill of a multi-model ensemble—made up of Canada's Earth System Model (CanESM), Australia's Commonwealth Scientific and Industrial Research Organisation (CSIRO), and the Max Plank Institute ESM (MPI)—outperforms each individual model for all of the 12 metrics. Ranking all five models, based on their relative errors, across all evaluation metrics and against each reanalysis product as a reference, yields the best-performing model for these AR metrics for this region: CSIRO, followed by CanESM, the Geophysical Fluid Dynamics Laboratory (GFDL) model, MPI, and the Model for Interdisciplinary Research on Climate (MIROC). The authors also note that each model ranks worst and best for at least one metric, emphasizing the need to approach model evaluation from a multi-variate perspective as done here.

All the studies mentioned above considered AR evaluation from a regional perspective, mostly driven by interest in the weather and climate effects of ARs in a given high-impact region (e.g., the US West Coast) and/or because the detection algorithms used to date are often region- or landfall-specific. The latter makes a global evaluation difficult. Leveraging the global detection algorithm of Guan and Waliser (2015) (see Chap. 4), Guan and Waliser (2017) evaluated 17 AR model performance metrics in 24 GCMs, with results summarized in the form of a portrait diagram (Gleckler et al. 2008) in Fig. 6.7. Note that this portrait diagram represents the typical information on relative model errors (shading) but also adds information on absolute model errors (circles). Of the 17 global AR performance metrics considered, the histograms of AR IVT magnitude and AR lengths—as well as the meridional IVT and its zonal mean—were all relatively well represented. On the other hand, three different measures of AR seasonality—fractional zonal circumference occupied by ARs and their contribution to meridional IVT (Zhu and Newell 1998; Guan and Waliser 2015), zonal IVT from ARs, and overall frequency patterns—were poorly represented across the models. The diagram also indicates the importance of model horizontal resolution to overall quality of AR simulation, with a resolution of about 150 km or better appearing to be somewhat of a necessary but not sufficient condition for relatively good performance.

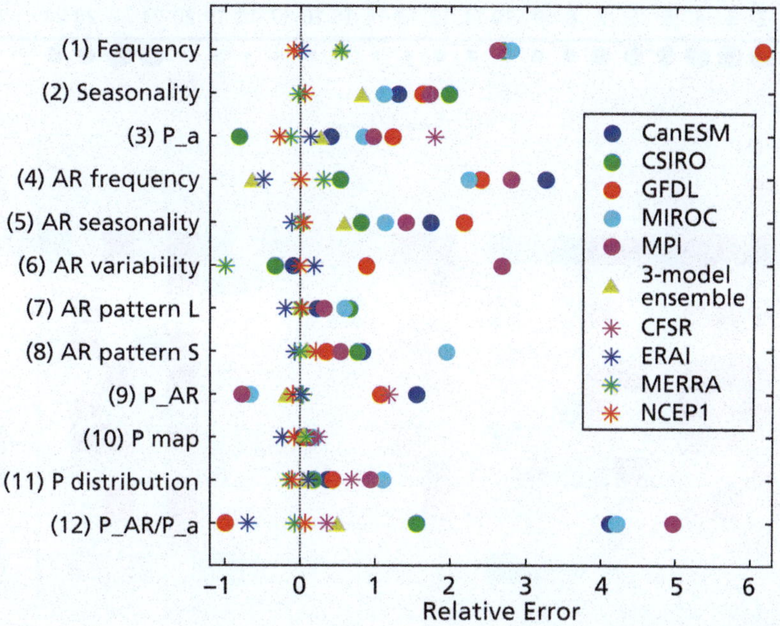

Fig. 6.6 Model performance diagram from Radić et al. (2015): Relative model and reanalysis error across all evaluation metrics used to evaluate model performance in simulating landfalling ARs in autumn (mid-August to December) for coastal British Columbia (BC) for the period 1979–2010. Performance measures include: (*1*) frequency and (*2*) seasonality of characteristic integrated water vapor transport (IVT) patterns identified via a self-organizing map (SOM) approach; (*3*) total autumn precipitation over BC, P_a; (*4*) frequency; (*5*) seasonality, and (*6*) inter-annual variability of AR-extreme events; IVT AR-extreme pattern over (*7*) large (northeast Pacific Ocean) and (*8*) small (BC) domain; (*9*) total AR-extreme precipitation over BC, P_AR; (*10*) its spatial map and (*11*) distribution; and (*12*) ratio between total AR-extreme precipitation and total autumn precipitation over BC (P_AR/P_a). Relative error = 0.5 means that the model error is 50% larger than the typical reanalysis error (TRE), based on the four reanalysis data sets; the three-model ensemble consists of Canada Earth Science Model (CanESM), Commonwealth Scientific and Industrial Research Organization (CSIRO), and Max Planck Institute (MPI) (Radić et al. 2015)

6.4 Climate Projections of ARs

Of all the sorts of climate modeling studies associated with ARs, the analysis of CMIP climate projections has garnered the most interest, with about a dozen studies in this area (see summary in Table 6.1). Except for the first study, which used CMIP3, all others have used CMIP5. All AR climate change studies to date, except one very recent one, have focused on a specific region: four on the UK/Europe and seven on the west coast of North America. As with the model-evaluation studies discussed above, focus on a specific region has typically derived from an interest in climate change effects on that region from an applications point of view, but has also been limited to some degree, until recently, by the lack of a uniform global detection algorithm that would support a global AR climate change study (e.g., Guan and Waliser 2015; Shields et al. 2018). In this section, a number of results from these studies are highlighted; North America is discussed first, followed by the European sector. The discussion of any given study is not comprehensive but rather highlights unique methodologies and/or results from all these studies. Note that two of the studies (Radić et al. 2015; Payne and Magnusdottir 2015) were discussed in Sect. 6.3 on the basis of the model-evaluation components of their studies.

Dettinger (2011) developed the first global change study on ARs, focusing on landfalling ARs in California using seven models from CMIP3. His analysis compared AR landfall characteristics between historical simulations for the period 1961–2000, and their projected changes under the CMIP3 A2 greenhouse-gas emissions scenario (i.e., emissions accelerating throughout the twenty-first century) for the periods 2046–2065 and 2081–2100. Because of the limited types of model output for CMIP3, the study developed a "GCM-friendly" approach to AR detection. The approach involved using daily values of 925-hPa wind speed and direction combined with IWV for a grid cell just off the central California coast. With these values, and targeting December–February, an AR was declared as occurring when there was an upslope wind speed of greater than 10 m s^{-1} and the IWV was greater than 2.5 cm. For the seven models Dettinger examined under the CMIP3 A2 scenario, the study found that the average AR statistics (e.g., number of AR days per 100 years) did not change appreciably (e.g., 0.3–7.2 additional AR days per year, depending on the model). However, there were notable changes in intensity (i.e., the product of IWV and upslope wind speed), with over half of the models exhibiting a + 5% increase over 100 years, and with two of the seven being ~10%. The analysis also showed that AR storm temperatures were expected to increase by about 1 °C

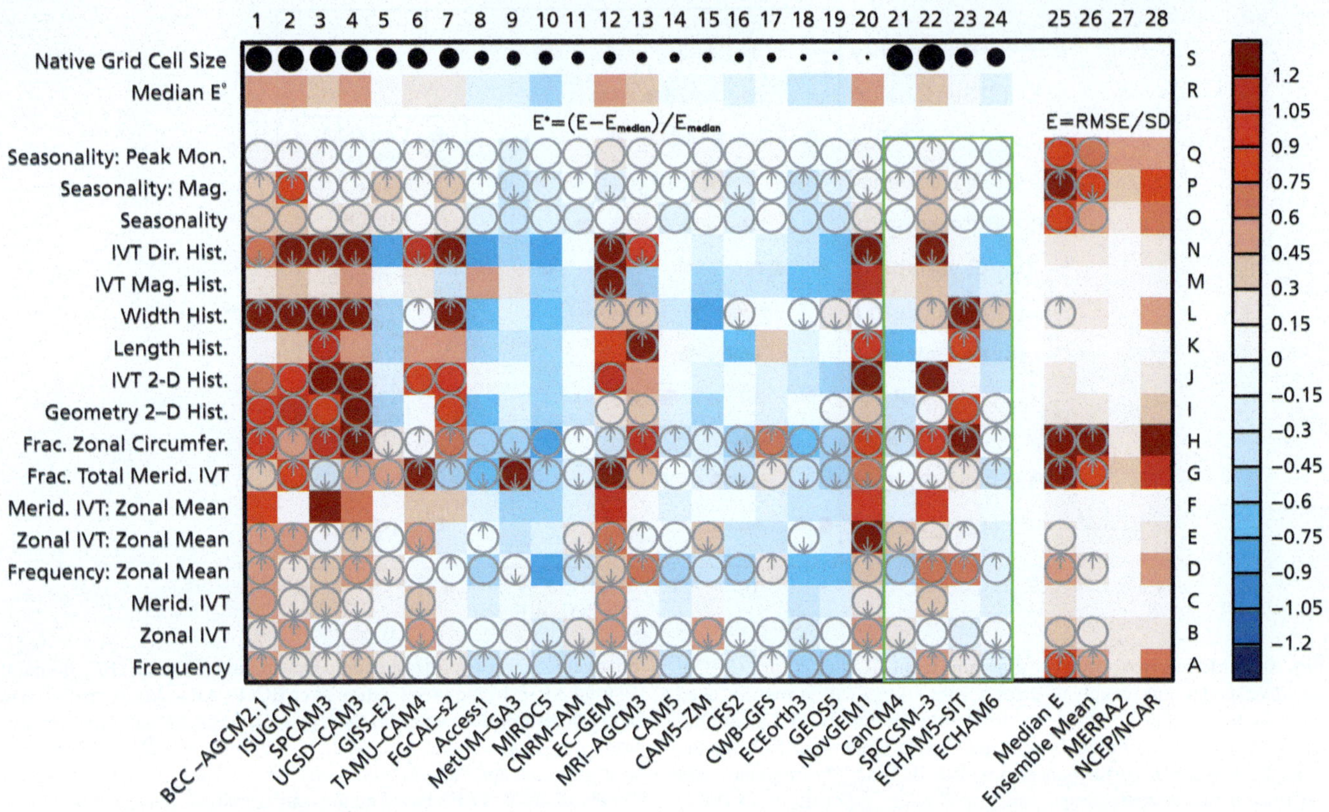

Fig. 6.7 Portrait diagram (Gleckler et al. 2008) showing the relative error $E*$ of each model for 17 AR performance metrics (*color shading*), with a few enhancements described below. $E*$ is defined as (E − Emedian)/Emedian, where E is the normalized root-mean-square error (RMSE) of an individual model relative to European Centre for Medium-range Weather Forecasts (ECMWF) Re-Analysis-Interim (ERA-Interim) for a given metric (*not shown*), and Emedian is the median E across all models for that metric. Emedian, as well as E for the ensemble mean of the models, Modern-Era Retrospective analysis for Research and Applications—Version 2 (MERRA2), and National Centers for Environmental Prediction-National Center for Atmospheric Research (NCEP-NCAR) relative to ERA-Interim are shown in the *four isolated columns on the far right*. For each model, the median $E*$ across all metrics is shown in the *isolated row near the top*. *Gray circles* indicate cases where E of a given model and metric is greater than 0.25 and meanwhile exceeds the primary measure of reanalysis uncertainty (MERRA2 relative to ERA-Interim). Wherever a *gray circle* is shown, an *arrow* additionally shows the sign of the bias (positive/negative if pointing upward/downward), but only if the magnitude of the normalized bias itself is greater than 0.05 and meanwhile exceeds the primary measure of reanalysis uncertainty. Biases in AR width (integrated water vapor transport [IVT] direction) are shown in the case of geometry (IVT) 2-d histogram. The *top row* shows the approximate size of native model grid cells for reference, based on which the models are sorted: *largest dot* is about 280 km; *smallest dot* is about 40 km. The four coupled models are placed together at the end whose $E*$'s are enclosed by the *green box*. For convenience, rows are labeled on the right by capital letters, and columns are labeled on the top by numbers (Guan and Waliser 2017)

by the mid-twenty-first century and by about 2 °C by the end of the twenty-first century—very close to the same pace of overall regional warming for the December–February season. Finally, there was evidence in most projections that the AR season would be expected to lengthen, indicating an extension of the flood-hazard season.

With largely the same objectives in mind, Warner et al. (2015) elaborated on the Dettinger (2011) study in a number of ways, namely by considering AR occurrence along the entire US West Coast and using a set of 10 models in the newer CMIP5 archive. The time-periods considered were 1970–1999 and 2070–2099, with the latter based on the Representative Concentration Pathways (RCP) 8.5 ("business as usual") scenario. With CMIP5's more elaborate model output (compared to CMIP3), IVT could be explicitly calculated—in their case integrating from the surface to 500 hPa. For each model, the authors identified extreme IVT days (i.e., ARs) by finding the 99th percentile in daily IVT during winter (October–March) for the 30-year historical and future periods at each of the offshore grid points. They verified this by NCEP–NCAR reanalysis. Their analysis showed that, when considering a multi-model mean across all 10 models, the number of AR days increased in the future period between ~200% and ~300% over the historical period, with the higher percentage increases associated with the northern part of the domain and the lower percentage increases with the southern part. Figure 6.8 shows a key result from their study, which illustrates the individual model

Table 6.1 Summary of climate change studies of ARs (adapted from Espinoza et al. 2018), including a comparison of mean changes in AR frequency (percent of time-steps) and integrated water vapor transport (IVT) (kg m^{-1} s^{-1})

Publication	Historical period	Projection period	Geographical region	AR FREQ (±%)	AR IVT (±%)
Dettinger (2011)	**1961–2000**	**2046–2065; 2081–2100**	**CA Coast**	**+10**	**+10**
Pierce et al. (2013)	**1985–1994**	**2060s**	**CA Coast**	**+25 to 100**	–
Warner et al. (2015)	**1970–1999**	**2070–2099**	**US West Coast**	**+230 to 290**	**+30**
Payne and Magnusdottir (2015)	**1980–2005**	**2070–2100**	**US West Coast**	**+23 to 35**	–
Gao et al. (2015)	**1975–2004**	**2070–2099**	**US West Coast**	**+50 to 600**	–
Hagos et al. (2016)	**1920–2005**	**2006–2099**	**US West Coast**	**+35**	–
Shields and Kiehl (2016)	**1960–2005**	**2055–2100**	**US West Coast**	**+8**	–
Espinoza et al. (2018)	**1979–2002**	**2073–2096**	**US West Coast/Global**	**+45**	**+30**
Lavers et al. (2013)	*1980–2005*	*2074–2099*	*W. Europe*	*+50 to 100*	–
Gao et al. (2016)	*1975–2004*	*2070–2099*	*W. Europe*	*+127 to 275*	*+20 to 50*
Ramos et al. (2016)	*1980–2005*	*2074–2099*	*Europe*	*+100 to 300*	*+30*
Shields and Kiehl (2016)	*1960–2005*	*2055–2100*	*N. Atlantic/W. Europe*	*+4*	–
Espinoza et al. (2018)	*1979–2002*	*2073–2096*	*Global*	*+60*	*+30*

Note: The **bold** *italic* region is previous studies focusing on the **US West Coast** (*Western Europe*). The studies are ordered from oldest to most recent within each geographic region (**bold** and *italic*). Note that each of the studies mentioned above differs in their methodologies, models used, and their study periods, limiting their compatibility

values and the ensemble means of IVT magnitude, IVW, 850-hPa wind speed, and precipitation for both periods, plotted versus latitude. The models represent the four quantities of interest relatively well during the historical period (left column—comparing green lines to blue). Most relevant to this discussion, however, are the climate change signals (right column—comparing red to blue), with demonstrative increases in IVT magnitude and IWV. Given virtually no change in the wind speed, the change in IVT magnitude evidently derives from the thermodynamic response of global warming leading to greater IWV. The extreme values of IVT magnitude and IWV increase by about 30%, independent of the latitude range studied. Precipitation values and their increases show markedly more inter-model variation, but with a multi-model mean increase of about 15–25%.

Hagos et al. (2016) used a 29-member ensemble of NCAR Community Earth System Model (CESM) simulations to investigate how global warming affects AR landfall in western North America. Given that the model exhibited notable biases in simulating the subtropical jet position and the relationship between extreme precipitation and moisture transport, they demonstrated a methodology for bias correcting these characteristics. After accounting for these biases, the model projects an ensemble mean increase of 35% in the number of landfalling AR days between the last 20 years of the twentieth and twenty-first centuries under the Representative Concentration Pathway (RCP) 8.5 scenario. Note that the large ensemble enabled the authors to present the projections in a clear probabilistic framework (see their Fig. 4). In this case, internal variability introduced an uncertainty of ±8% around the 35% increase in AR days. The significantly larger mean changes (i.e., 35%) compared to internal variability (i.e., 8%), as well as to the effects accounted for by correcting the model biases (i.e., 1%), highlight this study's findings of a robust AR response to global warming.

Although the studies above typically found the increase in landfalling ARs along western North America to be primarily associated with the enhanced atmospheric water vapor in accordance with the projected warming (i.e., Clausius–Clapeyron: warmer air can hold more water vapor), with little to no influence on IVT magnitude arising from increases in wind speed, Gao et al. (2015) illustrated a small but robust reduction in IVT magnitude in this region from changes in wind. They examined the climate change statistics in 24 CMIP5 models by contrasting the historical period 1975–2004 with the future period 2070–2099 under the Representative Concentration Pathway (RPC) 8.5 warming scenario. They adopted a model-dependent, IVT-based threshold approach to identifying ARs, similar in many respects to those mentioned earlier, and considered landfalls along nearly the entire North American coast, from Alaska to Baja. They used four reanalysis data sets for observation reference. Their results revealed a large increase of AR days by the end of the twenty-first century, with fractional increases between 50% and 600%, depending on the seasons and landfall locations. Although these increases were predominantly controlled by the Clausius–Clapeyron rate of increase of atmospheric water vapor with warming, the authors found that changes to the wind—i.e., the dynamical effect—countered some of the thermodynamical effect, particularly in spring and fall, limiting the increase of AR events in the future. They found that this negative effect from the changes in wind on the number of AR days could be linked to the robust poleward shift of the subtropical jet in the North Pacific basin.

For climate changes of ARs in the European sector, Lavers et al. (2013) examined five CMIP5 models, contrasting the period 1980–2005 with the period 2074–2099 from both the

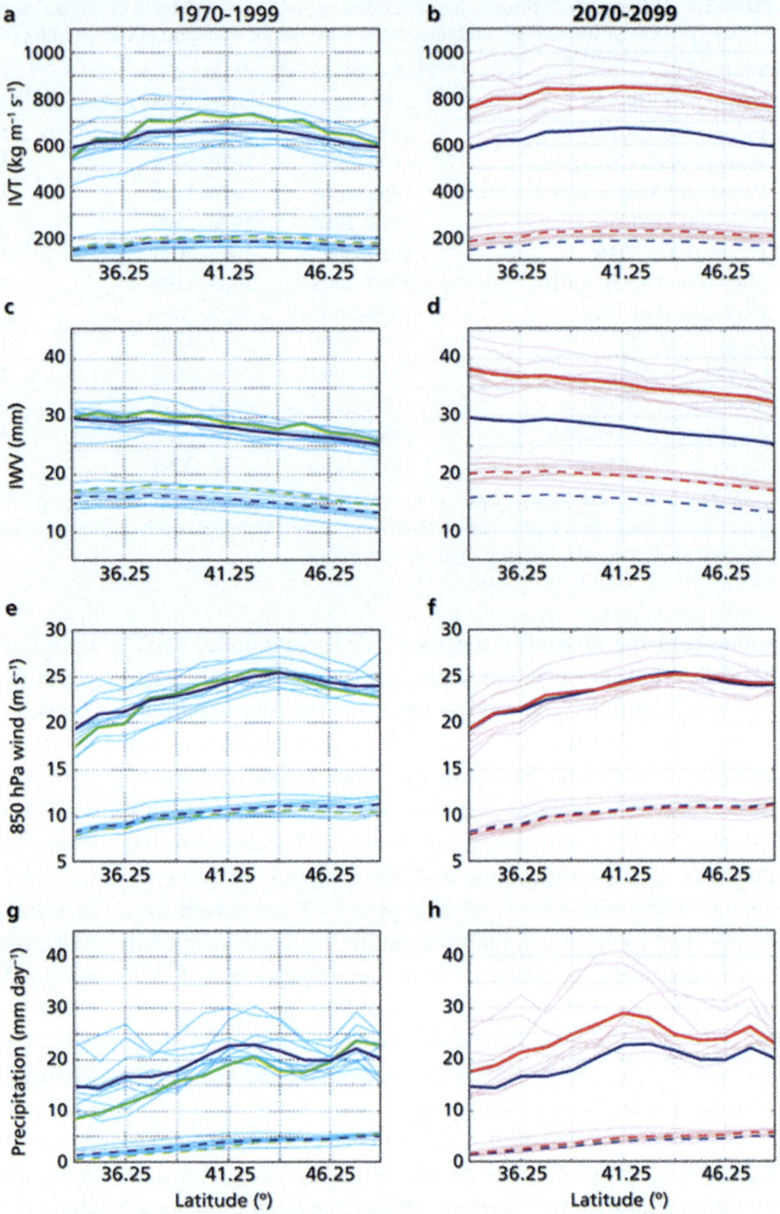

Fig. 6.8 (**a**, **b**) CMIP5 RCP 8.5 10-model means (*boldface lines*) for 99th percentile integrated water vapor transport (IVT) (upper values, *solid*) and winter mean (lower values, *dashed*) for *left* 1970–1999 (*boldface blue*) and *right* 2070–2099 (*boldface red*) along a 13-grid-box ocean transect near the US West Coast. Only October–March is considered. *Light blue* and *red lines* are individual models, and *boldface green lines* on the left are National Centers for Environmental Prediction-National Center for Atmospheric Research (NCEP-NCAR) reanalysis values for 1970–1999. Right-hand plots also show the multi-model means of the historical period for reference (*boldface blue lines*, same on *left* and *right*). Similar plots for (**c**, **d**) IWV, (**e**, **f**) 850-hPa total wind, and (**g**, **h**) daily precipitation are also shown (Warner et al. 2015)

RCP 4.5 and RCP 8.5 climate warming scenarios. Their observational references included five reanalysis products. In addition to using the coupled model simulations in the historical period, they also used each of the model's Atmospheric Model Intercomparison Project (AMIP) simulations (i.e., sea-surface temperatures [SSTs] prescribed from observations) to assess model credibility in replicating observed, present-day AR statistics. Establishing the methodology used in a number of the later studies described above, they computed IVT from the model wind and specific humidity fields, including levels from 1000 hPa to 300 hPa. They defined AR events by IVT values that exceeded a model- or analysis-dependent threshold, set to be the 85th percentile of IVT values, at one or more points along a north–south line just to the west of the UK (50°N–60°N, along 4°W). Figure 6.9 illustrates a major part of their findings, namely that North Atlantic ARs are projected to become more numerous (and stronger; see their Fig. 4), according to the projections from the five models examined. In the high-emissions scenario (RCP 8.5) for 2074–2099, AR frequency approximately doubles in the five model projections (i.e., comparing the gray "Hist" bars with the "RCP 8.5" bars). Based on Clausius–Clapeyron scaling of moisture from the historical runs using each model's projected temperature changes (lower right plot), the projected change in AR intensity is predominantly a thermodynamic response to warming rather than a result of changes to the winds (comparing hatched bars with corresponding gray bars). As with

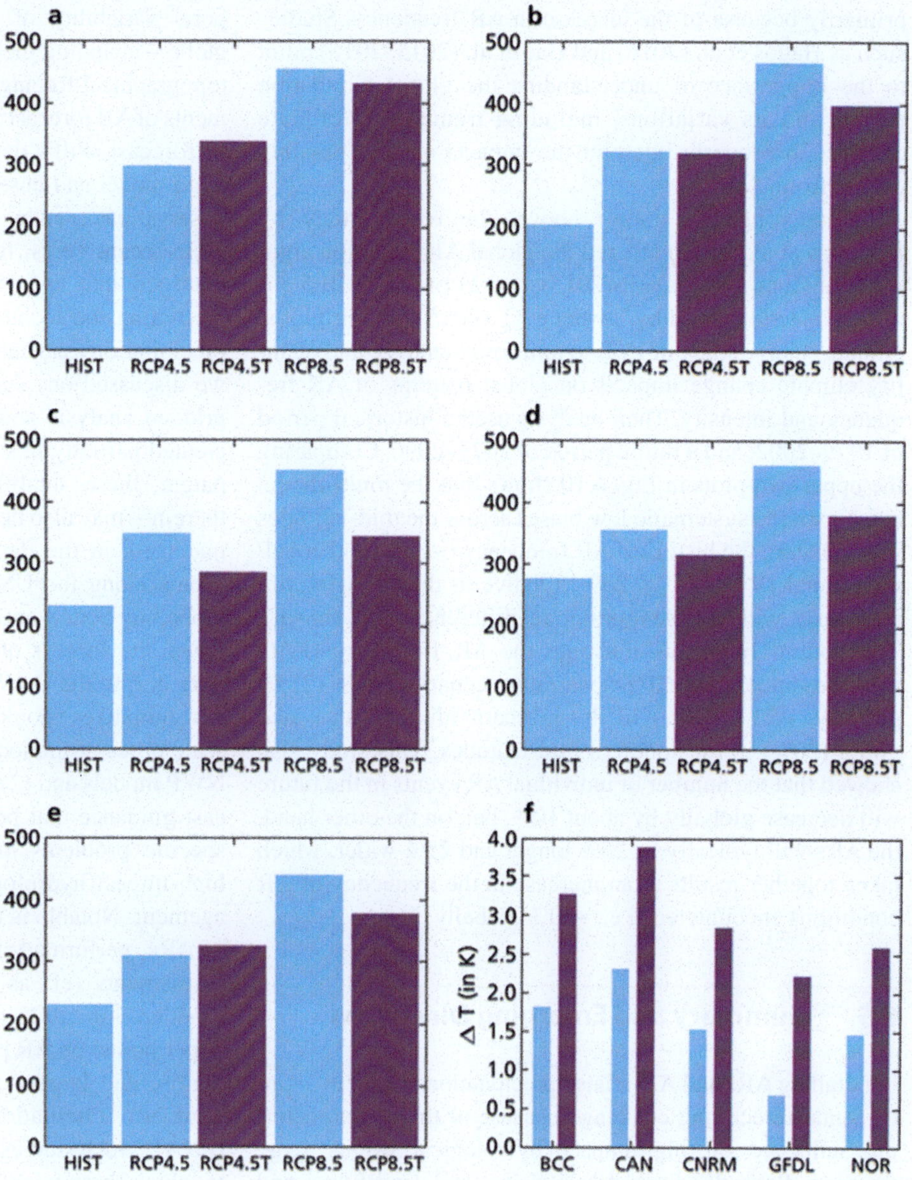

Fig. 6.9 The total number of winter ARs in the historical, and Coupled Model Intercomparison Project (CMIP5) 4.5 and Representative Concentration Pathway (RPC) 8.5 warming scenarios, and the scaled-up historical runs (RCP 4.5T and RCP 8.5T; *hatched bars*) for the following models: (**a**) Beijing Climate Center Climate System Model 1.1 (BCC CSM1.1), (**b**) Canada Earth Science Model 2 (CanESM2), (**c**) Centre national de la recherche scientifique Climate Model 5 (CNRM CM5), (**d**) Geophysical Fluid Dynamics Laboratory–Earth Systems Model 2G (GFDL-ESM2), and (**e**) Norway Earth System Model (NorESM 1-M). Panel (**f**) shows the rise in North Atlantic surface temperature (20° N–60° N; 60° W–0°) in the RCPs compared to the historical runs; *light gray* is RCP 4.5 and *dark gray* is RCP 8.5 (Lavers et al. 2013)

the results above for western North America, the study points to a greater risk of higher rainfall totals and, therefore, larger winter floods in Britain in the future.

With objectives and approaches similar to the Gao et al. (2015) study described above for western North America, Gao et al. (2016) have examined projected AR changes across the entire European sector to try to quantify changes in the number of AR days and heavy precipitation, as well as to compare thermodynamic and dynamic influences. They used 24 CMIP5 models, finding that the models generally captured the seasonal and spatial variations of historical landfalling AR days, with the inter-model variability in AR days strongly correlated with the inter-model spread of historical, near-surface, westerly jet position. Based on the RCP 8.5 warming scenario, they found that AR frequency is projected to increase significantly by the end of the twenty-first century, with a 127–275% increase in regions of peak AR frequency (i.e., 45°–55°N). Although Gao et al. (2016) found thermodynamic processes to play a dominant role in the future increase in the number of ARs days, as in all the studies above, they also found that changes associated with variability in wind speed and direction that related to shifts in the mid-latitude jet stream contributed significantly to changes in AR structure, resulting in a dipole change pattern associated with the wind (i.e., dynamical) effect in all seasons. This effect acts to reduce AR days south of the historical mean jet position, and increase to a lesser extent AR days north of the historical mean jet position, through the dynamical connections between the jet positions and AR days. Compared to present (modeled) conditions, both total and extreme precipitation induced by ARs in the future contribute more to the seasonal mean and extreme precipitation, primarily because of the increase in AR frequency. Studies such as Hagos et al. (2016) and Gao et al. (2015, 2016) point to the importance of understanding the global circulation pattern and its variations, including trends with climate change, to properly interpret the impacts of, and possibly effects from, ARs.

Augmenting the studies above, the recent study by Espinoza et al. (2018) utilized the global AR detection algorithm of Guan and Waliser (2015), and 21 of the CMIP5 climate projections used by Lavers et al. (2015) in their global climate change study of IVT, to uniformly analyze and quantify climate change impacts on global patterns of AR frequency and intensity. Their analysis used a historical period of 1979–2002, and a future period of 2073–2096. Comparing the upper two plots in Fig. 6.10 shows that the multi-model mean exhibits systematic low biases across the mid-latitudes in replicating the historical AR frequency (~10%) and zonal/meridional IVT (~15%/~25%) relative to the ERA-Interim reanalysis, with sizable inter-model differences (not shown). The bottom two plots illustrate the AR frequency (IVT) increases under the RCP 8.5 warming scenario: ~50% (25%) globally, ~50% (20%) in the northern mid-latitudes, and ~60% (20%) in the southern mid-latitudes. This study also showed that the number of individual AR events in the future will decrease globally by about 10%, but, on the other hand, the ARs will typically be 25% longer and 25% wider, which taken together results in an increase in the frequency of AR conditions encountered (i.e., ~50% globally).

6.5 Summary and Emerging Directions

Landfalling ARs and AR-related precipitation represent a fundamental forecasting challenge because of their overwhelming influence on high-impact hydrological events (e.g., flooding, flash flooding, debris flows, and landslides) and water resources (e.g., water supply, reservoir operations, drought mitigation) on the west coasts, and, in some cases, inland regions, of mid-latitude continents. This chapter highlights the important role of situational awareness and an ingredient-based methodology to better forecast the likelihood of extreme precipitation related to landfalling ARs. Although the synoptic-scale processes associated with mid-latitude winter storms and the ARs that can often accompany them have been shown to be forecast fairly well—up to ~7–10 days in many AR-dominated areas such as the western North American and European sectors—meso-scale processes related to AR intensity, width, landfall location, duration, and precipitation amount are not forecast well in advance. These meso-scale processes are key factors in determining which population centers and watersheds will ultimately be influenced by high-impact hydrological events or will face decisions related to water resources management. Because of the spatial and temporal variability of processes that shape ARs across the globe—including their meso-scale features imparted by local topographical forcing—additional broader and global assessments of AR forecast skill are needed—as is an understanding of forecast skill's dependencies on forecast system features (e.g., model and ensemble formulations, methodologies, and observations used by the data-assimilation system).

In recent years, field campaigns and forecast tools have emerged that specifically focus on better analyzing and predicting the location, intensity, landfall, orientation, and precipitation associated with landfalling ARs. In this chapter, we discussed one such tool: the "AR portal." Although the gridded analysis and forecast tools on the AR portal were created partially in support of the CalWater 2015 field campaign, these analyses and tools—or adapted versions therein—may also be useful to weather forecasters and water managers in the day-to-day analysis and forecasts of AR events along the US West Coast, or elsewhere around the globe, to better anticipate hydrometeorological extremes. Given the success of the synoptic-scale forecasts related to ARs during the CalWater 2015 field campaign, future forecast initiatives should focus on better illustrating: (1) meso-scale processes associated with ARs using higher-resolution NWP models and (2) probabilistic and ensemble-based forecast guidance that better serve stakeholders responsible for specific problems in decision support services related to high-impact hydrological events and water resources management. Notable is the need for more predictability studies of ARs, performed in a manner analogous to forecast skill assessments (cf. as done in Mani et al. [2015] for the Madden–Julian Oscillation [MJO]), to better quantify the upper bound on AR prediction and its sensitivities to season, region, and larger-scale climate variations (e.g., El Niño–Southern Oscillation [ENSO], Pacific/North American [PNA]). Such studies are also needed to determine the degree to which there might be S2S predictability of AR likelihood based on weather and climate patterns such as ENSO, PNA, Arctic Oscillation (AO), etc.

The study of ARs in the context of climate modeling is largely in its infancy. Studies to date have indicated that climate models represent observed statistics of ARs relatively well, including their seasonal and inter-annual variations. However, very few studies have examined modeled AR processes in detail, including their atmospheric or terrestrial water budgets, aerosol–cloud–precipitation processes, air–sea interactions, meso-scale circulation structures within the AR itself and associated with their interactions with topography, dependence of the above on latitude (i.e., subtropical, mid-latitude, polar ARs), and scale interactions between common climate patterns and AR frequency and evolution, etc. Such studies could be motivated by: (1) an interest in evaluating the fidelity of model representations of ARs to identify and reduce systematic biases and improve model forecasts and

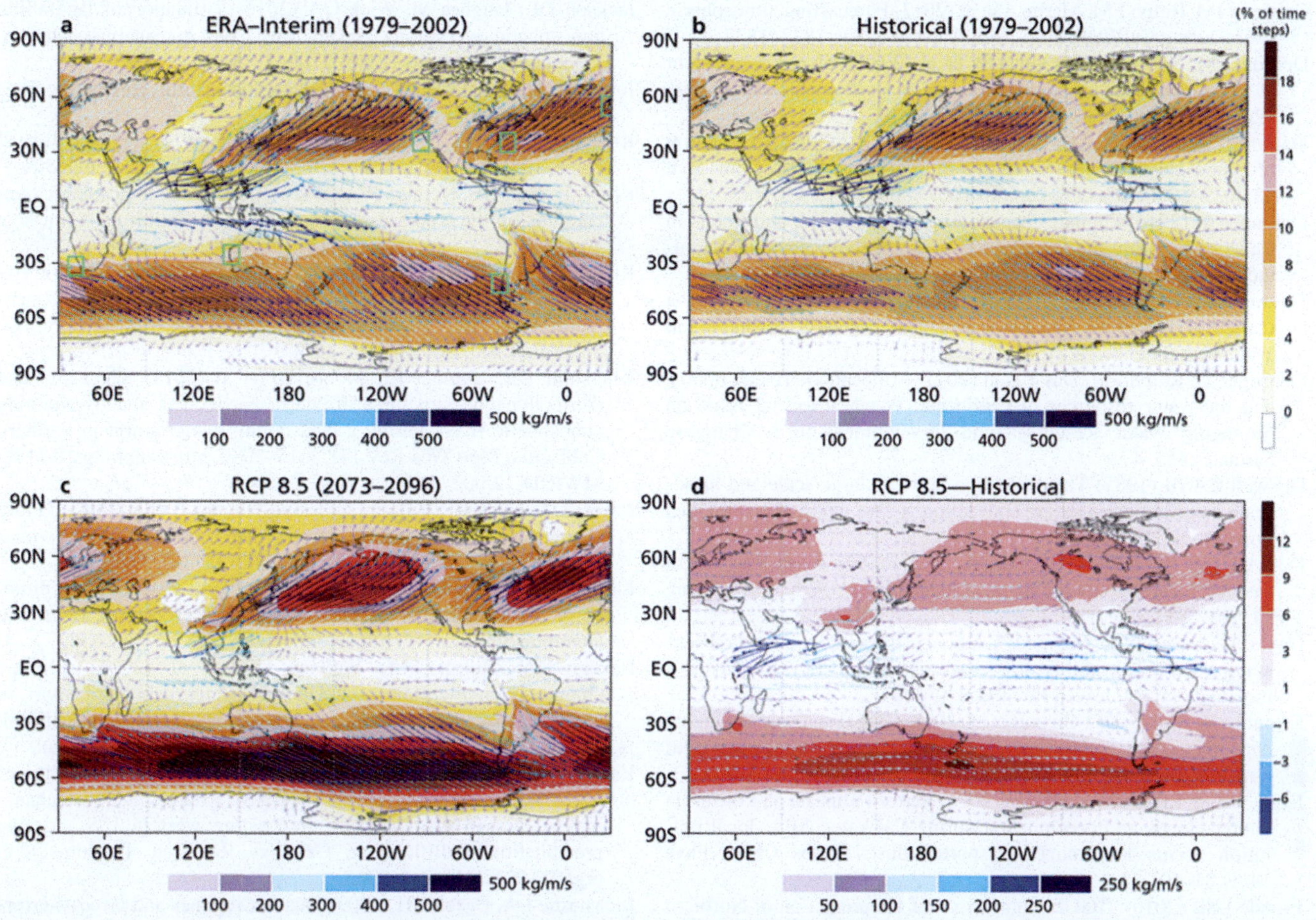

Fig. 6.10 AR frequency (*shading*; percent of time-steps) and integrated water vapor transport (IVT) (vectors; kg $m^{-1} s^{-1}$) for (**a**) European Centre for Medium-range Weather Forecasts (ECMWF) Re-Analysis-Interim (ERA-Interim) reanalysis for the historical period (1979–2002), (**b**) the multi-model mean for the 21 Coupled Model Intercomparison Project (CMIP5) models analyzed in this study for the historical period (1979–2002), (**c**) Representative Concentration Pathway (RCP) 8.5 warming scenario (2073–2096), (**d**) the difference between (**c**, **b**). Vector magnitudes are indicated both by their length and their color based on the *blue color bar*

simulations, or—in cases where the model is demonstrated to represent an AR process relatively well—(2) an interest to use the model as a "laboratory" to investigate and better understand the AR phenomena and the above sorts of processes and interactions. Given the importance of ARs in determining global water vapor and precipitation distribution—along with the character and patterns of precipitation and wind extremes—additional performance measures and more in-depth diagnostic studies on the global model representation of ARs and their effects are warranted.

As the meteorological community continues to consider and push for longer-lead-time S2S forecasting (NAS 2016), as well as robust projections of AR changes under global warming conditions, it is vital that climate models represent ARs as accurately as possible, accounting for the types of processes and interactions mentioned above. With this in mind, a broader range of performance metrics as well as more process-oriented model diagnostics for ARs need to be developed—diagnostics that probe beyond AR frequency, seasonality, intensity, etc. In particular, a better connection is needed between the analysis of S2S predictions and climate change projections of AR activity and the expected character of the precipitation, and related weather and hydrological impacts. With the above needs and gaps in mind, there may be a need for both offshore and onshore field campaigns to provide additional observational constraints to apply to model-evaluation and diagnostic studies.

References

Baggett CF, Barnes E, Maloney E et al (2017) Advancing atmospheric river forecasts into subseasonal-to-seasonal time scales. Geophys Res Lett 44:7528–7536. https://doi.org/10.1002/2017GL074434

Cordeira JM, Ralph FM, Moore BJ (2013) The formation and maintenance of two atmospheric rivers in proximity to western North Pacific tropical cyclones in October 2010. Mon Wea Rev 141:4234–4255

Cordeira JM, Ralph FM, Martin A et al (2017) Forecasting atmospheric rivers during CalWater 2015. Bull Amer Met Soc:1827–1837

DeFlorio M, Waliser DE, Guan B et al (2018) Global assessment of atmospheric river prediction skill. J Hydrometeorol 19:409–426. https://doi.org/10.1175/JHM-D-17-0135.1

DeFlorio MDE, Waliser B, Guan FM et al (2019) Global Evaluation of Atmospheric River Subseasonal Prediction Skill. Clim Dyn 52:3039–3060. https://doi.org/10.1007/s00382-018-4309-x

Dettinger M (2011) Climate change, atmospheric rivers, and floods in California – a multimodel analysis of storm frequency and magnitude changes. J Am Water Resources Assn 47(3):514–523

Dettinger MD, Cayan DR (2014) Drought and the California Delta—a matter of extremes. San Francisco: San Francisco Estuary and Watershed Science 12(2), 7 p.

Dettinger M, Ralph FM, Das T et al (2011) Atmospheric rivers, floods, and the water resources of California. Water 3(Special Issue on Managing Water Resources and Development in a Changing Climate):455–478

Doswell CA III (1987) The distinction between large-scale and mesoscale contribution to severe convection: a case study example. Wea Forecasting 2:3–16

Doswell CA III, Brooks HE, Maddox RA (1996) Flash flood forecasting: an ingredients-based methodology. Wea Forecasting 11:560–581

Doyle JD, Amerault C, Reynolds CA et al (2014) Initial condition sensitivity and predictability of a severe extratropical cyclone using a moist adjoint. Mon Wea Rev 142:320–342

Espinoza V, Waliser DE, Guan B et al (2018) Global analysis of climate change projection effects on atmospheric rivers. Geophys Res Lett. https://doi.org/10.1029/2017GL076968

Fan J, Leung LR, DeMott PJ et al (2014) Aerosol impacts on California winter clouds and precipitation during CalWater 2011: local pollution versus long-range transported dust. Atmos Chem Phys 14(1):81–101

Froude LSR (2010) TIGGE: comparison of the prediction of Northern Hemisphere extratropical cyclones by different ensemble prediction systems. Wea Forecasting 25:819–836

Fujita TT (1986) Mesoscale classifications: their history and their application to forecasting. In: Ray PS (ed) Mesoscale meteorology and forecasting. American Meteorological Society, Boston, MA

Gao Y, Lu J, Leung LR et al (2015) Dynamical and thermodynamical modulations on future changes of landfalling atmospheric rivers over western North America. Geophys Res Lett 42(17):7179–7186

Gao Y, Lu J, Leung LR (2016) Uncertainties in projecting future changes in atmospheric rivers and their impacts on heavy precipitation over. Eur J Clim. https://doi.org/10.1175/JCLI-D-16-0088.1

Gleckler PJ, Taylor KE, Doutriaux C (2008) Performance metrics for climate models. J Geophys Res Atmos 113:D06104

Guan B, Waliser DE (2015) Detection of atmospheric rivers: evaluation and application of an algorithm for global studies. J Geophys Res 120(12):12514–12535

Guan B, Waliser DE (2017) Atmospheric rivers in 20-year weather and climate simulations: a multi-model, global evaluation. J Geophys Res 122. https://doi.org/10.1002/2016JD026174

Guan B, Molotch NP, Waliser DE et al (2010) Extreme snowfall events linked to atmospheric rivers and surface air temperature via satellite measurements. Geophys Res Lett 37:L20401. https://doi.org/10.1029/2010GL044696

Hagos S, Leung LR, Yang Q et al (2015) Resolution and dynamical core dependence of atmospheric river frequency in global model simulations. J Clim 28(7):2764–2776

Hagos SM, Leung LR, Yoon J-H et al (2016) A projection of changes in landfalling atmospheric river frequency and extreme precipitation over western North America from the Large Ensemble CESM simulations. Geophys Res Lett 43(3):1357–1363

Jackson DL, Hughes M, Wick GA (2016) Evaluation of landfalling atmospheric rivers along the U.S. West Coast in reanalysis data sets. J Geophys Res Atmos 121(6):2705–2718

Johns RH, Doswell CA III (1992) Severe local storms forecasting. Wea Forecasting 7:588–612

Kim J, Kang H-S (2007) The impact of the Sierra Nevada on low-level winds and water vapor transport. J Hydrometeorol 8(4):790–804

Kim J, Waliser DE, Neiman PJ et al (2013) Effects of atmospheric river landfalls on the cold season precipitation in California. Clim Dyn 40:465–474

Kim HM, Zhou Y, Alexander MA (2017) Changes in atmospheric rivers and moisture transport over the Northeast Pacific and western North America in response to ENSO diversity. Clim Dyn. https://doi.org/10.1007/s00382-017-3598-9

Kingsmill DE, Neiman PJ, Moore BJ et al (2013) Kinematic and thermodynamic structures of Sierra barrier jets and overrunning atmospheric rivers during a land-falling winter storm in northern California. Mon Wea Rev 141:2015–2036. https://doi.org/10.1175/MWR-D-12-00277.1

Lavers DA, Allan RP, Villarini G, Lloyd-Hughes B et al (2013) Future changes in atmospheric rivers and their implications for winter flooding in Britain. Environ Res Lett 8(3):034010

Lavers DA, Pappenberger F, Zsoter E (2014) Extending medium-range predictability of extreme hydrological events in Europe. Nat Commun 5(5382):1–7. https://doi.org/10.1038.ncomms6382

Lavers DA, Ralph FM, Waliser DE et al (2015) Climate change intensification of horizontal water vapor transport in CMIP5. Geophys Res Lett 42(13):5617–5625. https://doi.org/10.1002/2015GL064672

Lavers DA, Waliser DE, Ralph FM et al (2016) Predictability of horizontal water vapor transport relative to precipitation: enhancing situational awareness for forecasting western U.S. extreme precipitation and flooding. Geophys Res Lett 43. https://doi.org/10.1002/2016GL067765

McMurdie LA, Casola JH (2009) Weather regimes and forecast errors in the Pacific Northwest. Wea Forecasting 24:829–842

McMurdie LA, Mass C (2004) Major numerical forecast failures over the northeast Pacific. Wea Forecasting 19:338–356

Miller JE (1946) Cyclogenesis in the Atlantic Coastal region of the United States. J Meteorol 3:31–44

Minder JR, Kingsmill DE (2013) Mesoscale variations of the atmospheric snow line over the Northern Sierra Nevada: multiyear statistics, case study, and mechanisms. J Atmos Sci 70(3):916–938

NAS (2016) Next generation earth system prediction: strategies for subseasonal to seasonal forecasts. National Research Council, National Academy of Sciences, Washington DC, p 290. ISBN-978-0-309-38880-1

Nayak MA, Villarini G, Lavers DA (2014) On the skill of numerical weather prediction models to forecast atmospheric rivers over the central United States. Geophys Res Lett 41:4354–4362

Neiman PJ, Ralph FM, Wick GA (2008) Meteorological characteristics and overland precipitation impacts of atmospheric rivers affecting the West Coast of North America based on eight years of SSM/I satellite observations. J Hydrometeorol 9:22–47. https://doi.org/10.1175/2007JHM855.1

Neiman PJ, Sukovich EM, Ralph FM et al (2010) A seven-year wind profiler-based climatology of the windward barrier jet along California's northern Sierra Nevada. Mon Wea Rev 138:1206–1233. https://doi.org/10.1175/2009MWR3170.1

Neiman PJ, Schick LJ, Ralph FM et al (2011) Flooding in western Washington: the connection to atmospheric rivers. J Hydrometeorol 12:1337–1358

Neiman PJ, Hughes M, Moore BJ (2013a) Sierra barrier jets, atmospheric rivers, and precipitation characteristics in northern California: a composite perspective based on a network of wind profilers. Mon Wea Rev 141:4211–4233. https://doi.org/10.1175/MWR-D-13-00112.1

Neiman PJ, Ralph FM, Moore BJ et al (2013b) The landfall and inland penetration of a flood-producing atmospheric river in Arizona. Part 1: observed synoptic-scale, orographic, and hydrometeorological characteristics. J Hydrometeorol 14:460–484

Neiman PJ, Ralph FM, Moore BJ et al (2014) The regional influence of an intense Sierra barrier jet and landfalling atmospheric river on orographic precipitation in Northern California: a case study. J Hydrometeorol 15:1419–1439. https://doi.org/10.1175/JHM-D-13-0183.1

Payne AE, Magnusdottir G (2015) An evaluation of atmospheric rivers over the North Pacific in CMIP5 and their response to warming under RCP 8.5. J Geophys Res-Atmos 120(21):11173–11190

Pierce DW, Cayan DR, Das T, Maurer EP, Miller NL, Bao Y et al (2013) The key role of heavy precipitation events in climate model disagreements of future annual precipitation changes in California. J Clim 26(16):5879–5896. https://doi.org/10.1175/JCLI-D-12-00766.1

Radić V, Cannon AJ, Menounos B et al (2015) Future changes in autumn atmospheric river events in British Columbia, Canada, as projected by CMIP5 global climate models. J Geophys Res 120:9279–9302

Ralph FM, Dettinger MD (2011) Storms, floods, and the science of atmospheric rivers. EOS Trans Am Geophys Union 92(32): 265–266

Ralph FM, Dettinger MD (2012) Historical and national perspectives on extreme west coast precipitation associated with atmospheric rivers during December 2010. Bull Am Meteorol Soc 93(6):783–790

Ralph FM, Neiman PJ, Wick GA (2004) Satellite and CALJET aircraft observations of atmospheric rivers over the eastern North Pacific Ocean during the winter of 1997/98. Mon Wea Rev 132:1721–1745. https://doi.org/10.1175/1520-0493(2004)132

Ralph FM, Neiman PJ, Rotunno R (2005) Dropsonde observations in low-level jets over the Northeastern Pacific Ocean from CALJET-1998 and PACJET-2001: mean vertical-profile and atmospheric-river characteristics. Mon Wea Rev 133:889–910

Ralph FM, Neiman PJ, Wick GA et al (2006) Flooding on California's Russian River: the role of atmospheric rivers. Geophys Res Lett 33:L13801. https://doi.org/10.1029/2006GL026689

Ralph FM, Sukovich E, Reynolds D, Neiman PJ (2010) Assessment of extreme quantitative precipitation forecasts and development of regional extreme event thresholds using data from HMT- 2006 and COOP observers. J Hydrometeorol 11:1288–1306

Ralph FM, Neiman PJ, Kiladis GN et al (2011) A multi-scale observational case study of a Pacific atmospheric river exhibiting tropical–extratropical connections and a mesoscale frontal wave. Mon Wea Rev 139:1169–1189. https://doi.org/10.1175/2010MWR3596.1

Ralph FM, Neiman PJ, Zamora RJ et al (2013a) Observed impacts of duration and seasonality of atmospheric-river landfalls on soul moisture and runoff in coastal in coastal Northern California. J Hydrometeorol 14:443–459

Ralph FM, Intrieri J, Andra D Jr et al (2013b) The emergence of weather-focused testbeds linking research and forecasting operations. Bull Amer Meteor Soc 94:1187–1210

Ralph FM, Dettinger M, White A et al (2014) A vision for future observations for Western U.S. extreme precipitation and flooding. J Contemp Water Res Educ 153(1):16–32. https://doi.org/10.1111/j.1936-704X.2014.03176.x

Ralph FM, Cordeira JM, Neiman PJ et al (2016a) Landfalling atmospheric rivers, the Sierra Barrier jet and extreme daily precipitation in Northern California's upper Sacramento river watershed. J Hydrometeorol 17:1905 1914. https://doi.org/10.1175/JHM-D-15-0167.1

Ralph FM, Prather KA, Cayan D (2016b) CalWater field studies designed to quantify the roles of atmospheric rivers and aerosols in modulating U.S. West Coast precipitation in a changing climate. Bull Amer Meteorol Soc 97:1209–1228. https://doi.org/10.1175/BAMS-D-14-00043.1

Ralph FM, Iacobellis SF, Neiman PJ et al (2017) Dropsonde observations of total integrated water vapor transport within North Pacific atmospheric rivers. J Hydrometeorol 18:2577–2596. https://doi.org/10.1175/JHM-D-17-0036.1

Ramos AM, Tomé R, Trigo RM, Liberato MLR, Pinto JG (2016) Projected changes in atmospheric rivers affecting Europe in CMIP5 models. Geophys Res Lett 43:9315–9323. https://doi.org/10.1002/2016GL070634

Rutz JJ, Steenburgh WJ, Ralph FM (2014) Climatological characteristics of atmospheric rivers and their inland penetration over the Western United States. Mon Wea Rev 142:905–921

Schultz DM, Schumacher PN (1999) The use and misuse of conditional symmetric instability. Mon Wea Rev 127:2709–2732. Corrigendum, 128, 1573

Shields CA, Kiehl JT (2016) Atmospheric river landfall-latitude changes in future climate simulations. Geophys Res Lett 43:8775–8782. https://doi.org/10.1002/2016GL070470

Shields CA, Rutz JJ, Leung L–Y et al (2018) Atmospheric River Tracking Method Intercomparison Project (ARTMIP): experimental design and project goals. Geosci Model Dev 11(6):2455–2474. https://doi.org/10.5194/gmd-11-2455-2018

Smith BL, S Yuter SE, Neiman PJ, Kingsmill DE (2010) Water vapor fluxes and orographic precipitation over Northern California associated with a landfalling atmospheric river. Mon Weather Rev 138(1):74–100

Sodemann H, Stohl A (2013) Moisture origin and meridional transport in atmospheric rivers and their association with multiple cyclones. Mon Wea Rev 141:2850–2867. https://doi.org/10.1175/MWR-D-12-00256.1

Sukovich EM, Ralph FM, Barthold FE et al (2014) Extreme quantitative precipitation forecast performance at the Weather Prediction Center from 2001 to 2011. Wea Forecast 29:894–911

Warner M, Mass CF, Salathé EP (2015) Changes in winter atmospheric rivers along the North American west coast in CMIP5 climate models. J Hydrometeorol 16:118–128

Weather Prediction Center (2012) The 2012 atmospheric river retrospective forecasting experiment: final experiment report. [Available online at: http://www.wpc.ncep.noaa.gov/hmt/ARRFEX_Final_Report.pdf]

Whan K, Zwiers F (2016) Evaluation of extreme rainfall and temperature over North America in CanRCM4 and CRCM5. Clim Dyn 46:3821–3843

White AB, Colman B, Carter GM (2012) NOAA's rapid response to the Howard A. Hanson Dam flood risk management crisis. Bull Amer Meteorol Soc 93:189–207. https://doi.org/10.1175/BAMS-D-11-00103.1

White AB, Anderson ML, Dettinger MD (2013) A 21st century California observing network for monitoring extreme weather events. J Atmos Ocean Technol 30:1585–1603. https://doi.org/10.1175/JTECH-D-12-00217.1

Wick GA (2014) Implementation and initial application of an atmospheric river detection tool based on integrated vapor transport. In: 2014 Fall Meeting. Amer Geophys Union, San Francisco, CA. Abstract A34E-06

Wick GA, Neiman PJ, Ralph FM (2013a) Description and validation of an automated objective technique for identification and characterization of the integrated water vapor signature of atmospheric rivers. IEEE Trans Geosci Remote Sens 51:2166–2176. https://doi.org/10.1109/TGRS.2012.2211024

Wick GA, Neiman PJ, Ralph FM et al (2013b) Evaluation of forecasts of the water vapor signature of atmospheric rivers in operational numerical weather prediction models. Wea Forecasting 28: 1337–1352. https://doi.org/10.1175/WAF-D-13-00025.1

Zhu Y, Newell RE (1998) A proposed algorithm for moisture fluxes from atmospheric rivers. Mon Wea Rev 126(3):725–735. https://doi.org/10.1175/1520-0493

Applications of Knowledge and Predictions of Atmospheric Rivers

7

Lawrence J. Schick, Michael L. Anderson, F. Martin Ralph, Michael D. Dettinger, David A. Lavers, Florian Pappenberger, David S. Richardson, and Ervin Zsoter

7.1 Introduction

The science and understanding of ARs has grown tremendously over the last 20 years. The application of this newfound knowledge is beginning to show benefits. For example, the connection between strong ARs and major flooding events for large drainage basins in the coastal western USA is undeniable. When there is a major AR forecast, everyone now takes notice. Water management professionals and the public want to understand the magnitude and duration of a forecasted AR and then prepare for its impacts. Real-time weather monitoring, forecasting, and understanding of ARs contribute to increasing confidence in decisions by water managers, emergency managers, and operational forecasters who ultimately serve public safety.

This chapter highlights where AR research is being applied to real-world problems. First, a 2006 event in Washington state is discussed from the perspectives of the US Army Corps of Engineers (USACE) in managing reservoirs for flood risk and gaining information from forecasts (Sect. 7.2). Next, Sect. 7.3 presents a developing application for reservoir operations that uses forecast information to provide both water supply benefit and reduced flood risk. A conceptual framework for the use of ARs in general flood planning follows (Sect. 7.4). The chapter concludes with the use of ARs as constructed extremes in the ARkStorm scenario, and with AR scaling methods related to extreme precipitation (Sects. 7.5 and 7.6).

Understanding flooding from strong ARs, and managing beneficial and hazardous aspects of ARs for integrated water management, are just some of many new applications of AR knowledge. California has particular interest because of its reliance on multi-purpose projects that balance water management objectives such as flood, water supply, and ecosystem services. Also challenging is how to communicate AR risk and benefit while confidently serving the user. Ultimately, application of AR research will enable informed decisions that affect people who live, work, and play in areas affected by these vast "rivers in the sky."

L. J. Schick (✉)
Retired, U.S. Army Corps of Engineers, Seattle, WA, USA

M. L. Anderson
State of California Department of Water Resources, Sacramento, CA, USA
e-mail: Michael.L.Anderson@water.ca.gov

F. M. Ralph
Center for Western Weather and Water Extremes, Scripps Institution of Oceanography, University of California–San Diego, La Jolla, CA, USA
e-mail: mralph@ucsd.edu

M. D. Dettinger
Retired, U.S. Geological Survey, Carson City, NV, USA
e-mail: mdettinger@ucsd.edu

D. A. Lavers · F. Pappenberger · D. S. Richardson · E. Zsoter
European Centre for Medium-Range Weather Forecasts (ECMWF), Reading, UK
e-mail: david.lavers@ecmwf.int; florian.pappenberger@ecmwf.int; david.richardson@ecmwf.int; ervin.zsoter@ecmwf.int

7.2 US Army Corps of Engineers: ARs and Flood Risk Management

7.2.1 A Case History

In early November 2006, the expected annual increase in stronger rainstorms for the Pacific Northwest arrived right on time—but this time the deluge was different. An intense AR pounded the North Cascades of Washington state with concentrated heavy rainfall. Initially, the heavy rain fell as was first forecast. But the conditions being forecast went from bad to worse—much worse. Rivers were rising rapidly. The latest forecast, issued while the storm was quickly ramping up, now predicted *double* the record river flow into the Upper Baker River reservoir. Never had USACE–Seattle District engineers seen such a large inflow forecast; this was no ordinary AR flood for the Pacific Northwest.

F. M. Ralph et al. (eds.), *Atmospheric Rivers*, https://doi.org/10.1007/978-3-030-28906-5_7

Heavy precipitation associated with ARs is a significant and necessary element for almost all serious flooding in western Washington (e.g., Neiman et al. 2011; Konrad and Dettinger 2017). Antecedent conditions such as previous high streamflows, saturated soils, or snowmelt can augment runoff, which amplifies flood size. But it was early in the flood season, so background streamflows were low. Also, there was very little snow in the Cascades to melt from rain falling on snow. But even so, given that 5–10 in. (12.7 cm–25.4 cm) of rain in 24 h was predicted in the Cascades, the extremely high river flow forecast was now plausible.

Initially, the reservoir flood pocket behind the dam—reserved for capturing and temporarily storing the peak floodwaters during a storm—was unfilled. This flood storage space began to fill quickly, however, which is typical of a flood risk management operation. But in this case, with the record high forecast, engineers determined the pocket would run out of room to store the water—something that had never happened before. The flood risk would be very difficult to manage as the floodwaters predicted in the updated forecast flood filled and then overtopped the dam. If water overflowed the dam and moved downstream, it would breach downstream levees. If that occurred, the river would be wild and free, and the entire valley would fill with water. Tens of thousands of people would be threatened, and several miles of the only north–south freeway between Seattle, Washington, and Vancouver, British Columbia, would be underwater and impassible.

At this point, people living near low areas by the river were alerted and told to prepare to evacuate.

Snow elevation levels soared (i.e., the altitude above which the precipitation falls as snow rather than rain), with the AR advecting warm air from the subtropics. Heavy rain fell, as a major AR flooding event unfolded. Relentless rains hammered the USACE Reservoir Control Center in Seattle, about 100 miles (161 km) south of the dam. It was raining throughout much of the region, but USACE personnel focused first on the North Cascades. Engineers continued to evaluate the changing weather, river flows, and dam operation options, considering observed and forecast river flows.

They instructed reservoir operations personnel to set calculated dam gate outflow settings, based on observations and forecasts. Typically, this involves closing the outflow dam gates and storing the peak floodwaters. The goal being to store the peak of the floodwaters behind the dam, again, a common operations procedure. Because reservoirs can usually only store the peak of the floodwaters—not the entire flood—flood storage space must be used wisely. But the revised forecast indicated that a difficult decision had to be made: Instead of impounding floodwater, the engineers believed the dam operators would have to initially *release* incoming water onto a rising river—the opposite of standard procedure.

This decision—the "dam operator's paradox"—means dam operators draft stored floodwaters more aggressively than they prefer, to make room for additional predicted high runoff. In effect, engineers intentionally allow a modest—or even large—flood to occur, by releasing amounts of stored floodwater to make room for even greater runoff. Usually highly undesirable because of its risk, this decision may be necessary during a very large AR event to avoid a catastrophic flood later in the storm cycle. In this case, it would be to avoid completely inundating the large and populated Skagit Valley below the dam.

After careful calculations, the USACE engineers directed the dam operators to adjust the dam outlet water flow settings to open—not close—the gates, choosing to increase downstream flood risk in the short term. But, again, this would allow more room in the reservoir for the heaviest part of the storm, which was yet to come. If the dam could no longer restrain the river, water would overflow the dam at the same rate it entered the reservoir. However, if the forecasted doubling of the record river inflow actually transpired, more flood storage space would be available when the worst part of the storm hit, because there would be space made available by these water releases early in the storm.

Rainfall was fierce in the mountains. June Lake, in the Cascades, broke the previous Washington state 24-h record rainfall with 15.20 in. (38.6 cm) of precipitation. The precipitation storm totals were far greater: At least eight Washington rivers set new all-time peak flow records. But in the end, the Upper Baker reservoir did not overtop the dam. Flows were very high, but not close to double the record, as predicted. In fact, the AR intense rainfall band defied the forecast and moved slightly south. The shift caused record massive flooding and damage in the Central Cascade basins, including major flooding in an adjacent basin, just to the south.

For this event, the forecast's *magnitude* was arguably very good, but the forecast's *location* was slightly off the original target of the North Cascades. This event in the Pacific Northwest highlights the uncertainties and consequences of extreme AR precipitation for reservoir engineers, emergency managers, and the general public (see Neiman et al. 2008 for further description of the AR and associated hydrometeorological conditions).

This case study was a testament to the power of a midlatitude storm effectively converting water vapor to heavy AR rainfall. In this case, when the storm collided with the Cascades in Washington state on the US West Coast, the mountains enhanced the rainfall process through orographic lifting. Finally, the runoff was concentrated, funneling down through steep valleys and drainages (see Chap. 5).

7.2.2 Real-World AR Issues

Where Will the AR Strike? Stalls, Shifts, Trajectories, and Drainage Basins

Rainfall from ARs occurs over large areas. The focused areas of high AR rainfall rates in the AR core are often the key to identifying where the worst flooding will occur. Often, the most intense rainfall band is narrow, and only covers a basin or two at any given time. At the same time, the AR airflow trajectory can line up with basin orientation to optimize precipitation. As a result, an entire region gets high water, but a few drainage basins get the worst of it. Adding to the challenge, ARs can stall, making flooding worse in a single location. At other times, the AR moves slightly from the region in which it was predicted to be, forcing water managers to pivot their expectations and adjust their levels of confidence about the storm. When the extreme AR rainfall band shifts, so do the worst flood impacts. Forecast guidance sometimes has difficulty capturing these minor, but important, changes. Forecasts must also grapple with the incoming AR trajectory, which is critical to understand, since this "angle of attack" often signals which basins are in the cross-hairs of the heaviest precipitation, and thus most vulnerable to flooding (Ralph et al. 2003; Neiman et al. 2011; Picard and Mass 2016).

Forecasts or Observations?

During a major flood event, it is vital to constantly verify a forecast's accuracy, because all forecasts have some level of uncertainty. Water management engineers typically prioritize real-time observations of actual rainfall and streamflow over forecast information, because only real-time data gives engineers the certainty needed for specific decisions. When a strong AR threatens, flood risk managers follow the forecast unless observations stray from the predictions. Decisions made as a storm unfolds offer the highest confidence for water managers who can watch how AR rainfall and runoff are tracking, and compare this to forecasts. If forecast trends and observations are similar, their confidence increases; if observations diverge from the forecast, their confidence in the forecast obviously decreases, and uncertainly will arise about the forecast's integrity. Sometimes, engineers try to analyze *which* weather element (timing, stalling, location, etc.) is driving the mismatch between observations and forecast. Increased uncertainty forces engineers to re-evaluate flood risk, essentially hedging their decisions—always on the side of safety and well-established engineering standards.

Drafting Floodwaters

When water is stored in a reservoir during a flood, the strategy is typically to store the peak of the floodwaters, reducing downstream impacts. When the storm is over and the immediate flood threat is past, water is released as quickly and as safely as possible, creating storage space for any near-future AR threat. Even as a flood is in progress, this strategy of prompt but prudent draft of reservoir floodwaters is important. During and after a flood event, the dam and reservoir are vulnerable to higher flood risk as long as water is in the reservoir's flood pocket, because, as just described in the Case Study, if additional runoff occurs, initially less flood space is available. It can take from a few days to well over a week to safely draft floodwaters after a big AR rainfall event; the draft rate depends on total storage, inflow, and release restrictions. As a result, back-to-back or stalling ARs produce major issues since the initial AR rainfall partially fills the allotted flood pocket storage space, making subsequent AR rainfall runoff difficult to manage, until the floodwaters are drafted out.

Flood Risks vs. Water Supply

Reservoirs in the western USA offer the best strategy for mitigating the effects of both major floods and extreme droughts. However, these two responsibilities can entail competing priorities for water managers, especially if the flood management season and water supply management refill season overlap. Flood protection requires reservoir storage space to be available when an AR flood threatens. But, to avoid drought, capturing abundant AR rainfall runoff to fill the reservoir may sometimes also be required. This conflict in storage priorities can become acute toward the end of the flood season (late winter) as ARs become less frequent and their magnitude decreases. However, even at the end of the flood season, uncertainty about AR magnitude remains. This uncertainty occurs while responsibility to refill becomes a higher priority.

AR Seasonality

AR seasonality is another issue relevant to reservoir management. Currently, there is often a clear difference between reservoir storage for the flood season (fall/winter) and for the water supply season (spring/summer). When can reservoir operators safely use flood storage space in reservoirs for water supply? Understanding the seasonal limits of major ARs—especially in the light of climate change—enables water managers to more confidently decide when to safely begin reservoir refill with little or no flood threat. When there is a conflict of water interests, flood risk management takes precedence over water supply. Thus, skillful AR forecasting, about which everyone is confident, is a key to managing flood control risk and water supply demands.

Probable Maximum Precipitation (PMP) Estimates

Since ARs are associated with the most extreme precipitation, another important issue is USACE's assessment of Probable Maximum Precipitation (PMP) estimates. PMPs are used in dam and levee planning to assist in design safety evaluations. A PMP is defined as "…the greatest depth of precipitation for a given duration meteorologically for a design watershed or given storm area at a particular location at a particular time of year…" (WMO 2009). PMPs are used to drive hydrological models to determine the greatest flooding a dam might be exposed to during its life-time. PMP estimates are enriched by more detailed studies of AR extreme rainfall events, which reveal the impacts of basin flooding. The November 2006 AR storm in the Case Study would be a prime candidate for inclusion into future PMP estimates for the Pacific Northwest.

Forecast Improvements

Improvements in short- and long-range rainfall, and hydrologic forecasts will continue to help water managers make better decisions. A good example of a valuable recent forecast improvement is the now common use of HRRR (High-Resolution Rapid Refresh). HRRR precipitation forecasting would have helped in November 2006 to understand if the precipitation forecast were tracking correctly. Ensemble forecasting techniques—using a set of forecasts to provide a likely range of outcomes and uncertainty—were also not used at that time. Those methods might have helped with the November 2006 storm, to put the extreme forecast into a more manageable context. General improvements in model resolution have also enhanced forecasts in the last decade, which also improves overall confidence. The advent of improved understanding, observations, and model forecasts of ARs has also helped in reservoir management decisions, as described by White et al. (2012). Hydrological forecasts have improved as well over the years, with ensemble forecasting and better resolution. However, no matter how much forecasts improve, there will always be some uncertainty. Unfortunately, difficult flood risk management decisions must be made with imperfect information.

Next Steps

A greater understanding of AR extreme precipitation, especially in mountainous terrain, combined with improved forecasts will provide the USACE with greater confidence in reservoir operations during flood risk management operations and with water supply decisions. With greater confidence, will come improved flood risk management decisions. As a result, there will be a valuable benefit for water management and public safety during major AR floods. Pushing forward on advancing the skill of AR forecasts will undoubtedly provide major benefits to flood risk management.

7.3 Forecast Informed Reservoir Operations, Lake Mendocino Pilot Study

Significant advances in the description of ARs and ability to forecast their location, strength, and associated hydrologic impacts have opened the door to explore the application of such information to forecast-informed reservoir operations (FIRO). An example of this potential application is a multi-agency collaborative project at Lake Mendocino in the Russian River watershed in California (see Fig. 7.1).

Recent modifications to the hydropower operating license that diverted water from the Eel River to the Russian River that flowed into Lake Mendocino resulted in significant declines in the volume of water transferred. As a result, the water supply reliability of the Lake Mendocino project has declined while requirements and uses for the stored water have increased. In an effort to restore lost reliability, the Sonoma County Water Agency (SCWA) partnered with the USACE and the Center for Western Weather and Water Extremes (CW3E) at Scripps Institution of Oceanography to convene a multi-agency group to explore the application of AR science to reservoir operations. The effort focuses on the wet season (October–April) when flood control limitations to reservoir storage are in place. The Russian River basin is nearly ideal for this application because (1) more than 80% of all historical floods have been caused by ARs (Konrad and Dettinger 2017), so that a focus on this almost-exclusive flood-generating mechanism—rather than on a wide range of possible hydrometeorological mechanisms—is possible, (2) the Russian River basin has the longest histories of AR Observatory (ARO) observations anywhere in the world (Chapter 3), so this basin is arguably where ARs are best known, and (3) the basin is home to the most current forecast and reservoir-operation technologies and agencies available.

Figure 7.2 describes the premise of the methodology. When a storm arrives and runoff increases inflows to the reservoir, storage in the reservoir may encroach into the flood reserve space according to the water control manual rules to mitigate downstream flows. (Each USACE-operated reservoir has a unique water control manual, which is the guiding document that specifies how that particular reservoir is to be operated.) If clear weather follows the storm, the methodology examines the forecast to determine the timing for releasing the water from the flood storage space. If more storms are in the forecast, the methodology follows the water control manual's guidance for releasing water from the encroached flood reserve space. If no storms are in the forecast, a decision process could be developed to hold onto a portion of the water for potential later use. Analogously, forecasts of storminess could trigger creation of additional flood reserve space in the reservoir by having water managers pre-release water to lessen downstream impacts that the water released from the conservation space would be refilled by storm runoff.

In 2014, the SCWA and USACE began a 3-year effort to explore the viability of such a strategy with the formation of a multi-agency steering committee. Membership on the committee included representatives from the USACE, the National Oceanic and Atmospheric Administration (NOAA), the US Geological Survey (USGS), the Bureau of Reclamation (Reclamation), and the California Department of Water Resources (DWR). The steering committee also had representation from the CW3E for their expertise in AR research,

Fig. 7.1 Map of Russian River Watershed and highlight of Lake Mendocino

and its Director co-chaired the steering committee with the SCWA Chief Engineer. Over the 3-year period, the steering committee reviewed the available and developing science of ARs and their forecasting as well as methods to evaluate the dynamic risk of carrying water in the flood reserve space.

The researchers at CW3E worked on characterizing and forecasting ARs and developing decision support tools to apply to reservoir management. While that occurred, two modeling efforts were explored on how to manage the dynamic risk of a limited encroachment for Lake Mendocino through the October–April wet season. The USACE Hydrologic Engineering Center led one effort, and the SCWA led the other. Although both groups considered both flood risks and supply reliability in their models, the USACE team focused its modeling method on determining whether flood-risk management would be adversely affected if FIRO were implemented. The SCWA team focused *their* modeling method on determining the extent to which water-supply reliability benefits could be obtained. Both groups used a common data set for forecasting from the NWS's California Nevada River Forecast Center. The data set runs from 1985–2010 and includes a number of large flood and drought events to test the two modeling methods. The two groups deployed relatively similar strategies for ingesting and using forecasts in their models of reservoir operations, and used existing operations to compare results to outcomes. Both modeling methods were able to improve supply reliability without increasing the risk of flood damage in the basin.

The results of the 3-year period of study were presented in a report titled "Preliminary Viability Assessment of Lake Mendocino Forecast Informed Reservoir Operations;" Jasperse et al. 2017). Three appendices accompanied the report that detailed work from the SCWA, the USACE, and CW3E. The report's policy findings included:

- Elements of FIRO at Lake Mendocino are currently viable for improving reliability in meeting water management objectives without impairing flood protection.
- SCWA should continue efforts with the steering committee to explore implementation of the concept through a formal USACE process used to make temporary modifications to the operating rules (referred to as the "deviation process" of the USACE).

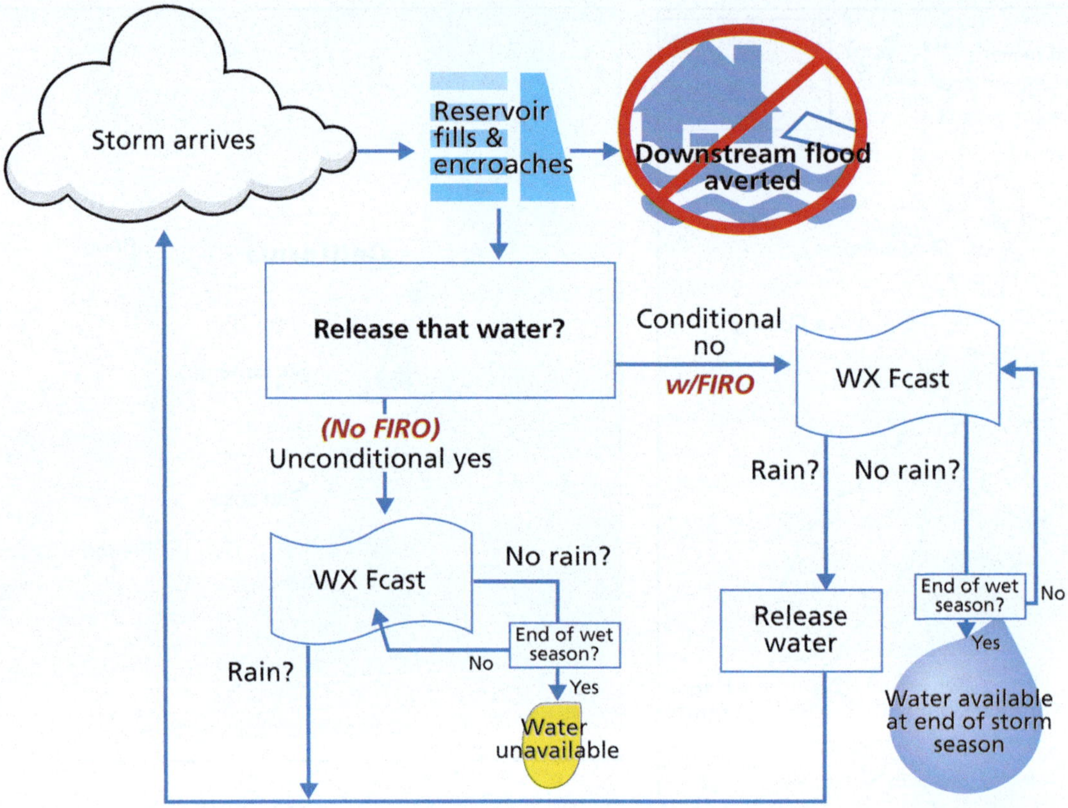

Fig. 7.2 Simplified schematic of a forecast-informed reservoir operations (FIRO) strategy based on a knowledge of how much lead time is needed to safely evacuate excess water behind a reservoir after a major storm that led to "encroachment" of reservoir levels into standard flood-mitigation space behind the reservoir, and forecasts of the arrival of major AR stortms that are skillful over at least that much lead time

- To realize the full potential of FIRO, further research on ARs and integrated hydrometeorological modeling, monitoring, scientific understanding, and prediction is needed, along with further decision support system development.

Its more technical findings included:

- When additional storage was maintained for the cases for which forecasts offered reasonable assurances of low flood risk, the SCWA study indicated that additional water could be stored and made available during nearly all historical years simulated. Increases in water that could be carried over from water year to water year ranged from about 8600 to 28,000 acre-feet (AF)—or as much as a 49% increase—depending on how good the AR forecasts are and how aggressively they are used in operations.
- The USACE study showed no significant deterioration of the ability of the reservoir system to manage flood risks for the river basin as a whole. Based on simulations of 60 years of historical flows, this study found no increase in annual average flood damages; a more wide-ranging flood-risk analysis using synthetic streamflows indicated a minor increase in damages if imperfect forecasts were used.
- These results ultimately depend on how well forecasts of major storms, specifically ARs, could be verified in the real world, and how much lead time forecasts would need to provide for operators to clear water from the flood-protection pools, once an approaching AR storm were indicated with confidence. Analysis of river channel geometry and hydraulics under flood flows showed that reservoir releases (to clear space in anticipation of a major AR storm) of up to 10,000 AF can safely be made in 2 days, with another 2–3 days needed to allow that water to safely exit river reaches downstream. Thus, to hold water in the reservoir until forecasts indicated the significant risk of a new AR arrival—particularly after an earlier event had caused the reservoir to fill beyond currently allowable levels—skillful forecasts of landfalling ARs at 5-day lead times would be required.
- The CW3E studies, along with the SCWA and USACE simulations, indicate that current forecasts already can be used to improve reservoir operations safely. The CW3E studies of real-world forecasts of major and minor AR landfalls also indicated that forecasts adequate to support FIRO are available or can likely be developed with additional research. In particular, the studies showed that forecasts of, and during, dry interludes between ARs are best,

whereas forecasting of the specifics of precipitation rates from individual ARs upon arrival includes significant errors. Thus, operations that rely on forecasting conditions with no ARs will result, in the near term, in actions about which personnel can take the most confidence and trust to be the safest.

Taken together, these findings support the FIRO concept in this setting wherein 80% or more of all historical floods are attributable to AR landfalls. Under these conditions, AR science has helped to focus analysis and operations on very specific storm conditions, and has focused research on the particular answers and skills with which engineers will need to be provided if forecasts are to form a reliable basis for managing flood and water supply jointly in the Russian River basin. The multi-agency Preliminary Viability Assessment (PVA) efforts have now led to a Final Viability Assessment (FVA) effort that strives to extend the skill of forecasts of ARs (and of non-AR conditions) and to put the particular tools, forecasts, and information flows in place to allow opportunities and risks to be comprehensively assessed. The methodology could greatly facilitate the challenge of meeting multiple water supply management objectives with existing infrastructure alternatives.

In summer of 2017, SCWA was preparing to submit a deviation request to the USACE to try out FIRO. Although the normal process for deviation requests in the case of a reservoir jointly operated by a local water agency (SCWA in this case) and USACE would entail SCWA submitting the deviation request, the USACE asked that it be submitted by the FIRO steering committee. In late 2018, the USACE formally approved (after thorough review) the major deviation request to try FIRO in winter 2018–2019. This decision marks a major milestone in the potential use of AR forecasts—and application of the underlying science and technology—to address major operational challenges in water management and flood control, with benefits to the restoration of endangered salmon that call the Russian River home.

7.4 ARs Use in Flood Planning in California

Perceived changes in flood magnitudes over the twentieth century in California's Central Valley rivers (see Fig. 7.3) raise concern that, with climate change, future floods will be even larger. Legislation enacted in the early part of the twenty-first century instructed the DWR to incorporate climate change considerations into its flood planning activities as part of the Central Valley Flood Protection Plan (CVFPP). The CVFPP was initially issued in 2012 and updated in 2017, and future updates are expected every 5 years. ARs provide a physical basis by which to better understand present-day storm characteristics; they can also inform our understanding of potential changes in future floods that will result from climate change.

The current standard of practice uses guidelines in Bulletin 17B (USGS 1982). The guidelines use statistical analysis of annual peak flows to determine the magnitude of flood peaks and volumes associated with a given annual exceedance probability of occurrence. Recent work by the USACE in the Central Valley Hydrology Study (CVHS) uses a flood-of-record approach to develop storm hydrographs with the requisite peak and volume information (DWR 2015). With this methodology, a 100-year design flood can be created by taking a historical flood and scaling its peak and volume to match the statistical estimates of peak and volume associated with the specified return interval (in this case, 100 years). A 100-year flood is a flood threshold that has a 1 in 100 chance of being met or exceeded in any given year. This concept deals only with historical data to develop the statistics for the different return-period thresholds of flood peak and volume.

As a means of incorporating climate change and using the latest research in ARs, work has begun to establish a methodology that relates flood thresholds (e.g., 200-year event) to a given construct of AR characteristics. Characteristics under consideration that may be influenced by climate change include moisture flux, duration, event clustering, and freezing elevation. To do this work, observations of AR characteristics and study of how these characteristics relate to flood events are needed. The hydrometeorology testbed (HMT) network (Chap. 3) provides the observations; these characteristics are now being related to flood peak and volume. Figure 7.4 shows an example illustration created by Benjamin Hatchett at Desert Research Institute. Figure 7.4, adapted from Fig. 5 in Hatchett and McEvoy (2018), presents overlays of data from the HMT network along with measurements of snow water equivalent, precipitation, and runoff for the winter of water year 2017, which produced notable flood peaks in January and February.

As the relationships between AR characteristics and flood peaks and volumes are built, design floods can be developed through constructed extremes methodologies to evaluate volume-management capacities for everything from parking lot drainages, which look at short-time rainfall intensities, to reservoir systems in large watersheds that fold together multiple lines of both observed and forecasted information.

With this framework for incorporating ARs into flood planning, climate change can be incorporated through physically meaningful parameters, and the nuances of interactions among different components of AR storms—such as moisture transport, freezing elevation, and antecedent conditions—can be explored.

Current efforts in AR research include the following entities and projects.

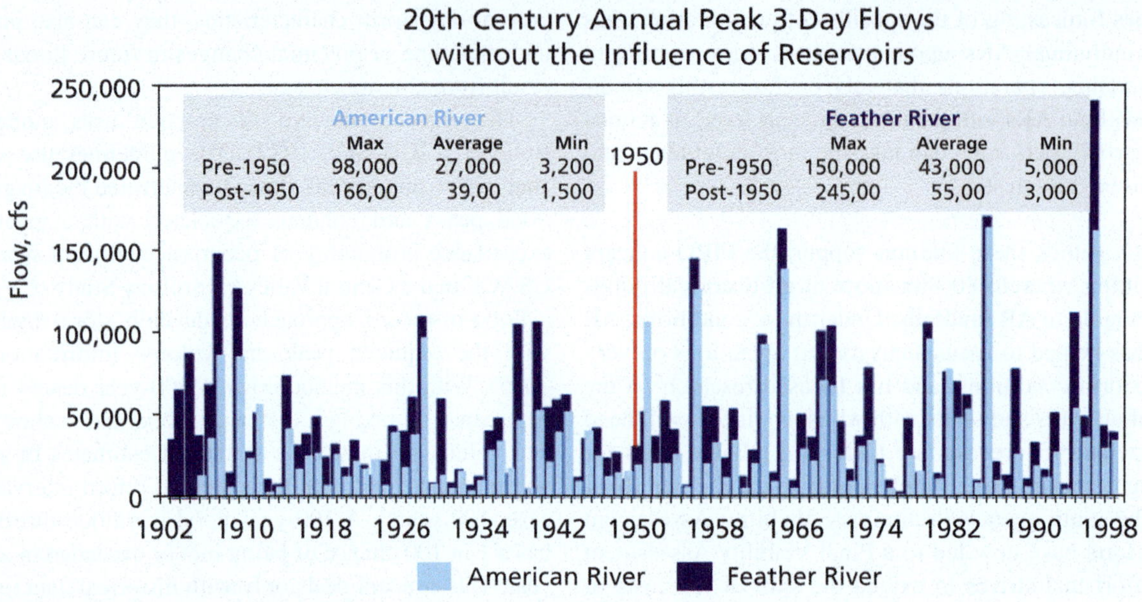

Fig. 7.3 Perceived changes in flood magnitudes over the twentieth century in Central Valley rivers

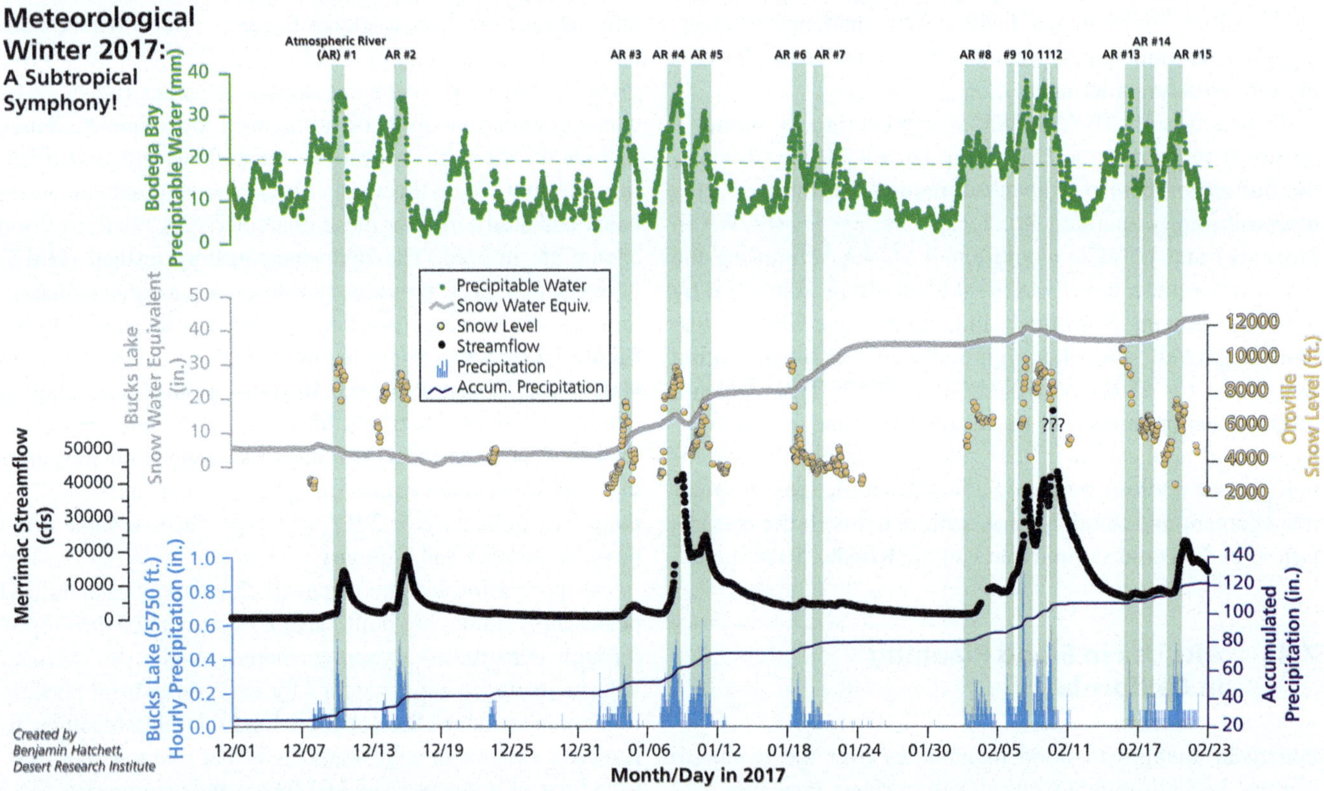

Fig. 7.4 AR characteristics matched to precipitation, snowpack, and runoff for the Feather River watershed for winter of water year 2017. (Adapted from Hatchett and McEvoy 2018)

- Scripps Institution of Oceanography and the CW3E, with collaborators at the Jet Propulsion Laboratory and elsewhere, are carrying out research to characterize past, present, and future ARs using characteristics such as freezing elevation and IVT (e.g., Ralph et al. 2016a, 2016b, 2018, 2019; Gershunov et al. 2017; Dettinger et al. 2018; Espinoza et al. 2018). To inform advance planning efforts during flood events, forecast tools (see

Cordeira et al. 2017) are helping to illuminate the magnitude and timing of events as represented in weather forecast models.

- The University of California Davis Hydrologic Research Laboratory (HRL) is using computer simulations to explore the interaction of extreme historical storm systems and the watersheds of the west-slope Sierra Nevada (Ishida et al. 2015, 2016; Chen et al. 2016; Trinh et al. 2016; Diaz et al. 2017; Toride et al. 2017; Ohara et al. 2017). The boundary condition shifting method—where atmospheric conditions are shifted relative to land surface—is being used to explore the character of different historical storms at the watershed scale. Changes in atmospheric conditions such as available moisture for transport also enables exploration of extreme conditions. The role of snowmelt in augmenting flood flows was studied as well.

The framework (i.e., organizational planning strategies, documentation requirements, decision processes) for incorporating ARs into flood planning can also enable forecasts of AR events to be associated with different classes of floods, which would motivate different levels of pre-event actions to mitigate flood damages. While the planning framework is being developed, complementary efforts are underway to develop a framework for embedding AR knowledge into forecast and warning activities. Longer lead times enable advance mitigative actions and volume management planning to occur, which improves integrated water management. Significant research remains to be done before this framework can be fully implemented. However, once implemented, planning and operations can work from a common perspective of AR characteristics to improve water management from the event time-scale out to planning horizons of decades.

7.5 AR Science, Natural Hazards Risk Reduction, and ARkStorm

In 2005, after Hurricane Katrina, the US Congress funded the USGS Multi-Hazards Demonstration Project to apply multi-disciplinary earth science to reduce the risk of natural hazards in southern California.

Now known as USGS Science Application for Risk Reduction (SAFRR), single, large, scientifically plausible scenarios are used to improve national security and reduce risks caused by natural hazards. This helps all involved to better understand cascading effects and interdependencies—such as effects on the social fabric and the economy as well as policy implications—and to unveil unknown vulnerabilities.

These scenarios include the following characteristics:

- To improve credibility, scenarios are not necessarily "worst-case," but are physics-based, quantifiable, and plausible natural hazard events.
- The events pair multiple scientific disciplines with social science and with the needs of external partners, such as the public and the community of emergency, resource, lifeline, and business continuity managers who would be responsible for physical effects on homes, buildings, and infrastructure.
- To be useful to these broad audiences, scenario efforts strive also to create the necessary tools and information to help this larger community visualize, plan, prepare for, and improve communications before, during, and after real emergencies.

Earthquakes were commonly understood to be California's most fearsome threat, so the goal of the highly successful first scenario, ShakeOut—based on a M7.8 earthquake in southern California—was to change California's culture of earthquake preparedness. (The success of this scenario has inspired other efforts worldwide: http://www.shakeout.org/.)

Given post-Hurricane Katrina awareness though, flood experts and the media raised the specter of California's flood risk. For instance, in an article titled "Is California Next?" Robert L. Reid, Senior Editor of *Civil Engineering* (the magazine of the American Society of Civil Engineers), pondered the decaying Sacramento-San Joaquin Delta levees that keep drinking water safe for 20 million people:

> ... But some experts wonder if the levees are leaving these regions vulnerable to a disaster far greater than the one inflicted on New Orleans by Hurricane Katrina.[1]

Also, then-Governor of California, Arnold Schwarzenegger, was quoted in The New York Times regarding "woefully inadequate flood prevention" in the Sacramento area:

> "Even New Orleans had a 250-year level of flood protection," Mr. Schwarzenegger wrote in a letter to members of Congress. "Sacramento only has about a 100-year level of protection."[2]

In 2008, ARkStorm thus became the second SAFRR scenario (the "AR" in ARkStorm represents "Atmospheric River").

West Coast storms, and particularly ARs, bring some of the highest 3-day rainfall totals to the US West Coast. AR storms are multi-hazard conglomerations of winds, rain, coastal erosion and inundation, landslides, and debris flows that have the demonstrated potential to cause widespread physical damages—with environmental, social, and economic implications. As such, AR storms were a good fit for

[1] Reid, R. (2005). "Is California Next?." Civ. Eng., https://doi.org/10.1061/ciegag.0000045, 39–47, 84–85.

[2] *The New York Times*, November 15, 2011

the multi-hazard team for a second scenario project, with the goal of improving preparedness and mitigation for—and resilience to—these extreme storms.

Because floods are difficult to conceptualize and gauge by a public usually unfamiliar with probabilities—the primary tool used in flood management—ARkStorm was also meant to elevate public awareness and understanding of pre-historic and historic storm potential. It would also leverage new knowledge of the physical aspects of ARs to make them more tangible and thereby useful to the public, emergency managers, and lifeline and business continuity experts. If this knowledge could complement AR forecasting, tying likely effects to the forecast, preparedness and mitigation efforts could be improved.

The ARkStorm effort involved participation and contributions from many federal, state, and academic organizations including NOAA's Earth Systems Research Laboratory. Using a new understanding of AR physics, combined with the down-scaled meteorological data that described two recent ARs (in 1969 and 1986), the ARkStorm team described and modeled a prolonged (23-day long) sequence of back-to-back storms (Dettinger et al. 2012), similar to the 40-day sequence that caused statewide flood damage in California in 1862 (Porter et al. 2011; Fig. 7.5).

The AR storm in January 1969 caused extensive damage to the southern half of California; the impacts of the February 1986 storm were focused more on the northern part of the state (Fig. 7.6). The research team developing the meteorology examined weather charts and found a meteorologically consistent structure to the atmosphere after the January 1969 storm and before the February 1986 storm. This provided the basis for stitching the two events together to create the extended flooding scenario. To incorporate the concept of an AR stalling over a region and amplifying impacts, the wettest day of the January 1969 storm was repeated in the constructed scenario time-series.

In 2010, the ARkStorm team—with details of the scenario in hand—engaged experts across California in a dozen panel discussions spread evenly among Pasadena, Sacramento, and Menlo Park. The discussions engaged local and state experts in water supply, flood management, power supplies and distribution systems, wastewater and environmental concerns, transportation, and emergency response. The discussions elicited plausible ARkStorm outcomes in considerable detail, likely responses, and practical proactive measures to mitigate expected impacts. Members of the team ultimately developed a coastal-storm inundation model (CoSMoS) and California's first landslide susceptibility map (California Geologic Survey; CGS) in response to questions raised in these panel discussions, and to better understand secondary meteorological and geophysical hazards (flood, wind, landslide, coastal erosion, and inundation) across California. In the years since 2010, scientists, engineers, and economists have studied the ARkStorm scenario, resulting in a special issue of the America Society of Civil Engineers' *Natural Hazards Review* (Volume 17, Issue 4, 2016). That issue described new methods and results for estimating landslide losses, evacuation resource planning, agricultural loss, environmental-health implications, and port-infrastructure operations.

Engineers and economists on the team estimated the physical damages to homes, infrastructure, agriculture, and the environment so the likely social and economic damages to California and the nation could be calculated. The scenario was the focus of a major ARkStorm Forum in Sacramento and of the 2010 *California Extreme Precipitation Symposium* titled *ARkStorm: Examining a Potential California Flood Disaster* at the University of California, Davis.

Across California, property damage from the ARkStorm scenario was estimated to exceed $300 billion, mostly from flooding. Additional demand surge, damage and losses, lifeline damages, and business interruptions brought the total cost of an ARkStorm-sized series of storms to nearly $725 billion—*nearly three times the losses estimated from the SAFRR scenario that described a M7.8 earthquake in southern California.*

Beginning soon after its creation, the ARkStorm scenario was used in preparedness exercises related to their agency needs and missions by NASA, the US Navy, and the County of Ventura and cities therein. ARkStorm II Emergency Planning and Response Exercise was a joint effort of the Ventura County Public Works Agency (VCPWA), the Federal Emergency Management Agency (FEMA), and the USGS to better identify critical community facilities and infrastructures along the Santa Clara River that are likely to be affected by a major flood event, as well as to identify areas of emergency service needs and supply deficiencies (Hosseinipour et al. 2013). The Ventura County Emergency Response Exercise illustrated that the framework could be a suitable template to be used by communities for emergency planning.

In 2012, ARkStorm was featured in the California Governor's Conference on Extreme Climate Risks and California's Future, and was used by the California Governor's Office of Emergency Service as the basis for the Northern California Catastrophic Flood Response Plan. ARkStorm research information was used throughout the plan, particularly impacts to infrastructure such as highways, power, wastewater, water, dams, telecommunications, and agriculture. This information was used to give planners an idea of what to expect and plan for in a major flood event.

The ARkStorm@Tahoe Project (Albano et al. 2016), similar to the Ventura County case above, developed new technical products to simulate and describe the scenario in the study area. In addition, the project convened six stakeholder

Fig. 7.5 K Street, Sacramento, California, looking east 1861–1862

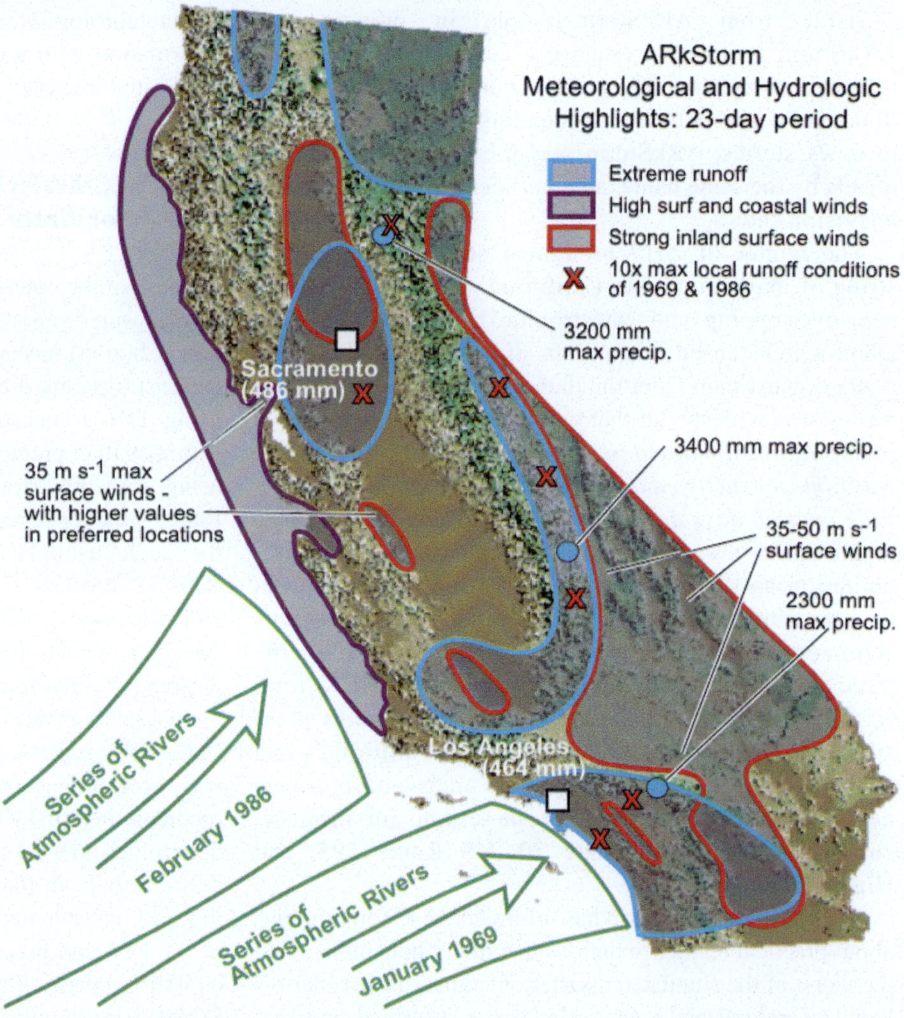

Fig. 7.6 Summary map of selected extreme weather conditions within the ARkStorm scenario (Dettinger et al. 2012)

meetings with over 300 participants, and ended with a tabletop emergency-response exercise that focused on the preparation, response, and recovery phases of the storm. ARkStorm@Tahoe described how over 185 miles of major roads would be inaccessible from flooding, including the Interstate 80 corridor west of Reno.

ARkStorm was applied to California's Lahontan Regional Water Quality Control Board's Climate Change Mitigation and Adaptation Strategy to identify vulnerable infrastructure. Several other agencies are using this as a reference point for a catastrophic flood planning scenario. The scenario is included in the Nevada County Local Hazards Mitigation Plan as a benchmark toward which future flood conditions should be managed. Beyond this, ARkStorm@Tahoe benefits included increasing awareness about weaknesses in inter-agency communications among utilities, stormwater managers, and emergency responders, which several participants have since sought to improve.

Five years after the release of ARkStorm, the Bay Area Council Economic Institute released Surviving the Storm (http://www.bayareaeconomy.org/report/surviving-the-storm/), which was based in part on ARkStorm, but with reduced storm impacts.

Aside from ARkStorm's role in preparedness, ARkStorm has been extensively covered in the media, thereby raising awareness of the potential for a massive flood and a new public understanding of ARs. In addition to news stories, ARkStorm was the inspiration for two novels by the same name, and has been featured in several television shows.

The winter of 2017 brought a seemingly unending string of major storms to California, setting the stage of near-overtopping and catastrophic releases at Oroville Dam, which caused evacuations of roughly 200,000 residents downstream from the dam. Flooding in the Central Valley was widespread that winter. In fact, *the amount of precipitation simulated in the 23-day ARkStorm scenario was almost exactly equal to the precipitation that fell in a little over 80 days during the 2017 storm season*. Thus, a single winter's worth of storms brought just as much precipitation as the ARkStorm scenario, albeit spread out over three to four times the duration, so that 2017 impacts were considerably less than those predicted based on the more concentrated ARkStorm onslaughts. The ARKStorm scenario rainfall has thus proved to be scientifically plausible, and the scenario's impacts around California—and in the Tahoe, Reno, and Carson City areas—have proven nearly prophetic. The 2017 storm season, for instance, cut off access to Interstate 80, US Route 395, and Highway 50.

The ARkStorm scenario has informed these communities about practical resilience options that might help them avoid the worst of the simulated disaster outcomes. The scenario's blend of natural and social science has informed and prepared Californians for possible AR-related catastrophes to come. For more information on ARkStorm, go to https://pubs.usgs.gov/of/2010/1312/.

7.6 Scales That Communicate AR Intensity and Impacts

With the advent of modern weather prediction, many preparations are made based on forecasts of extreme storms. For hurricanes (typhoons), Nor'easters, and tornadoes, there exist intensity scales that relate also to impacts. As imperfect as these scales are, they have proven invaluable in communicating weather event risks to the public, and are used by many people who must make decisions based on storm forecasts, e.g., emergency managers, transportation officials, energy providers. However, the west coasts of the USA, Europe, and South America, for example, do not generally experience these storm types. Instead, it is the AR-type storms that have the most impact. This section briefly summarizes two approaches aimed at quantifying AR storms in frameworks that should help non-meteorologists to assess the potential impacts of an incoming AR, as well as empower the meteorologists who are working to provide actionable information with a common language with which to communicate forecasts.

7.6.1 ECMWF's Extreme Forecast Index for Water Vapor Transport

Because of the connection between vertically integrated horizontal water vapor transport (IVT) and extreme precipitation (Chap. 5), research has assessed how well atmospheric forecast systems predict IVT versus precipitation to see if using IVT forecasts could provide earlier awareness of extremes than precipitation forecasts. Two studies—the first using forecasts from the European Centre for Medium-Range Weather Forecasts (ECMWF) (Lavers et al. 2014) and the second using NCEP's Global Ensemble Forecast System (GEFS) forecasts (Lavers et al. 2016a)—showed that the IVT forecasts had higher predictability than precipitation forecasts, which indicated that IVT may provide earlier awareness of hydrometeorological extremes.

In Lavers et al. (2014), the ECMWF Extreme Forecast Index (EFI; Lalaurette 2003; Zsoter 2006; Zsoter et al. 2014) was shown to be a possible product that could use the higher predictability of IVT to warn of extreme events. The EFI determines how the ensemble forecasts' probability distribution differs from that of the model climate, thus identifying times of possible extreme conditions. To investigate the skill of the IVT and precipitation EFI in capturing extreme precipitation events, Lavers et al. (2016b) evaluated them in ECMWF ensemble forecasts during three winters (2013–2014, 2014–2015, and 2015–2016) across Western Europe. Results suggested that the IVT EFI is more able than the precipitation EFI to detect extremes during early week 2 (i.e., forecast days 9 and 10) for forecasts initialized in a positive North Atlantic Oscillation (NAO) phase. Figure 7.7 shows the relative operating characteristic (ROC) curves on forecast days 9 and 10, and comparison of the solid lines highlights this result. This indicates the potential for earlier awareness of hydrometeorological extremes compared to using the precipitation EFI. During a positive NAO phase, extratropical cyclones are generally more frequent, which in turn means that extreme precipitation is more likely to be associated with the IVT within the AR of these storms. However, at shorter lead times, the large-scale IVT fields tended to result in a higher false alarm rate than when the precipitation EFI was used, which reduces its usefulness. For forecasts initialized on negative NAO days (i.e., negative phase), the precipitation EFI was more useful in capturing extremes, as shown by the dash-dot lines in Fig. 7.7.

In December 2015, an AR impacted northwestern Europe causing the largest daily rainfall total recorded in England; widespread flooding resulted. Figure 7.8a presents the IVT EFI at the global scale for storm Desmond on forecast day 9 from 00 UTC 27 November 2015, which shows the potential for extreme IVT over the North Atlantic Ocean. Figure 7.8b presents a global IVT EFI map for forecast day 1 from 00 UTC 5 December 2015. There are higher EFI values on day 1, which is expected at short lead times, and the EFI values of around one indicate that the ensemble forecasts are extreme compared to the model climatology; the AR responsible for extreme precipitation across northwestern Europe is clearly visible. As mentioned earlier, there will be a high false alarm rate for the IVT EFI on forecast day 1 (and at other short lead times), but, notably, the IVT EFI may be able to inform a user of the atmospheric processes that underlie the event.

Two further studies have highlighted that the IVT EFI has added usefulness (over the precipitation EFI) in capturing extreme precipitation in western North America (Lavers et al. 2017) and on the Iberian Peninsula (Lavers et al. 2018). Currently, the ECMWF is planning for the operational implementation of the IVT Extreme Forecast Index.

7.6.2 A Scale for Atmospheric River Strength and Impacts

The state of California observed an intense drought from 2011–2016, followed by record-breaking precipitation during the 2016–2017 cool season. This dramatic hydrological reversal played a key role in leading to the development of a scale (Ralph et al. 2019) to characterize the strength and impacts of ARs. First came the realization that ARs can be both hazardous and beneficial, depending on the intensity and duration of the storm, and antecedent conditions such as soil moisture and reservoir storage. Second, with a seemingly endless parade of ARs making landfall along the US West Coast, forecasters and researchers began looking for ways to differentiate between ARs that are primarily hazardous and life-threatening, and those that are primarily beneficial (e.g., by enhancing water supply and snowpack). Of course some ARs can be both hazardous in some ways and beneficial in others.

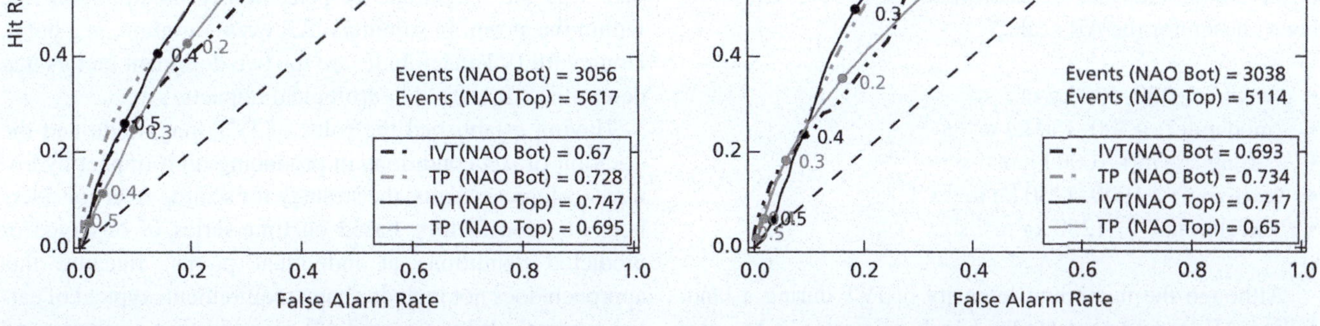

Fig. 7.7 The ROC curves on forecast days 9 and 10 (**a**, **b**, respectively) conditioned on the NAO index. *Solid lines* are for the 90 top-ranked NAO index days, and *dash-dot lines* are for the 90 bottom-ranked NAO index days (IVT in *black*; precipitation in *gray*). ROC areas are provided in the legends, as are the number of extreme precipitation events in each NAO category. For clarity, ROC curve points and numbering are given only for the solid lines. (Lavers et al. 2016b)

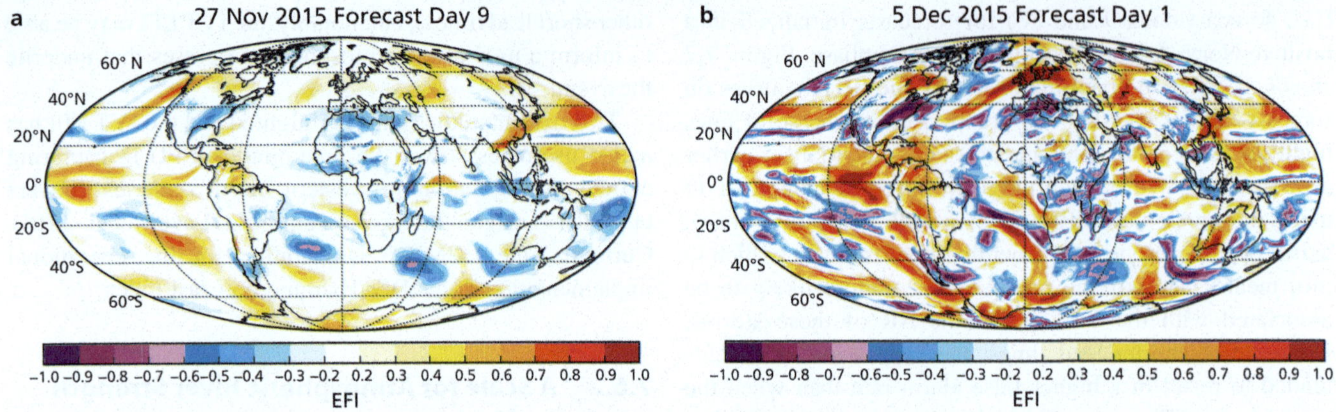

Fig. 7.8 The IVT EFI product at a global scale valid for storm Desmond on 5 December 2015 on (**a**) forecast day 9 initialized at 00 UTC 27 November 2015, and on (**b**) forecast day 1 initialized at 00 UTC 5th December 2015. (Lavers et al. 2016b)

Many AR identification and tracking methods define ARs as features with certain geometric characteristics and IVT magnitudes, such as at least 250 kg m^{-1} s^{-1} (and/or IWV values >2.0 cm) throughout. However, the range of IVT values along the US West Coast varies widely; observations, dropsondes, and reanalyses highlight many examples of landfalling ARs with IVT magnitudes >1000 kg m^{-1} s^{-1} (Ralph et al. 2004; Dettinger et al. 2018). For example, a landfalling AR in early January 2017 contained radiosonde-derived IVT and IWV magnitudes of 1102 kg m^{-1} s^{-1} and 3.5 cm, respectively, at Bodega Bay, California, where AR conditions (i.e., IVT ≥250 kg m^{-1} s^{-1}) persisted for ~36 h. In addition, dropsonde measurements collected from 21 offshore ARs over several field campaigns show that observed AR IVT intensities can exceed 1250 kg m^{-1} s^{-1} (Ralph et al. 2017; Guan et al. 2018), and, on average, an AR transports as water vapor about 25 times the discharge (as liquid, of course) of the Mississippi River in the Gulf of Mexico. In addition to observational data, forecasting experience with the Global Forecast System (GFS) and the GEFS shows that such forecasts contain many cases of ARs with IVT >1000 kg m^{-1} s^{-1}. Based on the observations and analyses cited above, the following IVT intensity thresholds (which can be applied to observational analyses of past events, or to forecasts) have been chosen for the AR scale:

- "weak": ≥250–500 kg m^{-1} s^{-1}
- "moderate": ≥500–750 kg m^{-1} s^{-1}
- "strong": ≥750–1000 kg m^{-1} s^{-1}
- "extreme": ≥1000–1250 kg m^{-1} s^{-1}
- "exceptional": ≥1250 kg m^{-1} s^{-1}.

Although the maximum intensity of IVT during a landfalling AR largely controls the hourly rain rates, it has also been shown that the storm-total precipitation (and hence runoff) at a given location is controlled heavily by the storm-total water vapor transport (Ralph et al. 2013). These authors found that 62% of variance in storm-total stream flow was explained by storm-total precipitation, and another 17% was explained by variations in antecedent soil moisture conditions. Another key result was that the top 10% longest-duration ARs in their climatology of nearly 100 AR landfalls had a duration that was *double* the average duration of all AR landfalls (i.e., 40 h vs. 20 h). Remarkably, instead of the streamflow simply doubling during the longer events, the top 10% longest events yielded seven times as much storm-total streamflow. This analysis was based on observations of AR conditions using an Atmospheric River observatory (ARO; see Chap. 3) and found that 74% of the variance in storm-total precipitation is explained by variance in storm-total upslope water vapor transport (upslope IVT; where the upslope direction is based on regional terrain orientation). This study found that an average AR lasted 20 hours at this location, a result confirmed by analysis from Rutz et al. (2014) using reanalysis methods. Rutz et al. (2014) extended this result from one coastal location to the entire western USA, and showed that AR duration varies widely as a function of location. Furthermore, Lamjiri et al. (2017) concluded that over the western USA, storm duration was a more important parameter in determining storm-total precipitation than was the magnitude of peak hourly precipitation rate within the storm. In summary, AR event duration, at a point, along with IVT magnitude, are the two dominant factors that determine AR-related hydrological impacts.

Having established the roles of IVT magnitude and the duration of AR conditions in producing high-impact hydrometeorological events, the strategy for scaling ARs considers both of these factors, based on time-series of observed or predicted conditions at individual points. Because this approach does not include shape requirements typical of earlier studies' object-oriented AR identification methods, and uses only time-series of IVT in an Eulerian framework, it greatly simplifies implementation of the scale in gridded data sets like reanalyses and forecasts.

The AR scale ranks AR events based on the maximum instantaneous IVT magnitude and the duration of the event (i.e., the duration of IVT ≥250 kg m^{-1} s^{-1}) at a given point (Table 7.1 and Fig. 7.9). Using Fig. 7.9, an AR event at a given location is ranked by locating the row associated with the maximum IVT and the column associated with the event duration. For example, a maximum IVT ≥750 and <1000 kg m^{-1} s^{-1} would be classified as being of "strong" intensity, and a duration ≥24 and <48 h would rank this as an AR 3 event. If the same event instead had a duration ≥48 h it would be upgraded to an AR 4 event, and a duration <24 h would downgrade it to an AR 2 event. (These duration-based upgrades and downgrades show how strongly duration contributes to impacts.) This system of classification works up and down the AR scale with two exceptions. First, for now, all ARs above the AR 5 thresholds are considered AR 5. Second, "weak" ARs (i.e., those with maximum IVT between 250–500 kg m^{-1} s^{-1}), but with a duration <24 h do not receive a ranking on the AR Scale, as represented by the gray area in Fig. 7.9 (even weak events require a minimum duration of 12 h).

Examples of AR 1–5 events from the 2016–2017 cool season, using both satellite-observed IWV (from Special Sensor Microwave/Imager [SSM/I]) and IVT (from the National Centers for Environmental Prediction's Global Forecast System [NCEP's GFS] model), are provided in Fig. 7.10. The synoptic-scale pattern associated with these events is characterized by a mid-latitude cyclone–anticyclone pair over the northeast Pacific and an AR located between these features. The AR, in each case, is readily identified as a long and narrow region of large IWV and IVT that extends poleward and eastward toward the US West Coast. Generally, the magnitude of IWV and IVT increases as a function of increasing AR ranking, but because IWV magnitude is not directly considered as part of the AR scale, it does not increase uniformly as a function of increasing AR ranking. For example, the IWV magnitude associated with the listed AR 5 event is smaller than that associated with the listed AR 3 and 4 events.

Table 7.1 *Top:* An AR intensity scale based on maximum instantaneous IVT magnitude and duration of AR conditions (i.e., IVT > 250 kg m^{-1} s^{-1}), *Bottom:* A subjective assessment of the potential for beneficial or hazardous impacts

	Duration of AR Conditions (h)		
Max IVT (kg m^{-1} s^{-1})	<24	24–48	>48
<250	Not an AR	Not an AR	Not an AR
250–500	Weak AR	AR 1	AR 2
500–750	AR 1	AR 2	AR 3
750–1000	AR 2	AR 3	AR 4
1000–1250	AR 3	AR 4	AR 5
> 1250	AR 4	AR 5	AR 5
AR category scale	Assessment of beneficial vs. hazardous impacts		
AR 1	Primarily beneficial		
AR 2	Mostly beneficial, but also hazardous		
AR 3	Balance of beneficial and hazardous		
AR 4	Mostly hazardous, but also beneficial		
AR 5	Primarily hazardous		

From Ralph et al. (2019)

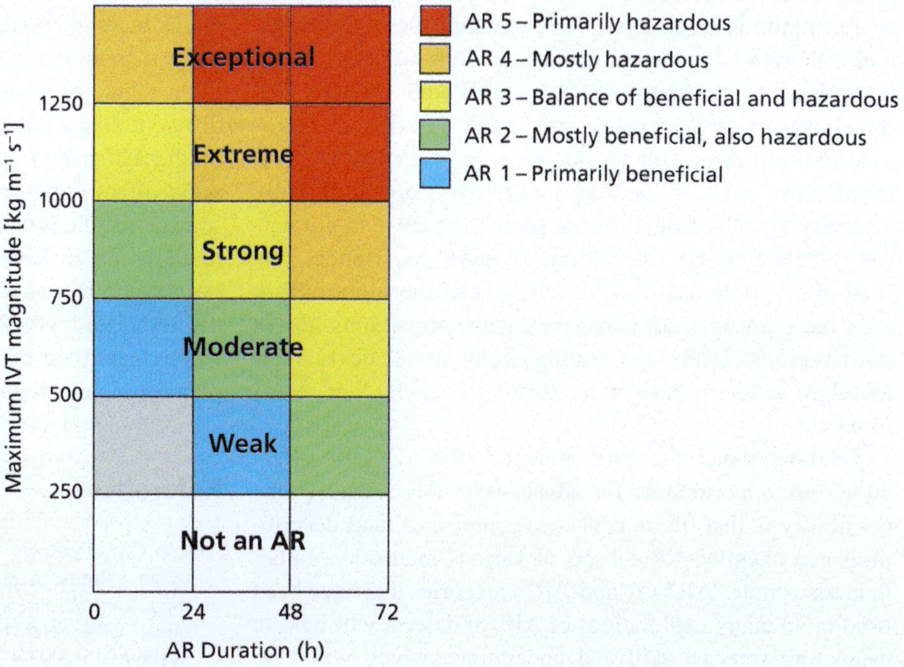

Fig. 7.9 A scale that ranks AR events based on the maximum instantaneous IVT associated with a period of AR conditions (i.e., IVT ≥250 kg m^{-1} s^{-1}) and the duration of those conditions at a point. (Ralph et al. 2019)

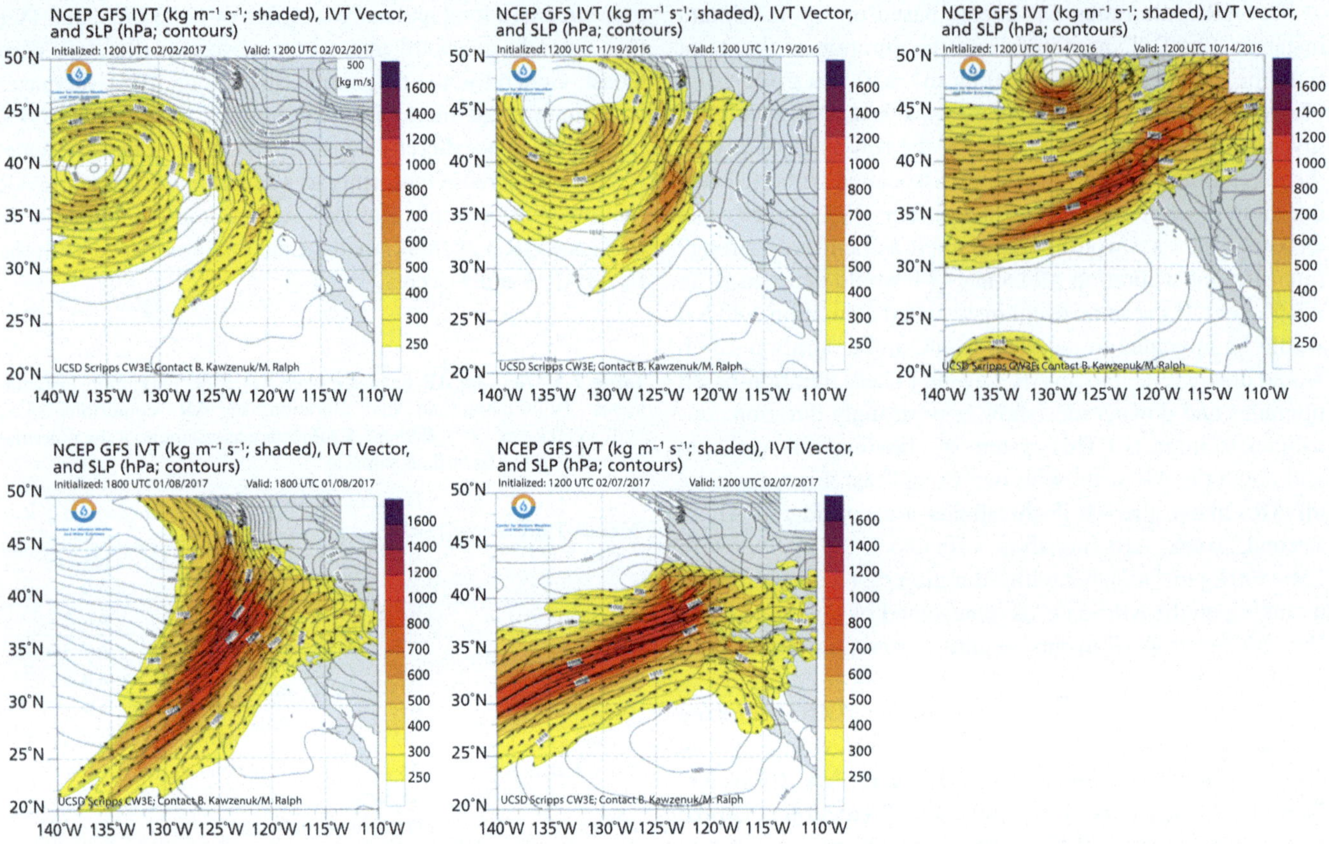

Fig. 7.10 Satellite-observed integrated water vapor (IWV) and Global Forecast System (GFS) -analyzed integrated water vapor transport (IVT) over the northeastern Pacific and western USA during examples of (**a**) AR 1, (**b**) AR 2, (**c**) AR 3, and (**d**) AR 5 events from the 2016–2017 cool season. (Ralph et al. 2019)

The categorization of ARs described here can be applied to in situ measurements (e.g., above an ARO), analyzed representations (e.g., from various reanalyses), forecasts, and other AR simulations and projections to allow storm conditions to be appropriately scaled and ranked. The thresholds of IVT intensity and storm duration that provide the framework for the AR scale (e.g., IVT>250, 500, etc. kg m^{-1} s^{-1}; 24- and 48-h durations) are admittedly arbitrary to some extent, but as such, they are also not narrowly tuned to US West Coast conditions. Hence, this scale may prove useful in many parts of the globe where ARs have impacts, allowing for inter-comparisons across such regions. Users can readily apply the scale to their location and calibrate it to various regional and local impacts.

Looking ahead, the development of the AR scale offers numerous opportunities for advances in AR science. One possibility is that future analyses, composites, and descriptions can examine AR subsets of various intensities, rather than the simple "AR" vs "non-AR" categories that have been used in so many explorations of ARs to date. It will help in evaluating forecast skill and uncertainty, which will help clarify the key forecast errors of importance. This should yield more nuanced descriptions of AR "life stages" such as intensification, maturity, and decay. Perhaps more practically, it should also lead to a better probabilistic assessment of the meteorological conditions and impacts associated with different stages of AR development, and different rankings on the AR scale. In addition to advances in AR science, this scale offers opportunities for better communication of AR science to scientific and non-scientific audiences alike, particularly by differentiating between weak and strong AR events, and by acknowledging the mix of hazards and benefits associated with ARs. This will enable forecasters to better augment their messages, and community leaders to better plan for major events.

References

Albano CM, Dettinger MD, McCarthy MI et al (2016). Application of an extreme winter storm scenario to identify vulnerabilities, migration options, and science needs in the Sierra Nevada mountains: USA. Nat Hazards 80:879–900. https://doi.org/10.1007/s11069-015-2003-4

Chen J, Kavvas ML, Ishida K et al (2016) Role of snowmelt in determining whether the maximum precipitation always results in the maximum flood. J Hydrol Eng 21(10):04016032
Cordeira JM, Ralph FM, Martin A et al (2017) Forecasting atmospheric rivers during CalWater 2015. Bull Am Meteorol Soc 98:449–459
Dettinger MD, Ralph FM, Hughes M et al (2012) Design and quantification of an extreme winter storm scenario for emergency preparedness and planning exercises in California. Nat Hazards 60:1085–1111
Dettinger MD, Ralph FM, Rutz JJ (2018) Empirical return periods of the most intense vapor transports during historical atmospheric river landfalls on the U.S. west coast. J Hydrometeorol 19:1363–1377. https://doi.org/10.1175/JHM-D-17-0247.1
Diaz AJ, Ishida K, Kavvas ML et al (2017) Maximum precipitation estimation for five watersheds in the southern Sierra Nevada. In: Dunn CN, Van Weele B (eds) Proceedings of the 2017 World Environmental and Water Resources Congress, Sacramento, California, May 21–25. American Society of Civil Engineers, Reston, pp 331–338
DWR (Department of Water Resources) (2015) Central Valley Hydrology Study. Prepared for DWR and Sacramento District USACE by David Ford Consulting Engineers. 121p
Espinoza V, Waliser DE, Guan B (2018) Global analysis of climate change projection effects on atmospheric rivers. Geophys Res Lett 45(9):4299–4308. https://doi.org/10.1029/2017GL076968
Gershunov A, Shulgina TM, Ralph FM et al (2017) Assessing climate-scale behavior of atmospheric rivers affecting western North America. Geophys Res Lett 44:7900–7908. https://doi.org/10.1002/2017GL074175
Guan B, Waliser DE, Ralph FM (2018) An intercomparison between reanalysis and dropsonde observations of the total water vapor transport in individual atmospheric rivers. J Hydrometeorol 19(2):321–337. https://doi.org/10.1175/JHM-D-17-0114.1
Hatchett BJ, McEvoy DJ (2018) Exploring the origins of snow drought in the northern Sierra Nevada, California. Earth Interact 22:1–13. https://journals.ametsoc.org/doi/pdf/10.1175/EI-D-17- 0027.1
Hosseinipour EZ, Trushinski B, Su Y (2013) ARkStorm II: a hydraulic Modeling and flood inundation mapping effort on Santa Clara River for emergency planning exercises by local responders in Ventura County, CA. In: Patterson CL, Struck SD, Murray Jr DJ (eds) World Environmental and Water Resources Congress 2013: Showcasing the Future. Cincinnati May 19–23, 2013, Reston: American Society of Civil Engineers. pp. 1669–1680. https://doi.org/10.1061/9780784412947
Ishida K, Kavvas ML, Jang S et al (2015) Physically based estimation of maximum precipitation over three watersheds in northern California: relative humidity maximization method. J Hydrol Eng 20(10):04015014. https://doi.org/10.1061/(ASCE)HE.1943-5584.0001175
Ishida K, Ohara N, Kavvas ML et al (2016) Impact of air temperature on physically-based maximum precipitation estimation through change in moisture holding capacity of air. J Hydrol 556:1050–1063
Jasperse J, Ralph M, Anderson M et al (2017) Preliminary viability assessment of Lake Mendocino forecast informed reservoir operations. Center For Western Weather and Water Extremes. 75 pp. Available from: https://pubs.er.usgs.gov/publication/70192184
Konrad CP, Dettinger MD (2017) Flood runoff in relation to water vapor transport by atmospheric rivers over the western United States, 1949–2015. Geophys Res Lett 44:11456–11462. https://doi.org/10.1002/2017GL075399
Lalaurette F (2003) Early detection of abnormal weather conditions using a probabilistic extreme forecast index. Q J R Meteorol Soc 129:3037–3057
Lavers DA, Pappenberger F, Zsoter E (2014) Extending medium-range predictability of extreme hydrological events in Europe. Nat Commun 5:5382. https://doi.org/10.1038/ncomms6382
Lavers DA, Waliser DE, Ralph FM et al (2016a) Predictability of horizontal water vapor transport relative to precipitation: enhancing situational awareness for forecasting western U.S. extreme precipitation and flooding. Geophys Res Lett 43:2275–2282. https://doi.org/10.1002/2016GL067765
Lavers DA, Pappenberger F, Richardson DS et al (2016b) ECMWF extreme forecast index for water vapor transport: a forecast tool for atmospheric rivers and extreme precipitation. Geophys Res Lett 43(22):11852–11 858. https://doi.org/10.1002/2016GL071320
Lavers DA, Zsoter D, Richardson DS et al (2017) An assessment of the ECMWF extreme forecast index for water vapor transport during boreal winter. Weather Forecast 32(4):1667–1674. https://doi.org/10.1175/WAF-D-17-0073.1
Lavers DA, Richardson DS, Ramos AM et al (2018) Earlier awareness of extreme winter precipitation across the western Iberian Peninsula. Q J R Meteorol Soc 25(4):622–628. https://doi.org/10.1002/met.1727
Lamjiri MA, Dettinger MD, Ralph FM et al (2017) Hourly storm characteristics along the U.S. West Coast: role of atmospheric rivers in extreme precipitation. Geophys Res Lett 44(13):7020–7028. https://doi.org/10.1002/2017GL074193
Neiman PJ, Ralph FM, Wick GA et al (2008) Diagnosis of an intense atmospheric river impacting the Pacific Northwest: storm summary and offshore vertical structure observed with COSMIC satellite retrievals. Mon Wea Rev 136:4398–4420
Neiman PJ, Schick LJ, Ralph FM et al (2011) Flooding in Western Washington: the connection to atmospheric Rivers. J Hydrometeorol 12:1337–1358. https://doi.org/10.1175/2011JHM1358.1
Ohara N, Kavvas ML, Anderson ML et al (2017) Characterization of extreme storm events using a numerical model–based precipitation maximization procedure in the feather, Yuba, and American river watersheds in California. J Hydrometeorol 18(5):1413–1423
Picard L, Mass C (2016) The sensitivity of orographic precipitation to flow direction: an idealized modeling approach. J Hydrometeorol 18:1673–1688. https://doi.org/10.1175/JHM-D-16-0209.1
Porter K, Wein A, Alpers C (2011): Overview of the ARkStorm scenario: U.S. Geological Survey Open-File Report 2010–1312, 183p. and appendixes
Ralph FM, Neiman PJ, Kingsmill DE et al (2003) The impact of a prominent rain shadow on flooding in California's Santa Cruz Mountains: a CALJET case study and sensitivity to the ENSO cycle. J Hydrometeorol 4:1243–1264
Ralph FM, Neiman PJ, Wick GA (2004) Satellite and CALJET aircraft observations of atmospheric rivers over the eastern North Pacific Ocean during the winter of 1997/98. Mon Wea Rev 132:1721–1745
Ralph FM, Prather KA, Cayan D et al (2016a) CalWater field studies designed to quantify the roles of atmospheric rivers and aerosols in modulating U.S. west coast precipitation in a changing climate. Bull Am Meteorol Soc 97:1209–1228
Ralph FM, Cordeira JM, Neiman PJ et al (2016b) Landfalling atmospheric rivers, the sierra barrier jet and extreme daily precipitation in northern California's upper Sacramento river watershed. J Hydrometeorol 17:1905–1914
Ralph FM, Wilson AM, Shulgina T et al (2018) Comparison of atmospheric river detection tools: how many atmospheric rivers hit northern California's Russian River watershed? Clim Dyn 52(7–8):4973–4994. https://doi.org/10.1007/s00382-018-4427-5
Ralph FM, Rutz JJ, JM C et al (2019) A scale to characterize the strength and impacts of atmospheric rivers. Bull Am Meteorol Soc 100:269–289. https://doi.org/10.1175/BAMS-D-18-0023.1
Ralph FM, Iacobellis SF, Neiman PJ et al (2017) Dropsonde observations of total integrated water vapor transport within North Pacific atmospheric rivers. J Hydrometeorol 18(9):2577–2596
Rutz JJ, Steenburgh WJ, Ralph FM (2014) Climatological characteristics of atmospheric rivers and their inland penetration over the western United States. Mon Wea Rev 142(2):905–921. https://doi.org/10.1175/MWR-D-13-00168.1

Toride K, Ishida K, Kavvas LM et al (2017) Maximum precipitation estimation over Shasta Dam watershed by means of atmospheric boundary condition shifting method. In: Dunn CN, Van Weele B (eds) Proceedings of the 2017 World Environmental and Water Resources Congress, Sacramento, California, May 21–25. American Society of Civil Engineers, Reston, pp 340–346. https://doi.org/10.1061/9780784480618

Trinh T, Jang S, Ishida K et al (2016) Reconstruction of historical inflows into and water supply from Shasta dam by coupling physically based hydroclimate model with reservoir operation model. J Hydrol Eng 21(9):04016029. https://doi.org/10.1061/(ASCE)HE.1943-5584.0001391

USGS (1982) Guidelines for Determining Flood Flow Frequency. In: Bulletin 17B of the Hydrology Subcommittee, Interagency Advisory Committee on Water Data. 194p

White AB, Colman B, Carter GM et al (2012) NOAA's rapid response to the Howard A. Hanson Dam flood risk management crisis. Bull Am Meteorol Soc 93:189–207. https://doi.org/10.1175/BAMS-D-11-00103.1

WMO (World Meteorological Organization) (2009) Manual on estimation of Probable Maximum Precipitation (PMP). Report No. 1045. Available from: https://library.wmo.int/pmb_ged/wmo_1045_en.pdf

Zsoter E (2006) Recent developments in extreme weather forecasting. ECMWF Newsl 107:8–17

Zsoter E, Pappenberger F, Richardson D (2014) Sensitivity of model climate to sampling configurations and the impact on the extreme forecast index. Meteorol Apps 22(2):236–247. https://doi.org/10.1002/met.1447

The Future of Atmospheric River Research and Applications

8

F. Martin Ralph, Duane E. Waliser, Michael D. Dettinger, Jonathan J. Rutz, Michael L. Anderson, Irina V. Gorodetskaya, Bin Guan, and William Neff

8.1 Introduction

In this chapter, the editors first give some of our own perspectives on what has been accomplished and where the science will lead us. Then emerging opportunities and needs for future AR observations, research, communications, and applications are discussed.

8.1.1 F. Martin Ralph (Western US Weather Science; Regional Water Applications Development)

Rivers in the sky. What a concept. The ideas have been received with everything from dismissive disdain for its oversimplification, to wonder at its intuitive nature—and to its ability to transcend disciplinary boundaries. And even to bridge boundaries between science and the public and policymakers. From being a topic of some controversy, ARs are now on the same page as the other most well-recognized and established types of storms, i.e., hurricanes, tornadoes, and winter storms. The term has literally been added to the earlier short list of three types of extreme storms that climate change scientists are now reporting on in the 4th US National Climate Assessment. After many years of angst-ridden debate in the halls of weather and climate science, in 2017, the community created a formal definition for the *Glossary of Meteorology*. Additionally, an AR scale, including an intensity scale, has been created and started being used in 2019.

AR science has led to better understanding of where many floods and much of the fresh water that people and ecosystems depend upon, comes from, and has revealed that ARs even help shape the cryosphere. These astounding impacts arise because ARs represent the largest fresh water "rivers" on earth: an average AR transports as much water, as vapor, as 2.6 Amazon Rivers discharge to the Atlantic, as liquid. This evocative term intuitively conveys what the phenomenon is. As noted, rather pithily, by a seasoned forecaster at the first International Atmospheric Rivers Conference (IARC 2016), he likes the term "thunderstorm" because it represents a storm with thunder and likes the term "atmospheric river" because it represents a river in the sky.

F. M. Ralph
Center for Western Weather and Water Extremes, Scripps Institution of Oceanography, University of California–San Diego, La Jolla, CA, USA
e-mail: mralph@ucsd.edu

D. E. Waliser
Jet Propulsion Laboratory, California Institute of Technology, Pasadena, CA, USA
e-mail: duane.waliser@jpl.nasa.gov

M. D. Dettinger (✉)
Retired, U.S. Geological Survey, Carson City, NV, USA
e-mail: mdettinger@ucsd.edu

J. J. Rutz
Science and Technology Infusion Division, National Weather Service, Salt Lake City, UT, USA
e-mail: jonathan.rutz@noaa.gov

M. L. Anderson
State of California Department of Water Resources, Sacramento, CA, USA
e-mail: Michael.L.Anderson@water.ca.gov

I. V. Gorodetskaya
Centre for Environmental and Marine Studies, University of Aveiro, Aveiro, Portugal

Department of Physics, University of Aveiro, Campus Universitário de Santiago, Aveiro, Portugal
e-mail: irina.gorodetskaya@ua.pt

B. Guan
University of California, Los Angeles, Los Angeles, CA, USA
e-mail: bin.guan@jpl.nasa.gov

W. Neff
Physical Sciences Laboratory, National Oceanic and Atmospheric Administration (NOAA), Earth System Research Laboratory, Boulder, CO, USA

Cooperative Institute for Research in Environmental Sciences, University of Colorado–Boulder, Boulder, CO, USA
e-mail: william.neff@colorado.edu

F. M. Ralph et al. (eds.), *Atmospheric Rivers*, https://doi.org/10.1007/978-3-030-28906-5_8

Specialized observations are now being collected, by radars, aircraft, and buoys. Ideas for new satellite capabilities are being proposed based on the need for better AR information. Sophisticated mathematical methods that now include the complex effects of water vapor phase changes have begun pointing directly to ARs offshore as the leading source of "initial condition" error that ruins forecasts of strong extratropical cyclones and their associated precipitation and impacts over the ensuing few days. An airborne AR Reconnaissance method is being developed and has brought together leaders of forecasting in the National Oceanic and Atmospheric Administration/the US National Weather Service (NOAA/NWS), the US Navy, and the European Centre for Medium-range Weather Forecasting (ECMWF) to work with researchers and Hurricane Hunter aircraft capabilities in the US Air Force and NOAA to explore this approach. It started in 2016, gained momentum in 2018, and has the attention of the US Congress and other policy-makers. In 2020 it became an operational requirement and over 30 flights were carried out to improve AR forecasts over 6 weeks. Climate model uncertainties in annual precipitation are now traced in heavily populated parts of the world to differences in how the models represent ARs. ARs are responsible for exposing hundreds of millions of people globally to flood risk, and are even associated, at many global locations, with many of the highest winds annually. Arctic pack ice melt events are associated with ARs penetrating into the Arctic; over 2–3 years of specialized field measurements, less than half a dozen ARs resulted in most of the snow that fell in the part of Antarctica where ice mass is increasing.

The public, non-meteorologists in technical decision-making roles, policy-makers, and the media now use the term extensively along the US West Coast. Although in some cases the perception is of risk, the reality is that many ARs (at least the weaker ones) are key to water supply and snowpack and are thus also (or even, primarily) beneficial. It has become clear that the role of ARs in day-to-day lives is extensive, and better information, forecasts, and climate projections are needed to address the myriad challenges faced by people, the economy, and the environment. There has been remarkable progress over the last 15–20 years, since Zhu and Newell's seminal article in 1998 and Ralph et al.'s (2004) article that rekindled the subject, and even more so over the 40 years since foundational work on low-level jets and orographic precipitation in the UK by Browning and Pardoe took place. Based on this, and on the ingenuity and energy of people, it is likely that the next 20 years will yield tools and ideas that help us cope with a complex and changing world—a world where science, data, and facts can help us thrive.

8.1.2 Duane E. Waliser (Global Atmospheric Science; Satellite and Reanalysis Applications)

It is interesting, although not surprising, that the rapidly growing recognition of the role and importance of ARs as a notable atmospheric and hydrometeorological phenomena has been almost entirely based on their regional impacts, and their connections to water availability, precipitation extremes, flooding, and other related hazards. In fact, nearly all the papers written on ARs in the last decade have been concerned with ARs in a regional context. This is despite the fact that the seminal finding by Zhu and Newell (1998) regarding ARs concerned a global consideration of the earth's water cycle and the dominant role of ARs in conducting the majority of the poleward moisture transport across mid-latitudes—with little or no mention of precipitation, flooding, water availability, etc. However, as the AR phenomena gets recognized across many regions of the Earth, there is growing interest in viewing and understanding ARs with a global view in mind. Such a perspective facilitates considerations of planetary-scale circulation impacts on ARs (e.g., El Niño modulations), AR water vapor source regions and extended-range forecasting, weather and climate model evaluations, and climate change projections of ARs.

Zhu and Newell not only identified a key phenomenon in the atmosphere that affects much of the globe's hydrometeorology and climate, they also gave it a simple, elegant, and intuitive name, that is also relatively sound scientifically. Although the term "atmospheric river" has taken time to catch on, now that it has greatly facilitated the discussion among scientists across disciplines, including forecasters and decision-makers. Moreover, it has helped in the dialog with the general public and media, allowing for a readier understanding of the phenomena than what might otherwise be the case with a more cryptic/scientific terminology (cf. El Niño, Madden-Julien Oscillation (MJO), potential vorticity, low pressure system). Within the next decade, it is likely that the term "atmospheric river" will become the basic scientific terminology for horizontal moisture transports in our atmosphere—particularly the extremes—much the same way that "deep convection" represents extreme vertical transports of moisture with direct connections to key regional and global water cycle processes and their extremes. Moreover, just as deep convection has its favored locations and environmental conditions for occurrence and impacts (e.g., tropics, monsoons, continental summertime thunderstorms), ARs have their favored contexts for development in the mid-latitude storm tracks. But, just as deep convection is important and studied in virtually all the contexts in which it occurs—including its integrated effects on the global water cycle—ARs are also likely to be recognized and studied as the extremes in large-scale horizontal water vapor transport that have both regional impacts and global ramifications on Earth's weather and climate.

8.1.3 Jonathan J. Rutz (Weather Forecast Improvement; AR Detection)

ARs have become part of the meteorological vocabulary, both scientifically and in terms of forecast messaging, for operational weather forecasters with the NWS. This did not happen overnight, but rather as the result of steadily advancing science that highlighted the role of ARs in high-impact weather events such as heavy rainfall, flooding, landslides, and debris flows. The adoption of ARs into the lexicon was also critically helped by the development of AR-related forecast tools, especially those leveraging ensemble data from numerical weather prediction (NWP) models to produce probabilistic information. Forecasters with the NWS, particularly across the western US, now routinely review forecast products based on integrated vapor transport (IVT), which allow them to better quantify and predict AR strength, timing, and intensity.

ARs have also proved to be an incredibly useful concept to forecasters who must communicate with the public, the media, and other non-specialist partners. This arises because the term "atmospheric river" is understood as neatly encapsulating many of the complex atmospheric processes that lead to heavy precipitation and other impacts over the western US. There is still much work to be done on properly serving these audiences, such as how best to communicate the strength of an AR, as well as whether its effects will be primarily beneficial or hazardous. The recent development of an AR Scale, which categorizes AR strength based on IVT magnitude and event duration, creates a foundation for future work in this area.

8.1.4 Michael D. Dettinger (Hydrologic Science and Applications; Climate Change Diagnostics)

As my colleagues have noted, over the past 20 years, understanding of atmospheric rivers—by scientists and increasingly by non-scientists—has expanded greatly. However, it is worth noting that for the first 5 years after Zhu and Newell (1998) coined the term, research on the topic languished. During the next 5 years, active research on ARs was mostly limited to a small group of scientists, often only at or associated with NOAA research groups. As a consequence, much of the explosion of information and understanding of ARs over the past 20 years has been restricted to the past 5–10 years. In recent years, the number of articles—in both technical and popular outlets—has grown almost exponentially, the number of research groups addressing ARs or incorporating them into core research efforts has expanded greatly, and academic theses and dissertations in most atmospheric departments have become commonplace, signaling the emergence of large numbers of early-career scientists who will carry AR research and applications forward. Thus, AR research and, especially, applications are growing now at rates and with momentum that have not been seen before.

All of this is to say that the AR science presented in this book is just the foundation for much more to come, a starting point for great advances that will surely emerge.

This chapter aims to provide perspectives on observational, research, and applications directions that are likely to be particularly productive for scientific advances and benefits to society and the environment in coming years. It is the result of invitations, from the editors, to various authors of this volume to identify specific gaps and opportunities for future AR research and applications. The sections that follow do not represent comprehensive lists, but just some of the more obvious gaps, as well as hints of the opportunities arrayed before the scientific community. If the past 20 years have taught us anything, it is that AR science is full of surprises. So many aspects and discoveries of AR science in the past 20 years have been right there in front of us, already present in more traditional meteorological concepts and framings but rarely recognized for their specialized significance and relevance to impacts, forecasts, and applications. Based on that experience of AR surprises, we anticipate that the collective efforts of the scientific community that is now focused on AR science and applications will surely yield discoveries and directions that we simply cannot guess on the basis of the work to date. We look forward to being as surprised and delighted as anyone in years to come.

8.2 Observational Gaps

This section uses experience gained largely in the western US to address the following:

- What are the key gaps in current observing capabilities to monitor ARs, both for applications and research?
- How do these relate to current capabilities?
- What ground-based, airborne, and satellite measurements are needed?

It is envisioned that much of this information would also apply to other heavily AR-impacted portions of the globe, especially the west coasts of mid-latitude continents. Undoubtedly, there are gaps on the global scale that are key, but they are not identified here, as the science behind them has yet to be substantially pursued.

8.2.1 Ground Based

California established a state-of-the-art network to observe AR conditions in 2009, with a collection of snow level radar, wind profilers, and GPS/MET stations (White et al. 2013; see also Ch. 3). While the over 100 sensors provide an abundance of information to track and understand ARs and their characteristics, several notable gaps remain. A notable issue that deserves further study is inland penetration of ARs or AR precipitation past various mountain barriers, which often occurs by flow through terrain gaps, but also in the form of lee effects of topography. These include "spill-over events" where locations on the lee side of mountains experience heavy precipitation because the depth of the AR exceeds the height of the terrain boundary. A related problem is the lack of direct measurements of the vertical profile of water vapor over land. Operational rawinsondes provide this information at spatial and temporal resolutions that are too coarse to track inland penetration of ARs, especially in complex terrain. A routine method for measuring water vapor profiles over land in all weather conditions is needed. The influence of topography on the freezing elevation of AR events is another area where further ground observations and analysis are needed. Finally, more study is needed on the role of topography in causing convergence of AR flow, which can lead to amplification of orographic precipitation. This would also benefit understanding of impact angle of ARs and the topographic barriers. Understanding how these processes are portrayed in numerical weather prediction (NWP), including the conversion of cloud condensate into precipitation, is required to help improve precipitation forecasts for mountainous regions.

Ralph et al. (2014), which includes over 20 co-authors from climate, hydrology, water management, meteorology, and forecasting across the western US, published a vision for how to tackle these issues and more by instrumenting the Intermountain West (IMW) US with state-of-the-art observing networks for extreme precipitation events and impacts. This vision has gained the support of the Western States Water Council and the Western Governors' Association. Although some of the identified gaps have been filled at least partially—e.g., water vapor aloft in some states, AROs along the northwest US coast, state funding for gap-filling scanning weather radars around the San Francisco Bay Area (AQPI; see below) has been approved, and serious testing of offshore airborne AR Reconnaissance (see Sect. 8.2.3) is underway based on water management applications—no national effort had been funded to fill these gaps as this book was being published. It is plausible that lessons from this region can be used to develop networks in other analogous regions around the globe.

In the San Francisco Bay Area of California, a regional enhancement of the statewide AR observing network is being developed to help focus emergency response and integrated water management. This Advanced Quantitative Precipitation Information project, or AQPI, is installing and operating an array of gap-filling radars, advanced algorithms, and a high-resolution, rapid-refresh NWP model to provide more spatial and temporal detail on precipitation conditions in the complex terrain of the region. As this project is built out and decision-support tools are developed to translate the observations and model output into information that supports response and management decisions, lessons learned will be used to determine how to translate this approach to other regions.

ARs are frequently causes of high streamflows and flooding in the Pacific Coast states of the US, but many ARs do not cause high flows. The intensity (vertically integrated water vapor transport [IVT]) of the ARs is an important determinant of whether or not intense precipitation and floods result, but the precipitation form (rain or snow) plays a particularly important role in many high-mountain settings. The roles of ARs in driving snowmelt, whether in winter-flood situations or spring seasonal snowmelts, have been partially addressed by Guan et al.'s (2016) analyses of the coincidence of ARs and rain-on-snow events in the Sierra Nevada. But—for the future—a better understanding of the meteorological and cryological conditions and processes that mediate connections between AR-provided rain, snow, and resulting snowmelt amounts would much improve the ability to forecast some of the worst mountain floods. Not all rain-on-snow events yield major snowmelt volumes; indeed, most do not. Thus, to improve understanding and the ability to forecast many AR-induced floods, a new generation of snow- and micro-meteorological observations is needed that focus on heat contents and transfers, and snowmelt processes, to augment the current observational focus on the overall volumes of water stored in snowpacks (mostly for seasonal water-supply forecast purposes). In tandem with those snow observations, more observational focus on sensible- and latent-heat deliveries into snowy catchments by ARs is needed.

Beyond this, Ralph et al. (2014) offered suggestions for additional monitoring at snow-instrumentation sites across the western US. These new monitoring directions were motivated by the following observations about major (AR) storms and floods from western mountains: Marks et al. (2001) have shown that, during some well-monitored and simulated rain-on-snow events in Idaho, the delivery of heat into snowpacks by the rain itself was a small contribution compared to the heat deposited on snow by rapid condensation of vapor onto the surface. The heat released by this condensation was the primary cause of melting. Even when melting snow is a minor contribution (as often happens at middle and higher altitudes), rain falling onto snowy surfaces can run off quite efficiently and rapidly, so that the rain itself can yield significant flooding. In warm storms, very large parts of river basins that normally would receive snowfall and generate little immediate runoff can instead receive rainfall, so that the areas contributing rapid runoff to downstream flows and flooding are greatly enhanced. For example, White et al. (2002) showed that for each of four watersheds studied in California, there existed an altitude range for which a 2000-foot rise in the snow level of a storm would triple the runoff. Thus, a vision of observations for extreme storms and flooding in the west necessarily includes consideration of the region's snowfields (Bales et al. 2006), and in particular a focus on surficial and meteorological observations of those snowfields. At present, high-altitude precipitation, snow-water contents, snow depths, and air temperatures are continually monitored at almost 1000 snow telemetry (SNOTEL) (or equivalent) sites in the western US, by the US Department of Agriculture's Natural Resources Conservation Service and by the California Cooperative Snow Surveys. These observations are used to track the seasonal storage and release of water for water-supply forecasts and management purposes across the region.

To broaden the usefulness of these instrumentation sites to address ARs and floods, several enhancements at selected SNOTEL sites (especially in the areas expected to shift to more frequent rain-vs-snow conditions) would be useful, including monitoring of wet-bulb temperature to complement the existing dry-air temperature measurements. Wet-bulb temperatures is a measure of humidity and availability/readiness of vapor for condensation and would reflect the conditions under which large amounts of water vapor are available and likely to condense onto and into snowpacks, which—as noted previously—can result in rapid melting of snow. The mass of vapor that condenses under such conditions also depends on how much moist air is passing over the snow surface, so winds should also be monitored. Under other conditions (when condensation is not the immediate cause of snowmelt), the amount of solar radiation absorbed by the snowpack is critical, and thus sites might be augmented with continual measurements of solar insolation and surface albedo. For the future, an important direction for monitoring snowpacks might be a greater focus on conditions in the interior of the snowpacks (i.e., between the ground surface and the top of the accumulated snow). There is an increasing understanding that dust and soot that falls on snow and that gets incorporated into snowpacks can greatly affect snowmelt timing and runoff amounts (Painter et al. 2011), and that aerosols in the atmosphere can influence the forms and amounts of precipitation, including snowfall, in the first place (Ault et al. 2011). Furthermore, it is well established that structures such as ice layers inside of snowpacks influence rates, locations, and amounts of snowmelt and the release of melt waters into streams (Pfeffer and Humphrey 1996).

Thus, monitoring enhancements at selected SNOTEL sites might add measurements of aerosols in the near-surface atmosphere or falling onto the snow surfaces and regular sampling of snow for measurements of stable isotopic compositions, dust, black-carbon, and aerosol contents. Monitoring physical conditions within snowpacks is also somewhat speculative but could include continual measurements of snow temperatures and densities at various heights above the soil surface in the snowpack and might (with more development) include monitoring liquid-water contents at the same heights.

Stable isotopic compositions of AR vapor and precipitation are being used to understand precipitation-formation processes within ARs, but more work is needed. For AR precipitation processes, precipitation and vapor sampling at more AROs and during more storms will be needed to discern geographic patterns and seasonalities of various micro-meteorological and synoptic-scale key storm processes. Work by Coplen et al. (2008, 2015) has shown how complicated the isotopes-in-precipitation story is within ARs, but the most immediate solution to some of that complexity is not to focus on just one or two locations but instead to begin observing at state- or region-spanning scales, so that a synoptic, Lagrangian view of the evolution of AR precipitation is possible. These scales will also enable analysis of samples from a very large number of storms, so that analyses can extend beyond the storm scale to whole-season and multi-decade scales. Many more observations of isotopic compositions of vapor *in* the ARs rather than just at the surface during AR precipitation will be needed to eliminate some hypotheses on the proper interpretation of compositional variations in precipitation during storms. Beyond the use of stable isotopic compositions to understand precipitation processes within ARs, stable isotopes of water are often used to track sources and fates of various terrestrial surface, plant, and groundwater flows and reservoirs. Only a few initial studies, mostly in the service of paleoclimatic reconstructions (e.g., McCabe-Glynn et al. 2016) have been conducted to date of the isotopic signatures of AR precipitation in these terrestrial waters. Isotopic monitoring of isotopic compositions of soil moisture, plant water, streams, and shallow groundwater have much to offer future analysts who study the role of ARs in land hydrology.

8.2.2 Airborne Physical Process Studies

The water vapor budget of ARs is based on balances between evaporation, water vapor convergence, and precipitation. The residual creates time tendency in the water vapor within an AR, i.e., either increasing or decreasing the water vapor content of the AR. Recent analyses, such as by Cannon et al. (2017), show that satellite-derived precipitation in ARs over the ocean differs greatly from that found in major reanalysis data sets, raising questions about the veracity of model-represented balance of processes. Additional analyses of many climate models has shown how sensitive the model representations of ARs are to the behavior of these physical processes within the models, and how these vary across models (e.g., Guan and Waliser, 2017). These differences are a source of climate projection uncertainty based on the fact that model variations in annual precipitation trends have been shown to depend critically on how the models are handling ARs (Chap. 6; Pierce et al. 2014). It is unlikely that anything other than aircraft-based measurements of each term can capture these differences in a way that allows identification of which climate models (and underlying parameterizations) most accurately represent the water budget in ARs. Section 8.2.4 provides more information on an approach to studying these key physical processes.

8.2.3 A Vision for AR Reconnaissance

The paramount importance of ARs in the western US, especially for water resources and flood management, has led to a vision for a next generation of airborne reconnaissance capacities to support forecasting high-impact weather in the cool season. This concept was described in a cross-disciplinary, multi-agency report (Ralph et al. 2014) on capabilities and gaps in observations to support western US weather, climate, and water information and decision-support needs requested by the Western States Water Council. Some of these needs were highlighted also in NOAA's "Water Cycle Science Challenge" report developed from an associated workshop and recommendations to NOAA leadership (Ralph and Davis 2012). These reports highlighted the key roles of ARs, which Lavers et al. (2016) demonstrated further by showing that IVT predictive skill is greater than precipitation forecast skill, i.e., prediction of ARs provides greater lead time in anticipating extreme precipitation than does prediction of precipitation directly.

A decade of research and development (R&D) focused on ARs has shown that there is useful skill in predicting ARs up to several days ahead of landfall on the US West Coast (Wick et al. 2013; Cordeira et al. 2017; Nardi et al. 2018), but improvements in predictability are required for lead times of 0–3 days. The details of landfalls of ARs (location, intensity, duration, orientation, snow level) have major implications on both beneficial and hazardous impacts (Sect. 5.6). This has been demonstrated recently in the Oroville spillway incident (White et al. 2019) and the deadly Montecito debris flow of 2018 (Oakley et al. 2018). New forecast skill metrics have been developed (Wick et al. 2013; Nardi et al. 2018) that show significant forecast errors in AR landfall. A scale for ARs has been developed (Ralph et al. 2019) and was used to show that just 2 days before the Oroville incident the forecast had been for an AR 2 event, when it instead reached AR 4.

Since 2016, an inter-agency, international team has developed a new method to target dropsondes for data assimilation, including input from a modern tool that independently found uncertainties in AR conditions offshore to be the leading source of forecast uncertainty for major US West Coast storms at 1–3 days of lead time (Doyle et al. 2014; Reynolds et al. 2019). While this new method has focused on improving west coast forecasts, it is highly likely that forecasts across the rest of the US will also benefit from observations over the northeast Pacific. AR Recon 2019 added input from ensemble methods used to target hurricanes into AR Recon flight planning. Results from this extensive testing of AR Recon will inform future cool-season airborne reconnaissance efforts to support the western US's unique information needs, as well as those of the rest of the contiguous US, which is downwind. In 2020 AR Recon became an operational requirement within the National Winter Season Operations Plan, which directs weather reconnaissance aircraft to conduct AR recon missions off the US West Coast using the methods developed by AR Recon in 2016, 2018 and 2019.

Key Observational Gap The gist of AR Recon efforts and results to date is that the position and structure of ARs over the northeast Pacific Ocean is not observed accurately enough to ensure the best possible forecasts of their landfall and associated precipitation for the US West Coast. ARs grow and mature over a large area of the northeast Pacific Ocean west of the US, where few direct observations are available. Satellite observations are not sufficient to resolve the fine-scale structure of winds and moisture that characterize and dictate the evolution and movements of ARs as they approach the West Coast. An AR Recon program would supply a western US equivalent of the hurricane observation program, yielding huge benefits to an economically important region by improving the management of water supply and flooding. Recent research at the US Naval Research Laboratory—Monterey has shown that details of wind and water vapor in ARs offshore are the largest source of forecast error in NWP for heavy US West Coast precipitation with 1- to 3-day lead times. (Reynolds et al. 2019), while Demirdjian et al. (2020) has shown the physical process basis for this finding.

Because of the relatively small size and complicated structure of watersheds, fine details of AR landfall are needed to determine exactly which watersheds flood. In other words, Forecast-Informed Reservoir Operation (FIRO) requires very site-specific skill from extreme precipitation forecasts. Because of the strength of the AR signal, it is possible that predictions of significant central and eastern US storms might be improved as well. For example, some precursors of the February 2016 eastern US storm—with its severe weather, blizzards, and flooding—appear to have formed over the northeast Pacific near the area sampled by a pilot project at the time.

Approach We envision a future observational program that identifies key offshore regions to sample based on NWP analysis, sends aircraft to observe the details of ARs in these regions, and introduces those data into NWP models to improve forecasts of AR landfall and inland evolution. This program would use existing NOAA, Air Force, and other aircraft to observe ARs over the northeast Pacific during the

October–March west coast AR season. In addition to improving existing forecasts, it would establish a core observation resource to enable research that would improve weather models, including ongoing evaluation and improvement of the models and reconnaissance methods. Initially, improvements of forecasts of landfalling AR location, strength, timing, duration, orientation and associated heavy precipitation, and snow line, at 1–5 days of lead times for the western US, could be targeted.

Current Performance and Gaps Extreme precipitation forecasts often have errors of a factor of two. Atmospheric river landfall forecasts on average have position errors of ±300 mile at 5 days of lead time. Most key watersheds are much smaller than that error. Flooding and water supply forecasts require far more accurate precipitation forecasts, which depend upon accurately predicting the AR landfall position, strength, duration, and orientation.

Science Needs Better understanding of mesoscale frontal waves that are a key cause of forecast errors and require better initial condition information (i.e., airborne AR Recon); data assimilation method development tailored to the US West Coast AR problem; assessment of weather and climate model representation of ARs—including identification of their errors in handling evaporation, condensation (including aerosols), dynamical forcing, and the effects of tropics and air–sea interaction and precipitation.

Region The program would target storms roughly in the region extending from Hawaii to Alaska in the west and from the Pacific Northwest to southern California. This area could be covered by C-130 s operated from Hawaii and from a US West Coast base. A pilot program had success operating from Hawaii (Hickam AFB), the Seattle area (McChord AFB), and the Bay Area (Travis AFB) and planned deployments from southern California (Point Mugu Naval Air Station). Depending on the results of the program and the observed benefits, the option of basing in Alaska could be attractive in some cases. In the pilot program, some storms warranted greater coverage than the two C-130 s alone could provide, and other aircraft, such as the NOAA G-IV, could be used to augment the sampling.

Proof of Concept: Data Collection

This section briefly describes the execution of AR Recon in 2016 (three storms were flown, with two aircraft per storm); 2018 (six AR storms were flown with one, two, or three aircraft per storm); and 2019 (six storms, with one to two aircraft per storm). The AR Recon pilot study continued in 2020 (it used 1-3 aircraft to simultaneously make observations on 17 days over 6 weeks). (See Fig. 8.1; Table 3.6 in Chap. 3 has further details.)

Proof of Concept: Data Assimilation and Modeling

Global modeling centers (National Centers for Environmental Prediction [NCEP], US Navy, ECMWF), and regional modeling efforts (Coupled Ocean/Atmosphere Mesoscale Prediction System [COAMPS®] and West-WRF [Weather Research and Forecasting]) have teamed up to collaborate on AR Recon. An AR Data Assimilation Steering Committee (ARDASC) has formed to bring together diverse expertise and substantial institutional capacity to carry out the collaboration.

This ad hoc committee formed as a grassroots effort. Members have agreed upon a Terms of Reference (ToR) designed to provide a framework and guiding principles for the collaborating organizations and individuals to work effectively together. Participating members have provided input into the ToR, and they concur with its principles and goals, provide their own support to carry out their components of the work, share information and results within the committee, and strive for consensus on results before presenting them externally from the committee. Members may leave the committee at their discretion. The team will use the peer-review process to publish results.

The ARDASC has led to several papers showing the impact of AR Recon, including Stone et al (2020) that concluded AR Recon missions typically had as much impact on daily forecasts of Northern hemispheric total energy as the entire North American Radiosonde network.

Anticipated Outcomes of AR Recon Data, Modeling, and Assimilation

- Improved DA methodology for dropsonde data assimilation such as improved error covariance statistics and other special requirements for enhanced use of dropsonde data, including use of full (temporal and vertical) resolution data
- Quantification of the impact of dropsonde DA on the analysis and the degree to which the data can be fit by the model
- Quantification of the impact of dropsonde DA on short-term and medium-range forecasts
- Potential improvements in parameterizations in models that lead to reduced errors in AR analyses and forecasts
- Development of effective strategies for future sampling of ARs through ensemble-based forecast sensitivity experiments

Vision of Potential Operational Implementation of AR Recon

Combining results from the AR Recon pilot study, with requirements documented through stakeholder needs assessments, it is envisioned that an operational AR Recon capability could be extremely useful. Two components and levels of

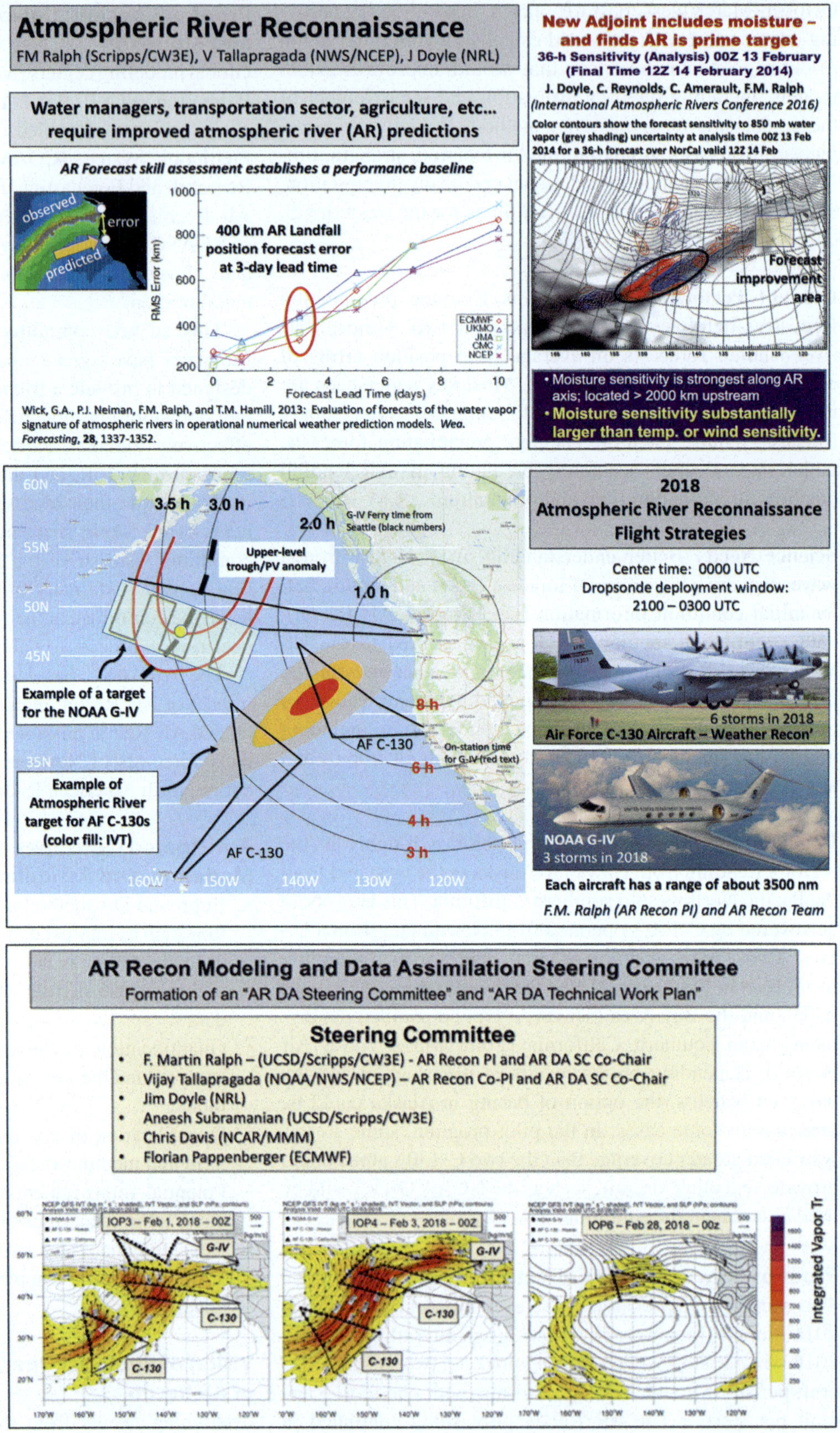

Fig. 8.1 Overview of AR Recon motivation, methods, tools, and execution and evaluation plans

investment could be developed: Priority (1) a west coast-based set of requirements supporting water management, flood control, emergency services, and transportation, built around predicting ARs, and Priority (2) a central and eastern US capacity focused on winter storm and severe weather events in those regions. Because it is very likely that the same areas targeted for Priority 1 will yield benefits for priority 2, it is plausible that the two efforts will end up being highly synergistic.

Implementation could focus on reconnaissance flights during November through March, which is a key season for ARs and conflicts very little with the hurricane season. This approach would benefit from leveraging existing aircraft and aircraft operations facilities and capacities that have traditionally been under-utilized in the cool season. However, storms of interest already approach Alaska as early as September, the Northwest in November, and the rest of the country to the east later into spring.

Ideally, some 20 AR storms per season over the northeast Pacific would be sampled using three aircraft in each storm, with roughly four storms sampled per month between 1 November and 31 March. This vision will evolve as experience is gained, and evaluations of methods take place. To ensure continual evaluation and improvement, an AR Recon science capability is needed that would provide independent, peer-reviewed assessments of plans and results, and support field operations annually.

8.2.4 Satellite

Satellite observations have played an important role in advancing AR science, predictions, and applications. Their most iconic contribution has been in the form of the global integrated water vapor (IWV) observations made with the Special Sensor Microwave/Imager (SSM/I), and other microwave imagers and sounders (e.g., Advanced Microwave Sounding Unit [AMSU], Atmospheric Infrared Sounder [AIRS]). The SSM/I IWV observations were key in the pivotal paper by Ralph et al. (2004), which combined them with research aircraft data and got the AR topic rolling again, after several years of minimal interest. Few, if any, other measurement systems so simply and elegantly delineate individual ARs and at the same time capture the global nature of their occurrences (see Sec. 8.5, Fig. 8.10 or Sect. 3.2, Fig. 3.4c).

Apart from SSMI/I and related sensors, many other seemingly relevant satellite products are of limited and sometimes confounding use. For example, scatterometers, such as from the long-running Quick Scatterometer (QuikSCAT) and Advanced Scatterometer (ASCAT) missions, provide surface wind measurements over the ocean which would be thought to be indicative of low-level convergence into the AR and/or along-river transport. However, because ARs generate internal boundary layers as seen in aircraft dropsonde data, the AR low-level jet (LLJ) is often decoupled from the surface winds and thus the two are not always well connected (Ralph et al. 2005, 2017). Similarly, cloud and infrared imagery sensitive to mid- and upper-level moisture might be thought to be a surrogate for AR flows and cloud structures. However, cloud structures above ARs vary dramatically (e.g., Ralph et al. 2004), and it is generally not possible to infer AR strength or location from traditional cloud imagery. In fact, steaks of high cloud that were initially thought to be representative of strong water vapor transport are often not associated with an AR beneath them. However, more sophisticated sensors and processing have provided satellite precipitation measurements that are essential for any sort of assessment of precipitation characteristics over the open ocean and for illustrating the connection between ARs and the distribution of global precipitation and its extremes (e.g., Matrosov 2013; Guan and Waliser 2015; Waliser and Guan 2017). On the other hand, satellites still struggle to provide robust observations of precipitation, including partitioning between snow and rain, over land, and in particular over high and variable topography (e.g., Behrangi et al. 2016; Cannon et al. 2017).

Along with the direct value of IWV and precipitation measurements to AR studies, the pathway for many satellite observations to contribute to AR research and applications is via their use and assimilation into NWP models. This includes contributions from the scatterometers, water vapor imagers and sounders, temperature sounders (e.g., AIRS, Infrared Atmospheric Sounding Interferometer [IASI]), radio occultations, and geostationary infrared (IR), and visible observations—including cloud-tracked winds. These, along with many other observational constraints, play a key role in the development of high-quality, global atmospheric reanalysis fields, which have become a critical observation resource for AR research (more discussion below). Nonetheless, serious gaps remain in sensing conditions in ARs with these systems. These include IR-based methods that cannot see through the typical high clouds often associated with ARs, or motion vector methods that cannot track winds in the lower altitudes where ARs exist beneath the high clouds, or surface wind measurements that do not represent the LLJ in the AR because of the internal boundary layer decoupling them. These gaps are part of the reason for both new satellite systems and potentially airborne AR Reconnaissance using dropsondes (described below).

The recent release of the 2018 Decadal Survey for Earth Science and Applications from Space (NASA's ESAS 2018) identifies a number of high-priority measurements that should be pursued for development and implementation in the coming decade. Observation targets that are particularly relevant to the study of ARs, and associated improvements in modeling and forecasting, include (1) clouds, convection, and precipitation—particularly adding better sampling, vertical motion, distinguishing between clouds and precipitation, and then, for the latter, distinguishing between snowfall vs rainfall, etc.; (2) vertical profiles of horizontal winds; and (3) information to better characterize the planetary boundary layer. Taken together, these three observations, along with high-fidelity water vapor profiles, provide a comprehensive characterization of AR conditions, and advancements to them would go a long way toward improving current model and forecasting capabilities of ARs. The 2018 Decadal Survey also identified aerosol (e.g., type, size, amount) and combined surface wind and ocean current measurements as key ESAS priorities, and these would also provide some secondary benefits to process studies, NWP, and climate modeling relevant to ARs.

8.2.5 Reanalyses: Evaluations and Gaps to Be Addressed to Support AR Science and Applications

The early studies that led to the concept of ARs (Newell et al. 1992; Zhu and Newell 1994) owe largely to the availability of ECMWF analysis products (specifically, wind and humidity profiles on multiple pressure levels in the troposphere). Their studies quantified some of the key characteristics and importance of these long and narrow features in the fields of integrated water vapor transport (IVT). One of the iconic notions about ARs—that they account for over 90% of the poleward water vapor transport in the mid-latitudes while occupying only about 10% of the zonal circumference of the earth (Zhu and Newell 1998)—was based on a few years of the ECMWF analyses, pointing to the key role of this type of global data product in AR studies. Even today, when high-quality imageries of IWV are routinely produced that can serve as a first-hand description of AR conditions nearly globally (e.g., Ralph et al. 2004; Neiman et al. 2008), the analysis and reanalysis products provide key information on vector winds, including their vertical structure, which is seriously lacking in current satellite capabilities and products. Furthermore, the reanalysis products, via their assimilation of satellite and other types of observations (as discussed in Sects. 8.1.1, 8.1.2, and 8.1.3), provide the means for these observation platforms to collectively contribute to better descriptions and forecasts of ARs. A key observational resource that can help evaluate AR simulations (e.g., Guan and Waliser 2017) and predictions (e.g., DeFlorio et al. 2018) and can be used to drive hydrological models to study hydrological effects (Paltan et al. 2017). In addition to the atmosphere, reanalysis products for other components of the Earth system provide opportunities for more comprehensive understanding of AR characteristics and effects. Among these, reanalysis products for mountain snowpacks have been crucial in quantifying the role of ARs in snow accumulation (Guan et al. 2010, 2012, 2013a, b) and variations along different elevations (Huning et al. 2017) in areas where the seasonal snowpack provides an important source of fresh water supply during the melt season.

A key shortcoming in current global reanalysis systems is the lack of direct assimilation of precipitation. Short-term forecast precipitation is often included in these global reanalysis products, but this should be used with caution because it depends heavily on model physics and is not constrained by precipitation observations. In mountainous regions, daily changes in snowpack could serve as a good proxy for snow precipitation, and therefore studies on ARs in those regions (e.g., Guan et al. 2010, 2012, 2013a, b) can benefit from the availability of snow reanalysis products. However, the general lack of reanalysis precipitation represents a notable gap in observational resources. Regional reanalyses, such as the North American Regional Reanalysis (NARR) and the European Reanalyses and Observations for Monitoring (EURO4M), shed light on improved representation of precipitation distribution, frequency, and intensity via direct assimilation of precipitation observations (Bukovsky and Karoly 2007; Jermey and Renshaw 2016).

Another shortcoming in the current generation of global reanalysis products is the relatively coarse horizontal and vertical resolutions relative to the complex topography in many mountainous regions affected by ARs, which may lead to inaccurate representation of topographic features associated with AR landfalls and inland penetration (Hughes et al. 2012). A factor of two increase in horizontal and vertical resolutions is expected to be achieved in the new and upcoming generation of global reanalysis products (e.g., ECMWF ReAnalysis [ERA5]), i.e., the resolutions will be on par with or exceed the current generation of regional reanalyses.

Besides resolutions, another expected key improvement in these upcoming global reanalysis products is a better balanced water budget, which is crucial for understanding the AR life cycle in terms of moisture sources and sinks, and evaluating the fidelity of their representations in global models. In the case of snow reanalyses, a major obstacle is the lack of in situ observations in many mountain ranges, such as the Andes where ARs frequently make landfalls. In mountain regions with relatively good in situ sampling, such as California's Sierra Nevada (although in situ sampling, there is still limited especially at high elevations), considerable advances have been made in snow reanalysis, which can now afford daily mapping of the snowpack at a 90-m resolution (Margulis et al. 2016). Extending similar snow reanalysis frameworks to other mountain regions where in situ observations are few or nonexistent could be beneficial for AR studies, although it means the snow product will rely heavily on the availability and quality of satellite remote sensing of snow in those regions. Filling the above gaps in reanalysis products needs advances in both observational networks and data assimilation systems.

To more carefully explore variations in snow level in ARs, and to compare the estimated rain rate from the Global Precipitation Mission (GPM) satellite with reanalysis estimates of precipitation rate, Cannon et al. (2017) used an independent catalog of ARs over the northeast Pacific to identify GPM overpasses of ARs (Fig. 3.1.3). The locations of the overpasses were then used as a basis for extracting the associated parameters from MERRA reanalyses. This revealed important differences between satellite-derived, and reanalysis-derived precipitation rates. Reanalysis indicated a much larger fraction of total rainfall was within very light rain rates compared to the satellite-based measurement. The difference could have important implications in terms of reanalysis representing AR water vapor budgets properly Sect. 8.3.2, for example.

8.3 Emerging Directions in AR Physical Processes Research

This section addresses the following questions:

- How far have we come?
- Why is it important to do more?
- What would the observational campaigns/required networks/ answers look like?
- What's a needed modeling component?
- Is there a model experimentation program or activity needed?

8.3.1 Polar Processes Associated with ARs

ARs are a well-described mechanism by which water vapor is transported through the atmosphere poleward of the tropics. Although the mid-latitude expression of ARs and their landfall have been well studied, their behavior in the high latitudes has only recently become the subject of studies—both in terms of the recent melting of the Arctic and the role of ARs in the surface mass balance of the Greenland and Antarctic ice sheets (e.g., Neff et al. 2014; Gorodetskaya et al. 2014; Bozkurt et al. 2018). Contrasts between the Arctic and Antarctic exist in major transport paths and synoptic controls. In the Arctic, major moisture pathways are focused in the North Atlantic and North Pacific (Fig. 8.2; Nash et al. 2018), and the continental land masses support an extensive observing network. In contrast, the Antarctic is surrounded by an open ocean, the ground-based observing network consists of scarce measurements from ships and very few Antarctic stations, and analyses depend on satellite observations and reanalysis products. Currently the major challenge in the high latitudes—over the Arctic Ocean and over the Southern ocean and the interior of Antarctica—is the more limited observational systems to characterize the life cycle and fate of AR transport events. In addition, understanding the behavior of ARs in the polar regions may require a broader view of the role of the tropics from a climatological perspective: for example, the recent warmings in both the Arctic in January–March 2016 and in the Antarctic in January 2016 (with large melt extent over the Ross Sea) were

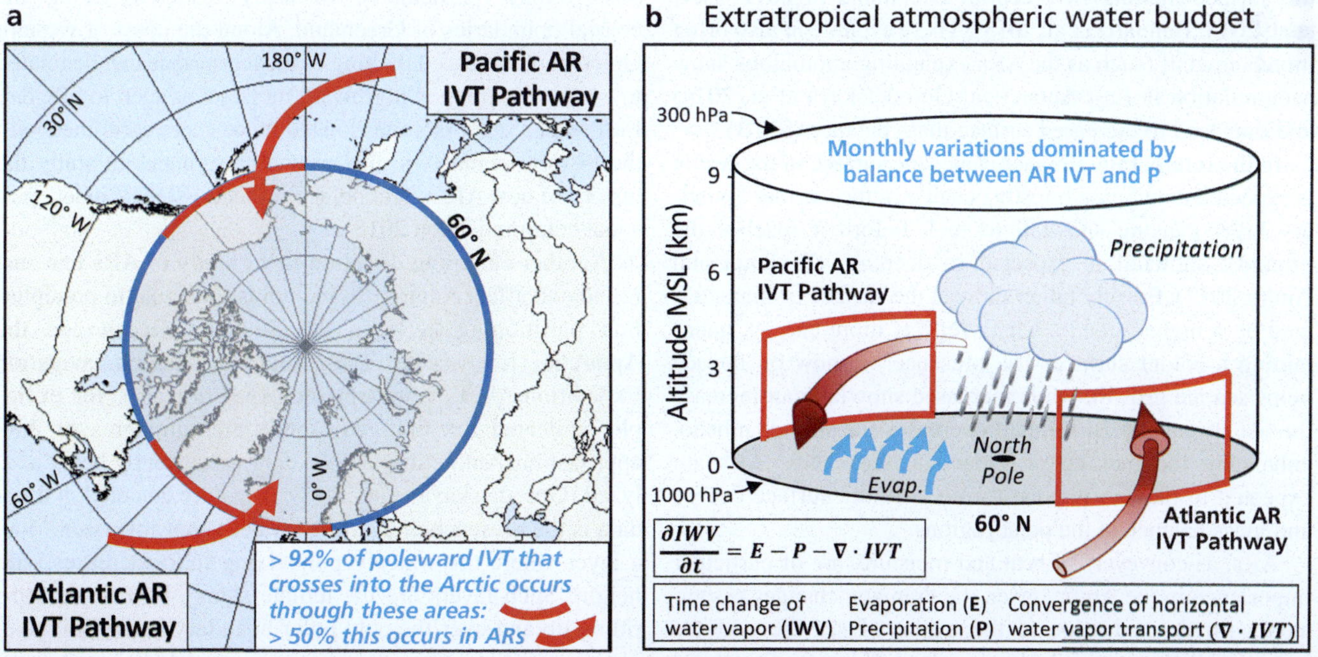

Fig. 8.2 The role of polar-penetrating ARs in the Arctic water vapor budget (Nash et al. 2018)

preceded by 6 months of weak planetary wave heat fluxes at 50 hPa into both polar regions, strong polar vortices, and subsequent rapid breakdowns (as seen in the MERRA reanalysis; see Fig. 8.3). The sudden breakdown of the polar vortices led to rapid warming in both hemispheres, (e.g., Nicolas et al. 2017; Binder et al. 2017). The question during such extreme events is whether the transport of heat and moisture is organized in AR-like structures.

There are already indications of the intensified global water cycle with more evaporation in the tropics and more precipitation in the mid and polar latitudes (Held and Soden 2006). ARs can thus become even more extreme in their intensity, associated with larger IVT and IWV because of the stronger moisture fluxes in the source regions, and also because of the higher air temperatures within the AR bands that leads to larger moisture-holding (and -carrying) capacity (e.g., Mattingly et al. 2018). ARs can also become more frequent because higher evaporation in the tropics will lead to the excessive amount of moisture that the tropics will need discharge more often into the global polar regions, which serve as energy and moisture sinks. While oceans act as an important source of water vapour to the polar regions, continental moisture sources, such as in North America and Siberia, can also play an important role (Komatsu et al. 2018; Neff et al. 2014) and should be investigated further.

Another issue for the polar regions particularly is the potential increase in precipitation (rain or snow), as well as an increase in warm advection and liquid-containing clouds. Warmer temperatures within AR bands can mean an increase in warm advection and melting (such as the AR event that brought warm air and liquid-containing clouds, resulting in the surface melting over central Greenland in 2012; Neff et al. 2014; Bennartz et al. 2013). These events can also bring more snowfall (such as the ARs explaining anomalous snow accumulation in East Antarctica; Gorodetskaya et al. 2014) and thus lead to increased surface mass balance (SMB).

In the future, rainfall that reaches the surface in the Arctic is expected to increase (mostly because of the warmer boundary layer causing snowfall to melt before it reaches the ground); snowfall is expected to decrease (Bintanja and Andry 2017). Freshly fallen snow at the surface is characterized by a higher albedo, which reflects more energy, maintaining a colder surface. The presence of snow on the ice helps sea ice growth, while increased snowfall can increase the ice sheets' SMB. Rainfall decreases the surface albedo, enhancing the melting of snow and ice. Thus, ARs are expected to play an important role in future surface energy and mass balance in the polar regions.

ARs, as conveyers of heat and moisture, are of particular importance in the Arctic, given the dramatic changes associated with the reduction in the extent of sea ice and the increased discharge of fresh water from Greenland's coastal glaciers. In particular, the melting of the Greenland ice sheet has accelerated since the early 1990s (Graeter et al. 2018). Such increases in the melting rate also coincide with (1) increasing moisture transport to the Greenland ice sheet associated with ARs (Mattingly et al. 2018), (2) warming in recent decades as seen in δ18O signatures in ice cores for northwestern Greenland (Masson-Delmotte et al. 2015), (3) potential sea level increases along the US East Coast associated with increasing Greenland melting (Davis and Vinogradova 2017), and (4) meltwater impacts on the oceanic ecosystem that lead to phytoplankton growth and increases in primary productivity in the Labrador Sea (Oliver et al. 2018). However, unlike the mid-latitudes, the Arctic is typically a low-moisture environment, so that detection of ARs—through normal diagnostic methods using reanalyses data as well as satellite observations over the complex mixture of sea ice, open water, and land areas—is complicated. At the same time, clouds and warm air associated with transport events can play a key role in melting sea ice via changes in the surface radiative balance. Fortunately, opportunities to study AR transport events and their fate deep into the Arctic may well result from comprehensive observational campaigns in the next few years, such as the Multidisciplinary drifting Observatory for the Study of Arctic Climate program (MOSAiC: https://mosaic-expedition.org/science/) and other campaigns such as the ones conducted at Svalbard observatory within the trans-regional German (AC)3 project (http://www.ac3-tr.de/). Exploiting long-term measurements at the existing Arctic Observatory stations (www.esrl.noaa.gov/psd/arctic/observatories/) is also the key to understanding long-term trends and evaluation of satellite and reanalysis products with larger spatial coverage. Another opportunity could also be explored with improved observations (particularly vertically resolved) along the coastal boundaries of Greenland. Along the coast of western Greenland, increased melting of the ice surface can be related to a coastally trapped northward moist jet as seen in Fig. 8.4. Last, water vapor isotopes lodged in ice cores combined with the historic reanalyses may provide the means to study the history of how ARs affect polar ice sheets (Steen-Larsen, personal communication 2018).

Another emerging direction is the study of ARs that surround and affect Antarctica, including their role in precipitation partitioning between the Southern Ocean and the Antarctic ice sheet. The Antarctic Circumnavigation Expedition (ACE, http://spi-ace-expedition.ch/), for example, collected the first radiosonde measurements of ARs approaching Antarctica, and the data are currently being analyzed (Gorodetskaya et al. 2018). A feature detected in these data is the presence of a strong "water vapor inversion," i.e., a layer where the water vapor mixing ratio increases with height. Such layers are the telltale signs of AR conditions riding up and over the cold polar boundary layer. This phenomenon is likely a source of error in models and reanalyses and yet is vital to polar surface energy balance.

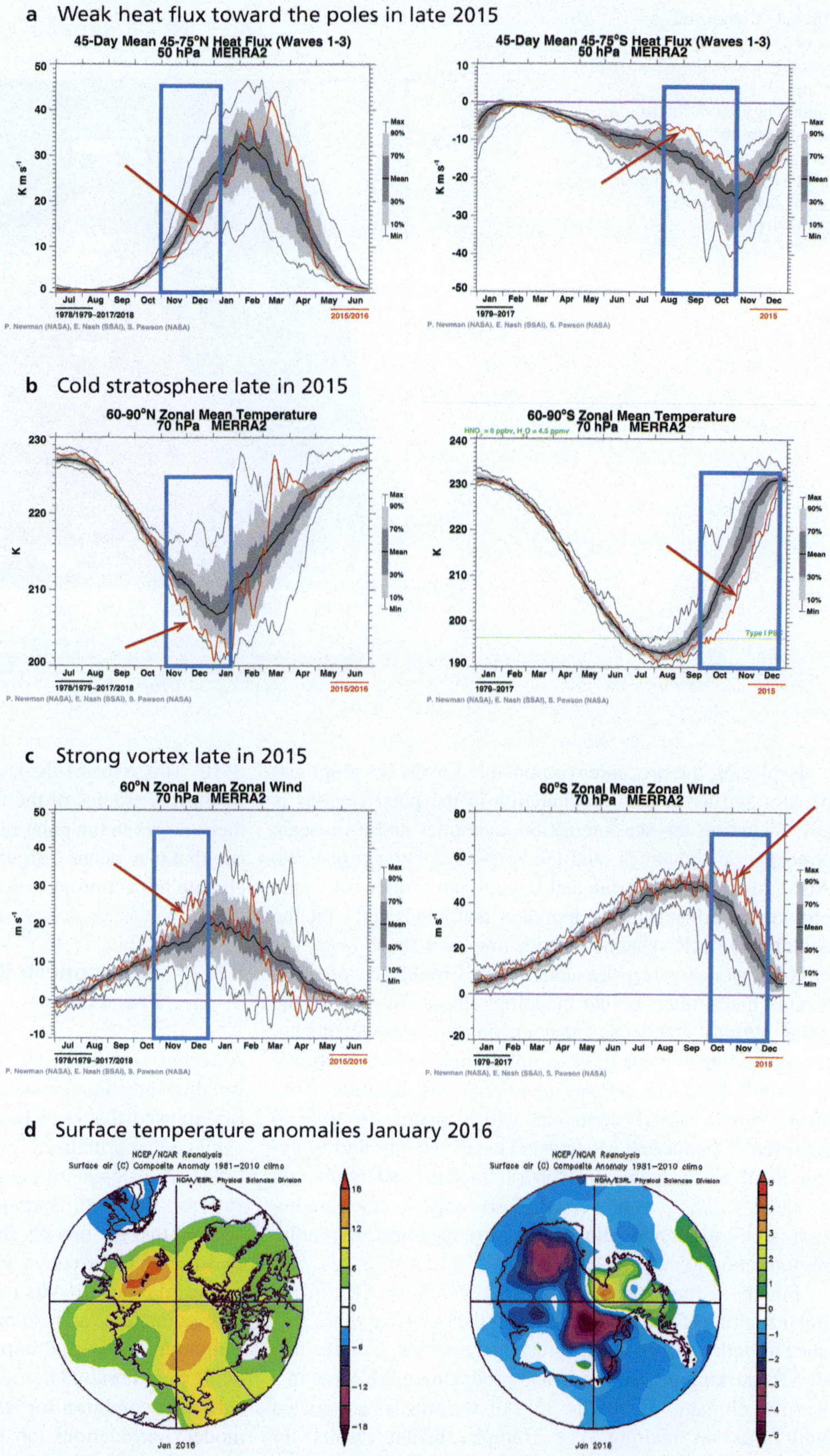

Fig. 8.3 (**a**) Heat flux toward the poles (waves 1–3 at 50 hPa) for the Northern Hemisphere (*left*) and Southern Hemisphere (*right*) showing weak fluxes in both hemispheres. (**b**) 60°–90° zonal mean temperature at 70 hPa showing cold polar stratospheres over both poles. (**c**) Zonal mean wind at 60° N and S at 70 hPa showing strong polar vortices that break down in both hemispheres in mid-December. (**d**) Surface temperatures showing strong warm anomalies in January 2016 in both polar regions. Upper atmosphere reanalysis results from Modern-Era Retrospective analysis for Research and Applications (MERRA)-2 (http://acd-ext.gsfc.nasa.gov/Data_services/met/ann_data.html). Surface temperature data from the National Centers for Environmental Prediction – National Center for Atmospheric Research (NCEP – NCAR) reanalysis data

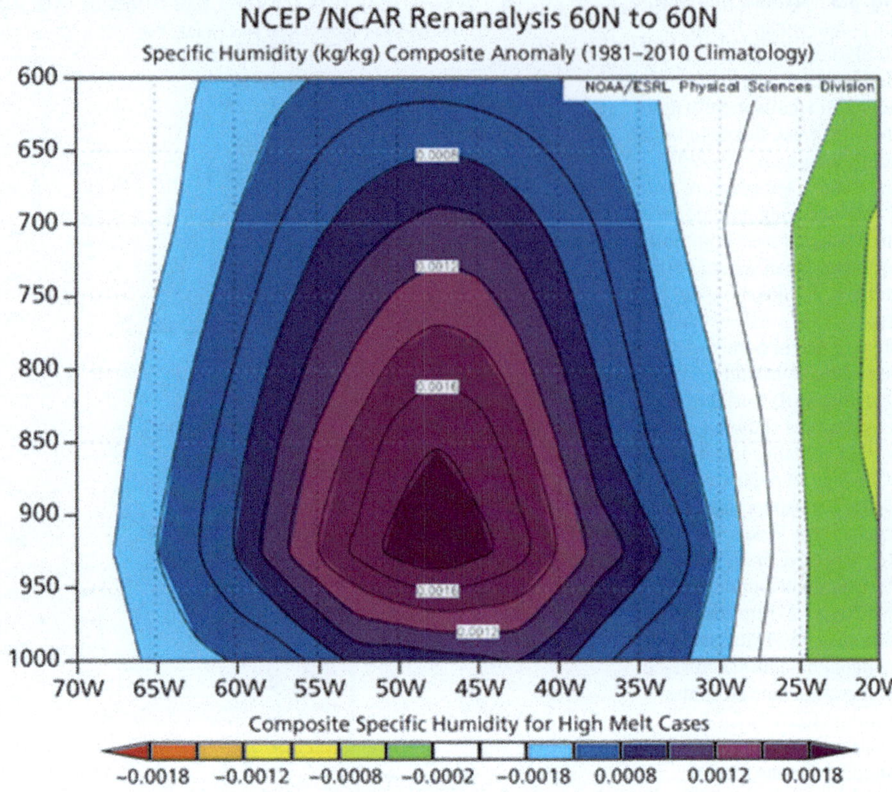

Fig. 8.4 Composite cross-section of specific humidity anomalies (relative to the 1981–2010 climatology) at 60°N for cases where the western portions of the Greenland ice sheet experienced high melt extent during the period 2000–2012

Exploring the processes responsible for the development of ARs and associated precipitation in the polar regions is key, including air–sea interaction, synoptic- and mesoscale atmospheric dynamics, and the role of diabatic processes. Also, combining Eulerian and Lagrangian frameworks may provide an advantageous approach that sheds light on the evolution of AR systems in both time and space (e.g., the ongoing work by Terpstra et al. 2018). Combined "process-based" measurements and modeling studies of ARs in the polar regions are needed that encompass various mechanisms responsible for their evolution and impact. ARs are frequently linked to extreme weather events, including both strong winds and precipitation, which models struggle to represent. Conducting AR-targeted measurements and applying them to evaluate atmospheric models and reanalysis products will help improve both short-range weather prediction and long-range climate projections—over the polar regions and globally.

Finally, a major implication of research on ARs in the polar regions is that future trends in the snow/ice mass balance of both Antarctica and Greenland may hinge on the fate of ARs striking the ice sheets and producing more snow in a warmer climate—versus the role of warmer air associated with ARs on melting. For example, recent results for Greenland (Mattingly et al. 2018) suggest that ARs result in melting in the ablation zone that exceeds the accumulation in the dry snow region of the ice sheet, providing a negative SMB. This issue of the ice sheet implications of ARs argues for further studies of the fate of ARs in a changing climate, their impact in the polar regions, and the amount of sea-level rise that may occur—arguably one of the key uncertainties in climate impact projections.

8.3.2 Atmospheric Water Budget and Moist Processes

A recent evaluation of a set of 24 contemporary global weather and climate models against 17 AR performance metrics showed that over half of the model metrics (17 × 24) exhibited a normalized root mean square error (RMSE) of at least 0.25 or greater (using ERA-Interim analysis as "observations"; Chap. 6; Guan and Waliser 2017). The evaluation showed that the models struggle most with basic patterns of seasonality and frequency of AR occurrence, the fraction of poleward moisture ARs transport (important to extratropical hydroclimate), and zonal moisture transport by ARs (important to landfall impacts). This evaluation showed that while finer horizontal grid resolution is important, it is not a sufficient condition for AR model fidelity; shortcomings in model formulations of key physical processes are also responsible for the systematic errors.

Assessments of the consistency of the impacts of global change on AR characteristics also exhibit considerable differences between the different models' projections despite using common future warming scenarios (Warner et al. 2015; Payne and Magnusdottir, 2015). For example, the first global assessment by Espinoza et al. (2018) shows that although all 21 model projections analyzed indicate an increase of +43% in the frequency of occurrence of AR conditions for the US West Coast, the models ranged between +2% and +95%. Similar uncertainty levels were also found for a number of other high-impact AR regions.

Shortcomings in model representations of ARs also limit the ability to accurately forecast ARs and their effects. Previous field campaigns focused on measuring AR widths and total water vapor transport (Neiman et al. 2014; Ralph et al. 2016, 2017; Cordeira et al. 2017) identified significant forecast errors in these quantities over the ocean as compared to dropsonde observations. In addition, average forecast errors in the AR landfall position along the US West Coast were found to be 600–900 km at 7–10 day lead times, with an average southward bias of 100–300 km (Wick et al. 2013). These forecast errors in AR location, strength, and structure result in large forecast errors in precipitation (Ralph et al. 2010).

One of the likely contributors to these uncertainties in AR analyses, predictions, and climate projections arises from variations in how the water vapor budget in ARs is represented. This water vapor budget largely determines the total water vapor transport in an AR, which averages about 25 times the flow of water in the Mississippi River (Ralph et al. 2017). The amount of water vapor content within an AR is determined by a balance of processes, including evaporation from the ocean, precipitation falling out of the atmosphere, and the convergence/divergence of water vapor into/out of a given region as a result of the combined structure of the water vapor and wind fields. These processes, and how they vary during the evolution of an AR and across its different meteorological sectors (Fig. 8.5), are central to the character, strength, and intensification of these storms and their delivery of precipitation over land. At present, an observed description of this balance—namely the water vapor budget of ARs (the needed quantities being beyond the reach and fidelity of available satellite and in situ observations)—does not exist.

8.3.3 Terrestrial Hydrology and Water Budget

The ties between ARs and hydrologic extremes are by now well documented, with their presence causing floods and their absence causing droughts. Further work is needed to articulate the relative contribution of freezing elevation, duration, integrated vapor transport rates, impact angles, and the timing and sequencing of AR arrivals in the development of floods. Understanding how each of these components can be expected to change in a warmer world, as well as their interplay, can facilitate planning and response capabilities for present and future extremes. In a warmer world with less snowpack and possibly larger storms, California (and many other regions) will have to transition water management capabilities from a seasonal paradigm that treats flood management as separate from water-supply and drought management, to event-based management that strives to manage flood and drought as interrelated challenges and opportunities. In this event-based management, understanding and predicting how large AR events and their likely effects are will play key roles. Increasing detail and lead times for fore-

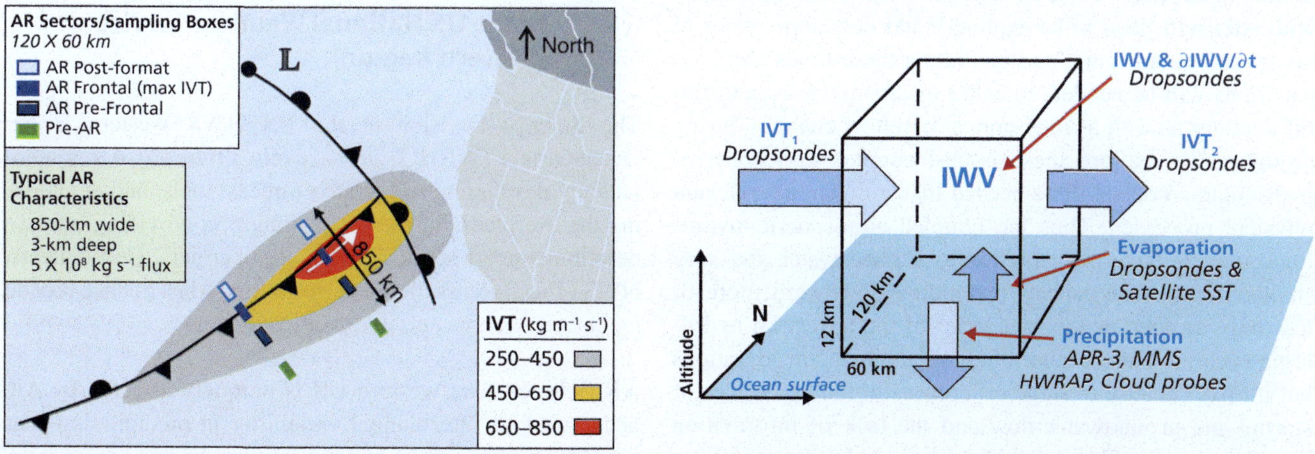

Fig. 8.5 *left panel* Plan view of an AR, with IVT shown as *color fill* and the AR's position shown with respect to weather fronts and the center of the main extratropical cyclone (L). *Small boxes* across the AR represent different sectors. The water vapor budget varies substantially from sector to sector. *Right panel* Terms of the water vapor budget in an AR, with hypothetical types of observations that could be used to measure them from aircraft. (Derived from Ralph et al. 2017)

casting AR conditions will also be key because this provides longer time-frames to manage and decide on different storage options for runoff.

Another less understood process associated with a warmer world is the expected changes that will come in the vertical distribution of atmospheric moisture in an AR. This will significantly affect hydrologic response as freezing elevations rise, and as storms that "spill-over" mountain ranges to impact even their lee sides become more common. The ARs that do manage to spill over are critical to the water balances of otherwise rain-shadowed areas. Moisture and vapor transport is centered higher in some ARs than in others. Early indications are that ARs with higher moisture profiles tend to interact with freezing elevations differently, and are more likely to spill over than ARs with lower moisture profiles. In a warmer atmosphere, vapor profiles will likely extend higher overall than in today's world. This aspect of ARs and their impacts is under-studied thus far. Most observations of vertical moisture profiles of ARs have been made in airborne campaigns far from shore; other studies have relied on the occasional weather balloon and, more so, various reanalysis data sources. But more detailed information and modeling in the areas where ARs interact with topography is needed to improve understanding and forecastability of this aspect of the storms.

Efforts are needed to understand any special roles that ARs may play in determining how (and providing that) water reaches and flows through subsurface and surface pathways from the mountain surfaces where so much of AR precipitation falls to the downstream locales where it serves as water-supply source or flood and inundation hazard. As the climate warms in the future, it will be particularly important to understand and predict how the movement of AR water in and from mountainous regions will change as snowmelt transitions to rain-fed runoff. The timing of subsurface recharge in the headwaters of mountainous watersheds associated with ARs will need to be explored, and new approaches to observing, quantifying, and predicting these flows and contributions will be needed, to better understand seasonal runoff declines as well as inter-annual baseflow changes during droughts. To explore these issues, coupled atmospheric–hydrologic models will be needed that simulate the relevant physical processes. Thus far, coupled atmospheric–hydrologic models have been limited to surface water and very shallow subsurface aspects (extending downward more or less only as deep as the root zone) of the terrestrial hydrologic cycle, in part because of computational considerations but mostly because of large differences in the timescales of storms and groundwater flow and the lack of information necessary to build models for deep groundwater flows beneath mountain settings. Nonetheless, in just the past decade, surface water and groundwater models have been fully coupled and applied in mountain environments (e.g., Huntington and Niswonger 2012). Thus, it is not too soon to envision and begin development of fully coupled atmospheric land–surface deep groundwater models. This coupling would greatly enhance both understanding and predictability of AR-mediated floods, droughts, and water resources.

8.4 Communicating and Applying AR Information

The section addresses:

- Why do we need to communicate this information? To whom?
- How is all this information used by the National Weather Service?
- How can we do better?

This section describes how AR information is being communicated, including recent developments at the NOAA/National Weather Service. In addition, a project designed to quantify the uncertainties in AR climatology and effects that result from different AR identification methods is discussed. It also describes the development of an AR intensity scale and impact categories developed by Ralph et al. (2019). The scale is envisioned for use as a situational awareness communication aid focused on the western US, and other such regions globally, where hurricane and tornado scales are not relevant. The identification of ARs as coherent features that make landfall at a certain location, also allows forecast errors to be quantified (Wick et al. 2013) analogously to hurricane landfalls, which can help focus efforts to improve this metric through research.

8.4.1 How AR Science Affects Operations in the US National Weather Service's Western Region

The Science Division at the US NWS Western Region Headquarters (WRHQ) has actively promoted AR science, recently developing AR-related forecast tools, and emphasizing the usefulness of the AR paradigm in providing effective decision-support services to its core partners. Consequently, NWS offices across the western US (and Alaska) have broadly embraced AR concepts and predictive tools.

AR Science The western US is uniquely affected by ARs because of its inter-annual variability in precipitation (particularly further south) and the fraction of this precipitation contributed by ARs (Dettinger et al. 2011; Ralph et al. 2013; Rutz et al. 2014; Guan and Waliser 2015). Studies have shown that vertically integrated water vapor transport (IVT), the defining characteristic of ARs, is strongly correlated with cool-season precipitation and streamflow across this region (Dettinger et al. 2011; Ralph et al. 2013; Rutz et al. 2014). These results have focused attention away from traditional parameters such as precipitable water or moisture flux at one level, which are less useful for diagnosing cool-season, heavy precipitation events over the western US.

Forecast Tools The recent development of several key probabilistic forecast tools has increased NWS forecasters' ability to assess the timing, location, intensity, and duration of upcoming AR events (Chap. 6). One of these is the NWS Situational Awareness Table, which highlights ensemble-mean forecasts of several atmospheric variables, including IVT, in terms of standardized anomalies, percentiles, and return intervals relative to a reforecast climatology. Another very useful tool is the GEFS QPF calibrated to the GEFS reforecast data set (sometimes referred to as model- or M-Climate). This too highlights regions where the GEFS is producing heavy precipitation relative to its own climatology at a given location and forecast lead time.

Additionally, a number of forecast tools developed by the Center for Western Weather and Water Extremes (CW3E) have become widely used among NWS forecasters, particularly the "AR Landfall Tool" (Cordeira et al. 2017). This tool displays the GEFS-based probability that IVT will exceed a chosen threshold as a function of forecast lead time and latitude along the US West Coast. An inland transect (located east of the Cascades and the Sierra Nevada) gives forecasters at interior offices a better sense of when an AR will penetrate into the interior western US.

Decision-Support Services The terminology associated with ARs and their impacts has caught on with NWS core partners in the western US, and NWS offices are leveraging it to improve services provided to emergency management, community leaders, and the public. The WRHQ Science Division provides training (both in person and through webinars) on AR science and the interpretation of AR-related forecast tools like those described above. In addition, the Science Division makes use of AR science as one aspect of the "Forecast Confidence" e-mail messages sent to regional forecast offices. These messages focus on leveraging quantitative, probabilistic forecast tools to diagnose confidence in NWP solutions at 5- to 10-day lead times.

8.4.2 Communicating ARs to Broader Technical and Lay Communities

ARs provide a paradigm for exposing key features, including threats, in meteorological and climatological data, and forecasts simply and efficiently for both technical and lay audiences. The extratropical water cycle and mid-latitude storm dynamics are extremely complex and chaotic, amounting to a boiling cauldron of visible and invisible fluxes and processes. ARs are threads that are visible in the right remote sensing imagery and analysis products—threads so dominant in generating precipitation, hydrology, and floods in areas such as the US West Coast that following and forecasting them is sufficient to explain 80% or more of all major events. They knit together the flood hazard and water-supply resource aspects of terrestrial hydrology into a single conceptual framework; they provide a basis for understanding and, ultimately, managing both droughts and deluges in the areas they dominate.

As such, in the midst of the remarkable complexities of modern meteorology, ARs provide a simplifying framework to rapidly extract important information from complex data streams and forecasts, to summarize storm conditions to enhance situational awareness, and, especially of late, to communicate storm conditions and forecasts to the technical community, public agencies, and the public alike. Initially just a curiosity, in recent years ARs have become a standard way to describe and explain approaching storms, flood risks, and even droughts.

However, to an extent, the inclusion of ARs into meteorological discussions that describe major storm and flood risks to the public has been *too* successful. Far too often, at present, when many in the media and in the public hear reports of the approach of ARs, they immediately assume that an AR is always a large destructive storm. Thus, by 2018, developing methods to better communicate the severity and specific challenges associated with individual AR storms to technical and public audiences was vital. An AR-storm severity scale (Ralph et al. 2019)—with rankings ranging from AR 1 (mostly beneficial) to AR 3 (mix of beneficial and hazardous) to AR 5 (mostly hazardous)—was developed based on the combination of the maximum intensity of the AR (in terms of IVT) and the duration of at least minimal AR conditions (AR duration had been found by Ralph et al. 2013 and Lamjiri et al. 2017 to be a key factor in determining impacts). The maximum AR "intensity" is broken into the following levels: 250–500 (weak), 500–750 (moderate), 750–1000 (strong), 1000–1250 (extreme), and >1250 kg m^{-1} s^{-1} (exceptional). Also, Dettinger et al. (2018) determined the long-term recurrence intervals associated with various extreme AR intensities that have been experienced in the past at any given location.

It is likely that communications of the risks, benefits, forecastability, and likely future changes in ARs and their impacts will always be in need of improvements, large and small. A natural process and feature with as important a role in the global climate system as ARs—and as important a set and range of impacts as these storms—is extremely important to communicate clearly. Most communications to date have been via technical media, media interviews, and some public-friendly graphics and summaries. In a few instances, ARs have been described in fiction (e.g., Davies 2014), albeit with considerable technical liberties taken in their description so far, and this avenue may have an increasing role in the future as well.

New forecast visualization tools showing likelihoods of landfall at selected locations, storm-total vapor transports, and storm freezing levels have been developed and are being used to communicate AR risks more and more completely, as described below in Sect. 8.4.3.

8.4.3 Developing New AR Forecast Methods and Displays

After years of development, CW3E (with several partners, and support primarily from California's Department of Water Resources), began creating experimental "Atmospheric River Updates," to address stakeholder needs as well as information requests from the public and media such as *"Is this storm an AR?"* The first of these was created on 3 March 2016, and quickly evolved by October 2016 to include the careful, quantitative interpretation of onset time, intensity, and duration at a sample landfall location (http://cw3e-web.ucsd.edu/? p = 5586). Figure 8.6 is an example. AR Updates generally include information from new tools and scientific understanding, and highlight elements of NWS forecasts. They focus on explaining the approaching storm in terms of AR attributes, especially, "Is it an AR?" If so, how strong it is, where it is, etc. The material is informed by an AR Forecast Products website: http://cw3e.ucsd.edu/iwv-and-ivt-forecasts/. These products have been developed through collaborations led by J. Cordeira and F. M. Ralph over several years, major elements of which are described by Cordeira et al. (2017). These products will evolve and grow with the science and with user needs. As just mentioned in Sect. 8.4.2 (and also briefly described in Chap. 7), an example of this type of product is the AR intensity and impacts scale, which Ralph et al. (2019) introduced; it is based on the intensity, duration, and typical impacts of various ARs. It is likely that the next several years will be marked by major innovations in AR prediction tools.

8.4.4 Intercomparisons of AR Tracking Methods

While the AMS recently adopted a qualitative definition of ARs into the *Glossary of Meteorology* (Ralph et al. 2018a), the recent surge of AR-related interest and research has led to the development of dozens of unique AR identification and tracking methods, the scope and variety of which are staggering, and which are described in Sect. 3.7 AR Identification. These differences in AR detection methods (ARDMs) arise largely because the criteria for the methods were developed to address a variety of science and practical goals for different regions, and they used differing data sets and time periods to do so. The use of these different methods throughout the published literature results in uncertainties about AR climatology (e.g., frequency, intensity, duration, and seasonality), the contribution of ARs to precipitation, and how ARs may be altered under future climate change scenarios. Because ARs play such a key role in high-impact weather events, seasonal water storage, and regional hydroclimate, a range of users—including weather forecasters, emergency response teams, and water resource managers—will benefit from improved knowledge and quantification of AR impacts. These applications hinge on how ARs are tracked, counted, and quantified in terms of strength, etc.

In response to the need to quantify the differences in AR identification and tracking, and to understand their origins, the AR community has recently undertaken two related efforts. One is based on using a single geographic location where ARs have been measured and studied extensively for more than a decade at Bodega Bay (BBY), on California's northern coast. This site boasts a state-of-the-art AR monitoring system called an AR observatory (ARO; White et al. 2013), which is incorporated into the analysis. Several ARDMs and different reanalysis data sets are used to answer the oft-asked question "How many ARs hit northern California annually on average?" This study, Ralph et al. (2018b), included many co-authors and concluded that the count annually is 19 (±7)/year, and shows that the variations largely result from the differing restrictiveness of methods based on use, or not, of geometric criteria, and on whether the thresholds are on IWV or IVT and are fixed or vary over time and space (Fig. 8.6). The counts were not as sensitive to the choice of reanalysis data set. Additionally, some methods use a much higher threshold for IVT and thus count only the most extreme of ARs, which average 2 (±1)/year. In spite of the variability in counts, the methods yielded very similar results for AR duration and IVT averages.

The second approach focuses on analysis of ARs globally using reanalyses. It established a broad international team who have developed a systematic method for carrying out intercomparisons, as is described by Shields et al. (2018), which explains the diagnostic approach and offers some preliminary results (Fig. 8.7). The first phase of this, for example, selected a single month, and provided various reanalysis data sets for that month to ARDM developers. Each ARDM was then applied to those reanalyses, yielding a variety of counts, much as was seen in the Ralph et al. (2018b) analysis of ARDMs at one location. These two approaches broadly represent the Atmospheric River Tracking Method Intercomparison Project (ARTMIP) (Fig. 8.8).

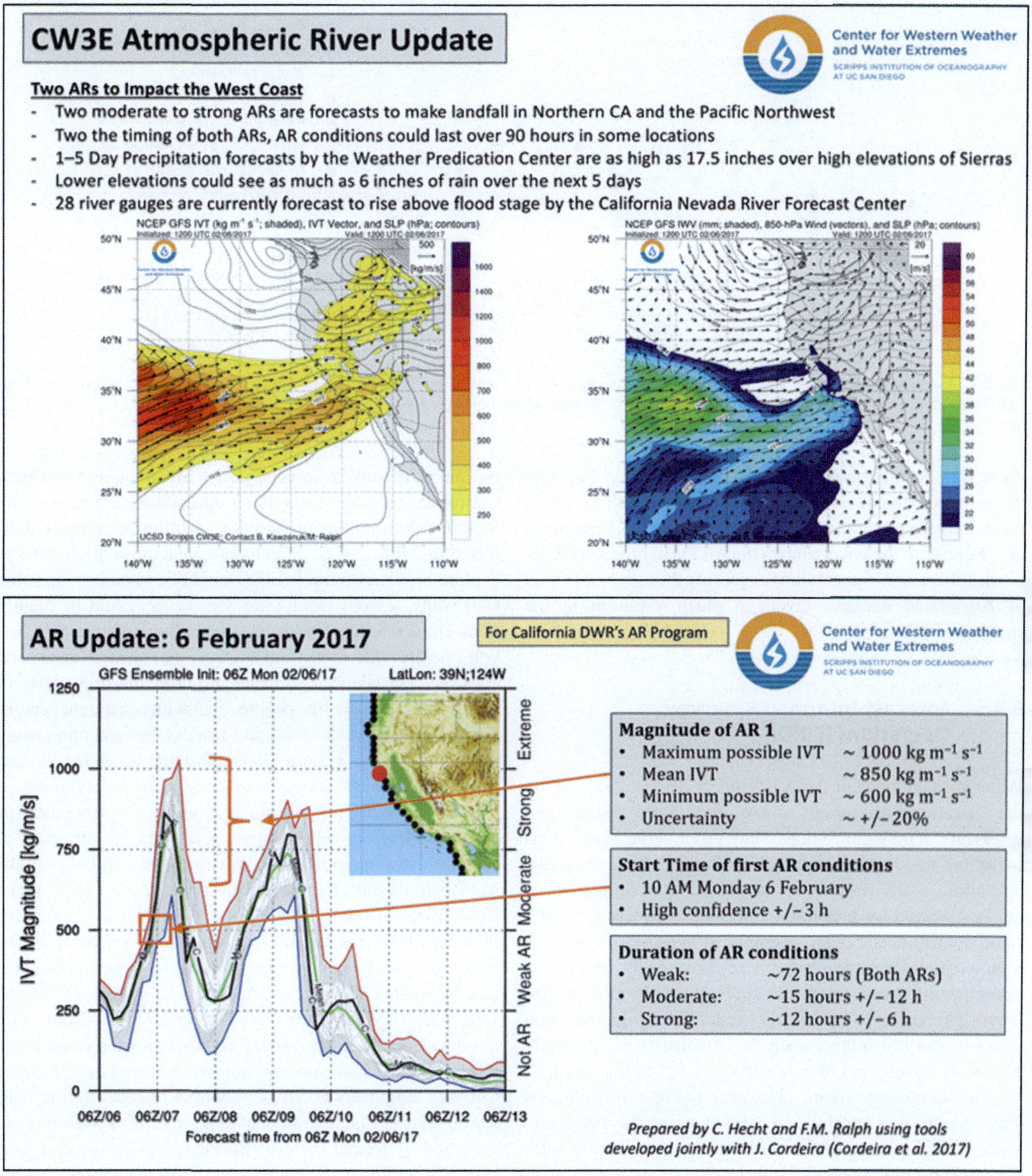

Fig. 8.6 An example of a CW3E AR Update, including details of AR characteristics of the predicted storm

The primary goal of ARTMIP is to quantify the uncertainties that arise purely as the function of a diverse set of AR identification and tracking methods. An international group of 25+ participants is currently working towards this goal, and the expectation is that the ARTMIP project will spur numerous avenues for AR-related research. For updated information on ARTMIP and related publications, please visit the ARTMIP website: http://www.cgd.ucar.edu/projects/artmip/.

Because so many studies and applications hinge on identifying, counting, and tracking ARs, and because the results

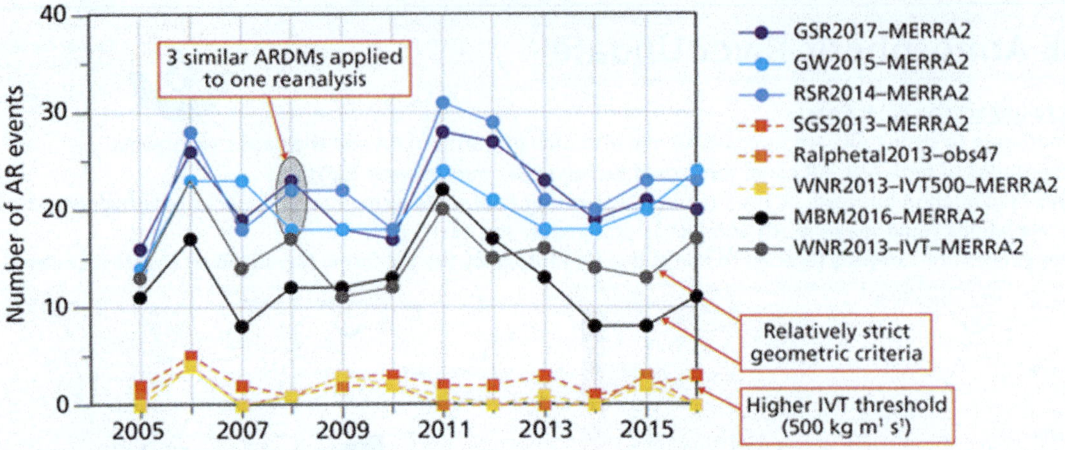

Fig. 8.7 Variation in the annual counts of ARs at Bodega Bay, on the coast of northern California, based on different AR detection methods (ARDMs) being applied to one reanalysis data set at one geographical location over 1 year (Ralph et al. 2018b)

differ significantly based on the ARDM used, it is likely that these intercomparisons will provide a foundation to assess future research results and develop applications opportunities. Related to the uncertainties the ARTMIP project is diagnosing, there is also uncertainty regarding the communication of AR-related hazards, given so much variation in the strength of ARs.

8.4.5 Forecast-Informed Reservoir Operations (FIRO)

Water management in the western US depends heavily on the operation of hundreds or even thousands of dams and reservoirs across the region. They often serve many uses, including flood control and water supply, which can make operation of the reservoirs complex. Water supply needs are best served by keeping a reservoir as full as possible; flood control needs them as empty as possible. To address this, each major reservoir has a set of rules by which it is operated. In many cases, the rules have not allowed for weather forecasts to be directly used in deciding how much water to store or release each day. Decisions are typically based on season, and how much water is "on the ground," i.e., in a rain gage, stream gage, or in the reservoir already. This system was established largely when the reservoirs were built—most of which were before precipitation forecasts had become reasonably accurate.

Thanks to progress in weather prediction, and to successful use of forecasts in a few unusual circumstances, a major effort is now underway to explore the use of precipitation forecasts in reservoir operations in a couple of key reservoirs. This "Forecast-Informed Reservoir Operations" (FIRO; http://cw3e-web.ucsd.edu/firo/) approach is showing remarkable promise. It is described briefly in Sect. 7.3, which focuses on the pilot study reservoir—Lake Mendocino in northern California. Here the focus is on what future directions may look like, beyond this preliminary viability assessment completed in 2017 for Lake Mendocino, including early efforts to explore how transferrable FIRO might be to other reservoirs. Ostensibly, lessons from Lake Mendocino could be readily transferred to other reservoirs; the reality is that each reservoir operates differently and exists in a context of needs and constraints that can vary dramatically from one to the next. In a way, reservoirs are like people, each with a different personality, strengths, weaknesses, and context. For example, many reservoirs in the western US serve both water-supply and flood control objectives; some focus only on one. Hydropower may be a primary purpose for one, while it is secondary, or even absent, from others. Some reservoirs are also part of efforts to restore endangered fish (e.g., salmon below Lake Mendocino), and other endangered bird species (e.g., Least Bell's Vireo in Prado Dam's flood basin near Los Angeles). Some reservoirs were built by and are now operated by the same agency; others may have been built by one and be operated by another. Ownership can be federal, state, local, or private, with varying mixes of operating responsibilities: e.g., Sonoma County Water Agency controls releases when water levels at Lake Mendocino are not too high, and the US Army Corps of Engineers (USACE) controls releases during high water. And as mentioned above, a reservoir's operation is governed by formal manuals developed when the dam was built (some manuals have been updated). Each manual includes a "rule curve," but these rules can vary greatly in their structure and complexity because of differences in the dams' purposes and potential vulnerabilities.

Reservoirs can also vary in terms of their geographic characteristics and climates, e.g., some are dominated by snow; others see little or no snow. A key factor that will determine if FIRO is viable at a reservoir is the amount of

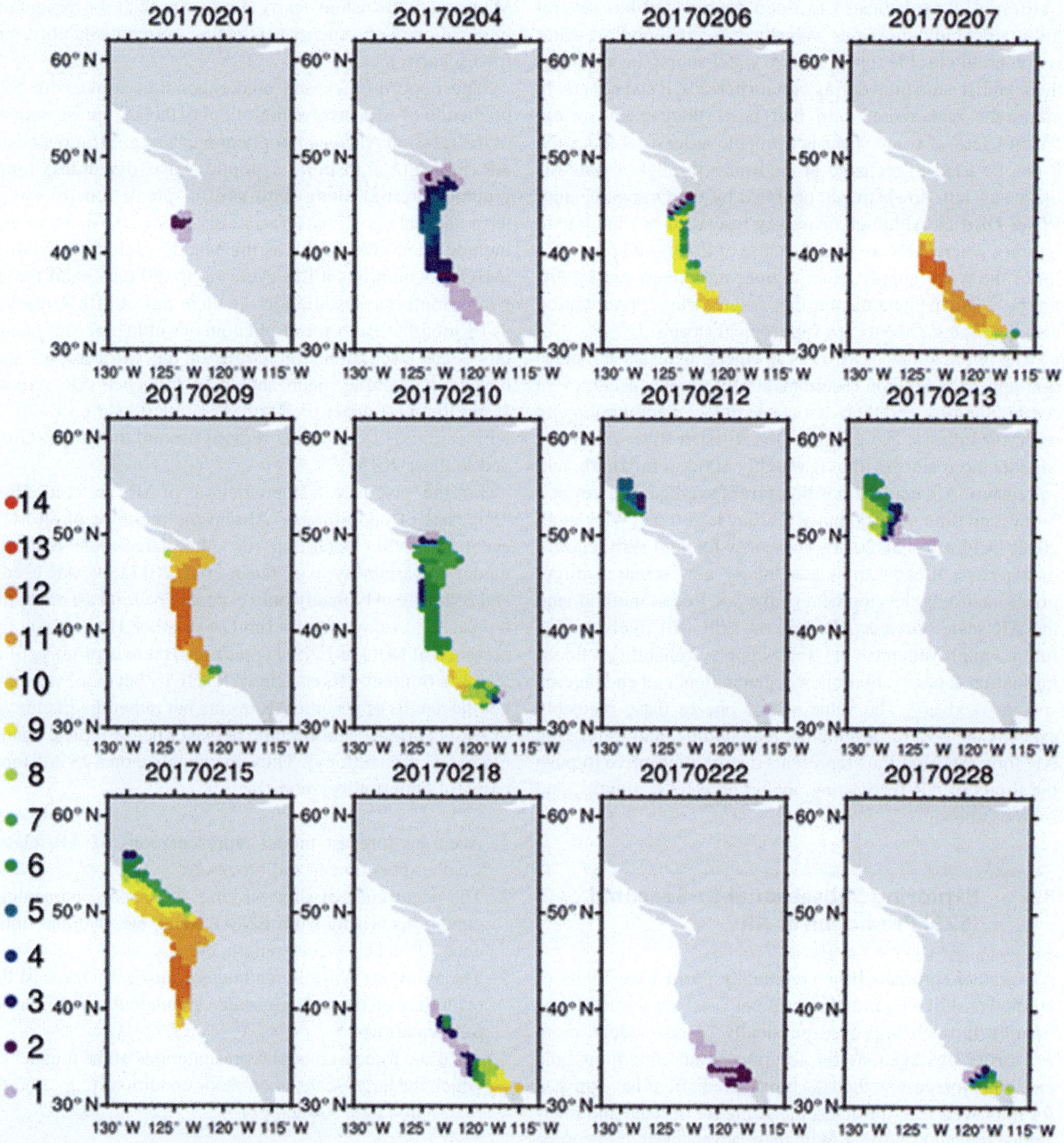

Fig. 8.8 Variation in the number of AR detection methods (ARDMs) identifying a landfalling AR along the US West Coast on twelve dates in the February 2017 trial period. (Shields et al. 2018)

lead time needed for accurate AR forecasts. At Lake Mendocino, this is 3–5 days; at Prado Dam it is likely 1–2 days because Prado is much closer to the ocean. Other reservoirs—with longer river reaches that flood or with endangered species downstream from the dam, or that need to release more water over a longer time—may require weeks of lead time.

The success of FIRO at Lake Mendocino will likely lead to serious evaluations over time at many other reservoirs. The first of these is happening at Prado Dam in southern

California. It is dedicated to flood control and has several million people living in its watershed. A small pool of water is potentially usable for short-term water supply needs, but if the pool rises too high during certain periods, it can adversely affect the endangered bird that nests there each spring. Nonetheless, if some water is available at least temporarily, it can be used to recharge groundwater through a state-of-the-art system downstream, operated by the Orange County Water District. Although many key reservoirs are in climate regimes where ARs are the key cause of floods and provide a lot of the water supply, other regions are dominated by different weather phenomenon that create heavy precipitation and runoff, e.g., hurricanes and tropical storms.

The implications of FIRO as an emerging reservoir operation approach include requirements for greater accuracy in AR predictions, and their associated heavy precipitation and reservoir inflows. For example, the Russian River that Lake Mendocino resides within is roughly 100 km in length, and yet current AR landfall position errors average 300 km at a 3-day lead time and 500 km at a 5-day lead time (Wick et al. 2013; DeFlorio et al. 2018). These new forecast skill requirements have already stimulated major new research directions, including development of the AR Recon method, and the AR scale. Greater AR forecast skill will likely enable future improvements in water-supply reliability, flood-mitigation capacity, hydropower generation, and endangered species recovery. The value of any one of these plausibly exceeds the cost of developing, maintaining, and improving AR forecasts, and thus represents a strong incentive to push the limits of AR forecasting, including weeks, months, and seasons ahead.

8.5 Exploring Subseasonal-to-Seasonal (S2S) Prediction of ARs

As weather forecasts have continually pushed the limits of predictive skill well into the period of 7–10 days of lead time (mainly through improved physically based models, more and better observations for accurate initial conditions, and ensemble forecast methods), climate predictions have pushed for improved skill for times, including for decadal and inter-annual down to seasonal. With these advances, it has become apparent that an intermediate time-scale exists that bridges these two prediction regimes, and runs roughly from 2 weeks to 3 months into the future. This has been termed "subseasonal-to-seasonal" or "S2S" timescales (e.g., Robertson et al. 2015; NAS 2016) and has become the focus of significant attention. One reason for this attention is the wide variety of decision-support applications that exist, for which even limited predictive skill could offer an advantage over the traditional inputs used currently in associated decision-making processes with these multi-week lead times. Many such decisions carry large financial or pragmatic implications (e.g., energy production, water availability, and food security).

The opportunities and challenges that come with S2S prediction of ARs involve high-fidelity model representations of the relevant AR/weather phenomena (e.g., frontogenesis, AR moisture transport, synoptic wave dynamics, topographic forcing) along with skillful predictions of short-term anomalous climate variations (NAS 2016). The latter include phenomena such as the Madden-Julian Oscillation (MJO)—which has a life cycle of 40–50 days, and the El Nino Southern Oscillation—which has a life cycle of 9–18 months. Both these phenomena influence the planetary-scale circulation, including in the extra-tropics and have, for example, been shown to influence AR activity along the west coast of North America (Guan et al. 2012; Kim et al. 2017) and other regions around the globe (Guan and Waliser 2015).

For the most part, S2S predictions of ARs are considered "forecasts of opportunity." Absent the presence of an MJO event or ENSO condition (or other large-scale patterns/modes of variability; e.g., Guan et al. 2013a, b), AR events and activity can typically only expect to be forecast out to the typical 10–14-day forecast limit of weather. However, in the presence of MJO or ENSO conditions, there is promise of at least intermittently forecasting AR activity beyond 2 weeks—not the details of specific AR events but rather the likelihood of AR occurrence and strength above or below typical conditions (i.e., climatology). Thus, to capitalize on S2S AR forecasts (of opportunity) requires:

1. Accurate forecast model representations of AR-related weather phenomena and processes.
2. The occurrence of slow-varying, large-scale, anomalous conditions arising from ENSO, MJO, etc. that can influence AR occurrence, strength, etc.
3. The ability to skillfully and deterministically forecast the evolution of these large-scale, anomalous conditions at S2S lead times.
4. Accurate forecast model representations of the manner by which the large-scale, anomalous conditions (e.g., ENSO, MJO) modulate AR activity.

With these in hand, there is hope for, at least intermittent, longer-lead (e.g., >2 weeks) forecasts of AR activity (i.e., more/less, stronger/weaker than usual) on the horizon.

This brief section, which builds on the AR forecast material reviewed in Chap. 6, discusses recent advances that suggest the needs and opportunities for S2S predictions of ARs and their associated weather and hydrologic conditions. It also provides some examples of the early development of S2S forecast products for the western US that incorporate ARs in their methods and/or outputs.

8.5.1 Western Water Management Requests Improved Precipitation Outlooks

Decisions regarding water supply, flood control, water usage, etc. occur over a wide range of lead times. One of the most challenging needs is for seasonal or annual projections of water supply part way through the water year, e.g., water year 2019 runs from 1 October 2018 to 30 September 2019 (WY 2019). California, where many advances in AR science and applications have occurred, has a complex water supply and flood control system. Each year in the late fall and early winter, agricultural users require information on water supply for the coming growing season to make decisions about crops, or fallowing land, etc. Water "transfers" (sales of water by one owner to a user) are affected as well. However, water delivery projections are fraught with uncertainty, for many reasons. One of which is that the western US experiences wider variations in annual precipitation than anywhere else in the US. This is especially pronounced in California and the Southwest, where the annual variability is on average more than triple that of the eastern US (Fig. 1e from Dettinger et al. 2011). This variability is largely a result of the fact that there is essentially one season that is wet (although the summer monsoon can help in parts of the Southwest), during which the annual precipitation occurs. The "cool season," from October through March, accounts typically for 85% or more of the annual precipitation. Thus, if that wet season is drier than normal, it sets back the entire year. No other season is normally wet enough to make up for that—unlike most of the rest of the US.

For the US West Coast, especially California, the wetness of the cool season overall is set largely by the fate and intensity of a relatively few storms (Fig. 1e in Dettinger et al. 2011). These storms provide the lion's share of annual precipitation in just a few events each year. For example, the Bay Delta watershed in central California, which is the major source of water supply for the region, experiences wide variations in annual precipitation. Eighty-five percent of the variability results from the variability of precipitation that occurs on the 5% of the wettest days each year (Dettinger and Cayan 2014). The silver lining in this difficult situation is that research has revealed a phenomenon that is the primary culprit for this variability—and that phenomenon exhibits useful predictability; it is the number and strength of landfalling ARs that strike the region. These number from just a few events in a drought year, to over three dozen in a very wet year (Ralph et al. 2018b, and Fig. 8.9). Each AR lasts about 1 day on average over a given location, though the AR itself may last a couple of days, and multiple ARs can hit an area over a several-day period. Recent research has shown that modern forecasting tools have at least some skill in forecasting AR event occurrence and landfall out to a week or more (e.g., Wick et al., 2013; DeFlorio et al. 2018), and forecast AR activity possibly out to week 3 (DeFlorio et al. 2019). Although there is some skill if the detailed location and intensity forecasts are less important, it should be noted that many applications require greater accuracies at shorter lead times than S2S addresses; for example, the AR that hit the watershed above Oroville in early 2017 (White et al. 2019) was predicted a couple of days ahead, but at even just 2 days of lead time it was under-forecast by two levels on the AR Scale and produced twice the runoff originally expected (Ralph et al. 2019)

The drought of 2012–2015 serves as a useful example of the role of extreme events. In the source region for northern California's major reservoirs, annual precipitation had been 60–85% of normal for WYs 2012–2015. Precipitation in this region is tracked by the "Northern Sierra 8-station index," which is shown in Fig. 1e in Dettinger et al. (2011). The day-to-day evolution of the precipitation in those years highlights the role of extreme precipitation events play in contributing to annual precipitation—and thus to whether meteorological drought conditions occur. Close examination of the precipitation time-series over those 4 years illustrates that:

- Most of the precipitation occurs from 15 October to 15 April (85% during WYs 2012–2015).
- 40% of WY 2012–2015 total precipitation occurred during the top 2 wettest 10-day periods of each water year. That is, 40% of the 4-year total precipitation fell in less than 5% of the days in those 4 years.
- Two of these 10-day episodes occurred during the CalWater field campaigns in February 2014 and February 2015 (Ralph et al. 2016). In each case, these studies documented the presence of landfalling AR conditions (Ralph et al. 2016). Other analyses have shown that the largest event of the eight episodes, which occurred in Nov/Dec 2012, was also associated with a series of landfalling ARs, as was the December 2014 case.

The patterns in precipitation seen in these 4 years represent patterns seen over much longer analysis periods where the fraction of annual precipitation from landfalling ARs in the region ranges from roughly 25–50% (Dettinger et al. 2011). In addition, the finding that 5% of the days each year provided roughly 40% of the annual precipitation is consistent with the analysis of much longer precipitation records by Dettinger and Cayan (2014) that showed how the inter-annual variability of the 5% wettest days each year explains 85% of the inter-annual variability in annual precipitation. It is also consistent with the results of Guan et al. (2010) that found 40% of the annual snowfall in the Sierra Nevada, on average fell in association with landfalling ARs.

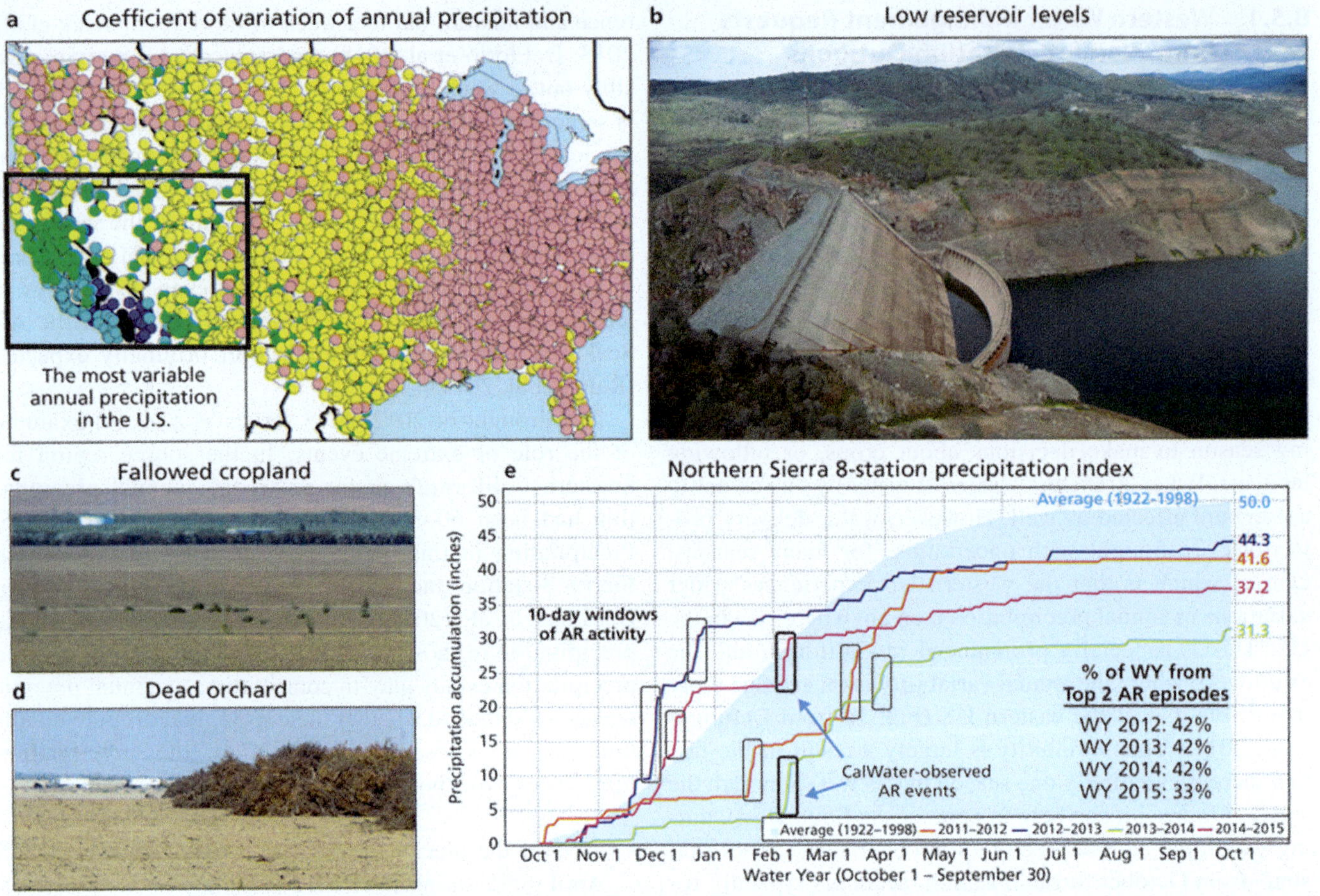

Fig. 8.9 (**a**) Inter-annual variability of annual precipitation based on Cooperative Observer Program (COOP) observations (standard deviation of annual precipitation, divided by annual mean precipitation; *pink* 10–20%, *yellow* 20–30%, *green* 30–40%, *blues and black* >40%). From Dettinger et al. (2011). (**b**) Example of a low reservoir during drought. (**c, d**) Agricultural impacts of drought. (**e**) Examples of the key roles of two 10-day wet periods (active AR episodes) in each of four water years. (Figure elements courtesy of J. Jones)

8.5.2 Large-Scale Processes/Short-Term Climate Variability that Modulate ARs

While ENSO has long been known to modulate the Pacific storm track, more recent studies have examined the role of short-period climate mode variability and the effects of either of these on AR activity and intensity. An early study (Ralph et al. 2011) described the dynamical linkages between a strong MJO, an extratropical Rossby wave packet, and the position of an AR and associated heavy precipitation in Oregon a few days later. They also found that a tropical easterly wave had displaced the northern edge of the tropical water vapor reservoir northward, which put it within reach to be "tapped" by an AR that formed on the southeast side of a breaking Rossby wave. Guan et al. (2012) explored MJO's role in modulating ARs and found strong correlations with AR-related snowpack on the West Coast. Additionally, Guan et al. (2013a, b) found that when the Arctic Oscillation (AO) was negative (i.e., southward cold-air outbreaks) and combined with a negative phase of the PNA "teleconnections" pattern (i.e., southern storm track position), this favored AR conditions striking the Sierra and causing near-record snowpack for the given month and water year. The MJO's ability to modulate AR position and behavior in the northeastern Pacific is carried out mainly through its forcing Rossby waves (with mid-latitude sources possible as well) that propagate northeastward. Depending on the planetary-scale zonal shear, these Rossby waves may break, either cyclonically, or anticyclonically, which then influences the structure and propagation of the associated extratropical cyclones and anticyclones, and hence the ARs. Recent studies by Ryoo et al. (2013), Payne and Magnusdotir (2014, 2016) and Hu et al. (2017) have explored the roles of Rossby wave breaking in modulating AR occurrence and characteristics and found a connection to duration and orientation of landfalling ARs. Figure 8.10 uses a specific case, from the early AR paper by Ralph et al. (2006) that first connected ARs to west coast flooding, where IWV coverage over the Pacific is complete,

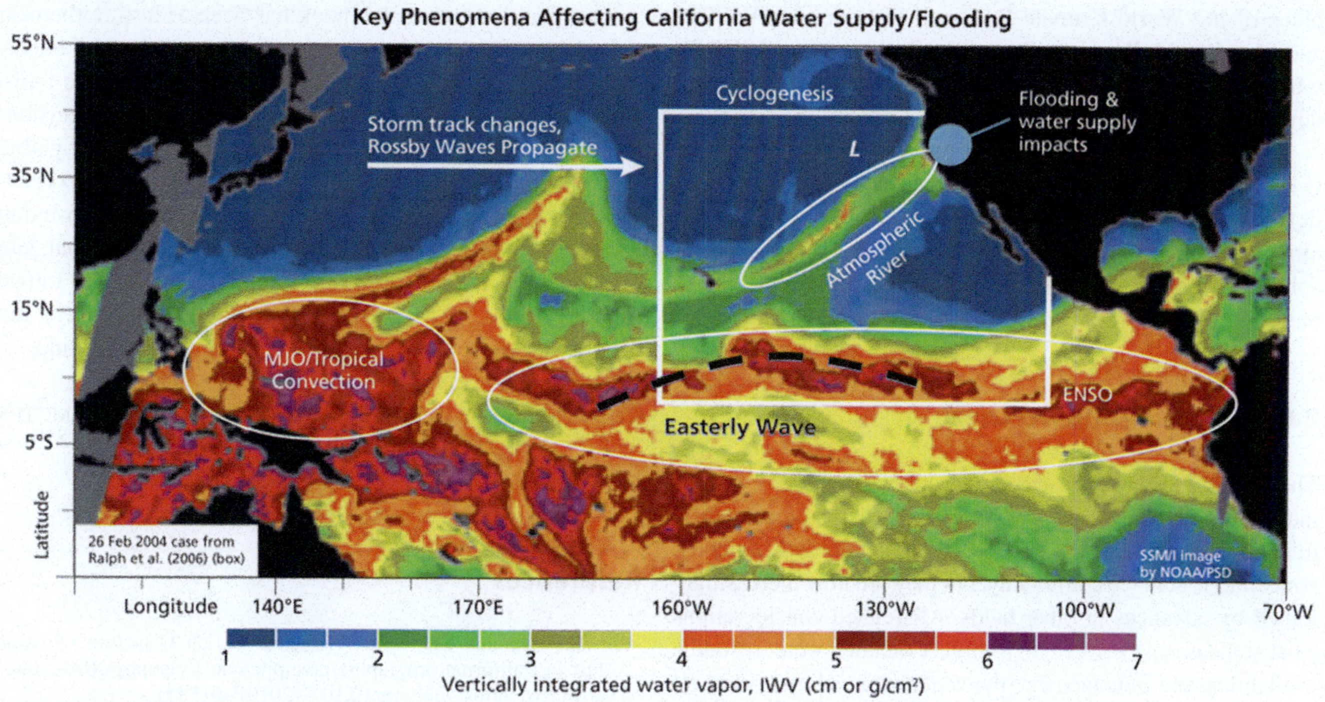

Fig. 8.10 Schematic summary of key physical processes that influence AR activity that affects the US West Coast. The processes are largely evident in structures seen in the integrated water vapor transport (IWV) field from satellite (*color fill*). The case used here caused flooding on northern California's Russian River and was the case documented by Ralph et al. (2006; study area is *outlined by white box*) that first recognized the connection between ARs and flooding

and many of these processes are manifested through structures seen in the IWV field.

Although several studies mentioned above have shown promising connections between MJO phases and AR landfalls in California, there are many cases where the connection seems absent. A study by Mundhenk et al. (2018) found an intriguing correlation between these AR–MJO teleconnections to US West Coast precipitation, and the phase of the quasi-biennial oscillation (QBO). The QBO is primarily a stratospheric circulation pattern that has roughly a 22-month cycle. This study found that the QBO's phase modulates the correlation between MJO and AR landfall. In one phase of the QBO, the connection is very weak; in the other, there is promising correlation. This result holds promise in the 2- to 6-week lead time-frame, which is within the normal range of the MJO's 30- to 60-day cycle.

On longer timescales, research into the impacts of climate variability from PDO, ENSO, and other climate modes has been explored by Gershunov et al. (2017) and Guirguis et al. (2018). These modes modulate the synoptic-scale environments of the ARs, as well as key AR characteristics, such as strength, orientation, and landfall duration. Some of these subseasonal regional climate modes are more influential than ENSO modes.

8.5.3 The Promise and Challenge of Creating S2S Precipitation Outlooks for the West

The studies discussed above suggest a promising path forward for research on what controls these larger-scale weather patterns and their regional impacts on AR locations and behavior. The degree to which these are potentially predictable out from 10 days to perhaps 30 days ahead will likely determine how useful S2S outlooks can be for water managers in the West.

Several efforts are underway to develop S2S outlooks involving ARs in the western US. A major set of these efforts is supported by California's Dept. of Water Resources (DWR). This includes development of novel methods from NASA/JPL, Scripps Institution of Oceanography's CW3E, Plymouth State University, Colorado State University, NOAA/Climate Prediction Center, and elsewhere. These methods are being brought together at CW3E into an AR S2S Outlook tool to explore their potential skill. An informal committee has formed to establish basic skill assessment criteria, and review the skill of prototype tools before they are posted online for experimental distribution to water managers and other interested individuals. Moreover, this activity represents one of a few application "pilot projects" being developed within the S2S Prediction Project, under the auspices of the World Climate Research Programme (WCRP) and the World Weather Research Programme (WWRP), which is (Vitart et al. 2017; s2sprediction.net) designed to demonstrate the decision-support value of S2S predictions.

The challenge of S2S prediction of precipitation in the western US represents an emerging area of study and tool development. While ARs will likely play a key role, as has already been seen, the occurrence of long-lasting atmospheric ridges will also likely need to be explored, in combination with the ARs they block.

8.6 Concluding Remarks

The past 2 decades of research and observations have greatly increased understanding of ARs, including their characteristics, impacts, and predictability. These gains came initially from the atmospheric science community, but they are now increasingly joined by advances in other fields. AR-related considerations exist at the nexus between high-impact weather, water resource availability and management, the water cycle, climate change, agriculture, economy, ecology, sustainability, and a host of other related topics. ARs evolve on a number of timescales, from their life cycle that varies with synoptic-scale meteorology day-to-day weather, to the variations they undergo associated with short- and long-term climate, offering important opportunities to improve understanding and prediction capabilities on these various timescales. All indications are that future changes in AR characteristics such as frequency, intensity, seasonality, and temperature will increase in importance to people and places around the world (Chap. 6). Taken together, these findings and advances indicate that ARs will become an ever-more important scientific topic in coming decades, with much more to be learned than has been so far, and with much more to be gained from applications of AR science than has been realized yet.

Advances in observational capabilities (both remote and in situ), NWP and climate modeling, and atmospheric reanalysis data sets have all contributed to this success and will continue to do so. Major field studies have brought scientists into the heart of these massive features, and findings from those visits are playing a key role in the implementation of model and forecast improvements. To date, the application of remote sensing methods and data sets has been limited in AR studies and so offers an important "new" emphasis.

Of course, all of these types of research activities require funding. As AR science continues to find applications with obvious benefits, continuation of support for AR-related observations, model development, integrated research activities, and applications will likely grow. In any event, given their place at the very foundation of much extratropical meteorology, climatology, and hydrology, continuation of such funding resources will be necessary and beneficial.

Our goal in writing this book has been to inspire the next generation of researchers to strive for an even greater understanding of ARs and to develop even better predictive tools. Our goal has been to lead the next generation of decision-makers to exploit this increasingly useful science in mitigating AR-related impacts. Our goal has been to guide the next generation of science communicators in effectively conveying AR-related information to the public. Above all, our goal, and our hope, is that the next generation of the public will follow us along the remarkable pathways of scientific discovery and benefit that we are confident ARs will continue to provide.

Acknowledgement Irina V. Gorodetskaya thanks FCT/ MCTES for the financial support to CESAM (UIDP/50017/2020+UIDB/50017/2020) through national funds.

References

Ault AP, Williams CR, White AB et al (2011) Detection of Asian dust in California orographic precipitation. J Geophys Res-Atmos 116(D16). https://doi.org/10.1029/2010JD015351

Bales R, Molotch N, Painter T et al (2006) Mountain hydrology of the western US. Water Resour Res 42:W08432. https://doi.org/10.1029/2005WR004387

Behrangi A, Guan B, Neiman PJ et al (2016) On the quantification of atmospheric rivers precipitation from space: composite assessments and case studies over the eastern North Pacific Ocean and the Western United States. J Hydrometeorol 17:369–382. https://doi.org/10.1175/JHM-D-15-0061.1

Bennartz R, Shupe MD, Turner DD et al (2013) July 2012 Greenland melt extent enhanced by low-level liquid clouds. Nature 496:83–86. https://doi.org/10.1038/nature12002

Binder H, Boettcher M, Grams CM et al (2017) Exceptional air mass transport and dynamical drivers of an extreme wintertime Arctic warm event. Geophys Res Lett 44:12028–12036. https://doi.org/10.1002/2017gl075841

Bintanja R, Andry O (2017) Towards a rain-dominated Arctic. Nat Clim Chang 7:263–267. https://doi.org/10.1038/nclimate3240

Bozkurt D, Rondanelli R, Marin JC et al (2018) Foehn event triggered by an atmospheric river underlies record-setting temperature along continental Antarctica. J Geophys Res-Atmos 123:3871–3892. https://doi.org/10.1002/2017jd027796

Bukovsky MS, Karoly DJ (2007) A brief evaluation of precipitation from the north American regional reanalysis. J Hydrometeorol 8:837–846. https://doi.org/10.1175/JHM595.1

Cannon F, Ralph FM, Wilson AM et al (2017) GPM satellite radar measurements of precipitation and freezing level in atmospheric rivers: comparison with ground-based radars and reanalyses. J Geophys Res Atmos 122:12747–12764. https://doi.org/10.1002/2017JD027355

Coplen T, Neiman P, White A et al (2008) Extreme changes in stable hydrogen isotopes and precipitation characteristics in a landfalling Pacific storm. Geophys Res Lett. 35(21):L21808. https://doi.org/10.1029/2008GL035481

Coplen T, Neiman PJ, White AB et al (2015) Categorization of northern California rainfall for periods with and without a radar brightband using stable isotopes and a novel automated precipitation collector. Tellus Ser A Dyn Meteorol Oceanogr 67:1

Cordeira JM, Ralph FM, Martin A et al (2017) Forecasting atmospheric rivers during CalWater 2015. Bull Am Meteorol Soc 98:449–459

Davies L (2014) Ark storm—a novel. Forge Press. 448 p

Davis JL, Vinogradova NT (2017) Causes of accelerating sea level on the East Coast of North America. Geophys Res Lett 44:5133–5141. https://doi.org/10.1002/2017gl072845

DeFlorio MJ, Waliser DE, Guan B et al (2018) Global assessment of atmospheric river prediction skill. J Hydrometeorol *19*(2):409–426. https://doi.org/10.1175/JHM-D-17-0135.1

DeFlorio MJ. Waliser DE, Guan B et al (2019) Global evaluation of atmospheric river subseasonal prediction skill. Clim Dyn 52:3039–3060. https://doi.org/10.1007/s00382-018-4309-x

Demirdjian R, Doyle JD, Reynolds CA et al (2020) A Case Study of the Physical Processes Associated with the Atmospheric River Initial Condition Sensitivity from an Adjoint Model. Journal of the Atmospheric Sciences 77:691–709. https://doi.org/10.1175/JAS-D-19-0155.1

Dettinger M, Cayan DR (2014) Drought and the California delta—A matter of extremes. San Franc Estuary Watershed Sci. 12(2):1–6. jmie_sfews_22330. Retrieved from http://escholarship.org/uc/item/88f1j5ht

Dettinger MD, Ralph FM, Das T (2011) Atmospheric Rivers, floods and the water resources of California (special issue: managing water resources and development in a changing climate). Water 3(2):445–478. https://doi.org/10.3390/w3020445

Dettinger MD, Ralph FM, Rutz JJ (2018) Empirical return periods of the Most intense vapor transports during historical Atmospheric River landfalls on the U.S. west coast. J Hydrometeorol 19:1363–1377. https://doi.org/10.1175/JHM-D-17-0247.1

Doyle JD, Amerault C, Reynolds CA et al (2014) Initial condition sensitivity and predictability of a severe extratropical cyclone using a moist adjoint. Mon Wea Rev 142(1):320–342

Espinoza V, Waliser DE, Guan B et al (2018) Global analysis of climate change projection effects on atmospheric rivers. Geophys Res Lett 45(9):4299–4308. https://doi.org/10.1029/2017GL076968

Gershunov A, Shulgina TM, Ralph FM (2017) Assessing climate-scale behavior of atmospheric Rivers affecting western North America. Geophys Res Lett 44:7900–7908. https://doi.org/10.1002/2017GL074175

Gorodetskaya IV, Tsukernik M, Claes K et al (2014) The role of atmospheric rivers in anomalous snow accumulation in East Antarctica. Geophys Res Lett 41(17):6199–6206. https://doi.org/10.1002/2014GL060881

Gorodetskaya IV, Silva T, Fish M et al (2018) Characteristics of enhanced moisture transport towards Antarctica from radiosonde measurements at coastal stations and over Southern Ocean. In: EGU Geophysical Research Abstracts, vol 20, Abstract 18988, EGU General Assembly 2018

Graeter KA, Osterberg EC, Ferris DF et al (2018) Ice core records of West Greenland melt and climate forcing. Geophys Res Lett 45:3164–3172. https://doi.org/10.1002/2017gl076641

Guan B, Waliser DE (2015) Detection of atmospheric rivers: evaluation and application of an algorithm for global studies. J Geophys Res Atmos 120(24):12,514–12,535. https://doi.org/10.1002/2015JD024257

Guan B, Waliser DE (2017) Atmospheric rivers in 20 year weather and climate simulations: a multimodel, global evaluation. J Geophys Res-Atmos 122:5556–5581. https://doi.org/10.1002/2016JD026174

Guan B, Molotch NP, Waliser DE et al (2010) Extreme snowfall events linked to atmospheric rivers and surface air temperature via satellite measurements. Geophys Res Lett 37(20):L20401. https://doi.org/10.1029/2010GL044696

Guan B, Waliser DE, Molotch NP (2012) Does the Madden-Julian oscillation influence wintertime atmospheric rivers and snowpack in the Sierra Nevada? Mon Weather Rev 140(2):325–342. https://doi.org/10.1175/MWR-D-11-00087.1

Guan B, Molotch NP, Waliser DE (2013a) The 2010/2011 snow season in California's Sierra Nevada: role of atmospheric rivers and modes of large-scale variability. Water Resour Res 49(10):6731–6743. https://doi.org/10.1002/wrcr.20537

Guan B, Molotch NP, Waliser DE et al (2013b) Snow water equivalent in the Sierra Nevada: blending snow sensor observations with snowmelt model simulations. Water Resour Res 49:5029–5046. https://doi.org/10.1002/wrcr.20387

Guan B, Waliser DE, Ralph FM et al (2016) Hydrometeorological characteristics of rain-on-snow events associated with atmospheric rivers. Geophys Res Lett 43:2964–2973. https://doi.org/10.1002/2016GL067978

Guirguis K, Gershunov A, Clemesha RES et al (2018) Circulation drivers of atmospheric rivers at the north American west coast. Geophys Res Lett 45(22):12576–12584. https://doi.org/10.1029/2018GL079249

Held IM, Soden BJ (2006) Robust responses of the hydrological cycle to global warming. J Clim 19:5686–5699. https://doi.org/10.1175/JCLI3990.1

Hu H, Dominguez F, Wang Z et al (2017) Linking atmospheric river hydrological impacts on the U.S. west coast to Rossby wave breaking. J Clim 30:3,381–3,399

Hughes M, Neiman PJ, Sukovich E et al (2012) Representation of the sierra barrier jet in 11 years of a high-resolution dynamical reanalysis downscaling compared with long-term wind profiler observations. J Geophys Res 117:D18116. https://doi.org/10.1029/2012JD017869

Huning LS, Margulis SA, Guan B et al (2017) Implications of detection methods on characterizing atmospheric river contribution to seasonal snowfall across Sierra Nevada, USA. Geophys Res Lett. 44:10445–10453. https://doi.org/10.1002/2017GL075201

Huntington J, Niswonger R (2012) Role of surface-water and groundwater interactions on projected summertime streamflow in snow dominated regions—an integrated modeling approach. Water Resour Res 48:W11524. 20 pp

Jermey PM, Renshaw RJ (2016) Precipitation representation over a two-year period in regional reanalysis. Q J R Meteorol Soc 142:1300–1310. https://doi.org/10.1002/qj.2733

Kim HM, Zhou Y, Alexander MA (2017) Changes in atmospheric rivers and moisture transport over the Northeast Pacific and western North America in response to ENSO diversity. Clim Dyn *52*(12):1–14

Komatsu KK, Alexeev VA, Repina IA et al (2018) Poleward upgliding Siberian atmospheric rivers over sea ice heat up Arctic upper air. Sci Rep 8:2872. https://doi.org/10.1038/s41598-018-21159-6

Lamjiri MA, Dettinger MD, Ralph FM et al (2017) Hourly storm characteristics along the U.S. west coast: role of atmospheric rivers in extreme precipitation. Geophys Res Lett 44(13):7020–7028. https://doi.org/10.1002/2017GL074193

Lavers DA, Waliser DE, Ralph FM et al (2016) Predictability of horizontal water vapor transport relative to precipitation: Enhancing situational awareness for forecasting western U.S. extreme precipitation and flooding. Geophys Res Lett 43 (5):2275–2282

Margulis S, Cortés G, Girotto M et al (2016) A Landsat-era Sierra Nevada (USA) snow reanalysis (1985–2015). J Hydrometeorol 17:1203–1221. https://doi.org/10.1175/JHM-D-15-0177.1

Marks DG, Link T, Winstral AH et al (2001) Simulating snowmelt processes during rain-on-snow over a semi-arid mountain basin. Ann Glaciol 32:195–202

Masson-Delmotte V, Steen-Larsen HC, Ortega P et al (2015) Recent changes in north-West Greenland climate documented by NEEM shallow ice core data and simulations, and implications for past-temperature reconstructions. Cryosphere 9:1481–1504. https://doi.org/10.5194/tc-9-1481-2015

Matrosov SY (2013) Characteristics of Landfalling atmospheric Rivers inferred from satellite observations over the eastern North Pacific Ocean. Mon Wea Rev 141:3757–3768. https://doi.org/10.1175/MWR-D-12-00324.1

Mattingly KS, Mote TL, Fettweis X (2018) Atmospheric river impacts on Greenland Ice Sheet surface mass balance. J Geophys Res Atmos 123(16):8538–8560. https://doi.org/10.1029/2018JD028714

McCabe-Glynn S, Johnson K, Strong C et al (2016) Isotopic signature of extreme precipitation events in the western US and associated phases of Arctic and tropical modes. J Geophys Res Atmos 121:8913–8924

Mundhenk BD, Barnes EA, Maloney ED et al (2018) Skillful empirical subseasonal prediction of landfalling atmospheric river activity using the madden–Julian oscillation and quasi-biennial oscillation. NPJ Clim Atmos Sci 1:20177. https://doi.org/10.1038/s41612-017-0008-2

Nardi KM, Barnes EA, Ralph FM (2018) Assessment of numerical weather prediction model reforecasts of the occurrence, intensity, and location of atmospheric rivers along the West Coast of North America. Mon Wea Rev 146(10):3343–3362. https://doi.org/10.1175/MWR-D-18-0060.1

NAS S2S (National Academies of Sciences, Engineering, and Medicine) (2016) Next generation earth system prediction: strategies for subseasonal to seasonal forecasts. The National Academies Press, Washington. https://doi.org/10.17226/21873

NASA–ESAS (National Academies of Sciences, Engineering, and Medicine) (2018) Thriving on our changing planet: a decadal strategy for earth observation from space. The National Academies Press, Washington. https://doi.org/10.17226/24938

Nash D, Waliser DE, Guan B et al (2018) The role of atmospheric rivers in extratropical and polar hydroclimate. J Geophy Res Atmos 123(13):6804–6821. https://doi.org/10.1029/2017JD028130

Neff W, Compo GP, Ralph FM et al (2014) Continental heat anomalies and the extreme melting of the Greenland ice surface in 2012 and 1889. J Geophys Res-Atmos 119:6520–6536. https://doi.org/10.1002/2014jd021470

Neiman PJ, Ralph FM, Wick GA et al (2008) Meteorological characteristics and overland precipitation impacts of atmospheric rivers affecting the west coast of North America based on eight years of SSM/I satellite observations. J Hydrometeorol 9:22–47. https://doi.org/10.1175/2007JHM855.1

Neiman PJ, Wick GA, Moore BJ et al (2014) An airborne study of an Atmospheric River over the subtropical Pacific during WISPAR: dropsonde budget-box diagnostics and precipitation impacts in Hawaii. Mon Wea Rev 142:3199–3223. https://doi.org/10.1175/MWR-D-13-00383.1

Newell RE, Newell NE, Zhu Y et al (1992) Tropospheric rivers? A pilot study. Geophys Res Lett 19:2401–2404

Nicolas JP, Vogelmann AM, Scott RC et al (2017) January extensive summer melt in West Antarctica favoured by strong El Nino. Nat Commun 8:15799. https://doi.org/10.1038/ncomms15799

Oakley NS, Cannon F, Munroe R (2018) Brief communication: meteorological and climatological conditions associated with the 9 January 2018 post-fire debris flows in Montecito and Carpinteria, California, USA. Nat Hazard Earth Sys 18(11):3037–3043

Oliver H, Luo H, Castelao RM (2018) Exploring the potential impact of Greenland meltwater on stratification, photosynthetically active radiation, and primary production in the Labrador Sea. J Geophys Res Oceans 123:2570–2591. https://doi.org/10.1002/2018jc013802

Painter H, Deems JS, Belnap J et al (2011) Responses of wind erosion to climate-induced vegetation changes on the Colorado plateau. Proc Natl Acad Sci 108:3854–3859

Paltan H, Waliser D, Lim WH et al (2017) Global floods and water availability driven by atmospheric rivers. Geophys Res Lett 44:10387–10395. https://doi.org/10.1002/2017GL074882

Payne AE, Magnusdottir G (2014) Dynamics of landfalling atmospheric rivers over the North Pacific in 30 years of MERRA reanalysis. J Clim 27:7133–7150. https://doi.org/10.1175/JCLI-D-14-00034.1

Payne AE, Magnusdottir G (2015) An evaluation of atmospheric rivers over the North Pacific in CMIP5 and their response to warming under RCP 8.5. J Geophys Res-Atmos 120(21):11173–11190. https://doi.org/10.1002/2015JD023586

Payne AE, Magnusdottir G (2016) Persistent landfalling atmospheric rivers over the west coast of North America. J Geophys Res-Atmos 121(22):13287–13300. https://doi.org/10.1002/2016JD025549

Pierce DW, Cayan DR, Thrasher BL (2014) Statistical downscaling using localized constructed analogs (LOCA). J Hydrometeorol 15(6):2558–2585

Pfeffer WT, Humphrey NF (1996) Determination of timing and location of water movement and ice-layer formation by temperature measurements in sub-freezing snow. J Glaciol 42:292–304

Ralph M, Davis B (2012) Understanding the water cycle Findings from NOAA's Water Cycle Science Challenge Workshop 28 August–September 2011, NOAA Earth System Research Laboratory, Boulder, Colorado NOAA internal report published 28 September 2012. Available from: https://nrc.noaa.gov/sites/nrc/Documents/Reduced%20size_water%20cycle%20workshop%20rpt_Sept%202012.pdf

Ralph FM, Neiman PJ, Wick GA (2004) Satellite and CALJET aircraft observations of atmospheric rivers over the eastern North-Pacific Ocean during the El Niño winter of 1997/98. Mon Wea Rev 132:1721–1745

Ralph FM, Neiman PJ, Rotunno R (2005) Dropsonde observations in low-level jets over the northeastern Pacific Ocean from CALJET-1998 and PACJET-2001: mean vertical-profile and atmospheric-river characteristics. Mon Wea Rev 133:889–910

Ralph FM, Neiman PJ, Wick GA et al (2006) Flooding on California's Russian River: role of atmospheric rivers. Geophys Res Lett 33:L13801. https://doi.org/10.1029/2006GL026689

Ralph FM, Sukovich E, Reynolds D et al (2010) Assessment of extreme quantitative precipitation forecasts and development of regional extreme event thresholds using data from HMT-2006 and COOP observers. J Hydrometeorol 11:1286–1304. https://doi.org/10.1175/2010JHM1232.1

Ralph FM, Neiman PJ, Kiladis GN (2011) A multiscale observational case study of a Pacific atmospheric river exhibiting tropical-extropical connections and a mesoscale frontal wave. Mon Wea Rev 139:1169–1189

Ralph FM, Coleman T, Neiman PJ et al (2013) Observed impacts of duration and seasonality of atmospheric-river landfalls on soil moisture and runoff in coastal northern California. J Hydrometeorol 14:443–459

Ralph FM, Dettinger MD, White A et al (2014) A vision of future observations for western US extreme precipitation and flooding. J Contemp Water Resour Res Educ 153:16–32

Ralph FM, Prather KA, Cayan D et al (2016) CalWater field studies designed to quantify the roles of atmospheric rivers and aerosols in modulating U.S. west coast precipitation in a changing climate. Bull Am Meteorol Soc 97:1209–1228

Ralph FM, Iacobellus SF, Neiman PJ et al (2017) Dropsonde observations of total water vapor transport within North Pacific atmospheric rivers. J Hydrometeorol 18:2577–2596

Ralph FM, Dettinger MD, Cairns MM et al (2018a) Defining "Atmospheric River": how the glossary of meteorology helped resolve a debate. Bull Am Meteorol Soc 99:837–839. https://doi.org/10.1175/BAMS-D-17-0157.1

Ralph FM, Wilson AM, Shulgina T et al (2018b) Comparison of Atmospheric River Detection Tools: How Many Atmospheric Rivers Hit Northern California's Russian River Watershed? Climate Dynamics. https://doi.org/10.1007/s00382-018-4427-5

Ralph FM, Rutz JJ, Cordeira JM (2019) A scale to characterize the strength and impacts of atmospheric Rivers. Bull Am Meteorol Soc 100:269–289. https://doi.org/10.1175/BAMS-D-18-0023.1

Reynolds CA, Doyle JD, Ralph FM et al (2019) Adjoint Sensitivity of North Pacific Atmospheric River Forecasts. Mon Wea Rev 147:1871–1897. https://doi.org/10.1175/MWR-D-18-0347.1

Robertson AW, Kumar A, Peña M et al (2015) Improving and promoting subseasonal to seasonal prediction. Bull Am Meteorol Soc 96:ES49–ES53. https://doi.org/10.1175/BAMS-D-14-00139.1

Rutz JJ, Steenburgh WJ, Ralph FM (2014) Climatological characteristics of atmospheric rivers and their inland penetration over the western United States. Mon Wea Rev 142:905–921

Ryoo J-M, Kaspi Y, Waugh DW et al (2013) Impact of Rossby wave breaking on U.S. west coast winter precipitation during ENSO events. J Clim 26:6360–6382. https://doi.org/10.1175/JCLI-D-12-00297.1

Shields CA, Rutz JJ, Leung L-Y et al (2018) Atmospheric river tracking method Intercomparison project (ARTMIP): project goals and experimental design. Geosci Model Dev 11:2455–2474. https://doi.org/10.5194/gmd-11-2455-2018

Steen-Larsen, Hans Christian, (June 20, 2018) Personal communication at POLAR 2018, Davos, Switzerland between Dr. Steen-Larsen (Geophysical Institute, University of Bergen, Bergen, Norway) and Dr. William Neff (Senior Research Scientist, CIRES, University of Colorado)

Stone RE, Reynolds CA, Doyle JD et al (2020) Atmospheric River Reconnaissance Observation Impact in the Navy Global Forecast System. Mon Wea Rev 148:763–782, https://doi.org/10.1175/MWR-D-19-0101.1

Terpstra A, Sodemann H, Gorodetskaya IV (2018) Dynamical mechanisms of anomalous moisture transport towards East Antarctica. In: Geophysical Research Abstracts, vol. 20, Abstract 18639, EGU General Assembly

Vitart F, Ardilouze C, Bonet A (2017) The subseasonal to seasonal (S2S) prediction project database. Bull Am Meteorol Soc 98:163–173. https://doi.org/10.1175/BAMS-D-16-0017.1

Waliser D, Guan B (2017) Extreme winds and precipitation during landfall of atmospheric rivers. Nat Geosci 10:179–183

Warner MD, Mass CF, Salathé EP (2015) Changes in winter atmospheric rivers along the north American west coast in CMIP5 climate models. J Hydrometeorol 16:118–128. https://doi.org/10.1175/JHM-D-14-0080.1

White AB, Anderson ML, Dettinger MD et al (2013) A 21st century California observing network for monitoring extreme weather events. J Atmos Ocean Technol 30:1585–1603

White AB, Gottas DJ, Strem ET et al (2002) An automated brightband height detection algorithm for use with Doppler radar spectral moments. J Atmospheric Ocean Technol 19(5):687–697

White AB, Moore BJ, Gottas DJ et al (2019) Winter storm conditions leading to excessive runoff above California's Oroville Dam during January and February 2017. Bull Am Meteorol Soc 100:55–70. https://doi.org/10.1175/BAMS-D-18-0091.1

Wick GA, Neiman PJ, Ralph FM et al (2013) Evaluation of forecasts of the water vapor signature of atmospheric rivers in operational numerical weather prediction models. Weather Forecast 28(6):1337–1352. https://doi.org/10.1175/waf-d-13-00025.1

Zhu Y, Newell RE (1994) Atmospheric rivers and bombs. Geophys Res Lett 21:1999–2002. https://doi.org/10.1029/94GL01710

Zhu Y, Newell RE (1998) A proposed algorithm for moisture fluxes from atmospheric rivers. Mon Wea Rev 126:725–735. https://doi.org/10.1175/1520-0493(1998)126<0725:APAFMF>2.0.CO;2

Index

F. M. Ralph et al. (eds.), *Atmospheric Rivers*, https://doi.org/10.1007/978-3-030-28906-5

The manufacturer's authorised representative in the EU is Springer Nature Customer Service Centre GmbH, Europaplatz 3, 69115 Heidelberg, Germany. If you have any concerns regarding our products, please contact ProductSafety@springernature.com

Printed and bound by CPI Group (UK) Ltd, Croydon, CR0 4YY
30/09/2026
02225269-0004